BIOLOGICAL SYSTEMS, BIODIVERSITY AND STABILITY OF PLANT COMMUNITIES

BIOLOGICAL SYSTEMS, BIODIVERSITY AND STABILITY OF PLANT COMMUNITIES

Edited by

Larissa I. Weisfeld, PhD, Anatoly I. Opalko, PhD, Nina A. Bome, DSc, and Sarra A. Bekuzarova, DSc

AAP | APPLE ACADEMIC PRESS

Apple Academic Press Inc. 3333 Mistwell Crescent Oakville, ON L6L 0A2 Canada	Apple Academic Press Inc. 9 Spinnaker Way Waretown, NJ 08758 USA

Exclusive worldwide distribution by CRC Press, a member of Taylor & Francis Group

Printed in the United States of America on acid-free paper

International Standard Book Number-13: 978-1-77188-064-0 (Hardcover)

Library and Archives Canada Cataloguing in Publication

Biological systems, biodiversity, and stability of plant communities / edited by Larissa I. Weisfeld, PhD, Anatoly Iv Opalko, PhD, Nina A. Bome, DSc, and Sarra A. Bekuzarova, DSc.

Includes bibliographical references and index.
ISBN 978-1-77188-064-0 (bound)
1. Biological systems. 2. Biodiversity. 3. Plant communities. I. Weisfeld, Larissa I., author, editor II. Opalko, Anatoly Iv, author, editor III. Bome, Nina A., author, editor
IV. Bekuzarova, Sarra A., author, editor

QH313.B55 2015 570 C2015-901985-0

CIP data on file with US Library of Congress

Apple Academic Press also publishes its books in a variety of electronic formats. Some content that appears in print may not be available in electronic format. For information about Apple Academic Press products, visit our website at **www.appleacademicpress.com** and the CRC Press website at **www.crcpress.com**

Cover photos by **Nina A. Bome**

ABOUT THE EDITORS

Larissa I. Weisfeld, PhD

Larissa I. Weisfeld is a senior researcher at the N. M. Emanuel Institute of Biochemical Physics, Russian Academy of Sciences in Moscow, Russia, and a member of the N. I. Vavilov Society of Geneticists and Breeders. She is the author of about 300 publications in scientific journals, patents, and conference proceedings, as well as the co-author of a work on three new cultivars of winter wheat. Her research interests concern the basic problems of chemical mutagenesis, cytogenetic and the other ecological problems. She has worked as a scientific editor at the publishing house Nauka (Moscow) and of the journals *Genetics* and *Ontogenesis*. In 2013-2015 she was a co-editor of the books *Ecological Consequences of Increasing Crop Productivity: Plant Breeding and Biotic Diversity* and *Biological Systems, Biodiversity, and Stability of Plant Communities* (Apple Academic Press).

Anatoly I. Opalko, PhD

Anatoly I. Opalko, PhD, is a Professor and Head of the Physiology, Genetics, Plant Breeding and Biotechnology Division at the National Dendrological Park "Sofiyivka" of the National Academy of Sciences of Ukraine, Uman, Cherkassy region, Ukraine, and a Professor of Genetics, Plant Breeding and Biotechnology Chair at Uman National University of Horticulture, Uman, Ukraine. He is also the Head of the Cherkassy Regional Branch of the Vavilov Society of Geneticists and Breeders of Ukraine. He is a prolific author, researcher, and lecturer. He has received several awards for his work, including the badge of honor for "Excellence in Agricultural Education" and the badge of honor of the National Academy of Sciences of Ukraine for professional achievement. He is member of many professional organizations and on the editorial boards of the Ukrainian biological and agricultural science journals. In 2013-2015 he was a co-editor of the books *Ecological Consequences of Increasing Crop Productivity: Plant Breeding and Biotic Diversity* and *Biological Systems, Biodiversity, and Stability of Plant Communities* (Apple Academic Press).

Nina A. Bome, DSc

Nina A. Bome, DSc in Agriculture, is a Professor and Head of the Department of Botany, Biotechnology and Landscape Architecture at the Institute of Biology at the Tyumen State University, Tyumen, Russia. She is the author of the monographs, articles, school-books, patents, and she is a lecturer. The is the Director and Founder of the Scientific School for Young Specialists. She is the author of about 300 publications. She participates in long-term Russian and international programs. Her main field of interest concerns basic problems of adaptive potential of cultivated crops, mutagenesis, possibility of conservation, enhancing biodiversity of plants, methods of evaluation of plants' resistance to the phytopatogens and other unfavorable environmental factors, and genetic resources of cultivated plants in the extreme conditions of the Western Siberia. In 2013-2015 she was a co-editor of the books *Ecological Consequences of Increasing Crop Productivity: Plant Breeding and Biotic Diversity* and *Biological Systems, Biodiversity, and Stability of Plant Communities* (Apple Academic Press).

Sarra A. Bekuzarova, DSc

Sarra A. Bekuzarova, DSc in Agriculture, is a Professor and Head of the Laboratory of Plant Breeding. She focuses on feed crops and long-term seed-growing of fodder crops at the North-Caucasian Research Institute for Mountain and Foothill Agriculture in the Republic of Northern Osetiya. She is a Professor at the L. N. Kosta Khetagurov North-Ossetia State University, Vladikavkaz, Republic of North Ossetia-Alania, Russia. She is also a prolific author, researcher, and lecturer. She is a corresponding member of the Russian Academy of Natural Sciences and a member of the International Academy of Authors of the Scientific Discoveries and Inventions; the International Academy of Sciences of Ecology, Safety of Man and Nature; the All-Russian Academy of Nontraditional and Rare Plants; and the International Academy of Agrarian Education. She is on the editorial boards of several scientific journals. In 2013-2015 she was a co-editor of the books *Ecological Consequences of Increasing Crop Productivity: Plant Breeding and Biotic Diversity* and *Biological Systems, Biodiversity, and Stability of Plant Communities* (Apple Academic Press).

CONTENTS

PART III: PROBLEMS OF GRAIN CROPS GROWING ON ACID SOILS OF THE EUROPEAN NORTH

PART IV: SCIENTIFICALLY SUBSTANTIATED SOIL AND CLIMATICAL REGIONS FOR THE INDUSTRIAL CULTIVATION OF CROPS

PART V: BIO-ORGANIC FARMING

PART VI: INNOVATIVE TECHNOLOGIES OF CULTIVATION OF AGRICULTURAL CROPS

PART VII: ADAPTIVE REACTIONS OF CULTURAL PLANTS TO STRESS FACTORS

LIST OF CONTRIBUTORS

Nurbek L. Adayev

Chechen Research Institute of Agriculture, d. 1, Lenin Street, Grozny District, Gikalo Settlement, Chechen Republic, 366021, Russia, Tel.: +7(8712) 21-20-04; E-mail: chechniish@mail.ru

Inaida M. Aleschenkova

Institute of Microbiology, National Academy of Sciences, d 2, Kuprevich Street; Minsk, 220141, Belarus

Galina A. Batalova

Behalf of N. V. Rudnicki Zonal Research Institute of Agriculture of North-East of the Russian Agricultural Academy, d.166-a, Vyatka State Agricultural Academy, d.133, October A. V., Lenin Street, Kirov, 610007, Russia, www.vgsha.info; E-mail: niish-sv@mail.ru; edaphic@mail.ru

Sarra A. Bekuzarova

Gorsky State Agrarian University, d. 37, Kirov Street, Vladikavkaz, Republic of North Ossetia Alania, 362040, Russia, Tel.: +7(963) 1778761; E-mail: bekos37@mail.ru

Anna A. L. Belozerova

Tyumen State University, d.10, Semakov Street, Tyumen, 625003, Russia, Tel.: +7(3452) 46-40-61, 46-81-69; E-mail: bomena@mail.ru; anna-bel@bk.ru

Olena V. Bilynska

V.Ya Yurjev Plant Production Institute of National Academy of Agricultural Sciences of Ukraine, d. 142, Moskovsky A. V., Kharkiv, 61060, Ukraine, E-mail: bilinska@ukr.net

Alexander Y. Bome

The Tyumen basing point of the N.I. Vavilov All-Russia Research Institute of Plant Growing, d. 42–44, Bol'shaya Morskaya Street, St. Petersburg, 190000; Russia, Tel.: +7(812) 3142234; E-mail: office@vir.nv.ru

Nina A. Bome

Tyumen State University, d. 1, Semakova Street, Tyumen, 625003, Russia, Tel.: +7(3452) 46-40-61, 46-81-69; E-mail: bomena@mail.ru

Mykola O. Bublyk

Institute of Horticulture of NAAS of Ukraine, 23, Sadova Street, Kyiv, Ukraine 03027; E-mail: mbublyk@mail.ru

Vera I. Bushuyeva

Belorussian State Agricultural Academy, Professor of the Department of Selection and Genetics, Doctor of Agricultural Sciences, Associate Professor, Gorki, h. 5, Michurin Street, 213407; Belarus, Mob: +375296910383; Fax: +375223359485; E-mail: vibush@mail.ru

Andriy V. Doronin

National Academy of Agrarian Sciences of Ukraine, 37 Vasylkivska Street, Kyiv, Ukraine 03022, E-mail: andredor@meta.ua

Volodymyr A. Doronin

Institute of Bioenergy Crops and Sugar Beet of NAAS of Ukraine, 25 Klinichna Street, Kiev 03141, Ukraine, E-mail: doronin@tdn.kiev.ua

Irina A. Dudareva
Tobolsk Complex Scientific Station, Ural Division of the Russian Academy of Sciences, d. 15, Yuriy Osipov Street, Tyumen Area, Tobolsk, 626152, Russia, Tel.: 89199234458; E-mail: sovet29@rambler.ru

Galina A. Evstratova
Education Establishment Belarusian State Agricultural Academy Michurin Street 5; Gorki, Mogilev region, 213407, Belarus

Vasiliy A. Fedotov
Voronezh State Agricultural University, d. 1, Mitchurina Street, Voronezh, 394087, Russia, Tel.: +7 (920) 424-20-80; E-mail: ovennn@mail.ru

Liudmyla A. Fryziuk
Institute of Horticulture of NAAS of Ukraine, 23, Sadova Street, Kyiv, Ukraine 03027; E-mail: mbublyk@mail.ru

Nikolay V. Glotov
Mari State University, 1, Lenina square, Yoshkar-Ola, Republic of Mari El, 424000, Russia, E-mail: nvglotov@inbox.ru

Galina N. Gordeeva
Khakas Scientific Research Institute of Agrarian Problems of Russian Academia of Agriculture Science, Abakan, P.O. Box: 709; Republic of Khakasia, 655019, Russia, Tel.: +89618958208; E-mail: gordeeva.gal2011@yandex.ru

Irina M. Haniyeva
V. M. Kokov's Kabardino-Balkar State Agrarian University, d. 1v, Lenin Avenue, Nalchik, Republic Kabardino-Balkaria, 360030, Russia, Tel.: +7(8662) 47-41-77; E-mail: imhanieva@mail.ru

Sergey M. Ivanov
Mari State University, 1 Lenin Square, Yoshkar-Ola, Republic of Mari El, 424000, Russia

Sabir V. Kadyrov
Voronezh State Agricultural University, d. 1, Mitchurina Street, Voronezh, 394087, Russia, Tel.: +7 (920) 424-20-80; E-mail: ovennn@mail.ru

Lesya M. Karpuk
Bila Tserkva National Agrarian University, pl. Soborna 8/1, Bila Tserkva, Kiev Region, 09117, Ukraine, E-mail: zuikes@ukr.net

Liliya E. Kartyzhova
Institute of Microbiology, National Academy of Sciences, d 2, Kuprevich Street, Minsk, 220141, Belarus, E-mail: Liliya_Kartyzhova@mail.ru

Vladimir F. Kazarin
The P.N. Konstantinov Volga Selective Breeding and Seed Production Research Institute, 76, Shosseynaya Street, Samara Region, Kinel, Village Ust-Kinel, 446442; Russia, Tel.: (84663) 46243; E-mail: KazarinVF@mail.ru

Alexandra V. Kazarina
The P.N. Konstantinov Volga Selective Breeding and Seed Production Research Institute, 76, Shosseynaya Street, Samara Region, Kinel, Village Ust-Kinel, 446442; Russia, Tel.: (84663) 46243

Svetlana S. Kazieva
Gorsky State Agrarian University, Vladikavkaz, Kirov Street, 37, Republic of North Ossetia Alania, 362040, Russia, E-mail: bekos37@mail.ru

Alan T. Kesayev
Gorsky State Agrarian University, d. 37, Kirov Street, Vladikavkaz, Republic of North Ossetia Alania, 362040, Russia

Fatima M. Khacajeva
K. L. Khetagurov North-Ossetian State University, Vladikavkaz, d. 46, Vatutin Street, Republic of North Ossetia-Alania, 362025, Russia, Tel.: 8 (867 2) 545108 / 8 906 4949597 / 8-918-8221990; E-mail: hacaevafm@mail.ru; tomaevvadim@mail.ru

Vladimir V. Kondratov
Voronezh State Agricultural University, d. 1, Mitchurina Street, Voronezh, 394087, Russia, Tel.: +7 (920) 424-20-80

Alexey L. Kornilov
Tyumen State University, 625003, Pirogova street, 3; Tyumen, Russian Federation, Tel.: 8(3452)64-07-24; E-mail: lelik_tgu@mail.ru

Natalia V. Korolyonok
Institute of Microbiology, National Academy of Sciences, d 2, Kuprevich Street; Minsk, 220141, Belarus

Liudmyla M. Levchuk
Institute of Horticulture of NAAS of Ukraine, 23, Sadova Street, Kyiv, Ukraine 03027, E-mail: mbublyk@mail.ru

Evgeniy M. Lisitsyn
Behalf of N. V. Rudnicki Zonal Research Institute of Agriculture of North-East of the Russian Agricultural Academy, d. 166a, Lenin Street, Russia, Vyatka State Agricultural Academy, d. 133, October A. V., Kirov 610017, Russia, Tel./Fax: (8332) 54-86-33; www.vgsha.info; E-mail: niish-sv@mail.ru; edaphic@mail.ru; info@vgsha.info

Svetlana M. Mostovshchikova
Tyumen State University, d.10, Semakov Street, Tyumen, 625003, Russia, Tel.: +7(3452) 46-40-61, 46-81-69; E-mail: bomena@mail.ru; anna-bel@bk.ru

Roman K. Nagorny
Institute of Microbiology, National Academy of Sciences, Belarus d. 2, Kuprevich Street, Minsk, 220141, Belarus, E-mail: roman19031988@mail.ru

Vladimir N. Obraztsov
Voronezh State Agricultural University, d. 1, Mitchurina Street, Voronezh, 394087, Russia, Tel.: +7 (920) 424-20-80; E-mail: ovennn@mail.ru

Anatoly I. Opalko
National Dendrological Park "*Sofiyivka*" of NAS of Ukraine, d. 12-a, Kyivska Street, Uman, Cherkassy Region, 20300 Ukraine, Uman National University of Horticulture, d. 1, Instytutska Street, Uman, Cherkassy region, 20305, Ukraine, E-mail: opalko_a@ukr.net

Olga A. Opalko
National Dendrological Park "*Sofiyivka*" of NAS of Ukraine, d. 12-a, Kyivska Street, Uman, Cherkassy Region, 20300 Ukraine

Ostap A. Paladich
Behalf of N. V. Rudnicki Zonal Research Institute of Agriculture of North-East of the Russian Agricultural Academy, 166-a, Lenin Street, Kirov 610007; Russia, E-mail: shikhova-l@mail.ru

Galina A. Petukhova
Tyumen State University, 625003, Pirogova street, 3; Tyumen, Russian Federation, Tel.: 8(3452)64-07-24; E-mail: lelik_tgu@mail.ru

Lyudmila V. Prokopyeva
Mari State University, 1 Lenin Square, Yoshkar-Ola, Republic of Mari El, 424000, Russia

Roman M. Puhachov
Education Establishment Belarusian State Agricultural Academy Michurin Street 5; Gorki, Mogilev Region, 213407, Belarus

Irina G. Puhachova
Education Establishment Belarusian State Agricultural Academy Michurin Street 5; Gorki, Mogilev Region, 213407, Belarus, E-mail: puhachova.irina@gmail.com

Elena I. Ripberger
Tyumen State University Ministry of Education of Russia, d.10, Semakova Street, Tyumen, 625003; Russia, Tel.: +7(3452) 46-40-61, 46-81-69; E-mail: bomena@mail.ru

Lyudmila V. Romanova
Institute of Microbiology, National Academy of Sciences, d 2, Kuprevich Street; Minsk, 220141, Belarus

Alissa S. Samsonova
Institute of Microbiology, National Academy of Sciences, Belarus d. 2, Kuprevich Street, Minsk, 220141, Belarus, E-mail: samsonova@mbio.bas-net.by

Irina V. Semyonova
Institute of Microbiology, National Academy of Sciences, d 2, Kuprevich Street; Minsk, 220141, Belarus

Diana I. V. Shchedrina
Voronezh State Agricultural University, d. 1, Mitchurina Street, Voronezh, 394087, Russia, Tel.: +7 (920) 424-20-80; E-mail: ovennn@mail.ru

Irina N. Shchennikova
Behalf of N. V. Rudnicki Zonal Research Institute of Agriculture of North-East of the Russian Agricultural Academy, d. 166-a, Lenin Street, Kirov, 610007, Russia, E-mail: niish-sv@mail.ru, edaphic@mail.ru

Lyudmila N. Shikhova
Vyatka State Agricultural Academy 133, Oktyabrsky Prospect, Kirov 610017, Russia, Tel./Fax: (8332) 54-86-33; www.vgsha.info; E-mail: niish-sv@mail.ru; info@vgsha.info

Georgy Yu. Sofronov
Department of Statistics, Macquarie University, Sydney, New South Wales, 2109, Australia

Yulia G. Suetina
Mari State University, 1 Lenin Square, Yoshkar-Ola, Republic of Mari El, 424000, Russia

Alexey A. Teplykh
Russian Center for Forest Protection, 83 Komsomolskaya Street, Yoshkar-Ola, Republic of Mari El, 424004, Russia

Vadim A. Tomajev
K. L. Khetagurov North-Ossetian State University, Vladikavkaz, d. 46, Vatutin Street, Republic of North Ossetia-Alania, 362025, Russia, Tel.: 8 906 4949597; 8-918-8221990; 8 (867 2) 545108; E-mail: tomaevvadim@mail.ru

Aleksey B. Trubyanov
Mari State University, 1, Lenina square, Yoshkar-Ola, Republic of Mari El, 424000, Russia, E-mail: a.b.trubyanov@gmail.com

Larissa I. Weisfeld
N. M. Emanuel Institute of Biochemical Physics RAS, d. 4, Kosygin Street, Moscow, 119334, Russia, Tel.: +79162278685; E-mail: liv11@yandex.ru

Ravil M. Zeleev
Kazan (Volga region) Federal University, d. 18, Kazan, Kremlevskaya St., 420018, Russia, Tatarstan region; E-mail: zeleewy@rambler.ru

LIST OF ABBREVIATIONS

ABU	absorption biochemical unit
AFLP	amplified fragment length polymorphism
AI	active ingredient
ANOVA	analysis of variances
AOP	antioxidant protection
APS	atomic power station
ARIF	All-Russian Institute of Feeds
ASP	active symbiotic potential
BOD5	biological oxygen demand for five days
bp	pair of base (sizes of fragments of specific amplicons)
BS	binder of sterility
BSAA	Belorussian State Agricultural Academy
CAB	Creative Association of Breeders
CAPS/dCAPS	cleaved amplified polymorphic sequences
CHP	combined heat and power
CIMMYT	Centro International de Mejoramiento de Maíz y Trigo
CLM	combined long-term method
CLTM	combined long-term method
CMEA	Council for Mutual Economic Assistance
CPI	carbon preference index
Cs	cesium
CSP	common symbiotic potential
CV	coefficient of variation
DB	digestible protein
DC	diene conjugates
DH	double haploid population
DIP	department for improving proficiency
DM	dry matter
DNA	deoxyribonucleic acid
DP	digestible protein
DUS	distinctness, uniformity, and stability
EC	enzyme classification
EDTA	ethylene diamine tetra acetic acid
EE	exchange energy
EE"BSAA"	educational establishment "Belorussian State Agricultural Academy"

EST	expressed sequences tags
FAO	The Food and Agriculture Organization of the United Nations
FER	farming energy ratio
FISH	method-Fluorescence *in situ* hybridization
FPU	feed protein unit
FT-IR	infrared spectroscopy
FU	feed unit
GE	gross energy
GENAN	GENetic ANalysis (computer software)
HC	hydrothermal coefficient
HTC	hydrothermal potential
IBCA	increased background of fertilizing—the common area
IBEA	increased background—the extended area of supply
icp	indicator of chemical pollution
ISSR	inter simple sequences repeats
kDa	kilodalton (unit to measure molecular weight)
LFSC	liquid fertilizing-stimulating composition
LPO	lipid peroxidation
LSD	small statistical distinction
MATE	multidrug and toxic compound extrusion gene family
MM	molecular mass
MPC	maximum permissible concentration
MS	male sterility
N	nitrogen
NAAS	National Academy of Agrarian Sciences
NAS	National Academy of Science
NBCA	normal background of fertilizing
NCA	nodulating competitive ability
NFE	nitrogen-free extractives
NSBC	nursery for study of biotypical composition
NSC	National Science Centre
OFU	oat feed unit
PABA	para-aminobenzoic acid
PAGE	polyacrylamide gel
PCR	polymerase chain reaction
QTL	quantitative trait locus
R	Roentgen
RAPD	Random Amplified Polymorphic DNA
RCA	rhizospheric competitive ability
RED	renewable energy directive
RFLP	restriction fragment length polymorphism

RIL	recombinant inbred line
RNA-asa	ribonucleasa
ROS	reactive oxygen species
RRG	relative root growth
SB	Schiff base
SCA	saprophytic competitive ability
SCAR	sequence characterized amplified region
SGP	complex-hybrid population
SNP	single nucleotide polymorphisms
SOD	superoxide dismutase
Sr	strontium
SSAP(S-SAP)	sequence-specific amplification polymorphism
SSD05	smallest significant difference
SSH	simple sterile hybrid
SSR	simple sequence repeats
STS	sequences tagged site
TCA	trichloroacetic acid
TEA	triethylamine
TOS	Creative Association of Breeders
Tr	ratio of radionuclide transition from soil to plants
UN	United Nations
UPOV	the International Union for the Protection of New Varieties of Plants
URC	units of regeneration coefficient
US	United States
VAT	value added tax
VNTR	variable-number tandem repeats
WPI	index of water pollution

LIST OF SYMBOLS

A	mineral horizon formed at or near the surface in the zone of leaching or eluviation of materials
A_h	horizon enriched with organic matter, contains no less than 17% organic compounds
A_{he}	A_h horizon that has undergone eluviations
A_p	plowed A soil horizon
Al^{+++}	ion of aluminum
C	concentration of Shiff bases
Ca^{++}	ion of calcium
C_{hum}	carbon of humus
C_{lab}	carbon of labile part of organic matter
C_{tot}	total carbon
Cu	energetic accumulation element
° C	degree Celsius
c/ha	center/ha, hundredweight
C_c	chemical substance concentrations
Ca^{++}	ion of calcium
d	degree of dominance
D_{233}	optical density of the sample for 233 nm
D_h	optical density of pure heptanes
D_s	optical density of sample
eV	electron volt
$F_1, F_2, F_3, F_4 \ldots$	generations of organisms from first to fourth and so on
g	gram
Gj	gigajoule
H	broad-sense heritability
H^+	ion of hydrogen
ha	hectare – is area unit that equal to 10,000 square meters
J	number of ontogenetic states
K	number of the model parameters
kD, kDa	kiloDalton
M_1–M_7	generations of mutants from first to seventh
mg	milligram
Mg^{++}	ion magnesium
mg/eV	milligram equivalent for 100 g of soil
MJ	megajoule

mkg	microgramm
mkl	microlitre
ml	milliliter
mln	million
mm	millimeter
mm, Mm	micromoles
mm per year	millimeters per year—atmospheric fallouts amount
MPa	megapaskal
N	amount of factors
n	haploid chromosomal set, chromosome complement
N	sample size
P_1	paternal parent of crossing
P_2	paternal parent of crossing maternal parent of crossing pH
Pb	weak grip elements
$P_{30}K_{30}$	mineral fertilizer
pH	hydrogen ion concentration
psc.	pieces
R	Roentgen
r	vector-column of the sample correlation coefficients
s_a^2	variance between samples
s_t^2	total variance
\$, USD	the United States dollar
V_{he}	final volume of heptane extract
V_{pe}	volume of plant extract
X_a	desirable level of the weather factor a and its virtual value
$\bar{x}F_2$	average values for F_2
Y_a	weight coefficient
z(.)	Fisher's z-transformation
$\hat{p}_{GLS}$	vector-column of the estimated correlation coefficients

PREFACE

"Better to be over thrift currently than to expose it to destruction what thousands and millions of years has been created by nature"

N. I. Vavilov

The authors of this volume are acknowledged scientists in the field of genetics, plant selection, ecology and agro-economy. They tried to find consensus between natural aspiration for the increase in welfare and overcoming of poverty in a number of countries and the not less natural wish to decrease excess loads on the environment in the majority of agro-developed countries.

The media's attention on the state of the environment at the turn of the millennium has increased significantly, which is manifested an irrational response from many entrepreneurs and officials in many countries of the world, resulting in a multidirectional search for ways of achieving consensus environmentally and socially for technical strategies of sustainable use of natural resources. Among the components of this strategy are arrangements directly connected with the minimization of anthropogenic loads on the one hand and the change of paradigms of the land tenure in the direction of the increase in anthropo-adaptive potential of the cultivated genotypes and technologies on the other hand, with the aim being the rise in agricultural production with simultaneous reduction of unwanted ecological consequences.

The maintenance and extension of biodiversity of cultivated plants on the territories with extreme edaphic-climatic conditions requires the development of new methodic approaches. The formation of the collections taking into account zonal peculiarities will allow providing selective-genetic programs with valuable initial material. Adaptation mechanisms of nutrient uptake and transport can be gained with the help of physiologically active compounds. The application of the contemporary genetic-biotechnical methods allows solving the problem of the extension of the plants biodiversity.

In our epoch, world agriculture tends to occupy moderate and subtropical areas. But in 1931, N. I. Vavilov noticed that 73.7% of the cultivated land worldwide belonged to a strip located between 55 and 23°30′ of north. To the beginning of 21st century, only 1/17 part of all arable land of the Earth was located on northern areas above 55°.

The change of biophysical and socioeconomic factors led to the increase in risks connected with unsatisfactory management of the natural resources, loss of biodiversity, pollution, soil and water degradation as well as vulnerability towards the climate change.

At present the growth of harvest of cultivated plant species is mostly gained through the use of chemicals, irrigation and mechanization of agriculture, which leads to the emergence of a number of problems.

To ensure food and environmental security on the Earth, by creating new types of raw materials, it is necessary to carefully study diversity use and enrichment of the plant and microbial world.

This can be ensured by developing methods for studying the biology, ecology and cultivation of plants; by maintaining the gene pool in gene banks, and by using natural populations of plants, microorganisms, and pure cultures of phytopathogenic fungi.

Throughout the world acid soils occupy large areas of land potentially suitable for agriculture. In the northeast of Europe the acidic soils are mainly located in the non-chernozem zone of Russia, where losses of yield in acid soils come to 15–16 million tons of agricultural production in terms of grain per year. Efficiency of mineral fertilizers on strong- and middle-acid soils goes down 20–25%. A separate section of this book is devoted to the study of genetic plasticity and adaptability of crops to acidic soils in the northeast of Russia.

One of the most important components of biotic diversity is the genetic resources of cultivated plants and their wild relatives. The application of the selective-genetic methods (hybridization, experimental mutagenesis) allows for plant varieties with the new or improved characters and provides for a high level of adaptation to the ambient conditions. Scientists, graduate students, and students of Tyumen State University (Russia) conducted long-term experimental and methodical research in the field of biotic variety of plants and microorganisms.

Information about the composition the chemical elements in the soil-plant system are of interest to study their migration through the food chain. The composition of the soil and the distribution of elements in the soil-plant system are manifested as adaptations of plant organism to environmental conditions. The influence of physiologically active compounds also is an actual problem.

This volume presents a section of the Mari State University's proposed new model for measuring fluctuating asymmetry (FA). Correlation between different attributes can't be arbitrary and depend on the size of the FA. Theoretical studies have been devoted to the analysis of bilateral characters on the example of a plate of plant leaves and on the example of the heterogeneous plant population development spectrum; as an example natural populations of lingonberry and lichens were used.

Everyone knows that ontogenetic spectra vary within a population, but it is hard to summarize these data, not having the appropriate statistical methods. One chapter in this book not only provides valid comparison of ontogenetic spectrum of populations, but also provides a way to estimate the proportion of the differences between the spectra within and between populations. The proposed methods are tested on natural populations of cowberry and lichens.

A chapter from Kazan Federal University examines the causes of nonconformity of the hierarchical model of assessment of biodiversity to the needs of practical systematics.

Another section of this volume presents works on selection and methods of estimation of breeding material of *Galega* east and red clover with the possibility of applying this material in plant-breeding practice. The main directions of the research into sampling and estimation methods of breeding material of *Galega* eastern and red clover are presented in Belarus with evaluation of the possibilities of this material usage in breeding practice. The role of nodule bacteria and degradation of trimethylamine were widely studied with application of their achievements in agriculture.

The work of the North Ossetian State University in Chapter 19 offers ways to reduce disaster risk. The greater part of the book is devoted to works of Ukrainian researchers on genetics and selection of technical and fruit crops.

A consequence of a retrospective analysis of the historical background of the "green revolution" interconnected with current ecological and social problems was suggested to consider anthropo-adaptability of new cultivars as the aim of contemporary breeding and a basis component of a new biotechnological wave of the "green revolution."

The book presents the scientific works from 24 different research centers of Russia (16), Ukraine (8), Belarus (3), and Australia (1). They are

- National Dendrological Park "Sofiyivka " of NAS of Ukraine
- Uman National University of Horticulture, Ukraine
- Institute of Bioenergy Crops and Sugar Beet of NAAS of Ukraine, Kyiv, Ukraine
- Bila Tserkva National Agrarian University, Ukraine
- Yurjev Plant Production Institute of National Academy of Agricultural Sciences of Ukraine, Kharkiv, Ukraine
- Institute of Horticulture of NAAS of Ukraine
- Institute of Microbiology of NAS of Belarus, Minsk, Belarus
- Belorussian State Agricultural Academy, Gorki, Belarus
- Behalf of N.V. Rudnicki Zonal Research Institute of Agriculture of North-East, Kirov, Russia
- Vyatka State Agricultural Academy, Kirov, Russia
- N.M. Emanuel Institute Biochemical Physics RAS, Moscow, Russia
- Voronezh State Agricultural University, Voronezh, Russia
- Gorsky State Agrarian University, Vladikavkaz, Republic of North Ossetia Alania, Russia
- North Caucasus Research Institute of Mountain and Foothill Agriculture, Village Mikhailovskoe, Republic of North Ossetia Alania, Russia
- V.M. Kokov's Kabardino-Balkarsky State Agrarian University, Nalchik, Kabardino-Balkaria, Russia

- K.L. Khetagurov North-Ossetian State University, Vladikavkaz, Republic of North Ossetia Alania, Russia
- Tyumen State University, Tyumen, Russia
- Tyumen basing point of the N.I. Vavilov All-Russia research institute of plant growing, St. Petersburg, Russia
- Tobolsk complex scientific station, Ural Division of the Russian Academy of Sciences, Tyumen area, Tobolsk, Russia
- Research Institute of Agrarian Problems Khakassia, Russian Academy of Agricultural Sciences, Abakan, Khakassia, Russia
- Mari State University, Yoshkar-Ola, Republic of Mari El, Russia
- Macquarie University, Sydney, New South Wales, Australia
- Russian Center for Forest Protection, Mari El Republic, Yoshkar-Ola, Republic of Mari El, Russia

The results of the research presented in this volume can serve as a theoretical-practical base for the ecosystems optimization, designing the variety “model” on the basis of valuable quantitative characters.

The authors, editors, and other participants associated with volume hope that the materials offered to the reader will help him address in the most difficult and multiplane problem of biodiversity.

— Larissa I. Weisfeld, Nina A. Bome

INTRODUCTION

This book discusses theoretical approaches to the taxonomy of biological systems and theory and mathematical approaches to the problem of plant diversity. Particular attention is given to theoretical and practical problems of soil and the environmental sustainability of phytocoenosis.

This volume provides information on the distribution of chemical elements in the soil-plant system and on migration of chemical elements in the food chain. The composition of the soil and the distribution of elements in the soil-plant system are manifested as adaptations of plant organism to environmental conditions.

This book also presents precision farming techniques, including the introduction of differentiated agrochemicals considering variability of soil fertility and crop conditions. An important element for the conservation of and adaptation to plant organism to environmental conditions is the use of physiologically active compounds.

The book contains articles on theoretical and applied problems of biological systems, of environmental sustainability phytocoenosis, and on their biodiversity to enhance the productivity of agricultural crops: cereals, legumes, vegetables and fruit.

This volume organizes the chapters in a rather different in order to prepare scientifically based fundamentals of compromise between the natural wish to increase food production and the aspiration for preserving the habitat of humans as *Homo sapiens*, not only for the current generation, but also for future ones.

Their authors are well-known scientists in genetics, selection, horticulture, as well as in agrarian marketing.

This volume is based on the understanding that proper selection of agro-climatic zones for the cultivation of particular varieties and cultivars of varieties cultures, optimization of mineral nutrition, improvement *in vitro* of the technologies, as well as deep analysis of agricultural production marketing outlets, decrease the anthropogenic load on the environment, and environmental pollution. At the same time the forecast of production distribution perspectives minimizes risks of overproduction, which remains in the 21st century as it did in the previous one.

— **Anatoly I. Opalko**

PART I

INTERACTION OF ANTHROPOGENIC PRESSURE ON THE ENVIRONMENT: GLOBAL WARMING AND BIOLOGICAL STABILITY

CHAPTER 1

ANTHROPO-ADAPTABILITY OF PLANTS AS A BASIS COMPONENT OF A NEW WAVE OF THE "GREEN REVOLUTION"

ANATOLY I. OPALKO[1,2] and OLGA A. OPALKO[1]

[1]National Dendrological Park "Sofiyivka" of NAS of Ukraine, d. 12-a, Kyivska St., Uman, Cherkassy region, 20300 Ukraine

[2]Uman National University of Horticulture, d. 1, Instytutska St., Uman, Cherkassy region, 20305, Ukraine; E-mail: opalko_a@ukr.net

CONTENTS

The destiny of world civilization depends upon providing a decent standard of living for all mankind.
— *Norman Borlaug*

ABSTRACT

The results of a retrospective analysis of the historical background of the "green revolution" interconnected with current ecological and social problems were presented. Key events and processes, affecting global changes of the environment, in particular those which concern the ways of overcoming the lack of foodstuff in the developing countries, rational use of land resources, introduction of new cultivars including the ones developed by biotechnological methods, were emphasized. The beginning of radical changes, which took place in agriculture of the developing countries in the 40s–70s of the previous century and were classified in the world literature as the "green revolution" was analyzed as well as the role of moral-ethical life principles of Norman Borlaug in its successful accomplishment. It was suggested to consider anthropo-adaptability of new cultivars as the aim of contemporary breeding and a basis component of a new biotechnological wave of the "green revolution."

1.1 INTRODUCTION

The preservation of any kind of living organisms as a biological species in its entirety, including *Homo sapiens*, is due to a survival of some of its representatives, their physical existence in a highly organized and improved system of propagation and symbiosis. From a biological point of view, man does not differ much from other highly developed mammals. The criteria of *Homo sapiens* concerning brain size, mental abilities, speech, social way of life, suggested in school textbooks, are far from being undeniable; however, in a complex of the above-mentioned criteria a behavioral component is most likely to be a fundamental element. Animals are impelled to action mostly by hunger and an instinct of propagation. The major components of the inhibition of the actions initiated by the instincts mentioned earlier, in a simplified interpretation, range within: "want—do not want," "want—cannot," "want—can—but feel scared." It does not matter what "scares"—water, fire, a flock leader, a wolf, a master's stick and many-many other factors. The same inhibitors affect human behavior, but "want and can, even do not fear—but feel ashamed" should be added. It is "feel ashamed" that expresses morale, the consequence of abstract thinking (in particular, perception of distant consequences of their actions) and memory of ancestors (history) which are indigenous only to man. Other differences are rather nominal. Direct correlation between cognitive abilities of an individual and brain mass was not identified, positive correlation was observed in large samplings though. Social animals are well known all over, including dolphins, primates, dogs, bees, ants and others. A scientific subject "Zoosemiotics" is resulted from the studies about bio-communication of animals [1–3]. In turn, a distance between the most developed animal and man is determined mostly by an individual's morality, but public ethics defines civilization of a state. The provision of decent living standards for all its citizens should be considered the main criteria of society civilization, but neither chemization, nor electrification, nor computerization, nor sport achievements, nor space exploration success and definitely not military superiority can play this role.

In this context, radical changes associated with the name of Norman E. Borlaug and which took place in agriculture of the developing countries in the 40–70s of the previous century require special analysis. Emphasizing an exceptional significance of these changes for world economy in general and for the solution of food problem in particular, the then head of the USAID William Gaud called them the "green revolution." Norman Borlaug's personal contribution to the modernization of world agriculture received world recognition. High appraisal of his achievements was confirmed by awarding him Nobel Peace Prize in 1970 [4], and in 2014 scientific societies and world community marked his 100th anniversary, as an outstanding scientist and founder of biological and agrarian science, who prevented over 3 billion people from malnutrition [5]. In the biography of Norman E. Borlaug, as a famous scientist and a first-rate organizer to introduce researches into world agricultural production, it is appropriate to underline his moral-ethical life principles brought up in the family and competition spirit of a trained athlete which could be developed and mastered exclusively by his ability to concentrate on the main issue, to control mental and emotional activity based on his passionate devotion to his work. From his early age he was characterized by impartial attitude to people of different nationalities, skin color, religion and political beliefs—without any signs of xenophobia. Keen sense of a novelty of the useful ideas for mankind which he developed in his student's years, his understanding that the best scientific discovery would be useless without proper efforts for its introduction motivated his activity. Life principles of Norman Borlaug combined with titanic hard work and persistence resulted in the scientist's success. The name of Norman Borlaug will always be associated with the name of the Father of the "green revolution;" not only he outlived its triumph but, without any bias, he facilitated the beginning of a new biotechnological wave of the "green revolution."

1.2 HISTORICAL BACKGROUND OF THE "GREEN REVOLUTION"

Homo sapiens have always been aimed at saving from famine since prehistoric times. In various time-periods man as a biosocial being appeased hunger in different ways, but he received food in combination with nature. In Paleolithic, Old Stone Age, when population density was low, nomadic (sometimes seminomadic) way of life prevailed, man's needs slightly differed from the needs of the animals of the similar size; thus the principles of self-regulation of "man-nature" relations met the requirements of a stable coexistence of people with environmental bio-sphere. To survive in those conditions, it was enough to match the rules of existence and evolution of biosphere nature [6]. Under this pure consuming (taking) type of natural resource use, man could afford himself not to worry about the restoration of the destroyed area without tangible negative outcomes. There was no problem to leave poor hunting grounds and to find new, better ones. After that, thanks to natural re-cultivation, destroyed localities self-regenerated but not as fast as they were ruined. It took from several decades to two-three centuries for this process [7].

In late Paleolithic (35–10 centuries ago), due to global warming which caused dramatic climate changes, the number of large mammals reduced, and some species disappeared. This resulted in a considerable decrease of the population and most likely initiated the change in man's household life [8].

In the middle of Mesolithic, Middle Stone Age (10–9 centuries BC), hunters and fishermen appeared, they did not lead a roving life, and they spent most of the time at one place. As paleontological studies prove, sites in Palestine were the first known examples of the transfer from roving to settled life, which stimulated considerable changes in machinery and forms of running households occurring in Neolithic (8–3 thousand years BC). Further reclaiming of new areas by man and their more efficient use were classified as "Neolithic revolution" in the history [8]. Plant domestication, which continues till present time, goes back to Neolithic. It was then that breeding began—when an ancient man scratched the ground and sowed the seeds of a wild plant whose fruits he liked most of all [7, 9].

Gradually, instead of walking around forests and steppes in search of food, man learned how to grow plants and store their fruits and also to keep animals. The feasibility to save wealth without spending time on necessary hunting, fruit and root harvesting of wild plants facilitated self-improvement of man. There appeared settlements where various crafts began to develop including those, which were not connected with arable farming, hunting and fishing. Instinctively ancient growers directed the evolution of domesticated plants towards the satisfaction of their various needs exactly the same way as nature, based on the vectors of moving natural selection, moved evolution in accordance with the environmental changes. Sometimes the courses of breeding and evolution coincided when artificial selection according to hardiness and higher propagation coefficient took place, but sometimes their courses disagreed and even became opposite. Small grain, rachis fragility in wheat, barley and other cereals, small multiseeded fruits and similar features enhance the adaptability of wild plants, facilitating the increase of their propagation coefficient; however, the same features make cultured plants less attractive for man. They were not supported by artificial selection even if they occurred. On the contrary, the evolution guided by man (breeding) favored the accumulation of other characteristics, valuable for arable farming and horticulture, in the genotypes of cultivated plants, that is, signs of anthropo-adaptability which we understand as the ability of a new changed plant to satisfy any needs of man (utilitarian and esthetic) [7, 9–13].

Priority in using wild plants in food and their selective domestication are not clear yet. It can be assumed that fruits and leaves edible without being cooked were used first; they saved from hunger when hunting failed. When time progressed self-seed plants brought from forests and steppes could be seen near caves (or other shelters of man). Several centuries later, conscious cultivation of plants to be used in food and medicine began. Further development of breeding had a great impact on general arable farming. New cultivars required better care for their growth, which initiated the appearance of new agricultural tools. The yield increased and reduced the dependence on hunting and fishing; the need to rob neighboring tribe did not exist any more. The lack of mechanisms explained the necessity for a grower to have a constant contact with plants; this factor favored the becoming of breeding very much. The selection of better plants was made mostly according to the results of organoleptic evaluations. A grower was encouraged by natural curiosity supported by a desire to find new food sources. Crop production, vegetable production and fruit production as well as dual-purpose cattle breeding and poultry production, developed as a result of animal domestication, gave man a unique opportunity to live a civil (contrary to military) life, without killing anyone in the competition for food. Hence, it can be stated that the achievements of ancient breeders created a material foundation of the civilization [7, 9], which unfortunately we failed to use properly till now.

We can suppose that on the territory of present-day Ukraine a gradual transition from gathering wild plants to growing of some crops took place during Cucuteni-Trypillian culture. Archeologists found melon seeds (and possibly cucumber ones) during excavation of Chersonese Tavricheskiy dated from IV–II centuries BC. There is evidence of Kievan Rus period concerning cabbage, turnip, onion and garlic consumption. Arabian writer Ibn Fadlan described the rite of onion offering to Gods which he watched being among of Old Russian merchants [9].

As the history of our planet proves, living organisms, including man, and at present man plays a key role, influence the environment, which changes considerably under a constantly growing anthropogenic pressing/stress. The effect is exercised directly on other organisms through existing changes of habitat, the ones, which appeared earlier, that is, before the habitat was changed [14]. In this connection, we can agree with the opinion of D.E. Goldberg [15], as the grounds for understanding of interspecific competition of plants (and we think that not only plants), the consequences of such competition should be divided into two interdependent components—"reaction" and "effects." Within the competition among species, "effects" of the influence of one species are seen in such change of the environment that genotypes of another species are to "react" to it. Explaining this concept in other words, species A "builds" habitat, in which species B must live, but species A will have to survive in new habitat partly "built" by species B and other changed species and/or in the habitat which was changed as a result of the disappearance of this or some other species. The consequences of excessive influence of *Homo sapiens* on "neighbors" in habitat were vividly expressed and became a real threat for the survival of man himself beginning from the third epoch of globalization, when one species disappears every 20 minutes, which is a thousand times faster than ever during a larger part of the history of our planet [16, 17]. Similar process take place and they concern atmospheric, water, land, mineral and other natural resources. Therefore, presently a very complicated and vulnerable ecosystem, which formed and very slowly evolved during centuries, can be destroyed by irrational actions of man for several years, months and even days [18].

In a great diversity of a natural resource complex and with different approaches to their classification solar energy, intraplanet warmth, atmospheric, water, land, mineral resources, flora and fauna, drinking water, wild plants and a great deal of other gifts of the planet, man's well-being depends on, and eventually the survival of the very species of *Homo sapiens*, it is the land resources that present a special interest of business circles in other countries and a growing concern of the world community.

The concern is caused by both their condition and the increased interest of business people to land resources with their biological derivatives, and first of all to plant and animal world. The problem of rational use of the natural resources is not new. However, at present in social-political conditions of the third epoch of globalization information pressing around ecological problems became stronger, and it is often observed in irrational forms causing responsive irrational reaction of numerous entrepreneurs and some officials in all the countries of the world. A great number of publications in foreign [14, 19–23] and domestic editions [18, 24–28] prove an extraordinary significance of this problem.

Wise application of the achievements of breeding can help slow and in prospect reverse the damaging processes of land resource [12, 13, 19, 20, 25]. However, taking into account selfishness of a particular individual [29], and group egoism of every state, and self-interest of almost every national, political, clerical and other grouping, it is difficult to expect any voluntary important self-restraint which in fact is due to real fears of depletion, pollution or damage of land or any other natural resources. On the contrary, the observation of the last decades demonstrates planned development of irrational needs. These needs are associated with neither the survival, nor the satisfaction of physiological and social, spiritual and material and other rational needs, including self-preservation, self-affirmation and self-actualization [30, 31]. Attempts to limit public consumption were not effective in the past, they bring no results nowadays, and it is hopeless to believe that they will be more efficient in the near future. These artificial needs are implanted to please the commercial interests of some unfair producers, corrupted officials, and they are automatically supported by the majority of undereducated consumers.

Unfortunately, Ukraine (as well as other former Soviet Union countries) failed to avoid the above-mentioned and other global problems concerning national use of land resources and difficulties in restoration of the damaged environment. The more so, the system, which controls how the requirements of good ecological legislature of Ukraine are observed does not ensure the expected results [24, 32]. In most cases ecological expertise and estimation of the industrial effect on the environment are carried out in view of the projects dealing with the creation of new companies as well as modernization and reconstruction of the existing ones, whereas the control over functioning businesses, as a rule, consists in periodic inefficient checks, performed by the inspections of the authorized bodies of the Ministry of ecology and natural resources of Ukraine. Managers, holders and owners of ecologically dangerous businesses and entities do not show any initiative; instead they use every possible and impossible way to ignore ecological measures, which require investment. There are no real leverages to stimulate nature-protecting measures. However, monitoring of the farm producers' performance confirms that there are many violations, in particular, excessive erosion, which quite often leads to complete soil damage without any compensation of destroyed soil fertility [24, *26*, 33]. Some researchers consider their expectations, concerning the solution of ecological problems, in the context of Ukraine's intention to enter European and world community. It is assumed that this can encourage local entrepreneurs to attract foreign investors to reconstruct and reequip their businesses, which in turn will make them ecologically safer, and mainly they will meet international requirements and will be competitive at world markets [24].

More affirmative suggestions are presented in the often-quoted article of Matson et al. [22], there it is stated that the extension of cultivated lands and intensification of farm production are among predominating global changes of today, their advantages cannot be ignored. During the last 50 years the world growth of foodstuff production is ensured mainly due to the use of high-yielding cultivars, application of fertilizers and pesticides, irrigation and other components of the intensification of crop production.

However, intensification goes along with disproportions of biotic interactions in the ecosystem; this can have serious local, regional and global ecological consequences. The application of ecologically grounded sociotechnical management strategy can enhance the stability of farm production along with the decrease of allowed limits of undesirable consequences. In this context, it concerns minimization of tillage and innovative crop production [14, 21], which uses anthropo-adaptive cultivars.

Land resources are the foundation of Ukraine's national economy: 95% of food fund and 2/3 of consumer goods fund are formed on their use. The share of land resources in the country's productive forces exceeds 40%. And agricultural reclamation of dry land is 72.2%, the share of arable land in total agricultural land area is equal to threatening 79%. This is much more than agrarian-developed countries have. The share of arable land in Great Britain, France and Germany ranges from 28 to 32%, dry land reclamation is 40–58%, while in the U.S.A. (an ecologically safer country) arable land covers 15.8% and agricultural land constitutes 35.9% [24]. Among numerous ingredients of innovative crop production, *achievements* of plant *breeding* should be recognized as a base component of the rational use of land resources. By 2050 owing to plant breeding, the increase of the yielding capacity of leading food crops by 1.5 times is predicted worldwide, which will make it possible to provide enough food for future 9-billion population of the planet without a serious increase of arable lands [34]. The purpose of new *breeding* programs is to develop new cultivars which will be both adapted to standard natural changes of growing conditions and able to ensure expected productivity and quality of the yield in anthropogenically changed habitat [9, 10, 12, 13, 35–37]. A priori it is expected that extensive introduction of these anthropo-adaptive cultivars can improve ecological situation without derogating from the interests of producers and consumers of agricultural produce [12, 13, 35] worldwide and in every state taken separately.

It should be mentioned that earlier, in the epoch of industrial development, man and nature had enough time to adapt to new phenomena of smoothly progressing social transformations. Intensification rates of agriculture in industrialized countries began to speed up in the XIX century along with the growth of city population and the decrease of the people employed in agrarian production. Owing to the introduction of new highly productive cultivars, mechanization, chemization and alike, modernization of the agrarian sector of the economy occurred rather regularly—with stable growth of gross yields and yielding capacity of the main farm crops [38]. In most modern European states relative stability of the growth of agrarian production was interrupted mainly in the war years, in particular World War I, and in the U.S.A.—in the years of "Great depression" 1929–1933, when the demand for American food at European markets fell down due to the rebirth of agriculture in Great Britain, France, Belgium, Holland (the Netherlands) and in some other countries which went out of war. On the other hand, in Latin America, South and South-East Asia, and Africa small agricultural business prevailed at traditionally small farms: draft cattle instead of machinery, lack of proper technical means for cultivation and harvesting, no chemical fertilizers and pesticides, no irrigation, local cultivars only.

1.3 TRIUMPH OF THE "GREEN REVOLUTION"

On the background of some political changes at the end of the XIX and the beginning of the XX centuries low yielding capacity and poor labor productivity in the regions mentioned were the reason for food deficit, constant malnutrition of a greater part of the population, and in off-years they led to famine. Tragically repeated grain crop failures, which Mexico experienced at the beginning of the 40s of the last century forced the government of this country to ask the US administration for help. On the initiative of the US Vice-President Henry A. Wallace a delegation of American scientists funded by charitable Rockefeller Fund was sent to Mexico. Young plant pathologist and breeder Norman Borlaug was part of the delegation [5, 39].

When Norman Borlaug found out that diseases, in particular rust, were the main reason for unstable wheat yields at the farms in Mexico, he began to look for initial material for a purposeful breeding program. The scientist discovered that among many cultivars brought from other countries only four of them were resistant to rust. Crossing of the selected immune cultivars with local ones and including them in hybridization with Japanese cultivar Norin 10, which was characterized with a short firm stem, allowed to single out a short-stem, resistant to lodging so-called "dwarf" wheat, which was resistant to diseases and pests and adapted to specific aspects of Mexican climate; it gave super high yields (up to 40 metric centner per hectare (centner/ha) when moisture and mineral nutrition were sufficient. It is worth mentioning that Mexican farmers harvested 8–10 centner/ha of grain in those years.

The use of "shuttle breeding" beginning from 1945 was a great contribution to Norman Borlaug's success—hybrid F_2, F_3 and consecutive generations were studied by turns in various (summer and winter) seasons in two zones with different (contrast) agro-meteorological conditions, the distance between them was 1000 km and the difference in above-sea level was over 2.5 km [40]. As a result, in 1956 Mexico shifted from a regular grain importer to a grain exporter having satisfied the country's needs in grain completely. Mexican cultivars of wheat Sonora 63, Sonora 64, Lerma Rojo and Mayo 64 got world recognition. In 1963 the *International Maize* and *Wheat Improvement Center* (CIMMYT) was established; till now the institution actively promotes the "green revolution" worldwide [41–43].

From the very beginning the "green revolution" had rather a centered and wave character than a total one [44]. Having started in the 40s of the previous century in Mexico, it spread gradually across the countries of Asia, in particular in India, Pakistan and Philippines at the end of the 60s and at the beginning of the 70s (faster than in Mexico though). In Afro-Asian countries the "green revolution" did not naturally result from previous social-economic evolution. It was brought to life as denial of all social forms, being based on the traditional technical foundation; it became an "industrial surge," quite an adequate after-sound of the epoch of scientific-technical revolution which was taking place in developed countries [45]. Before the "green revolution" was introduced there existed a great gap between the methods of agriculture used in developed and developing countries. In South and South-East Asia, 6 kg/ha of chemical fertilizers were applied, whereas in West Europe—about 100 kg, and in Japan—250 kg/ha. Due to the results of the CIMMYT performance, Norman Borlaug's activity and purposive investment, food deficit decreased considerably in India and Pakistan; and such courtiers as South Korea, Taiwan, Singapore, Hong-Kong, Malaysia, Indonesia, Thailand and Philippines in South-East Asia as well as Mexico, Brazil, Argentina and Chile in Latin America showed an outstanding economic and even social progress.

Similar processes but much later (the 70–80s of the last century) occurred in Ukraine as well as in other republics of the former Soviet Union but under other slogans of "chemization" and "intensive" or "industrial" technologies of agricultural crop cultivation [46]. In the 40–60s of the last century 8–11 centner/ha of grain crops were harvested in the USSR (Mexican farmers harvested as much in the 30–40s). In Ukraine these indicators were 1.5 times higher [47, 48], yet they were not sufficient. Thanks to the work of outstanding breeders V.N. Remeslo, P.P. Lukianenko, F.G. Kirichenko, V.V. Morgun, D.A. Dolgushina and others, as well as the introduction of new intensive cultivars, developed by them, together with intensive cultivation technologies, the yielding capacity of winter wheat doubled within a 20-year period.

Triumphal progress of the "green revolution" was ensured by wide borrowing of new cultivars and advanced technologies from the most developed agro-systems with stable yielding capacity to bring them to the fields of the developing countries.

All this helped achieve better food safety in South America, South-East Asia and many districts of Africa. The unbelievable increase of wheat and rice yielding capacity gave all grounds to hope that the threat of famine remained in the past and a new ear of agricultural development started on the planet. Norman Borlaug was among five people in the world history who were awarded Nobel Peace Prize, President's medal of Freedom and Gold medal of the US Congress. The four others bestowed such honor were Mother Teresa, Martin Luther King, Nelson Mandela, and Elie Wiesel [49].

1.4 NEW WAVE OF THE "GREEN REVOLUTION"

Investments into an agrarian-industrial complex from the developed countries and transnational companies promoted the "green revolution," but mostly it was due to extensive activity of the governments of the developing countries, which favored the creation of the additional infrastructure and usually supported high purchasing prices of agricultural produce. As a result of a goal-oriented national economic policy, by 2000 the share of Asian developing countries in world agricultural production caught up with their share in the number of the world population, they were 53.1 and 53.7%, correspondingly [44, 45].

Ambiguity of so-called intensive and industrial technologies of farm crop cultivation, which were widely advocated in those years in the USSR, was commented by academician Zhuchenko [36] (although in a veiled manner, he was one of the first who spoke in favor of adaptive potential of cultivated crops). The point is that super intensive cultivars appeared in those years in our country, and one-two decades earlier in many other countries, could give all-time high yields when industrial technologies were applied. Since energy carriers were cheap, the cultivation of these cultivars guaranteed high economic efficiency. However, in current conditions of real market economy, which spread all over the post-Soviet states, when energy carriers become more expensive, it gets more difficult to meet the requirements of industrial technologies. Besides, because of insufficient resistance of many super intensive cultivars to pests and diseases, their cultivation requires additional expenses for purchasing pesticides, which in turn worsens the environmental ecology.

In view of this, last decades do not prove the stability of the results of the "green revolution" initiated by Norman Borlaug. Thus, in 1950–1984 the increase of gross grain yielding capacity exceeded the growth of the world population, and average grain production per capita increased by the third—from 247 to 342 kg per year, but by the end of the 90s this indicator decreased to 317 kg. Norman Borlaug had to admit that his success in preventing famine was temporary, further achievements could be expected from the introduction of biotechnological methods into plant breeding and from the improvement of a demographic situation. The scientist stated: agriculture is a unique type of human activity which can be considered as art, science and trade of managing plant and animal growth for man's needs. The main purpose of this activity has always been the increase in agricultural output to feed constantly growing population of the planet [45, 50, 51]. Besides, he expressed his belief that today (the year of 2000 was meant) mankind would have technologies totally ready for the introduction, or the ones which were at final stages of their development, that their application would allow to safely feed future 10-billion population of the plant. The matter was whether food producers all over the world would have access to these technologies [52], whether farmers in every corner of the world would have access to the most advanced farm practices to cultivate the most highly productive cultivars of cultivated crops. Norman Borlaug warned about one of the biggest problems of the twenty-first century—education renewal and widening access to it for all social groups in all regions of the planet, and education updating should go with science. Opposition to the use of biotechnological agricultural crops observed in Europe and in many other regions of the world could be avoided, provided more people received better education about genetic variety and its variability; using inadequate education of people, extreme ecologists seem to do their best to break scientific progress in biology and agrarian production [52]. In addition they use abbreviation GMO, which is not very much clear but quite threatening for undereducated citizens. This journalistic-routine and not very accurate slang moves slowly from mass media to biotechnological science in Ukraine and not only. Present-day spread of biotechnological crops in the world as well as the extension of their grown areas, despite "GMO fright," can be considered as biotechnological realization of Norman Borlaug's idea [21, 30]—a new wave of the "green revolution" [44].

This explained the revision of the strategy aimed at further yielding capacity increase of agricultural plants and that of agro ecosystem productivity. To substitute traditional concepts of plant breeding, a new approach based on the development of the cultivars with anthropo-adaptive features is suggested. Anthropo-adaptive cultivars can guarantee high (though not super high) yields under favorable conditions and in prolific years, and slight decrease of the yield and its quality in unfavorable years, that is, they can stabilize agrarian production. According to a new concept of breeding programs yielding capacity is considered as a derivative of two components—productivity and complex hardiness/resistance of new cultivated plants [7, 9].

1.5 CONCLUSIONS

Thus, anthropo-adaptivity as the ability of a cultivar to satisfy man's needs – to ensure yearly yields and quality, resistance to diseases, pests, unfavorable soil-climatic conditions, adaptability to mechanized cultivation and harvesting of the crops, the ability to efficiently accumulate sun energy, to grow on polluted backgrounds without absorbing harmful substances (pesticides, nitrates, salts of heavy metals, radio nuclides), resistance to unfavorable effect of man' business performance—can be the aim of plant breeding and a base component of a new biotechnological wave of the "green revolution."

KEYWORDS

- **biotech crops**
- **domestication**
- **Father of the "green revolution"**
- **food production**
- **Nobel Peace Prize**
- **Norman Borlaug**
- **plant breeding**
- **wheat cultivars**

REFERENCES

1. Hare, B., Call, J., & Tomasello, M. (1998). Communication of Food Location between Human and Dog (Canis Familiaris), Evolution of Communication, *2(1)*, 137–159.
2. Janik, V. M., Sayigh, L. S., & Wells, R. S. (2006). Signature Whistle Shape Conveys Identity Information to Bottlenose Dolphins, Proceedings of the National Academy of Sciences, *103(21)*, 8293–8297.
3. Slabbekoorn, H., & Smith, T. B. (2002). Bird Song, Ecology and Speciation, Philosophical Transactions Biology Sciences, *357(1420),* 493–503.
4. Brown, L. R. (1970). Nobel Peace Prize Developer of High-Yield Wheat Receives Award (Norman Ernest Borlaug), *Science, 170(3957),* 518–519.
5. Quinn, K. M., & Dr. Norman Borlaug (2009). Full Biography the World Food Prize URLhttps://www.worldfoodprize.org/en/dr_norman_e_borlaug/full_biography/

6. Man of Technical-Industrial Civilization in the Conditions of the Information Support of Society Monograph [Eds. Viktor, E. Ivanov et al. (2011)] Kaliningrad Kaliningrad State Technical University, Kaliningrad Department of the International Academy of Information Support, 504p (in Russian).
7. Opalko, A. I. (2013). Anthropo Adaptability of the Plants as a Base Component of the Rational Use of Land Resources, The Development of the Regions in XXI century Proceedings of the 1st International Scientific Conference North-Ossetian State University Named After Hetagurov, K. L. (October 31–November 2, 2013) [Under the Editorship of Sozanova, V. G.] Vladi Caucasus North Ossetian State University Press, Part I C. 348–354 (in Russian).
8. Kuznetsova, O. D., Permiakova, L. I., Lisovskaia, Ye G. et al. (2002). History of Economics a Textbook [Kuznetsova, O. D. & Shapkina, I. N. Eds] Moscow INFRA-M Publishing House, 384p (in Russian).
9. Opalko, A. I., & Opalko, O. A. (2012). Breeding of Fruit and Vegetable Crops Part 1 General Principles of Vegetable Plant Breeding [Anatoly Opalko, I. V. Ed.] Uman National Dendrological Park "Sofiyivka" of NAS of Ukraine Press, 340p (in Ukrainian).
10. Opalko, A. I., & Kotko, I. K. (1989). The Effect of the Exogenous DNA on Winter Wheat Anthropo Adaptivity, Selective-Genetic Methods of Productivity Increase of the Cultivated Plants, Uman State Agrarian University Press, 11–15 (in Russian).
11. Opalko, A. I., & Kotko, I. K. (1992). The Use of the Foreign DNA to Provide Winter Wheat Anthropo Adaptive Mutations, Plant Breeding and seed Production (Branch Bulletin), *2*, 1–4 (in Polish).
12. Opalko, A. I., & Opalko, O. A. (2004). The Problems of enhancing the Anthropo Adaptive Potential of the Cultivated Plants, Urgent Problems How to Save Hardiness of Living Systems, Proceedings of the VIII International Scientific Ecological Conference (Belgorod, September 27–29, (2004)), Belgorod State University Press, 152–153 (in Russian).
13. Opalko, A. I. (1996). Peculiarities of the Breeding Concerning Anthropo Adaptivity, Summaries of the Research Work in (1991–1995) Uman State Agrarian University Press, 66–67 (in Ukrainian).
14. Vandermeer, J. (2011). The Ecology of Agro Ecosystems Sudbury Massachusetts et al. Jones & Bartlett Publishers, LLC, 387p.
15. Goldberg, D. E. (1990). Components of Resource Competition in Plant Communities Perspectives in Plant Competition [Grace, J., & Tilman, D. Eds] N. Y. Academic Press, 27–49.
16. Friedman, T. L. (2009). Hot, Flat, and Crowded Why We Need a Green Revolution and How it Can Renew America London Penguin, 528p.
17. Opalko, A. I. (2012). Problems of Saving Genetic Diversity of Plant Resources in the Third Epoch of Globalization, The State of the Mountainous and Slope Lands in North-Caucasian Federal District and the Problems ofPreserving Soils Fertility, the Development of Arable Farming Systems, Meadow-Pasture Areas, Livestock Production on Landscape and Resource-Saving Basis, Proceedings of the Meeting of Russian Agricultural Academy, Department of Arable Farming, North-Caucasian Research Institute of Mountainous and Piedmont Agriculture, Vladicaucasus Tsopanova, I. P., 146–147 (in Russian).
18. Melnyk, L. H. (2006). Ecological Economics, Sumy University Book, 367p (in Ukrainian).

19. Arent, K. P. (2011). Economic aspects of Ecological Development of the National Economy Monograph, Moscow State University of the Environment Press, 193p (in Russian).
20. Connor, D J., Loomis, R. S., & Cassman, K. G. (2011). Crop Ecology Productivity and Management in Agricultural Systems, Cambridge University Press, 562p
21. Coughenour, C. M., & Chamala, S. (2000). Conservation Tillage and Cropping InnovationConstructing the New Culture of Agriculture, Ames Iowa State University Press, 360p.
22. Matson, P. A., Parton, W. J., Power, A.G., & Swift, M. J. (1997). Agricultural Intensification and Ecosystem Properties, *Science*, *277*, 504–509.
23. Romanova, T. A., Chervan, A. N., & Andreeva, V. L. (2011). Theoretical Basis and Practical Significance of Investigations into Soil Cover Patterns, Eurasian Soil Science, *44(3),* 272–280.
24. Havrylenko, O. P. (2008). Eco-Geography of Ukraine Kyiv Knowledge, 646p (in Ukrainian)
25. Hulinchuk, R. M. (2012). The Problems of the Formation and Use of the Potential of Agricultural land, *Innovative Economy, 6(32),* 102–105(in Ukrainian).
26. Medvedev, V. V., & Laktionova, T. M. (2003). Current Situation in Soil Monitoring in Ukraine Ecological gazette Kyiv All-Ukrainian *Ecological League*, 8–10 (in Ukrainian).
27. Oliynyk, Ya B., Shyshenko, P. H., & Havrylenko, O. P. (2012). Principles of Ecology Kyiv Knowledges, 558p (in Ukrainian)
28. Khodakivska, O. V. (2012). Natural-Resource Potential of the Agricultural Areas as to their Stable Developments of the Economy, *4(12)*, 160–163 (in Ukrainian)
29. Berger, P. L., & Luckman, T. (1991). The Social Construction of Reality, a Treatise in the Sociology of Knowledge London Penguin Books Ltd, 251p.
30. Dumont, F. A. (2010). History of Personality Psychology Theory Science and Research from Hellenism to the Twenty-first Century, Cambridge University Press, 559p
31. Hergenhahn, B. R. (2013). An Introduction to the History of Psychology, Belmont, C. A. et al. Cengage Learning, 720p.
32. Buhera, S. I. (2011). Legal Regulation of the Output of Farm Produce, Problems of Adaptation to the Requirements of the European Union, Our Law, *4*, 157–161 (in Ukrainian).
33. National Report about the State of the Environment in Ukraine in (2000). KyivMinistry of Economic Resources (2001) 184p (in Ukrainian)
34. Jaggard, K. W., Qi, A., & Ober, E. S. (2010). Possible Changes to Arable Crop Yields by (2050) Philosophical Transactions of the Royal Society, *Biological Sciences, 365,* 2835–2851.
35. Balabak, A. F., & Opalko, A. I. (1999). Efficiency of Rhyzogenesis in Increasing Anthropo Adaptive Potential of Small Fruit Production, Bulletin of Agrarian Science (Special Edition, September), Kyiv Agrarian Science, 65–69 (in Ukrainian).
36. Zhuchenko, A. A. (1988). Adaptive Potential of Cultivated Plants, Kishinev Science, 766p (in Russian).
37. Zhuchenko, A. A. (2007). Mobilization of World Genetic Resources and Phyto-Technologies Which Improves the Environment, Moscow Peoples' Friendship University of Russia Press, 149p (in Russian).
38. Potapov, N. A., Salitskiy, A. I., & Shakhmatov, A. V. (2008). The Green revolution Economy of Present-Day Asia Moscow International Relations, 37–44 (in Russian).

39. Weasel, L. H. (2009). Food Frayinside the Controversy over Genetically Modified Food, New York et al. AMACOM, 256p.
40. Miller, H. I. (2012). Norman Borlaug the Genius Behind the Green Revolution, Forbes URL http://www.forbes.com/sites/henrymiller/2012/01/18/norman-borlaug-the-genius-behind-the-green-revolution/
41. Borlaug, N. E. (1971). Man kind and Civilization at another Crossroad, McDouga, l. L. Memorial Lecture, Conference Food and Agriculture Organization of the United Nations, Sixteenth Session, 73p.
42. Borlaug, N. E. (1965). Wheat Rust and People, *Phytopathology*, *55,* 1088–1098.
43. Swaminathan, M. S. (2009). Obituary Norman Borlaug, E. (1914–2009). Plant Scientist who Transformed Global Food Production, Nature, *461, 7266*, 894.
44. Kabachevskaia, K. V. (2007). A Wave Component of the "Green Revolution"the Culture of the Peoples of the Area Nears the Black Sea a Scientific *Journal Crimean Scientific Center of the National Academy of Sciences*and Ministry of Education and Science of Ukraine (Simferopol), Taurida National Vernadsky V. I. University Inter-University Center "Crimea" (Simferopol), Simferopol Inter-university Center "Crimea" Press, *106*, 49–54 (in Russian).
45. Savitskaia, M. T. (2012). Social-Philosophical Aspect of the "Green Revolution"of the Mid of XX century URL http://www.rgazu.ru/db/vestnic/2012(3)/gum/003.pdf (in Russian).
46. Iofinov, S. A., & Lyshko, G. P. (1983). Industrial Technologies of Crop Cultivation, Moscow Ear, 191p (in Russian).
47. Agriculture of the USSR (Statistic Reference Book), Moscow State Plan Edition (1939) Part V. Sown Areas and Yields, 136–247 (in Russian).
48. National Economy of the USSR in (1960) (Statistical Year Book), Arable Farming Moscow National Statistic Edition of the USSR (1961) 384–447 (in Russian).
49. Phillips, R. L., & Norman Borlaug, E. (2013). A Biographical Memoir Washington, NAS, 27p.
50. Borlaug, N. E. & Dowswell, C. R. (2004). Prospects for World Agriculture in the Twenty-First Century Sustainable Agriculture and the International Rice-Wheat System [La, R., Hobbs, P. R., Uphoff, N., & Hansen, D. O. Eds.] Madison Marcel Dekker Inc., 1–18.
51. Borlaug, N. E. (2007). Sixty-Two Years of Fighting Hunger Personal Recollections, Euphytica, *157(3)*, 287–297.
52. Borlaug, N. E. (2000). Ending World Hunger the Promise of Biotechnology and the Threat of Anti Science Zealotry, Plant Physiology, *124,* 487–490.

CHAPTER 2

OPTIMIZATION OF THE STRUCTURALLY FUNCTIONAL CHANGES IN THE CULTURED PHYTOCOENOSES IN THE AREAS WITH EXTREME EDAPHIC-CLIMATIC CONDITIONS

NINA A. BOME[1], LARISSA I. WEISFELD[2], SARRA A. BEKUZAROVA[3], and ALEXANDER Y. BOME[4]

[1]Tyumen State University, d. 1, Semakov St., Tyumen, 625003, Russia;
E-mail: bomena@mail.ru

[2]N.M. Emanuel Institute of Biochemical Physics RAS, d. 4, Kosygin St., Moscow, 119334, Russia; E-mail: liv11@yandex.ru

[3]Gorsky State Agrarian University, d. 37, Kirov St., Vladikavkaz, Republic of North Ossetia – Alania, 362040, Russia; E-mail: bekos37@mail.ru

[4]The Tyumen basing point of the N.I.Vavilov Institute of Plant Industry, 44, Bolshaya Morskaya Str., St-Petersburg, 190000, Russia;
E-mail: office@vir.nv.ru

CONTENTS

ABSTRACT

The formation of fundamental knowledge and practical approaches for the impact assessment of the modern agricultural land use on structurally functional state of domestic plants biological variety given weather-climate and soil factors is a paramount objective of our complex research.

On the basis of cultivated plants gene pool assessments totality analysis qualitatively new information about the changeability and variety of plant species in extreme conditions of the ambient with the revelation of valuable genotypes has been received. The problem of search techniques to minimization effects search on phyto- and microbiocoenosis is under consideration. New ways of phenotypic correction of practically meaningful plants characters are suggested providing for expand genetic diversity accessible for selection. The possibility of experimental mutagenesis application for the extension of genetic variety is being estimated; the possibility of genetically heterogeneous material receipt on the basis of complex of traditional and nontraditional techniques was proved.

2.1 INTRODUCTION

For the supplying with food, bioresource, ecological safety of the country, creation of new raw materials for the industry reliable maintenance, enrichment, study and careful use of the plants variety. At the same time, the variety of plants on Earth is under the threat of "genetic erosion" [1, 2]. The part of cultivated plants in the world does not exceed 3% from the general number of higher plants of the world flora, genetic variety is characteristic of them [3]. Genetic diversity of the plants enlarges the possibilities of choice and provides defense from the future unfavorable conditions such as extreme and changeable environment; drafts; growing threaten for ecosystems entailed by the climate change and pollution with biogenic substances. The human activity led to a considerable and mostly irreplaceable reduction of the life forms on the Earth [4].

In accordance with the Biodiversity Convention [5], techniques of genetic resources components assessment holding based on usage of complex databases and leading to their complementation with new data exist or they are being developed.

Along with the available information about resources (soil, water, vegetative cover, climatic and agro-climatic zones), additional evaluations are necessary, for example, with respect to microbial genetic resources, functions of agriculture biodiversity. The understanding of underlying causes and consequences of agricultural biodiversity loss is extremely restricted.

A vast diversity of native zones requires the creation and selection of genotypes adapted to definite edaphic-climatic conditions. The elaboration of genetic diversity of plants maintenance is indispensible to life, the sustainable use of vegetative genetic resources is not less important.

According to the FAO [6] data, there are 1750 gene banks in the world maintaining 7,03 million samples. Among them the main gene banks are opened in the USA, China, India, Russia, Japan. An incontestable role belongs to collections created for concrete ecological conditions [7].

2.2 RESULTS AND DISCUSSION

Our research aimed to create the database of samples of different ecological and geographic origin and to detect the role of ontogenetic mechanisms in the formation of phenotypic characteristics of cultivated plants will allow formulating new understanding of mobilization, study and preservation of genetic resources of cultivated plants.

The advantage of Tyumen gene pool that we are creating relatively to working collections of other research institutions consists in considerably higher diversity of organisms, presence of equipment, which allows to perform deep research of plant objects submitted to the effect of the complex of factors including stress ones, both in modeled, as well as in natural growth conditions. In the region there is no analogs of herbarium samples of cultivated plants, microbiological herbarium, collection of microorganisms.

International treaty on plant genetic resources for food production and agriculture of UN FAO [8] defines priority ways of conservation of plant genetic resources: *in situ*, i.e. in the nature, within natural plant communities, and ex situ, that is, in collections, including seed and genetic banks. Low-temperature methods of genetic material storage are now applied in the research institutions of Russia (e.g., N.I. Vavilov Institute of Plant Industry; and Kuban genetic bank of seeds).

Taking into account the insufficient knowledge about the reaction of genotypes on the long storage of biological material, in particular seeds, as well as their reaction on depressive, the methods of the preparation of seeds cultivated on different territories, in particular with cold climate, to the storing have scientific foundations, in our study we focus the attention on the elaboration of new methodological approaches to the long-term preservation of genotypes without loss of vitality.

Introduction and reintroduction are placed among perspective measures of restoration of endangered plant populations [9, 10]. A considerable experience is accumulated about plant reintroduction. Long-term experiments were performed in natural reserves and botanical gardens [11–13]. Success of reintroduction is determined by detailed research of wild populations with the detection of genetic and ecological peculiarities. The methods of cenopopulations observation [14] represents practical and theoretical interest. It is necessary to take into account that reintroduction works require considerable financial expenses. Reintroduction is recommended only if all possible ways of conservation and restoration of populations *in situ* showed to be ineffective [15].

For example, gene pools of mountainous meadows plant communities keep the highest adaptive properties, which determine their resistance to unfavorable factors [16]. Mobilization of genetic resources is closely related with the identification of each species, ecotype, sample according to main economically valuable characters with the evaluation of the possibility of its application in selection [17–19]. The results achieved and the research planned will give the new understanding of the re-introduction, successfulness indicators selection and of the duration of populations monitoring in new conditions [20].

The reaction of the variety on environmental factors is an important component in the maximum realization of its potential productivity. In the study of single species and their groups the method of their transmission from the different parts of species areal with contrasting ecological conditions in the homogenous environment of experimental field, garden or greenhouse is applied.

The transmission of plant populations from one environment to another can lead to the display of many different characters. It was shown on the example of winter wheat populations that small or big deviations detected can be accumulated and fixed by selection [21]. Zheleznov and Zheleznova [22] observed the variation of a set of characters in barley (seed productivity, plant height, vegetation period, protein content and some others. The transmission of populations into new ecological conditions leads to the change of their structure as a result of elimination ones biotypes or preferable development of others [22, 23]. According to opinion of the authors, it is necessary to take into account the presence of such variability when doing plant introduction and selection, formation of collections and when performing studies related with use of material from other regions.

The issue of adaptive strategies, allowing to populations and species to sustain their quantity, is an important topic of biological research. The complexity of characters, inherent in one adaptation direction or another, considerably complicates their research. Consequently, approaches detecting strategy types basing on the substance and energy cost of functions realized, especially on the resources distribution between vegetative and generative parts of plants according to competitive and reproductive functions, become more and more recognized. The determination of quantitative ratios of elements allows in some cases to imagine the mechanism of possible adaptive changes of generative sphere in ecological gradient [24].

The topsoil is indispensable component of biosphere in the processes of energy and mass exchange, where the main role belongs to the complex and dynamic system soil – plant. The distribution of substances in this system depends on multiple factors [25]. Among them physical and chemical properties of soil, the substance concentration in soil, elements mobility, forms of compounds and interactions determining the level of substance migration and availability to plants, as well as climatic conditions of environment are the most important [26]. These factors may not correspond to plants needs, forming conditions for stress and suppressing their growth. In its turn, plant organism has its selective specificity of absorbing of substances presented in the environment, its own reciprocal reactions to the shortage or excess of nutritional and exchange elements, mechanisms and stability level [27]. The application of physiologically active compounds influencing of adaptation to soil chemical composition in particular climatic conditions can be perspective in the given aspect [28–33]. Besides, the information about the content and distribution of chemical elements in the soil-plant system represents the interest for the basis of new developing line of investigation of structural and functional migration of chemical elements in trophic chains [34]. Our field and laboratory experiments in TBGU foresee the research of physicochemical properties and element composition of soil, as well as peculiarities of presence and distribution of elements in the soil-plant system, which are reflected in the plant organism adaptation to environmental conditions under the effect of physiologically active compounds.

The role of physiologically active compounds in the plant preadaptation to possible stress situations was shown [35]. The induction of accumulation of osmoprotector proline in barley leaves [36] or accumulation of lectin in the roots of wheat germs [37] participating in nonspecific protective reactions may be attributed to such reactions. The data on the decrease of negative effect of salinization on the wheat germs roots growth under the effect of epibrassinolide (brassinosteroid) may be an additional illustration of its antistress effect [28]. In the field experiments of Orekhova et al. [38] the positive effect of epibrassinolide on the gluten content and quality in winter wheat seeds was detected.

High amount of publications concerns the research of activation of plant growth and development processes, photosynthesis, nitrogen assimilation, increase of plant resistance to stress factors, increase of productivity and quality of agricultural production under the effect of epibrassinolides [39–42]. The effects of abscisic acid on the content of polyamids and proline in haricot plants under salinity stress [43] and on the expression of cysteine proteinase genes under cold adaptation of wheat plants [44] were shown.

The earliest indications on the possibility of induction of plants immunity to damaging biological agents (viruses, bacteria, fungi) by means of plants vaccination were made by Vavilov [45]. Now it was proved that immunomodulators are able not only increase resistance of susceptible species to damages but also considerably increase the resistibility of constitutionally stable plant genotypes to the damaging effect of phytopathogens and phytophagans by means of protective (biochemical, morphological and growth) reactions complex induction [46–49].

The problems of plants biological diversity conservation, activation of mechanisms of adaptation to stress factors are annually discussed at forums and conferences, e.g., at the International conference devoted to the 115th anniversary of N.I. Vavilov Institute of Plant Industry "Role of Vavilov collection of plant genetic resources in the changing world," Sankt-Petersburg, 2009; at the International conference "Problems of environment and food security under changing climate and land use," Puschino, 2013; at VII Moscow International congress "Biotechnology:

state and development perspectives," Moscow, 2013; International scientific conference "Fourth International Scientific Symposium "Agrosym-2013," Moscow; at VIII International conference "Factors of experimental evolution of organisms," devoted to150th anniversary of V.I. Vernadsky and 95th anniversary of Ukrainian National Academy of Sciences foundation (Alushta, Autonomian Republic of Crimea, Ukraine).

Our aim is to create models for study of genetic stability of plants relatively to environmental factors and to elaborate long-term program of cultivated plants genetic diversity maintenance in the conditions of Western Siberia and Northern Caucasus. The attainment of this aim is performed by the solution of priority scientific task with the application of modern methods and approaches.

1. Improvement of theoretical foundations and methods of cultivated plants resources study.
2. Detection and creation of genetic sources of resistance to biotic and abiotic stresses on the basis of application of recombinational and mutational plants variability.
3. Creation of gene pool aimed to rational use of local and distant plant resources. Elaboration of plants gene pool management system including genotypes identification; creation of passport and evaluation databases resulting from ecologically geographic and laboratory study of plant samples from the world collection of N.I. Vavilov All-Russian Research Institute of Plant Growing, other research institutions and own collections.
4. Elaboration of phyto- and microbiocoenoses rehabilitation technology. Under changing climate and increasing anthropogenic stress the increase of amount of plant diseases is forecasted, therefore samples with resistance genes are particularly valuable. Their detection requires the research of underlying mechanisms of interorganism interactions and the conduction of main cultivated plant species gene pool screening from the point of view of resistance to pathogenic fungi.
5. Optimization of symbiotic activity of legumes under seed treatment by *rhizobia* cultures (at the example of *Pisum sativum* L.). Mountain ecosystems degradation disturbs natural processes of energy exchange, decreases soil fertility, and impoverishes plant species composition. The additional sowing of perennial herbs, especially legumes, with powerful root systems, high protein content, capacity of biological nitrogen accumulation in the soil is one of the methods of disturbed landscapes restoration. Wild species of Northern Caucasus natural phytocoenoses have such characters and represent unique initial material for selection genetic programs and for biodiversity restoration by means of reintroduction.
6. The research of mechanisms of the effect of biologically active compounds applied in agriculture. Elaboration of the increase of cultivated plants productive and adaptive properties by means of application of para-aminoben-

zoic acid (PABA), appin, and zircon. The PABA reparagenic properties allow reduce damaging effect of substances with mutagenic activity on plant organism.

7. Mastering of technologies of save and long-term preservation of plants genetic resources. Under frequent reseeding of samples biological clogging up occurs, the probability of technical errors increases, the natural change of relationship of biotypes in polymorph samples is possible as well, which leads to the loss of valuable genotypes. At the same time it was shown in the world practice that under low temperatures seeds remain viable during a long time. The diversity of collection material under condition of its secure storage and rational use is able to provide the development of priority directions aimed to the ecological contamination control of territories and conservation of their biodiversity.

Potential feasibility of scientific task is determined by existing scientific basis. The research of mutational and combinational variability increasing probability of rise of plant forms with transgressive display of valuable and unique characters was performed; reparagenic effect of PABA, reducing damaging effect of chemical mutagens was studied.

The separation and creation of genetic sources of resistance to biotic and abiotic environmental factors is conducted.

The study aimed to cultivate wild forms in agrocoenoses of southern Tyumen region is conducted in order to preserve biodiversity and create the possibility of herbal medicine production. The results of comparative evaluation of plants growth and development in cultivated and natural ecosystems allow to consider this direction as perspective [50].

By present time collection stock of Tyumen University Biology Institute (NUBI) includes around 1800 plant samples (cereals, legumes, medicine herbs, ornamental) from 58 countries of the world and 33 regions of Russia and about 150 cultures of microorganisms.

Growth regulators application methods and concentrations were determined on the base of morphological characters and physicochemical soil properties, pigments concentration in the cellular sap of leaves, elemental composition of caryopsides and productivity characters in order to provide more complete realization of ecological-biological potential of spring wheat in the condition of subtaiga zone of Tyumen region.

Methods of plant organisms metabolism activation by means of weak electromagnetic radiation and chemical stimulators were elaborated and optimized. Optimal parameters of low-intensity laser, magneto-infrared-laser and X-ray emissions of red clover seeds were justified. For the first time the effectiveness of magneto-infrared-laser emission, PABA, "Alanit" clay and their joint effect on sowing qualities of red clover seeds with long conservation terms was shown. The results of the study are protected by three invention patents: "Method of Presowing Treatment of

Legume Herb Seeds" [51], "Method of Selection of Clover Plants with Elevated Sugars Content" [52] and "Method of Presowing Treatment of Cereals Selection Samples Seeds" [53].

Joint research of the following research collectives is conducted now and planned for future: Tyumen State University (DSc of Agriculture, professor N.A. Bome); Gorsky State Agrarian University (DSc of Agriculture, professor S.A. Bekuzarova); N.I. Vavilov Institute of Plant Industry, DSc of Biology Loskutov I.G., PhD of Agriculture, Al.Ya. Bome); N.M. Emanuel Institute of Biochemical Physics RAS, senior researcher L.I. Weisfeld), Tobolsk complex research station of Ural branch of RAS (PhD I.A. Dudareva).

The results of joint Russian-German study consisted in ecological testing of spring wheat and soya varieties and the research of effectiveness of bio-preparations application on legume plants (at the example of soya) [54, 55] showed as advisable the involvement of German colleagues Hochschule (University of Applied Sciences Osnabruck, Germany, professor D. Trautz, PhD, graduate students I. Kuehling and E. Ripberger) to the implementation of priority scientific task.

In order to solve the proposed scientific tasks further development of research in the Center of Plants and Microorganisms Genetic Resources on the base of scientific laboratories, herbarium and seed funds of Institute of Biology of Tyumen base station of N.I. Vavilov Institute of Plant Industry and "Kuchak lake" Biological station of Tyumen State University.

The teams have modern equipment for biological study: climatic chambers for plants cultivation, refrigerators for low-temperature storage of biological objects, thermostats, autoclave, lamellar box, laboratory scales, microscopes, and spectrometers.

Deepening of research and obtaining of more precise results in short-terms can be reached if additional laboratory equipment will be obtained, namely spectrophotometers, temperature and climatic equipment, chambers for plants cultivation, complex field laboratory, laboratory for biotechnological studies, laboratory microscope Carl ZEISS Axio Lab. A1 for the research using the method of bright field in the transmitted light with color photo/vide or chamber (5 Mp) and software.

The risk analysis shows, that the existing material and technical basis of our scientific collectives is sufficient for obtaining scientific results within the terms up to 5 years. In order to decrease the risk of insufficient qualification of research team the question of junior specialists education in the given field by means of net magister study foundation ("Biology" branch, "Biotechnology" magister program) is under consideration.

The expected results of presented tasks solution may be considered in two aspects: scientific and applied results.

The scientific results are given below:

1. Study of the reaction of cultivated cereals to stress factors and detection of forms with high ecological plasticity, resistant to pathogens basing on the

laboratory express-diagnostics in early ontogenesis and on the methods of field evaluation.
2. Obtaining new data on the development biology of wide-spread pathogens aimed to the understanding of mechanisms of phytopathogenic fungi effect on plant organism and of plants resistance system functioning in extreme conditions.
3. Inventory of accumulated genetic resources, providing of collections replenishment by new plants and microorganisms samples, their guaranteed secure storage.
4. Optimal methods of introduction, cultivation and reintroduction of valuable and unique plants.
5. Determination of experimental mutagenesis method effectiveness for the plants genetic diversity extension; development of plants morphological characters phenotypic correction methods, providing the increase of volume of variability available for selection and increasing mutational selection effectiveness.

Applied results (production and services of Center of plants and microorganisms genetic resources):

- technology of secure low-temperature storing of biological objects;
- initial material (seeds and other planting material) for selection and genetic programs;
- valuable plant species multiplication technology, in particular *in vitro* (medicinal, ornamental, rare and endangered);
- methods of identification and description of plants and microorganisms taxonomic units, as well as plant diseases;
- performance of consultations and training seminars on the topics of conservation and research of biodiversity;
- character collections, herbarium and seed fund for the education of specialists in biology and agriculture;
- involvement of scientific and technical potential in the solving of problems concerning resources of plants and microorganisms.

The following stages are provided in order to achieve the aims:

1. Preparation stage – research plan and planned schedule of works.
2. Field studies: evaluation of collection fund according to a set of characters, research of physiologically active compounds application effectiveness, expedition in state variety-testing plots of West-Siberian region, formation of cultivated plants herbarium, microbiological herbarium (Tyumen State University, Biology Institute, Tyumen base station of N.I. Vavilov Institute of Plant Industry); collection of material in natural and cultivated phytocoenosis of Northern Caucasus foothills (Gorsky State Agrarian University); soil and plants sampling for chemical analysis (Tobolsk Complex Research Station of RAS Ural Branch); ecological varieties testing, increase of legume

plants symbiotic activity in the conditions of southern Tyumen region and Lower Saxony (Germany) – Tyumen State University, Biology Institute, Tyumen base station of N.I. Vavilov Institute of Plant Industry; Institute of Applied Sciences (Osnabruck, Germany).
3. Laboratory studies: processing of plant objects by chemical mutagens, cytological studies (N.M. Emanuel Institute of Biochemical Physics RAS); sample preparation and determination of chemical elements content in plant-soil system (Tobolsk Complex Research Station of RAS Ural Branch); determination of pigments in plants, elaboration of biological objects long-term storing technology, microorganisms and plants cultivation, research of seeds biological properties (scientific laboratories of Biology Institute of Tyumen State University, Institute of Applied Sciences (Osnabruck, Germany), Gorsky State Agrarian University).
4. Cameral stage: research report, scientific publications, patents, and computer database of collection fund.
5. Foundation of plants and microorganisms genetic resources Center – annual replenishment of collections, identification, organization of storage and vitality maintenance.

2.3 CONCLUSIONS

Commercialization of expected results is possible via administrations of Tyumen region, Northern Caucasus, Germany; agricultural companies interested in the increase of productive and adaptive properties of cultivated varieties; research institutions working in the field of plants adaptive selection, state variety-testing plots; universities and other educational institutions.

The final goal of the complex research work determines storage and efficient use of plant resources in the conditions of modern agricultural land-use with taking into account regional weather and soil factors.

KEYWORDS

- **appin**
- **biological variety**
- **epibrassinolide**
- **mutagenesis**
- **para-aminibenzoic acid**
- **stresses biotic and nonbiotic**

REFERENCES

1. Alexanyan, S. M. (2007). Interaction Strategy of Gene Banks of Gene Banks of World in Conditions of Globalization, Proceedings on Applied Botany, Genetics and Selection, *164,* 11–33 (in Russian).
2. UN FAO Commission in the Field of Food and Agriculture, Rome (2010) http://unyearbook.un.org/1985yun/1985_p2_ch3.pdf.
3. Hammer, K. (2004). Resolving to Challenge Posed by Agrobiodiversity and Plant Genetic Resjurses an Attempt, *Journal of Agriculture and Rural Development in Tropics and Subtropics,* Addition, 76–184.
4. Evaluation of the State of Earth Ecosystems (2001–2005). http://rpp.nashaucheba.ru/docs/index-35466.html (in Russian).
5. The United Nations Convention on Biological Diversity, the International Secretariat Geneva, Executive Center (1992). http://www.un.org/ru/documents/decl_conv/conventions/biodiv.shtml.
6. The Second Report on State of the World's Plant Genetic Resources for Food and Agriculture of (26 October 2010) FAO Organization of the United Nations, Comission on Genetic Resourses for Food and Agriculture (2010) 350p. http://www.fao.org/docrep/013/i1500e/i1500e_brief.pdf
7. Bome, N. A. (2012). Applied and Theoretical Aspects of Cultivated Plants Gene Pool Formation, Bome, N. A., Bome, A. Y., & Kolokolova, N. N., Fruit and Berry farming in Russia, *XXXIV,* Part 1, 106–113 (in Russian).
8. The International Treaty on Plant Genetic Resources for Food and Agriculture from (29–06–2004) 28p http://www.fao.org/newsroom/en/news/ (2004)/47027/index.html.
9. Modern Methods and International Experience of Wild Plants Gene Pool Conservation (at the Example of Wild Fruiters) Alexanyan, S. M., Gorbunov, Yu. N., Dolgikh, S. G. et al. (2011) Author Collectives from Russia, Kazakhstan, Great Britain, Almaty, 188p.
10. Standards Gene Bank for Plant Genetic Resources for Food and Agriculture Rome (2013). The FAO Commission on Genetic Resources, 182p http:// www.f ao.org/f ileadmin/ templates /agphome/ documents/PGR genebank/GeneBank_RU_low_res_01.pdf (in Russian).
11. Rysina, G. P. (1984). Experiences of Protected Plants Populations Restauration in Moscow Region, Bulletin of Main Botanical Garden, *133,* 81–85(in Russian).
12. Belovodova, N. N., & Tikhonova, V. L. (2001). Plants Re Introduction as Method of Protected Species Quantity Restoration, Agrarian Russia, *2*, 31–33 (in Russian).
13. Gorbunov, Yu N., Dzybov, D. S., Kuzmin, Z. E., & Smirnov, I. A. (2008). Methodic Recommendations on the Introduction of Rare and Endangered Plant Species (for Botanical Gardens) International Council of Botanical Gardens on Plant Protection (BGCI), Tula Grif & K, 56p (in Russian).
14. Denisova, V. L., Nikitina, S. V., & Zaugolnova, L. B. (1986). Program and Methods of Monitoring of Ceno Populations of Plant Species from USSR Red book, Moscow, All-Russian Research Institute of Nature Protection, 34p (in Russian).
15. Valee, L., Hogbin, T., Monks, L., Makinson, B., Matthes, M., & Rossetto, M. (2004). Publishers Sundder Land, Guidelines for the Translocation of Threatened Plants in Australia Second Edition *Australian Network for Plant Conservation*, Australia, Canberra, 80p.
16. Zhuchenko, A. A. (1990). Adaptive Plant-Farming, ecological and Genetic Foundations, Kishinev, Moldavia, Shniintsa, 587p (in Russian).
17. Bekuzarova, S. A. (2006). Selection of Red Clover, Bekuzarova, S. A., Vladikavkaz, 176p (in Russian).

18. Bekuzarova, S. A. (2011). Breeding Adaptive Legumes Cultivars for Grassland Agriculture, Autochthonous and Alien Plants, Collection of Proceedings of the National Dendrological Park "Sofiyivka" of NAS of Ukraine, *7,* 72–76 (in Russian).
19. Bekuzarova S.A. (2011). Reintroduction is a Method of Biodiversity Conservation. Ancient Parks and Botanical Gardens-Research Centers of Plant Diversity Preservation and Protection of Historical and Cultural Heritage. Proceedings of the International Scientific Conference Devoted to 215 Anniversary of the Founding National Dendrological Park "Sofiyivka" of NAS of Ukraine (in Uman, Ukraine, from the 5-7 October 2011). Uman. 56–58 p. (in Ukrainian).
20. Standards Gene Bank for Plant Genetic Resources for Food and Agriculture, Rome, The FAO Commission on Genetic Resources (2013) 182p. http://www.fao.org/fileadmin/templates/agphome/documents/PGR/genebank/GeneBank_RU_low_res_01.pdf (in Russian).
21. Sinskaya, E. N., & Vorobyeva, F. M. (1961). Winter wheat Populations Analysis within the Process of their Transformation into Variety with Spring Mode of Life, Populations Problem in Higher Plants, Leningrad, *1*, 106–140 (in Russian).
22. Zheleznov, A. V., & Zheleznova, N. B. (1994). Problems of Conservation and Use of Plant Genetic Resources, Genetic Collections of Plants, Novosibirsk, *2*, 6–32 (in Russian).
23. Zheleznov, A. V., & Zheleznova N. B. (2000). Inbreeding and Heteros is in Amaranth, Abstracts Mendel Centenary Congress, Vortage für Pflanzenzuchtung, Bd, *47*, S 171.
24. Magomedmirzayev, M. M. (1990). Introduction in Quantitative Morphogenetics, Moscow, Nauka Publishing, 230p (in Russian).
25. Afanasyev, R. A., & Merzlaya, G. E. Transformation of Mobile Phosphorus in Agro Ecosystem Soils in Long-Term Experiments, Present Collection.
26. Kabata-Pendias, A. (1989). Microelements in Soil and Plants, Kabata-Pendias, A., & Pendias, H., Moscow, Mir publishers, 439p (in Russian).
27. Alexeyev, Yu V. (1987). Heavy Metals in Soils and Plants, Alexeyev, Yu V., Agropromizdat, Leningrad, 142p (in Russian).
28. Shakirova, F. M. (2001). Nonspecific Plant Resistance to Stress Factors and its Regulation, Shakirova, F. M., Ufa Gilem, 160p (in Russian).
29. Prusakova, L. D. (2005). Application of Brassinosteroids in Extreme Conditions, Prusakova, L. D., & Chizhova, S. I., Agrochemistry, *7,* 87–94(in Russian).
30. Bome, N. A. (2012). Increasing of *Triticum aestivum* L. Seed Vitality and Plant Resistance to Changing Environmental Conditions, Bome, N. A., & Cherkashina, I. A., Natural and Technical Sciences, Sputnik Publishers, *2(58)*, 108–113.
31. Cherkashina, I. A. (2012). Increasing of *Triticum aestivum* L Varieties Resistance to Stress Environmental Factors by Means of Physiologically Active Substances, Cherkashina, I. A., & Bome, N. A. Fruit and Berry Farming in Russia, T. *XXXIV*, *2*, 362–368.
32. Dudareva, I. A., & Bome, N. A. (2014). Basic properties of Sod-Podzolic Soil of North Territories of Tyumen Region, Ecological Consequences of Increasing Crop Productivity Plant Breeding and Biotic Diversity, Toronto Apple Academic Press, Inc.
33. Dudareva (Cherkashina), I. A. (2013). Content of Chemical Elements in Soil-Plant System under the Effect of Biologically Active Substances, Dudareva (Cherkashina), I. A., & Bome, N. A. Scientific Bulletin of Belgorod State University, Natural Sciences Series, *10(153)*, *(23)* 116–124 (in Russian).
34. Ilyin, V. B. (2006). Heavy Metals in Soil-plant System, Soil Science, *8,* 56–64 (in Russian).
35. Gimalov, F. R. (2006). Effect of 24-epibrassinolid on Cabbage Germs under Cold Stress, Gimalov, F. R., Matniyazov, R. T., Chemeris, A. V., & Vakhitov, V. A., Agrochemistry, *8,* 34–37 (in Russian).

36. Bokebayeva, G. A. (1991). Protective Effect of Brass in Osteroids on Barley Plants under Salinization, Abstract of Thesis of Candidate of Biological Sciences, Moscow, Lomonosov, M. V. Moscow State University, 25p (in Russian).
37. Shakirova, F. M. (1998). Change of ABK and Lectin Levels in the Roots of Wheat Germs under the Effect of 24-epibrassinolid and Salinization, Shakirova, F. M., & Bezrukova, M. V., Plant Physiology, *45,* 451–455(in Russian).
38. Orekhova, A. N. (2007). Effect of Epibrassinolid on the Formation of Reserve Protein Complex and Grain Quality of Winter Wheat, Orekhova, A. N., Maxyutova, N. N., Neshin, I. V., & Dudenko, N. V., *Agrochemistry, 11,* 36–41 (in Russian).
39. Kulayeva, O. N. (1989). Brass in Steroids in the Regulation of Protein Synthesis in Wheat Leaves, Kulayeva, O. N., Burkhanova, E. A., & Fedina, A. B., Proceedings of AS of USSR, *305(5),* 1277–1279 (in Russian).
40. Korablyova, N. P. (1995). Biochemical Aspects Pf Plant Dormancy and Immunity, Korablyova, N. P., & Platonova, T. A., Applied Biochemistry and Microbiology, *31(1),* 103–114 (in Russian).
41. Shakirova, F. M. (1999). Participation of Wheat Phytogormons and Lectin in the Plant Response to Stress Effects, Shakirova, F. M., Abstract of Thesis of Doctor of Biological Sciences, Sankt-Petersburg State University Publishers, 44p (in Russian).
42. Kripach, V. (2000). Twenty Years of Brass in Osteroids Steroidal Plant Hormones Warrant Better Crops for the XXI century, Kripach, V., Zhabinski, I. V., & de Groot, A., Annual Bot., *86,* 441–447.
43. Shevyakova, N. I. (2011). Physiological Peculiarities of Salt-Resistant Plants, *Plant Physiology, 56(3),* 321–336 (in Russian).
44. Talanova, V. V. (2009). Phytogormons as Regulation of Plant Resistance to Unfavorable Environmental Factors, Abstract of thesis of Doctor of Biological Sciences, Talanova, V. V., Moscow, 32p (in Russian).
45. Vavilov, N. I. (1935). Teaching on plant Immunity to Infectious Diseases, Moscow, 100p (in Russian).
46. McCloud, E. S. (1997). Herbivory and Caterpillar Regurgitate Amplify the Wound-Induced Uncreases in Jasmonic Acid but Nicotine in Nicotiana Sylvestris, McCloud, E. S., & Baldwin, I. T., Planta, *203,* 430–435.
47. Mithell-Olds, T. (1998). Chemical Ecology in Molecular Era, Mithell-Olds, T., Gershenzon, J., Baldwin, I., & Boland, W., Tredds *Plant Science, 3(9),* 362–365.
48. Agrawal, A. A, Gorski, P. M. & Tallamy, D. W. (1999). Polymorphism in Plant Defense Against Herb Ivory Constitutive and Induced Resistance in Cucumis Sativus, Agrawal, A. A., Gorski, P. M., Tallamy, D. W., *J.Chem. Ecology, 25(10),* 2285–2304.
49. Wittstock, U. (2004). Successful Herbivore Attack Due to Metabolic Diversion of a Plant Chemical Defense, Wittstock, U., Agerbir, N., Stauber, C. E., Olsen, M., Hippler, T., Mithell-Olds, J. & Gerschenzon, H., Vogel, PNAS, *101(14),* 4859–4864.
50. Mikhailova, A. N. (2012). Research of Medicinal Plants Reaction on the Changing Environmental Conditions Based on the Duration of Phonological Phases, Mikhailova, A. N., & Bome, N. A., Natural and Technical Sciences Sputnik+ Publishers, *6(62),* 139–144 (in Russian).
51. Khegaturova, L. G., Bekuzarova, S. A., & Beliayeva, V. A. (10–01–2010). Methods of Legume Herbs Pre-sowing treatment, Patent of Russian Federation, 2377752, Bulletin, 1.
52. Bekuzarova, S. A., Beliayeva, V. A., Kharchenko, Yu. A., & Bushuyeva, V. I. (10–02–2010). Method of Sampling of Clover Plants with Elevated Sugars Content, Patent of Russian Federation, 2380885, Bulletin, *4.*

53. Bekuzarova, S. A., Bome, N. A., Weisfeld, L. I., Tsimartoba, F. N. & Luschenko, G. V. (2012). Method of Pre-sowing treatment of Cereals Selection Samples Seeds, Patent of Russian Federation, 2461185, Bulletin, *26*, 5p (in Russian).
54. Kuehling, I. (2013). Relationship Between Soj Bean Varieties, Rhizobia Inoculation and Spad-502 chlorophyll Meter Readings in Western Sibiria, Kuehling, I., Huesing, B., Bome, N., & Trautz, D., IV International Symposium, "Agrochym-2013" Saraevo, 71–76.
55. Kuehling I. (2014). Readings and Rhizobia Inoculation in Western Siberia, Bome, N. A., & Trautz, D., Ecological Consequences of Increasing Crop Productivity, Plant Breeding and Biotic Diversity, Apple Academic Press.

PART II

DIRECTIONS OF SELECTION AND EVALUATION METHODS OF BREEDING OF GALEGA EASTERN AND RED CLOVER IN BELARUS

CHAPTER 3

SOURCE MATERIAL FOR BREEDING VARIETIES OF RED CLOVER FOR DIFFERENT WAYS TO USE

VERA I. BUSHUYEVA

Belorussian State Agricultural Academy, Gorki, h. 5, Michurin St., 213407, Belarus; E-mail: vibush@mail.ru

CONTENTS

ABSTRACT

This chapter presents the characteristic of variety samples of red clover in the collection nursery on the main economically valuable traits and their division into five types of ripeness. In each group of ripeness the most significant sources of quantitative and qualitative traits for breeding varieties of red clover for different areas of use were identified.

3.1 INTRODUCTION

Red clover (*Trifolium pratense* L.) is the most significant and widely cultivated forage crop in Belarus and many countries in the world. It is of fundamental importance for the development of the field and meadow-pasture forage, biological farming, increasing soil fertility, environmental conservation and production of environmentally friendly products.

Effectiveness of its cultivation depends on the quality of varieties and their conformity to the areas of use. Each area of use requires a particular model of the variety with a corresponding set of economically useful traits. In connection with this worldwide purposeful breeding work on creating specific varieties for field, meadow and pasture use is carried out. Special breeding programs have been developed as to the areas of use. First such programs have been adopted abroad in 1985 at a conference on selection of legumes in the United States, and in 1986 at the XI General Meeting of the European Federation in Portugal [1].

There are different requirements to the varieties depending on the area of use.

Field grass cultivation requires intensive varieties of different maturity that are able to form a high yield of fodder for 2–3 years, are highly resistant to cold, have rapid regrowth of vegetation in early spring and after cutting, are resistant to diseases and pests and fit in quality for the production of mono-fodder, grass meal and silage.

For varieties of meadow and pasture grass cultivation the main requirements are longevity, high competitive ability in grass mixtures, different earliness and rhythm of growth, rapid regrowth and high resistance to grazing and trampling. Uniformity of the seasonal regrowth over a long grazing period is of practical importance for grazing use, and for the varieties of meadow grass cultivation intended for the procurement of roughage it is ability for rapid regrowth in the spring and in the subsequent cycles of cutting [2].

Priorities and objectives of breeding of the red clover in different countries depend not only on the purpose of the use of varieties, but also features of the soil and climatic conditions for their cultivation [3].

In the United States where early and late-ripening types of clover are successfully cultivated, varieties of different maturity groups are created in which along with the general requirements special attention is given to improving immunity to nematodes and viruses.

In New Zealand the main requirement in the selection of red clover is to create competitive perennial varieties with fibrous root system, high phosphate-mobilizing capacity and high resistance to drought.

In the UK and Germany more attention is paid to improving productive longevity of the varieties for intensive use on meadows and pastures, improving the quality of food and its palatability.

In Switzerland there are perennial varieties of red clover with different maturity, with the potential use of 4–5 years in the meadows and pastures [4].

The problem of increasing productive longevity is most relevant and common to all the countries cultivating red clover. Efficient selection work is carried out in this direction in the UK [3]. It was found that rapid thinning of red clover is associated with the morphology of the plant, its erect form especially vulnerable to adverse environmental conditions, causing physiological stress of the root system. In this regard, most of the available varieties of red clover are sensitive to autumn grazing. New varieties of red clover with semierect and semihumifuse forms of plants capable in moderate mode of use produce high yields for 6 years have been created. Among them is variety Norzman whose yields in a long-term use in the UK are 69% higher than those of the famous European varieties Redhead and Hangaropoli. Perennial varieties are successfully used in the development of hilly land. According to many-year experiments done at the Welsh Plant Breeding Station tetraploid varieties of clover Astra, Aled, Aa-22, Hangaropoli cultivated on hilly lands mixed with timothy formed yearly hay yields on average of 8 t/ha during 6 years.

Among the best varieties of perennial hay-pasture use intended for radical improvement of meadows in Western European countries are early maturing diploid varieties Start, Hlumetsky, Radan and early maturing tetraploid variety Temara, as well as mid-and late- maturing tetraploid varieties Quarta, Yavoryna, Radegast. The best and most valuable varieties of red clover for field grass cultivation are considered early maturing tetraploid varieties Redhead, Deben, Barfiola, Norzman, Maro, Valley, the diploid ones: Kuhn, Granta, Kean, Sabtoron, Trifomo, Merviot, Markom and late tetraploid varieties Astra, Britta, Power, Palna. The most high-yielding of early maturing varieties is diploid Merviot, of the late ones – tetraploid Palna. Late varieties are valued for high yield in the first mowing and are used for fodder. The presence of varieties of red clover of different maturity enhances the effectiveness of its use [3].

As a result of the implementation of the programs adopted in Australia, the UK and the U.S. a successful breeding work has been done to improve the quality of forage varieties whose consumption does not cause animal tympanites and are characterized by low estrogen active substances such as phormononetin. In the UK a method for determining the operational phormononetin has been developed which is used for the selection of promising varieties with its low content. It is assumed that the allowable rate of phormononetin in the diet should be less than 0.3%, the common varieties of red clover contain 0.7% of it. Varieties of red clover not causing tympanites have been genetically engineered [4]. The created varieties represent a new gene pool of red clover that carries the most important features and properties for varieties of different areas of use.

In Russia and Belarus clover selection was carried out on a single program the implementation of individual directions of which is still being done by joint efforts. This is confirmed by the joint cooperation of the Creative Association of Breeders (CAB) "Clover" organized in 1988 on the initiative of the All-Russian Research Institute of feeds named after V.R. Williams comprising representatives of 14 scientific institutions in Russia and Belarus. The Academic institutions, members of the CAB "Clover," are located in different soil-climatic zones covering all regions of Russia and Belarus. This allows the targeted use of promising breeding material, to conduct its environmental testing for a year in 10–12 locations with sharply different soil and climatic conditions and thus accelerate the process of developing new varieties more sought after by production. In 10 years of working together nine varieties of red clover were created, such as: Sozh, Trio, Altyn, Orlik, Ratybor, Pamyati Burlaki, Stodolich, Pamyati Lisitsyn and TOS-870 [5, 6].

More than 90 selected varieties, 19 of them are of Belarusian selection, are currently allowed in production for cultivation conditions of Russia and Belarus. The main direction of red clover breeding in the current conditions, as well as throughout the world, is creation of high-yielding varieties for the field, meadow and pasture grass cultivation with a corresponding set of economically useful traits.

One of the most pressing problems, both home and abroad, is the productive longevity of varieties and in this regard productive breeding work has been done. Variety Atlanta created in Russia may persist in herbage for 5–6 years and form hay yields of 8–9 mc/ha [7].

The produced varieties must be of high quality forage. The task of selection of red clover on quality improvement is to create a breeding material with a high content of protein in the later phases of development when there is maximum dry matter accumulation. Selection for resistance to diseases, such as cancer, anthracnose, fusarium, etc., remains relevant. Considerable attention is paid to improving seed yield. Among the new directions of breeding of red clover is to create varieties with high nitrogen-fixing and phosphate-mobilizing ability [8].

Methods for creating original material are being improved to increase the efficiency of breeding. Along with the methods of selection and hybridization in breeding of red clover they began to apply the methods of polyploidy, mutagenesis, multiple heterosis, gamete, edaphic and symbiotic selection. Using the method of polyploidy tetraploid varieties VIC, VIC-84, Salute, Mars, Durable, Ustoylivy, Amber having high potential of green mass yield have been created. Increase of their seed production is carried out by gamete selection.

Using chemical mutagenesis it was possible to overcome biological barrier of negative correlation between early ripening and winter hardiness in the created varieties resulting in short-season, hardy varieties Early 2, Trio, Orlik [8, 9].

Using method of multiple heterosis early maturing and high-yielding in fodder mass and seed production variety Altyn was created. As a result of edaphic selection Topaz cultivar resistant to excessive soil acidity and aluminum toxicity was created. Using symbiotic selection allowed to create original material of red clover with increased phosphate-mobilizing and nitrogen-fixing ability. High symbiotic efficiency is achieved by selecting complementary pairs of macro- and microorganisms showing effect of heterosis in interaction. Biotechnological methods of cell and gamete selection are promising, the use of which allows to carry out mass regeneration and rapid reproduction of the most valuable genotypes, perform hybridization on embryonic level, effectively do selection for resistance to high soil acidity and stressful environmental factors (drought, the return of spring frosts, etc.) [10, 11].

However, the urgent problem is the establishment of effective clover varieties that combine complex of specific features and properties that are to be taken into account when planning models for each variety use. In addressing this problem, along with the methods of selection the quality of the gene pool of red clover, its genotype, ecological-geographical and morphobiotypical diversity, its containing the necessary sources of the most important economically useful traits and properties are of exceptional importance [12, 13].

Therefore, the aim of our research was to examine the existing gene pool of red clover in the collection nursery, to determine to which kind the studied variety samples belong, to give them a comprehensive assessment and identify sources of economically valuable traits needed to breed new varieties of field, meadow and pasture areas of use.

3.2 MATERIALS AND METHODOLOGY

The studies were conducted in the experimental field of the Breeding and Genetics Department of the Belorusian State Agricultural Academy. The soil is sod-podzolic, light-loamy, underlaid at a depth of 1 m with moraine loam. Basic agrochemical properties of the arable layer are: humus (by I.V. Tyurin) – 2.0–2.2%; pH KCI – 6.3–6.5; mobile forms P_2O_5 and K_2O (by A.T. Kirsanov) – 252 and 206 mg per 1 kg of soil, respectively.

Meteorological conditions during the years of studies differed significantly in rainfall and temperature regime from each other, as well as from the long-term averages. This allowed more objectively and fully to appreciate the material sources and define the most important economically useful features and properties.

Sixty three variety samples of red clover of different ecological-geographical and breeding origin that were evaluated in the collection nursery were objects of study. Sowing was carried out manually with row spacing of 15 cm. The plot size was 1 m^2; the location was randomized with double replication. Sown on May 22, 2003 the variety samples of red clover were studied on the main economically valuable traits and properties over 3 years of life. One replication was used for phenological observations and establishment of the length of the growing period, approbation and determining the average number of internodes on the main stem. In the second replication type of the clover the number of leaves, dry matter content, plant height in the phase of cutting maturity, yield of green mass and dynamics of its accumulation in mowing and at different times of life were studied.

In the first year of life the variety samples were studied by morphobiotype of the bush and their relation to vernalization. In the second and third years of life the maximum possible number of cuts and distribution pattern of harvest among them, the first mowing flowering time, the number of leaves and dry matter content were determined. By the third year of life trend of variability of the yield level of feed mass depending on the duration of the herbage use were studied.

3.3 RESULTS AND DISCUSSION

According to the results of phenological observations significant variation in the length of the growing period of the studied variety samples was found. By the variance analysis of variability in the length of the growing season and the length of time from the beginning of spring regrowth until the phase of budding, flowering and maturing, grouping of variety samples of red clover on precocity was done (Table 3.1).

TABLE 3.1 Results of Grouping of Variety Samples of Red Clover by Types

Type of red clover or maturity group	Date of the beginning and duration of development phases of variety samples in days from the start of spring regrowth to:					
	Budding		Flowering		Ripening	
	Date	Length	Date	Length	Date	Length
Early	14.06–18.06	61–65	21.06–25.06	68–72	31.07–8.08	106–114
Mid-early	19.06–23.06	66–70	26.06–1.07	73–76	9.08–17.08	115–123
Mid	24.06–28.06	71–75	2.06–6.07	77–82	18.08–23.08	124–129
Mid-late	29.06–3.07	76–80	7.07–11.07	83–87	24.08–4.09	130–139
Late	4.07–10.07	81–85	12.07–17.07	88–92	3.09–12.09	140–149

This allowed for a preliminary separation of them into five types: early, mid-early, mid, mid-late and late ripening. Variability ranges of all the traits characterizing each type within the populations considerably reduced. Differences in the length of the growing season among the types of ripeness were 9 days, the duration of the period from the beginning of spring regrowth until the phase of budding and flowering was 5 days.

Spring regrowth in all variety samples of red clover in the second and third year of life was observed on April 14. Subsequent growth and development of the variety samples of different maturity groups differed significantly. The budding phase in the variety samples of early maturing group began on June 14. Duration of forming the first mowing was 61–65 days; the length of the growing period was 106–114 days. The variety samples of each subsequent group of ripeness entered the budding phase 5–8 days later than the previous one, and the latest-ripening forms reached cutting maturity almost a month after the early maturing ones. Difference in the length of growing season in the variety samples of early and late maturing groups was 43 days. Correctness of the separation of the studied variety samples to the types of ripeness was checked by the number of cuts and time of mowing conducted strictly at the budding stage – beginning of flowering (Table 3.2).

TABLE 3.2 Terms of Mowing of Variety Samples of Red Clover of Different Types

Mowing	Early ripening	Mid-early ripening	Mid ripening	Mid-late ripening	Late ripening
First	18.06	23.06	29.06	7.07	13.07
Second	25.07	3.08	13.08	20.08	29.08
Third	15.09	25.09	no	no	no

Variety samples of early and mid-early ripening groups formed three green mass cuts during growing season, mid and mid-late ripening produced two cuts, late ripening- one cut plus the afterfeed.

Duration of formation of the first mowing from the beginning of spring regrowth in early maturing variety samples was 61 days. Starting from June 18 every 5–10 days harvesting maturity of the variety samples of the following group began, first of the first cut, then of the second and third, resulting in harvesting green mass until September 25. It was the experimental confirmation of the possibility of cultivation of varieties of different maturity groups in agricultural production to effectively form a continuous green summer conveyor of high protein; energy resource saving and environmentally friendly feeds.

Evaluation of the variety samples in the collection nursery allowed to reveal their degree of variation by the length of the vegetation period, the height of the grass before mowing, the yield of green mass at different times of life, foliage and dry matter content (Table 3.3).

TABLE 3.3 Characteristic of Variety Samples of Red Clover in the Collection Nursery

Variety sample		1st year of life		2nd year of life		3d year of life				
	Vegetation, days	Plant height, cm	Yield green mass, kg/m^2	Plant height, cm	Yield green mass, kg/m^2	Plant height, cm	Yield, green mass, g/m^2	Yield for 3 years, green mass, kg/m^2	Foliage, %	Dry matter content %
Early ripening										
Altyn	106	70	4.5	82	9.2	65	2.7	16.4	39.3	25.4
T-46	108	75	5.2	80	8.6	70	4.5	18.3	41.4	23.8
Milvus	110	66	4.2	90	8.6	67	2.9	15.7	39.5	24.5
Early 2	110	74	5.0	91	8.8	65	3.1	16.9	44.0	26.1
TOS-early	112	68	4.5	73	9.5	74	5.5	19.5	38.0	25.2
SL-38	112	75	5.0	87	7.0	70	4.3	16.3	40.6	25.0
T-30	112	79	5.8	80	8.8	70	4.3	18.9	42.8	21.7
SL-43	112	76	5.0	89	9.5	75	5.4	19.9	42.9	21.5
SL-58	112	71	4.0	90	9.8	72	5.5	19.3	51.8	22.5
S-20	112	72	4.3	85	8.9	70	4.8	18.0	45.2	21.0
Dolgoletny	114	72	4.8	81	9.2	78	5.2	19.2	44.1	21.5
Ustoylivy	114	73	6.5	92	10.9	78	5.4	22.8	42.2	22.2
T-54	114	75	4.7	84	9.3	74	4.2	18.2	44.3	21.8
T-62	114	75	5.0	83	9.3	70	4.4	18.7	42.0	22.0
T-70	114	76	4.8	85	10.1	68	4.5	19.4	49.9	23.0
SL-20	114	73	4.5	78	8.3	68	3.5	16.3	47.9	21.6
SL-31	114	73	5.0	90	7.8	70	4.1	16.9	47.5	23.0
SL-36	114	74	5.0	83	8.0	70	5.0	18.0	47.2	21.1
SL-56	114	76	5.0	85	8.9	70	4.0	17.9	44.6	20.3
Mid-early ripening										
Mars	116	75	6.0	87	10.2	80	6.8	23.0	42.5	22.5
T-58	116	74	4.5	90	9.0	76	4.1	17.6	40.0	23.8
T-100	116	74	4.7	80	8.9	70	4.4	18.0	47.2	22.6
SL-21	116	77	3.8	85	9.1	72	5.2	18.1	36.5	21.3
SL-26	116	79	4.5	90	9.2	65	3.5	17.2	41.8	22.3

TABLE 3.3 *(Continued)*

SL-27	116	69	4.3	88	9.4	75	4.2	17.9	43.9	22.2
SL-47	116	75	4.8	88	7.7	70	3.2	15.7	46.5	21.7
VIC-84	116	74	4.5	88	8.5	65	2.7	15.7	47.8	22.5
S-22	116	71	4.3	88	8.7	65	3.8	16.8	44.2	22.0
TOS mid-early	118	71	4.9	80	9.6	70	5.9	20.4	42.5	22.7
Mid ripening										
Srednespely	124	67	4.5	86	8.2	65	3.0	15.7	40.8	23.2
BAA-31	124	64	5.5	92	11.0	85	7.0	23.5	46.8	24.1
SD-24	126	71	3.7	115	12.2	70	5.3	21.2	35.3	22.5
Vitebchanin	126	86	4.5	95	9.9	72	4.3	18.7	34.7	22.0
VIC-133	126	74	3.9	102	11.2	72	4.7	19.7	33.5	23.2
Berezka	126	75	3.7	105	10.8	82	4.5	19.0	36.3	21.6
Mid-late ripening										
Mereya	130	80	4.5	98	10.6	76	4.6	19.7	37.0	24.2
Minsky	130	75	3.8	95	10.2	77	4.7	18.7	37.0	23.5
TOS-870	130	76	3.9	102	12.9	70	4.8	21.6	38.2	24.7
VIC-865	130	75	3.7	100	11.3	71	4.7	19.7	41.5	23.2
VIC-159	130	70	3.3	98	11.5	70	4.6	19.6	36.4	22.8
Sozh	130	66	3.7	105	12.0	82	4.9	20.6	33.0	23.1
Edeysmo	130	87	4.0	98	14.1	82	4.1	22.2	36.2	21.8
Yaskravy	130	95	4.5	108	15.3	87	5.3	25.1	36.3	22.5
Martum	130	90	3.8	100	11.8	93	3.7	19.3	36.6	22.6
SL-44	130	87	4.0	100	13.1	90	4.0	21.1	34.8	23.2
VIC-7	132	85	4.3	88	10.0	78	3.5	17.8	40.5	23.3
Vega	132	69	3.5	115	12.4	80	5.7	21.6	37.3	21.3
Pradi	132	74	4.2	100	13.9	78	5.3	23.4	43.2	20.7
Eken	132	82	3.5	100	13.1	86	4.6	21.2	38.2	21.0
BM-7	132	85	4.2	115	12.9	85	5.2	22.3	35.7	21.1
BN-15	132	91	4.5	100	13.0	82	5.2	22.7	36.5	22.2
SGP-12	132	90	5.4	112	15.0	97	5.4	25.8	38.2	24.1
Falensky 86	139	76	3.1	110	16.5	100	3.5	23.1	42.4	20.5
Orfey	139	74	3.4	118	12.8	89	3.7	19.9	38.3	22.0
Dymkovsky	139	82	3.4	108	12.7	90	4.3	20.4	40.0	20.9
Tetravic-84	139	90	5.0	115	14.7	90	3.5	20.4	39.7	21.1

TABLE 3.3 *(Continued)*

Late ripening										
MOS-1	140	70	3.7	105	12.2	75	3.6	19.5	41.8	23.2
Mut-19–1-1	141	63	3.0	102	11.3	77	5.2	19.5	44.0	21.1
Tetravic	142	80	4.0	112	12.8	87	4.2	20.8	36.5	22.5
Kirovsky-159	147	71	3.6	120	13.5	102	4.2	21.0	41.8	22.5
Vityaz	147	75	3.4	116	12.1	98	6.3	21.8	42.5	21.5
SGP-6	147	83	4.4	115	15.0	99	4.2	23.6	37.7	22.5
Mut-6–4-48	149	74	3.2	107	11.6	73	4.0	18.8	41.6	22.3

The studied traits were taken into account when planning the breeding process and making models of new varieties for various uses. Thus, the length of the growing season that varies considerably among the studied variety samples was taken into account in dividing them into groups of ripeness for use in the selection of complementary varieties. Variety samples with the length of growing season of 106–114 days were included in the early maturing group, and in mid-early ripening – 116–118 days, in mid ripening – 124–126 days, in mid-late ripening – 130–139 days and late ripening – 140–149 days. The best variety samples characterized by a shorter growing season and high forage productivity were defined in each group of ripeness.

Variety samples Altyn (106 days) and T-46 (108 days) had the shortest growing season in the early maturing group, and in the late maturing – MOS-1 (140 days). Variety samples in each group of ripeness differed in terms of growth and productivity potential of forage mass at different times of life. Variety samples of early maturing, mid-early and mid-season groups had the highest growth rate of plants in the first year of life and in a short period reached 66–79 cm of height and entered the phase of flowering, and the highest plant height (95 cm) was observed in variety samples of Yaskravy from the mid-late ripening group reaching budding phase at this stage. Variety samples from the late maturing group were the shortest, their height was 63–70 cm, and branching phase was observed in the plants.

In the second year of life variety samples from the mid-late ripening group were the tallest and produced the highest yields, their height ranged from 88 to 115 cm. Maximum yield was obtained from variety sample Falensky-86 (16.5 kg/m^2). In the third year of life the plant height compared to the second year of life in the variety samples from all maturity groups was lower, at the same time mid-late and late ripening variety samples were taller compared to the other groups.

Plant height in mid-late samples was 70–100 cm, in the late ones – 75–102 cm vs. 65–85 cm in the early, mid-early and mid-season variety samples. Variety samples Ustoylivy and Dolgoletny were the tallest in the early maturing group (78 cm),

Mars (89 cm) – in mid-early group, BAA-31 (85 cm) – in mid-season group, Falensky-86 (100 cm) – in mid-late group, and Kirovsky-159 (102 cm) – in late group.

Productivity of green mass also changed over the years. The highest yield was obtained in the second year of life for all the variety samples. In the third year of life the highest-yielding samples were from: early maturing group – Ustoylivy (5.4 kg/m^2) and SL-58 (5.5 kg/m^2), Mars (6.8 kg/m^2) – mid-early, BAA-31 (7.0 kg/m^2) – mid, Vega (5.7 kg/m^2) – mid-late and Vityaz (6.3 kg/m^2) – late maturing groups.

In total during the three years the best yields of forage were produced by: early maturing samples TOS (19.5 kg/m^2), Ustoylivy (22.8 kg/m^2); mid-early TOS (21.4 kg/m^2), Mars (23.0 kg/m^2); mid – BAA-31 (23.5 kg/m^2) and SGP-12 (25.8 kg/m^2); late – SGP-6 (23.6 kg/m^2).

Foliage of the variety samples varied on the whole in the nursery ranging from 33% in the mid-late sample Sozh to 51.8% in the early maturing SL-58. Within each group of ripeness the limits of variation of foliage indicators in the variety samples decreased and amounted to: in early ripening – 38.0–51.8%, in mid-early – 36.5–47.8%, in mid – 33.5–46.8%, in mid-late – 33.0–43.2%, and in late – 36.5–44.0%. There is a general pattern of increasing foliage from late forms to the early ones. However, in each group of ripeness we isolated sources of high foliage, among which the best are: in early maturing group sample SL-58 (51.8%), in mid-early – VIC-84 (47.8%), in the mid – BAA-31 (46.8%), in the mid-late – Pradi (43.2%), and in the late – Mut-19–1-1 (44%).

The dry matter content also varied significantly in the samples from 20.3% in SL-56 to 26.1% in Ranny-2. The highest percentage of dry matter was in the following samples: from early maturing group – SL-38 (25.0%), TOS-early (25.2%), Altyn (25.4%) and Ranny-2 (26.1%); from mid-early – T-59 (23.8%); from mid – SGP-12 and BAA-31 (24.1%); from mid-late Mereya (24.2%), TOS-870 (24.7%), and from late – MOS-1 (23.2%). Both parameters, the number of leaves and dry matter content, are of practical importance in the selection of red clover. A greater number of leaves enhances the protein content, and high dry matter content increases winter hardiness. Therefore, the correct selection of sources of the given traits in different groups of maturity is necessary to breed new varieties of different areas of use.

Belonging of the variety samples of red clover to one or another type has been studied by other parameters as well, such as the average number of internodes, the number of formed cuts and distribution of harvest among them, in terms of flowering, and also by bush morphobiotype and relation to vernalization (Fig. 3.1). To do this, we have separated from each group of ripeness the best and most typical variety samples created as a result of the joint selection of the Belorusian State Agricultural Academy and the All-Russian Research Institute of Feeds named after Williams on the TOS program "Clover." Among these samples are TOS-early, TOS mid-early, BAA-31, TOS-870 and SGP-6.

Given that the number of internodes on the main stem in the phase of beginning of plant flowering is one of the main features to determine the type of red clover testing of crops was conducted. At the same time, along with the average number of internodes the amplitude of fluctuation of the trait and the coefficient of variation were defined. A comparison of the indicators of the average number of internodes of all the studied samples of red clover confirmed their belonging to different types. A strict pattern of consistent increase in the average number of internodes from early maturing to the late types was revealed. So, in the early maturing sample TOS-early this figure was the lowest, the average number of internodes was six, and in each subsequent samples by the degree of ripeness it increased by one unit, and was, respectively, in TOS mid-early – 7, BAA-31–8, TOS-870–9, and SGP-6–10 internodes. The amplitude of fluctuation of the number of internodes on the main stems of plants of the same samples was not so significant.

The coefficient of variation for all samples had an average value, except the variety sample TOS-870 in which it was 8.5% and showed to a weak variation. The obtained results indicate that the studied variety samples are well selected, are characterized by relative uniformity of the height of stalks, uniform flowering and ripening.

Features	Variety sample, type of clover				
	TOS early maturing	TOS mid-early maturing	SGP-31 mid-season maturing	TOS-870 mid- late maturing	SGP-6 late maturing
The average number of internodes Amplitude of fluctuation Coefficient of variation, %	6 4–8 14,6	7 5–9 16,3	8 6–10 15,6	9 6–12 8,5	10 8–12 11,1
Number of formed cuts	3	2–3	2	2 or 1 plus afterfeed	1 plus afterfeed
Distribution of harvest among cuts, %	30:40:30	70:30 42:33:25	63:37	75:25	85:15
Mean perennial flowering time in the North- Eastern part of the Republic of Belarus	2nd decade of June 15.06–18.06	3d decade of June 23.06–26.06	end of 3d decade of June beginning of 1st decade of July 28.06–2.07	1st decade of July 7.07–10.07	2nd decade of July 12.07–17.07
Biotype of bush by International Classifier *	3 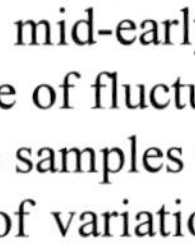	4 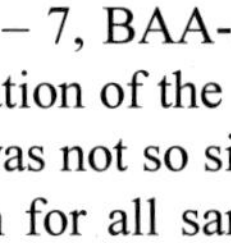	4–5 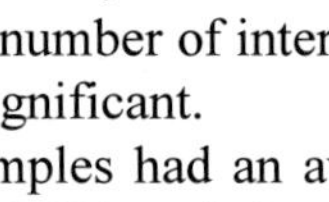	5 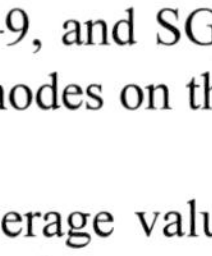	6 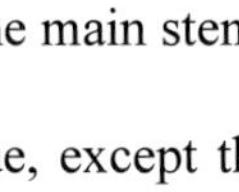
Type of vernalization	spring biotypes	spring and spring-winter biotypes	spring-winter biotypes	winter and partially spring-winter biotypes	winter biotypes
Using for seeds	from the 2nd cut	from the 2nd cut	from the 1st cut	from the 1st cut	from the 1st cut

Note * - 3 –the bush is formed by the main and lateral flowering stems and a shallow crown; 4 – the bush is formed by flowering and non-flowering stems and a medium crown; 5 – the bush is formed by stems that are not blooming and a large crown; 6 – the bush is formed only by a large crown that differ significantly.

FIGURE 3.1 Characteristic of the traits of variety samples of different types of red clover.

Samples from different maturity groups differed significantly in the number of generated cuts. Early maturing samples TOS-early and TOS mid-early in the second and third year of life in the growing season formed three cuts, BAA-31 – mid-season and TOC-870 – mid-late types – two cuts, and late sample SGP-6 produced one cut and afterfeed. Distribution of the green mass yield among cuts largely depended on the supply rate with heat and moisture during the most critical periods of growth and development of plants. The highest yields in early maturing samples of TOS-early maturing was obtained in the second cut, whose share in the total yield was 40%, in the remaining samples the highest yield was in the first cut, which amounted to in TOS mid-early – 42%, in BAA-31–63%, in TOS-870–75% and in SGP-6–85%. Flowering of all the samples in the first cut began differently. The most precocious TOS-early reached mowing maturity in mid-June 15.06–18.06, and then every 8–10 days samples of subsequent maturity groups were flowering. In the late ripening samples of SGP-6 flowering phase occurred in the second decade of July 12.07–17.07.

Characteristic traits of the types of red clover are morphobiotype of the bush and type of vernalization. Within the same sample there are spring, spring-winter and winter forms of plants that differ significantly. There are differences that are manifested in morphobiotype of the bush and precocity in the variety samples depending on the ratio of different spring planting.

The variety sample TOS-early consists only of early maturing spring plant forms, in the first year of life it formed a bush of main and lateral blooming stems and a shallow crown. In the sample TOS there occur spring and spring-winter plants, morphobiotype of the bush consisted of nonflowering and flowering stems with an average crown. Mid-season variety sample BAA-31 consisted of spring-winter forms and had two different morphobiotypes, one of which was formed from flowering and nonflowering stems and an average crown, the other one from nonflowering stems and a large crown. In TOS-870 sample winter forms of plants dominated and there were partially spring-winter forms. The bush was formed from nonflowering stems and a large crown. The late ripening variety sample SGP-6 included only winter forms of plants, the bush was formed in a very large crown, the height of which reached 45–50 cm by the end of the growing season in the first year of life.

Thus, a comprehensive study of red clover samples allowed us not only to confirm that they belong to one type or another, but also to use the studied traits as a marker in the allocation of sources of valuable properties in the selection of varieties for different areas of use.

3.4 CONCLUSIONS

1. In the breeding of red clover varieties for different use of practical importance is the diversity of the gene pool and the presence in it of the necessary sources of quantitative and qualitative traits. For the varieties of the field grass cultivation the most significant traits are different earliness, high parameters of forage productivity within 2–3 years, foliage, dry matter content, the rapid rate of spring and after cut regrowth, resistance to diseases and pests. For the varieties of meadow and pasture grass cultivation of practical significance is longevity, high competitive ability in grass mixtures with their prolonged use in hayfields and pastures, different ripening and different rhythm of growth, fast regrowth and high resistance to grazing and trampling.

2. The studied gene pool of red clover is characterized by a considerable variety of quantitative and qualitative attributes and contains the most valuable sources for breeding varieties of different areas of use.
3. For breeding complementary varieties of field, meadow and pasture areas of use different sources of precocity were selected which are divided into five types: early ripening, including 19, mid-early – 10, mid-season – 6, mid-late-21, and late – 7 samples. The most typical representatives of the early type is TOS-early, of mid-early – TOS-mid-early, of mid-season – BAA-31, of mid-late – TOS-870, and of late – SGP- 6.
4. The most valuable sources for breeding varieties of red clover characterized by a complex of economically useful quantitative and qualitative attributes are: variety samples of early maturing group – TOS – early maturing, Ustoylivy, Ranny-2, SL-38 and T-70; mid-early – TOS mid-early and Mars; mid-season – BAA-31; mid-late – Mereya, TOS −870, SGP-12, Yaskravy and Falensky-86; late – SGP-6.
5. For field use more suitable samples are: TOS early maturing, TOS mid-early, BAA-31, SGP-12, TOS-870 and SGP-6.
6. For meadow and pasture clover sowing the best are perennial samples Ustoylivy, Mars, SGP-12, Falensky-86, SGP-6 and Mut-19–1-1.

KEYWORDS

- **green conveyor**
- **group of ripeness**
- **longevity**
- **productivity**
- **selection**
- **variety sample**

REFERENCES

1. Yield and Quality in Herbage Seed Production, Proceeding of the Third International Herbage Seed Conference, Halle (1995) Germany, 476 (in German).
2. Shlapunov, V. N. (2005). Forage Crops of the Green Conveyor, Shlapunov, V. N., & Lukashevich, T. N. Modern Technologies of Crop Production in Belarus, Collection of Scientific Materials, Institute of Agriculture and Breeding of the National Academy of Sciences of Belarus, Minsk, Unitary Establishment of Information Computer Centre of the Ministry of Finance, 266–270 (in Russian).
3. Pozdnyakov, V. A. (2000). State of Research on Breeding and Seed Production of Forage Crops in Russia and Abroad, Pozdnyakov, V. A., Ways to Increase the Yield and Quality of

Crops in the North-West of Russia the Works of the Russian Academy of Agricultural Sciences, St. Petersburg, 147–151 (in Russian).
4. Novoselov, Yu M. (1999). Breeding of Red Clover (*Trifolium Pratense* L), Novoselov, Yu M., Moscow, 184p (in Russian).
5. Bushuyeva, V. I. (2007). Achievements and Challenges of Red Clover Breeding in Russia and Belarus, Bushuyeva, V. I., Agriculture and Plant Protection, *4(53)*, 56–59 (in Russian).
6. Novoselova, A. S. (2003). The Main Achievements and Directions of the Research in Selection of Red Clover, Novoselova, A. S., Novoselov, Yu M., Development of Scientific Ideas of Academician Lisitsyn, P. I., Collection of Works, Moscow Agricultural Academy named after Timiryazev, K. A., & Pylnev, V. V. Moscow, All-Russian Scientific-Research Institute of Meat Processing Industry named after Gorbatov, V. M., 45–52 (in Russian).
7. Polyudina, R. I. (2005). Basic Methods and Results of Selection of Red Clover (*Trifolium Pratense* L) in Western Siberia Author's Abstract of dis. Dr. of Agriculture Science (06.01.05) Plant Breeding and Seed Growing, Polyudina, R. I., Siberian Scientific Research Institute of Feeds, Moscow,34p (in Russian).
8. Novoselova, A. S. (2007). Scientific Bases and Results of Ecological Breeding of Red Clover (*Trifoliumpratense* L.), Novoselova, A. S., & Novoselov, Yu M., Feed Production, Problems and Solutions, Scientific Works of All-Russian Scientific-Research Institute of Feeds named after Williams, V. R. Ed. Kosolapov, V. M., the Printers FGNU "Rosinformagroteh" Lobnya, 278–283 (in Russian).
9. Tumasova, M. I. (1998). Ecological and Geographical Method in the Selection of Red Clover, Tumasova, M. I., Gripas, M. N., & Svetlakova, V. Y., New Methods of Selection and Creation of Adaptive Crop Varieties, Results and Perspectives theses of Reports of Scientific Session July (1–3–1998), Kirov, 81–82 (in Russian).
10. Novoselova, A. S. (2005). Breeding and Seed Production of Perennial Grasses, Novoselova, A. S., Voronezh Regional Printer named after Bolkhovitinov, E. A., Moscow, 375p (in Russian).
11. Novoselov, Yu M. (2002). Main Directions of Selection and Efficient Methods for Creating Adaptive Varieties of Red Clover, Novoselov, Yu M. et al. Adaptive Feed Production Problems and Solutions, for the 80th Anniversary of All-Russian Scientific Research Institute of Feeds named after Williams, V. R., Ros informagro teh, 252–270 (in Russian).
12. Bushueva, V. I. (2006). Gene Pool of Red Clover and its Application in Breeding Varieties of Different Areas of Use, Bulletin of the Belorusian State Agricultural Academy, *3*, 66–72 (in Russian).
13. Bushueva, V. I. (2007). Assessment of Variety Samples of Red Clover of Different Maturity Groups, Bushuyeva, V. I., Science and Innovation, *8(54)*, 37–42 (in Russian).

CHAPTER 4

RESULTS OF ELECTROPHORETIC ANALYSIS OF SEED STORAGE PROTEINS OF VARIFTY SAMPLES OF RED CLOVER AND GALEGA ORIENTALIS

VERA Iv. BUSHUYEVA

Belorussian State Agricultural Academy, d. 5, Gorki, Michurin St., 213407, Belarus; E-mail: vibush@mail.ru

CONTENTS

ABSTRACT

The paper analyzes the effectiveness of new methods for assessing the source material in breeding. The greatest attention is paid to methods of electrophoresis of proteins and protein markers, as well as methods of molecular genetic marking RFLP and PCR based on the use of DNA. Experimental data of the electrophoretic analysis of seed storage proteins in the selection samples of red clover and *Galega orientalis* are given. It was found that breeding material of red clover and *Galega orientalis* created in the Belorussian State Agricultural Academy is characterized by genetic variability. Variety samples BSAA-3, SL-38, TOS-early, BSAA-4, BSAA-2 and variety TOS-870 of the red clover are characterized by the greatest variability of polypeptide spectrum. In *Galega orientalis* differences of polypeptide spectrum were observed in variety samples SEG-1a, SEG-2, SEG-3 and BSAA-3. The obtained results will be used for the selection of genotypes in the breeding process to create patentable varieties characterized by novelty and distinguishability.

4.1 INTRODUCTION

One of the most important conditions for successful breeding of high-yielding varieties is proper selection process at all stages, starting with the creation of the source material and completing with the selection of the best breeding forms that correspond to the parameters of the planned variety model. No matter by what methods of selection the source material was created, the breeding success of the work depends on the effectiveness of the methods of its evaluation and capabilities of recognition of potentially more productive genotypes by phenotype [1]. As phenotypic traits in plants are the result of complex interactions between genotype and environment, hence the identification of necessary genotypes according to them is of probabilistic nature and is associated with a certain degree of error [2]. It should be noted that the selection of the most valuable genotypes by phenotype among the large number of studied forms in the early stages of the selection process in the practical implementation is very time consuming, of little effectiveness and not always resultant. It was possible to increase the efficiency of selection of genotypes due to the improvement of assessment methods of breeding material based on biochemical, electrophoretic and molecular genetic studies. These methods were especially effective in crop breeding aimed at improving the quality characteristics of varieties [3].

4.1.1 ANALYSIS OF LITERATURE SOURCES

N.Iv. Vavilov once called breeding for quality and chemical composition of plants a priority and urgent task in the science of selection. He then noted: “The time has come to begin serious research genetic work on mastering the chemistry of plants, on improving the chemical composition of plants. The geneticist and chemist must unite in this direction" [2]. The idea of N.Iv. Vavilov has been developed in the numerous studies conducted afterwards by the joint efforts of breeders, biochemists, geneticists, physiologists and technologists. N.N. Ivanov made a significant contribution to the development of biochemical methods for plant breeding [4]. He was the first in the world who fundamentally and in different aspects – varietal, agronomic, soil-climatic and geographical studied variability of major crops by many biochemical parameters. The results of these works have been published in a multivolume edition of "Biochemistry of cultivated plants" where patterns and limits of variability of biochemical traits of each crop and ways of their selection improvement are shown. Introduction of biochemical methods in practical plant breeding let him select such valuable crops as alkaloid-free lupine, melilot with little coumarin content and successfully introduce them into production. However, paying attention to the improvement of biochemical methods he thought it was necessary to use proteins "for accurate distinction of varieties."

Productive research in this direction began under the leadership of Konarev [3] and got a real development with the advent in the 60s of the twentieth century of electrophoretic, chromatographic and immunochemical methods for the identification of proteins, revealing their polymorphism and specificity. Using these methods, all the proteins were divided into three groups according to the specificity which is revealed in one group at the genus level, the tribe, the family, in the other one-at the level of species, subspecies, genetic groups, in the third one – at variety, biotype or line level. Accordingly, the proteins were used as genetic markers unique only to a particular genus, species, variety, and even biotype. Proteins inherent, for example, to representatives of one variety and absent in others were named protein markers. Each protein marker is characterized by its own biological specificity of the structure of the component spectrum. Seed storage proteins have the most pronounced specificity, the spectrum of components of which is defined by electrophoretic analysis. This method has proven effective and is widely used not only in plant breeding, but also in solving actual problems of applied botany, genetics and seed breeding.

Since the second half of the last century methods of molecular genetic marking of restriction fragments of DNA and polymerase chain reaction – PCR have been developed. After the discovery made by D. Botstein in 1980 [5] of phenomenon of restriction fragment length polymorphism (RFLP) and the subsequent development of PCR by K. Myulles in 1985 [cited by Ref. 6], they began a deep study of the structure of DNA and biological processes occurring in the cell with DNA (replication, repair and recombination), the basic mechanisms of transcription, translation and gene expression. An opportunity appeared to study the inheritance of certain phenotypic traits in their relation to specific DNA sequences, to identify the individual characteristics of organisms at the molecular level.

Methods of analysis of RFLP and PCR are now successfully used in various areas of breeding science, for example, to create molecular-genetic maps of plants, to identify donors of economically useful traits, to certify varieties, to diagnose diseases, to create DNA-marker of concomitant breeding.

The method of PCR-analysis is more valuable for studies of plant genomes for which the following primers are used: RAPD (Random Amplification of Polymorphic DNA), SSR (Simple Sequence Repeats), and in recent years SNP (Single Nucleotide Polymorphisms) [7].

The advantage of DNA-markers compared to the protein ones is that DNA of any organ of plant and at any stage of development can be used for marking.

It is known that for plant breeding of greatest practical importance is the opportunity to mark such genomes and genetic systems, which are of economically valuable traits. Therefore, protein and DNA markers are used in selection to identify the sources of economically valuable traits, such as high content and quality of protein, resistance to diseases and environmental stress, etc. Using DNA marker technology allows the researcher to work directly with the carrier of genetic information and in a short time to create forms of plants with the necessary characteristics and properties. All this testifies to the significant progress in the development of breeding science. Currently available for breeders theoretical and instrumental framework provides an opportunity not only to put into practice the idea of N.N. Ivanov to provide "accurate distinction of varieties," but also to accelerate the selection process for the creation and revealing of genetically variable forms of plants and carry out the selection process more purposefully and effectively.

New marking techniques of breeding material can be successfully used in combination with any of the methods of selection and at all stages of the selection process from searching economically important traits to developing varieties [3]. At the same time it should be remembered that new methods only help the breeder to intensify breeding process, and field experience has been and remains the basis allowing to select the best genotypes that are more adaptive to the specific soil and climatic conditions. With regard to improving varieties with targeted genes, the carriers of the most valuable traits, the greatest effect can be obtained by "embedding" them in the best varieties previously studied by conventional selection and adapted to specific areas [8].

It should be noted that the molecular genetic methods for creating and evaluating source material are already used in practical breeding of certain crops. Successful work in this direction is carried out in the Institute of Genetics and Cytology of the National Academy of Sciences of Belarus where a genetic map of rye has been made and certification of Belarusian varieties of wheat, potato, tomato, flax and beet has been done. Wheat genes responsible for baking qualities and resistance to brown rust were identified. RAPD-analysis method revealed intervarietal polymorphism of flax (*Linum usitatissimum L.*) used for fiber and linseed oil production; molecular analysis of inbred lines and hybrids of sunflower for predicting heterosis effect was conducted [9].

As for the selection of perennial legumes, including red clover, the most efficient molecular-genetic studies are conducted at the All-Russian Research Institute of Feeds named after V.R. Williams where together with the scientific center of agricultural research of Hokkaido and DNA Research Institute (Japan) two genetic maps of red clover on the basis of restriction fragment length polymorphism (RFLP) and amplification products of simple sequence repeats of DNA (SSR – Simple Sequence Repeats) were made [7].The genetic map created on the basis of RFLP includes 167 markers covering 535.7 cm. The map created on the basis of amplification products SSR includes 1305 markers and covers 868.7 cm. Both maps are presented in the form of seven linkage groups that have been identified relating to chromosomes of red clover using FISH-method. Three valuable for breeding traits of red clover (date of flowering, winter hardiness and resistance to sclerotinia) were put on the genetic map and the sequence of sites that correlate with these traits was defined. In order to improve breeding and seed operation and integration into the international system of UPOV (International Union for the Protection of New Varieties of Plants) the All-Russian Scientific Research Institute of Feeds developed a method for the identification of varieties. It is based on amplification products of DNA specific for a variety and isolated from the five-, seven-day seedlings and genotyped by the set of RAPD-primers [10]. However, the key problem at present remains the question of creating a collection of protein and DNA-marked forms by the main breeding valuable traits that are of complex quantitative character. With the advent of such collections protein and DNA markers can be successfully used in breeding varieties of red clover and *Galega orientalis* having the most favorable combination of economically useful traits. This can be achieved due to capability of giving a comprehensive assessment of the source material at the genetic level. With the help of protein and DNA markers it will be possible practically to establish the differences among the samples of different selection or geographical origin, to determine a biotypical composition of the population and to select the best biotypes [7].

In our studies the method of electrophoretic analysis of seed storage proteins was used to evaluate the best selection samples of red clover and *Galega orientalis* by genotype.

The aim of the research was to carry out electrophoretic analysis of seed storage proteins of the variety samples of red clover and *Galega orientalis* and to identify genetic differences among them.

4.1.2 RESEARCH OBJECTS

The objects of the study were the best by the complex of economically useful traits variety samples of red clover and *Galega orientalis* created at the Department of Breeding and Genetics at the Belorussian State Agricultural Academy. Thirty variety samples of different maturity groups of red clover have been studied, among which: (a) early ripening variety Dolgoletny – standard, variety samples BSAA-

1, BSAA-2, BSAA-3, SL-38, TOS-early, (b) mid-early ripening: variety Mars – standard, BSAA-4, BSAA-31, TOC mid-early (c) mid-season ripening: variety Vitebchanin – standard, BSAA-5, SGP-12 (d) mid-late ripening: varieties Minsk – standard, Mereya, TOS-870, variety samples BSAA-6, BSAA-7, BSAA-8 and (e) late ripening: variety MOS-1 – standard, variety samples BSAA-9, SGP-6, SOZH, BSAA-10, BSAA-11, BSAA-14, BSAA-15, BSAA-12, BSAA-13, BSAA-870. According to the results of the field evaluation the variety samples of red clover differed in morphobiotype of the bush, color of flowers, vegetative organs and seeds, foliage and other phenotypic traits.

Twenty variety samples of different selection and ecological origin were studied in *Galega orientalis*. Among these samples are SEG-1, SEG-1A, SEG-2, SEG-2A, SEG-3, SEG-4, SEG-4A, SEG-5, SEG-6, SEG-7, SEG-8, varieties Moscow, Gale, Estonian, Minsk, Fast Regrowing, Caucasian Branets, variety samples BSAA-1, BSAA-2, BSAA-3, BSAA-4, BSAA-5 and variety Nesterka – standard. According to the results of the preliminary assessment in the field more contrasting differences in color of flowers, vegetative organs and to a lesser extent of seed color were identified among the variety samples of *Galega orientalis*. The variety samples also differed in morphobiotype of the bush, the length of the growing season, foliage and yield.

4.2 MATERIAL AND METHODOLOGY

Seed protein electrophoresis was performed on the device SE-250 produced by the company GE Healthcare (England). Proteins from seeds of the studied crops for electrophoresis were extracted by Tris-HCl buffer, pH 7.0 containing saccharose, EDTA and 0.01% ascorbate. Electrophoresis of proteins was performed in 15% polyacrylamide gel (PAGE) in denaturing conditions according to the method of Laemmli [11]. Protein was determined by the method of Bradford [12]. Bovine serum albumin was used as a standard.

4.2.1 PREPARATION OF SEPARATING GEL

To prepare 10 mL of a 15% polyacrylamide gel (volume which is required for a single plate), 2 mL of deionized or twice-distilled water mixed with 4 mL of 30% acrylamide and with 2 mL of 0.8% solution of bis-acrylamide (29.8 g of acrylamide and 0.8 g of bis-acrylamide of firm Sigma, Bio-rad, are dissolved in 100 mL of deionized water) and added 2 mL of 1.5 M Tris-HCl buffer, pH 8.8.

4.2.2 PREPARATION OF STACKING GEL

Gel of 6% and volume of 1 mL was prepared by mixing 0.54 mL of deionized water, 0.2 mL of 30% acrylamide and 0.8% bis-acrylamide and 0.25 mL of 0.5 M Tris-HCl pH 6.8.

4.2.3 PREPARATION OF SAMPLES PRIOR TO ELECTROPHORESIS

Equal volume of buffer was added to 100–200 mkg of protein in the samples comprising: 10 mL−5.55 mL of deionized water, 1.25 mL of 0.5 M Tris -HCl pH 6.8, 3.0 mL of glycerol, 0.2 mL 0.5% of bromphenol blue 5% SDS, 2% mercaptoethanol. Samples were heated in a water bath at 100 °C within 1 min. Thereafter, the samples were applied onto the gel in the bulk of 20 mkl.

4.2.4 ELECTROPHORESIS AND ANALYSIS OF THE RESULTS

At the beginning of electrophoretic separation the mode of electrophoresis: voltage was 40 V, current – 10–12 mA. After the samples entered the separating gel, the voltage should be 140–200 V, current – 20 mA.

The electrode buffer consisted of 25 mM Tris-HCl pH 8.3 containing 198 mM of glycine. Ovalbumin, 45 kD; RNA-asa, 25 kD; β-laktatglobulin, 18.4 kD; lysozyme, 14.4 kD. Gels were fixed with 20% TCA (trichloroacetic acid). Electrophoregram processing was performed by a computer program TOTAL/LAB, Control Centre V. 2,1.

4.3 RESULTS AND DISCUSSION

We have received 4 electrophoregrams with the spectra of components of seed storage proteins of the 30 studied variety samples. Analysis of the spectra showed that they differed from each other in varying degrees. It has been found that differences in the spectra of the proteins are more pronounced in those variety samples in which they were manifested with greater contrast by phenotypic traits when being evaluated in the field.

An example is electrophoregram of the spectrum of protein components of red clover varieties Dolgoletny and Mars, and variety samples BSAA-1, BSAA-2, BSAA-3, SL-38, TOS-early maturing and BSAA-4, presented in Fig. 4.1.

In this group of the studied variety samples differences were revealed in the polypeptide composition of proteins, for example, in the variety Dolgoletny at the range of 36 kDa. The polypeptide of this variety has proved to be less pronounced in comparison with the other samples. Depending upon the sample to be examined different composition of the polypeptide spectrum is observed both in the area of 56 and 112 kDa, in particular in the samples of BSAA-3, SL-38, TOS-early, BSAA-4, and in variety Mars protein 75 kDa appeared that is virtually absent in other samples. In the sample BSAA-2 a distinctive feature of the spectrum is the appearance of the polypeptide in the area of 69 kDa. All this shows to the presence of a polymorphism of seed storage proteins in the red clover variety samples. However, numerous studies have found that the polypeptide composition of proteins in red clover is relatively stable at the species level, so it is not always possible to identify polymorphism at the variety level. Clover varieties can vary considerably by economically valuable traits while remaining unchanged in the structure of the protein components.

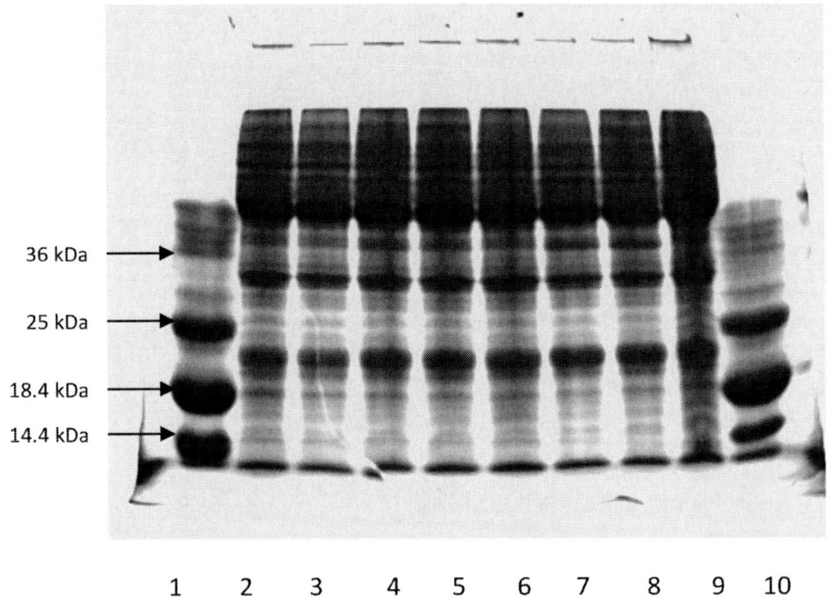

FIGURE 4.1 Electrophoregram of separation of extracts of red clover seeds: 1 and 10 – protein molecular weight markers (25; 18.4; 14.4 kDa). Samples of red clover seeds: 2 – Dolgoletny; 3 – BSAA-1; 4 – BSAA-2; 5 – BSAA-3; 6 – SL-38; 7 – TOS-early; 8 – Mars; 9 – BSAA-4.

This is confirmed by electrophoregram of the spectra of seed storage proteins of the variety samples of red clover BSAA-31, TOS-mid-early, BSAA-5, SGP-12, BSAA-6, varieties Vitebchanin, Minsk-late local, Mereya; the differences among them were more of quantitative content than of the polypeptide composition. Differences in electrophoretic spectrum of seed storage proteins were not so contrasting among red clover variety samples belonging to mid-late and late groups, such as the BSAA-870, BSAA-7, BSAA-8, MOS-1, BSAA-9, SGP-6, SOZH and BSAA-10. As for the group of variety samples BSAA-11, BSAA-14, BSAA-15, BSAA-12, BSAA-13 and variety TOS-870, differences in their polypeptide composition have been identified according to the results of the electrophoretic analysis of seed storage proteins. From the practical point of view, the variety TOS-870, which is largely different from the others in electrophoretic protein spectrum, is of the greatest significance for selection in this group of variety samples. So, in the variety TOS-870 (track 7) the appearance of polypeptide 42kDa was observed and was not noticed in the other samples. Furthermore, there was an increase of 64kDa protein in the same variety. It should be noted that the variety of TOS-870 according to the results of the biochemical analysis also positively differed from the other variety samples. The reason, in our opinion, is the genetic structure of the variety which is a composite-hybrid heterotic population formed on the basis of selected by combining ability highly productive biotypes from 4 variety populations of different breeding and ecogeographical origin.

We studied all the diversity of the variety samples of *Galega orientalis* that was at our disposal. Among these were varieties of different ecogeographical origin and our new variety samples of mutant and hybrid origin. It is known that varieties of *Galega orientalis* that are currently being commercially cultivated, including the variety of our selection Nesterka, differ to a greater extent by quantitative traits as they all have the same phenotype characterized by the blue color of flowers and dark-green vegetative organs. New variety samples bred at the Department of Breeding and Genetics of the Belorusian State Agricultural Academy differ not only by quantitative, but also clearly distinguishable qualitative characteristics (by color of flowers, vegetative organs). In order to establish the differences among them at the genetic level we carried out electrophoretic analysis of seed storage proteins. We studied three electrophoregrams in all including spectra of the components of seed storage proteins of 20 variety samples. The studies allowed us to identify the differences both among crops of red clover and *Galega orientalis* at the genus level, and also among variety samples at the species and cultivar level. As a result of the preliminary comparison of the electrophoregrams of red clover and *Galega orientalis* clear differences between the crops were revealed. Analysis of the spectra of electrophoregrams of each group of variety samples showed differences in the spectrum of the protein components depending on the genotype of the studied samples. In this connection phenotypic variation with the polypeptide composition of the protein was identified.

On the electrophoregram shown in Fig. 4.2 variety samples of *Galega orientalis*, such as SEG-1, SEG-1a, SEG-2, SEG-2a, SEG-4, SEG-4a, SEG-5 and SEG-6 differing by contrast in phenotype were studied.

All variety samples are of hybrid origin and belong to different species. The differences in the polypeptide composition are observed in a white flower sample SEG-1a (Track 3), where there is no band 41 kDa and there is increase of 44, 39 kDa bands. In the lilac flower variety sample SEG-2 (track 5) there are also differences in comparison with the other studied variety samples, namely polypeptide doublet of 45, 43 kDa, which are almost not present in the other samples of the electrophoregram. It should be noted that the differences among these variety samples were also found [13] at the molecular genetic level by PCR-analysis with RAPD-primers.

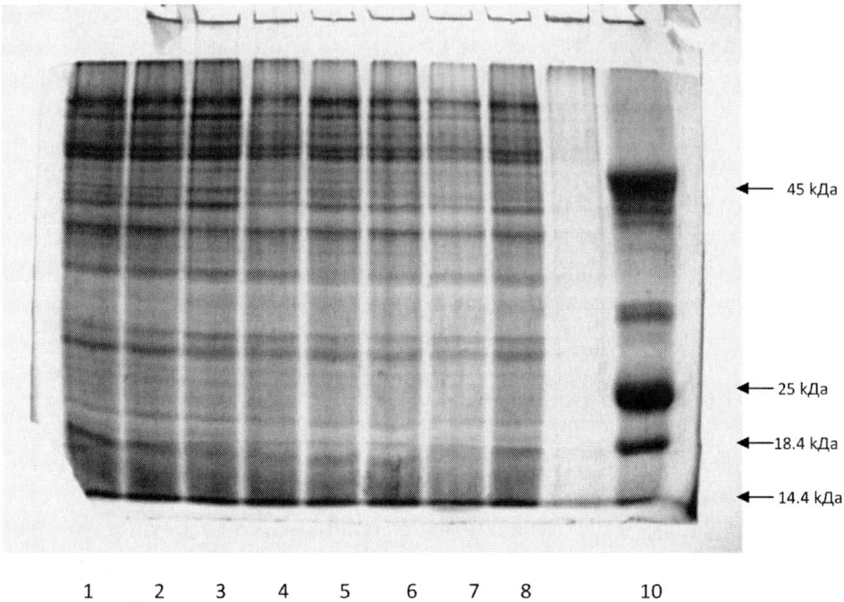

FIGURE 4.2 Electrophoretic separation of the soluble proteins of seeds of variety samples of *Galega orientalis*: 1 – SEG 1, 2 – SEG-1a, 3 – SEG-2, 4 – SEG-2a, 5 – SEG-4, 6 – SEG-4a, 7 – SEG-5, 8 – SEG-6, 10 – molecular weight markers (45, 25, 18.4 and 14.4 kDa).

A distinctive feature of the seed protein electrophoregrams of the variety samples of *Galega orientalis* of the varieties Minsk, Moscow, Estonian, Gale, Nesterka, samples SEG-7, SEG-8, SEG-3 was the emergence of a new major band at 73 kDa in a blue flower sample SEG-3 (track 8).

On the electrophoregram of seed proteins of the variety samples Fast Regrowing, Caucasian Branets, BSAA-1, BSAA-2, BSAA-3, BSAA-4 and BSAA-5 differences were observed in the sample BSAA-3 (track 5). The peculiarity of this sample was increased content of polypeptides 25, 30, 32 kDa.

With regard to the varieties and variety samples with a phenotype typical to the varieties commercially cultivated today and characterized by a blue color of flowers and dark-green vegetative organs, they all had the same peptide and protein composition, did not differ in the spectrum. The obtained results let us assume, on the one hand, that the genotype of the varieties developed mainly by selection has not changed fundamentally, and, on the other hand, it is likely that the protein of *Galega orientalis* itself has a relatively stable structure.

4.4 CONCLUSIONS

Variety samples characterized by genetic variability established on the basis of the results of electrophoretic analysis of seed proteins were identified in the studied breeding material on red clover and *Galega orientalis*.

Variety samples BSAA-3, SL-38, TOS-early, BSAA-4, BSAA-2 and variety TOS-870 of red clover were characterized by the greatest variability of polypeptide spectrum.

Variety TOS-870 according to electrophoretic analysis differs from the other variety samples by polypeptide 42 kDa, which either did not occur in the other samples or was present in minor amounts. Furthermore, an increase of protein 64 kDa content was observed in this variety.

In *Galega orientalis* polypeptide spectrum differences were observed in the following variety samples: white flowering SEG-1a, lilac flowering SEG-2, blue flowering SEG-3 and violet flowering BSAA-3, which proves that they also have genotypic variability.

The obtained results of the electrophoretic analysis of seed proteins of the variety samples of red clover and *Galega orientalis* are an important scientific basis for the effective selection of variety samples with genetic variation and their use in the future selection process to establish patentable varieties characterized by novelty and distinguishability.

KEYWORDS

- **electrophoresis**
- **genetic variability**
- **polypeptide spectrum**
- **protein markers**
- **variety samples**

REFERENCES

1. Obraztsov, A. S. (2001). Potential Productivity of Cultivated Plants, Obraztsov, A. S., Moscow, Ros informagroteh, 504p (in Russian).
2. Vavilov, N. I. (1987). Theoretical Basis of Selection, Vavilov, N. I., Moscow Nauka (Science), 512p (in Russian).
3. Konarev, V. G. (2007). Molecular-Biological Studies of the Gene Pool of Cultivated Plants in the all-Russian Institute of Plant Growing (ARIPG), Konarev, V. G. (1967–2007) 2nd Edition Expanded, St. Petersburg, ARIPG, 134p (in Russian).
4. Ivanov, N. N. (1935). Biochemical Basis of Plant Breeding, Ivanov, N. N., Theoretical bases of Selection, *1,* Moscow-Leningrad, 991–1016 (in Russian).

5. Kartel, N. A. (2007). DNA Markers in Genetics and Plant Breeding, Kartel, N. A., From Classical Genetics and Breeding Methods for DNA-based Technologies, Proceedings of the International Scientific Conference on the 95th Anniversary of Academician Turbine, N. V., IX Congress of the Belarusian Society of Geneticists and Breeders, Gomel, (October 2–5, 2007), Minsk Law and economics, 157 (in Russian).
6. Edwards, K. A. (1991). Simple and Rapid Method for Preparation of Plant Genomic DNA for PCR Analysis Edwards, K., Johnstone, C., & Thompson, C. Nucleic Acids Resolution, *19(66),* 1349 (in English).
7. Kozlov, N. N. (2002). Results and Prospects of Research of DNA Polymorphism of Feed Plants, Kozlov, N. N., Pribytkova, T. F., Komkova, T. N. et al. Adaptive Feed Production, Problems and Solutions, Scientific Works of the All-Russian Research Institute of Feeds named after Williams, V. R., Lobnya Rosinformagroteh, 365–374 (in Russian).
8. Kadyrov, M. A. (2005). Selection of Major Crops in Belarus State, Problems and Priorities, Kadyrov, M. A., Principles and Methods for Optimizing the Selection Process of Crops, Proceedings of the International Scientific-Practical Conference (July 14–15, 2005, Zhodino) Minsk Publishing Centre of the Ministry of Finance of Belarus, 3–14 (in Russian).
9. Kilchevsky, A. V. (2005). Genetic Priorities of Modern Plant Breeding, Kilchevsky, A. V., Principles and Methods for Optimizing the Selection Process of Crops, Proceedings of the International Scientific-Practical Conference (July 14–15–2005), Zhodino Minsk Publishing Centre of the Ministry of Finance of Belarus, 14–18 (in Russian).
10. Shamsutdinov, Sh Z. (2007). Results and Current Priorities in the Selection of Feed Plants, Shamsutdinov, Sh Z., Piskovatsky, Y. M., & Novoselov, M. Y. Feed Production Problems and Solutions, Scientific Works of the All-Russian Research Institute of Feeds named after Williams, V. R., Lobnya Rosinformagroteh, 241–256 (in Russian).
11. Laemmli, U. K. (1970). Cleavage of Structural proteins during the Assembly of Heat of Bacteriophage, 4, Laemmli, U. K., *Nature, 2127,* 89–99 (in English).
12. Bradford, M. M. (1976). Rapid and Sensitive method for Quantitation of Microgram Quantities of Protein Utilizing the Principle of Protein Dye Binding, Bradford, M. M., *Anal Biochemistry, 8,* 248–254 (in English).
13. Bushuyeva, V. I. (2008). The Results of PCR-analysis of Variety Samples of *Galega Orientalis* by RAPD-primers, Bushuyeva, V. I., Genetics and Biotechnology of the XXI century, Fundamental and Applied Aspects, Proceedings of the International Scientific Conference Dedicated to the 60th Anniversary of the Department of Genetics of the Belorusian State University, Minsk December (3–6–2008) Minsk Publishing Center of the Belarusian State University, 53–55 (in Russia).

CHAPTER 5

USE OF GENOTYPIC VARIABILITY OF *GALEGA ORIENTALIS* FOR IDENTIFICATION OF VARIETIES

VERA I. BUSHUYEVA

Belorussian State Agricultural Academy, d. 5, Gorki, Michurin St., 213407, Belarus;
E-mail: vibush@mail.ru

CONTENTS

ABSTRACT

The characteristic of the new breeding material of *Galega orientalis* by morphological and economically useful traits is given. Variability of quantitative and qualitative characteristics and the possibility of their use for identification of varieties are shown. The criteria of distinguishability, uniformity and stability of *Galega orientalis* by qualitative and quantitative criteria are given. Characteristic of variety samples-standards and the list of traits for testing varieties of *Galega orientalis* for patentability are presented.

5.1 INTRODUCTION

Successful development of the agricultural sector in the Republic of Belarus or any other country in the world is impossible without the use of selection achievements such as new high-yielding varieties of plants. Under the current conditions in Belarus and in other countries with a highly developed selection science the potential of variety yield is very high and exceeds in cereals 10 t/ha. The further increase of the productivity potential and qualitative improvement of such varieties in the selection process requires the use of more efficient and very expensive methods of creating original material, needs deep and comprehensive knowledge, significant financial and human resources and many years of painstaking work of the breeder. In this regard, there is a need for legal protection of new varieties and legislative protection of copyright of the breeder. Therefore, in 1995 the Republic of Belarus adopted the law "On patents for plant varieties" guaranteeing copyright protection of breeders at the state level [1]. Since 2002 after passing the law on accession to the International Convention and the accession of the Republic of Belarus in the UPOV (International Union of Countries for the Protection of New Varieties) such protection also became possible at the international level [2]. Belarus became the 52nd state – member of the International Convention signed in Paris on December 2, 1961 and entered into force in 1968. The UPOV Convention is periodically reviewed and improved which is associated with the emergence of new scientific discoveries in biological science, with the improvement of breeding methods and changing in this connection requirements for variety identification. Genetic engineering and biotechnology methods using tissue culture that appeared between1961and 1991 have fundamentally changed the process of creating new varieties and their reproduction, which caused the need for improvement of legal protection within UPOV. Therefore, the UPOV Convention since its inception in 1961 has been revised three times in Geneva: November 10, 1972, October 23, 1978 and March 19, 1991 [3]. At the same time, along with the improvement of methods of legal protection of varieties the affiliation of new states is taking place, and the range of protected crops is expanding. The Act of the UPOV Convention of 1991 provides for ensuring the legal protection of all plant species in each acceded state. At the moment there are already countries (USA, France, etc.) where all cultivated crops are guaranteed legal protection. In accordance with article 3 of the 1991 Act of the UPOV Convention all member-states of the UPOV are given time period to 10 years from the date of their accession to the UPOV to perform these obligations. It follows from this that in Belarus legal protection to all kinds of cultivated plants must be provided by 2012 [4]. However, the UPOV highly approves of the states' desire to accelerate the introduction of new plant species in the number of protected, especially in the case of creating new varieties of them. In the Republic of Belarus *Galega orientalis* belongs to such type of plants (*Galega orientalis Lam.*) of which new varieties have already been created and are being cultivated commercially. The provision of legal protection to new varieties of *Galega orientalis* needs special method for testing varieties for distinguishability, uniformity and stability the development of which requires the presence of the gene pool of the plant characterized by a wide range of variability of quantitative and qualitative traits.

Galega orientalis is a perennial leguminous grass with cross-pollinated flowers. Its varieties have a complex genetic structure characterized by a certain set of quantitative and qualitative traits. However, currently commercially cultivated varieties of *Galega orientalis* comprising the included in the State Register of the Republic of Belarus variety Nesterka of our selection differ to a greater extent by quantitative traits as they all have the same phenotype characterized by blue colored flowers and dark-green vegetative organs.

In recent years at the Belorusian State Agricultural Academy a new gene pool of *Galega orientalis* was created whose forms differ not only in quantitative, but also clearly distinguishable qualitative characteristics (by color of flowers and vegetative organs). The available gene pool includes forms with white, lilac, light-blue, blue and purple color of flowers; with green, dark-green and anthocyanin vegetative organs. This is an important condition for conducting research on the nature and extent of variability of quantitative and qualitative traits in *Galega orientalis*, compiling table of traits and developing on their basis criteria of distinguishability, uniformity and stability.

The purpose of this research is to study the nature and extent of variability of quantitative and qualitative traits in the gene pool of *Galega orientalis* and on their basis to compile a table of traits to identify varieties, to select and propagate constant variety samples-standards with different combination of morphological characters and give them a comprehensive description.

5.2 MATERIALS AND METHODOLOGY

The research was carried out at the experimental field of Breeding and Genetics Department of the Belorusian State Agricultural Academy for 12 years from 2001 to 2012. New forms of *Galega orientalis* and variety samples of different types whose quantitative and qualitative characteristics were studied were the objects of research: SEG-1 – with white colored flower and light-green leaves and stems; SEG-2 – with lilac colored flower and dark-green leaves and stems; SEG-3 – with light-blue colored flower and dark-green leaves and stems; SEG-4 – with blue colored flower and dark-green leaves and stems; SEG-5 – with purple colored flower and anthocyanin leaves and stems. One hundred plants of each species were studied which were planted individually with the feeding area of 70×70 cm^2 in the nursery to study biotypical composition. As to the quantitative traits we studied duration of the phases of development and plant height at different periods of growth and years of life, leaf length, the number, length and width of leaflets, the average number of internodes and weight of 1000 seeds.

The study of the traits was done according to the technique recommendations and requirements of the UPOV on variety identification. At the same time, we took into account genetic structure of the populations of varieties of perennial leguminous grasses and their inherent interpopulation variability, methods of creating and reproducing varieties, genetic variability of their quantitative and qualitative characteristics.

Observation of traits of plants and their evaluation were performed according to the criteria of novelty, distinguishability, uniformity and stability. To establish criteria for distinguishability among variety samples by quantitative traits we used a combined long-term method (CLM) recommended by the Act of the UPOV Convention in 1991 [4]. The method is based on the principle of a long-term analysis of the varying characteristics and comparison of the average statistical characteristics of annual and long-term observations. In compiling the table of traits we took into account the limits of their variability in whole in the crop in the past 6 years. In describing the quantitative traits they were divided into the opposite pairs depending on the degree and nature of their expression, e.g., weak or strong, short or tall, small or large, narrow or broad. Within varying from minimum to maximum the degree of trait manifestation was assessed on a 9-point scale. For example, the height of the plants with the minimum degree of the trait expression was characterized as very short and had 1 point, and at the maximum manifestation – as very high and had 9 points.

Qualitative characteristics, color of flowers and vegetative organs were visually evaluated [5].

5.3 RESULTS AND DISCUSSION

Providing legal protection to the varieties is done by the results of patent examination. A variety is granted legal protection if according to the results of the patent examination it has novelty, distinguishability, uniformity and stability. Novelty of the variety is, thus, confirmed by the fact that prior to the date of application to be granted legal protection it is not sold or used for cultivation. At the same time, the new variety must be clearly different from any other, at least in one trait, be homogeneous and characterized by relatively stable limits of variation of quantitative traits or qualitative characteristics showing their stability or constancy after repeated propagation. Both qualitative and quantitative traits of varieties serve as criteria of distinguishability, uniformity and stability; they vary greatly depending on the genetic structure of the variety, methods of its creating and reproducing. We took all this into account in doing research.

The studies found that variety samples differed significantly in the duration of the phases of plant development. The period from the beginning of spring regrowth before the budding phase varied in variety samples within 37–44 days, of flowering phase – 44–50 days, the ripening phase – 90–102 days (Table 5.1).

TABLE 5.1 Variability of Quantitative Traits in the New Gene Pool of *Galega orientalis*

Trait	Parameters of the main statistical characteristics of quantitative traits of *Galega orientalis*.						
	X_{min}	X_{max}	$X_{average}$	S	V%	S_x	S_x%
Duration of the phases of development (days) from the start of regrowth to:							
budding	37	44	40.7	1.7	4.2	0.34	0.8
flowering	44	50	47.3	1.5	3.3	0.3	0.6
maturation	90	102	95.4	2.5	2.6	0.5	0.5
Plant height in the first year of life	50	95	73.1	11.1	15.2	2.2	3.0
Plant height in the second year of life: during the budding phase	80	150	116.7	17.2	14.7	3.4	2.9
during the flowering phase	90	180	130.7	18.8	14.4	3.8	2.9
during the maturation phase	83	174	130.7	21.2	16.2	4.2	3.2
In the second year of life during the flowering phase of plants: leaf length, cm	15	30	22.1	3.6	16.2	0.7	3.2
number of leaflets on a leaf, pcs.	9	15	12.4	1.5	12.0	0.3	2.3
width of leaflets, cm	2	5	3.7	0.7	19.7	0.15	4.0
length of leaflets, cm	4	8	6.1	0.99	16.3	0.19	3.1
length of the main stem with blossoming cluster, cm	93	185	135.8	22,0	16.2	4.4	3.0
number of internodes on the main stem, pcs.	7	15	10.6	1,7	16.0	0.34	3.2
Weight of 1,000 seeds, g	5.0	9.0	7.1	0,97	13.6	0.2	2.8

Plant height in the first year of life varied among variety samples within 50–95 cm. In the second year of life this figure varied during phases of development and differed significantly depending on the variety sample. Its variation during the budding phase was 80–150 cm, during the flowering phase – 90–180 cm, during the maturation phase – 83–174 cm. The variety samples also differed in the leaf length: the shortest – 15 cm, and the longest – 30 cm. Variety samples of the studied gene pool were characterized by a wide range of variability in the number of leaflets on a leaf, their width and length. Minimum height of the main stem of plants was 93 cm, and the highest – 125 cm. The average number of internodes varied among variety samples within 7–15 pcs. The lowest rate of the weight of 1,000 seeds was 5.0 g, and the highest – 9.0 g.

Thus, from the study of quantitative traits in the new gene pool of *Galega orientalis* differences in the degree of their expression were established according to which a table of traits for identification of varieties was compiled. Qualitative traits are more effective for the identification of varieties, they can be found visually and with a high degree of probability. As a result of the observations significant differences of morphological character were revealed among variety samples, among them were forms with white, lilac, light-blue, blue, violet and dark-violet colored flowers and green, dark-green and violet-green leaves and stems. The selected variety samples also differed in other qualitative characteristics, such as size and shape of the leaves, the presence or absence of pubescence on the leaflets, venation, a spinelet, as well as the shape of the bush (Table 5.2).

TABLE 5.2 Variability of Qualitative Traits in the New Gene Pool of *Galega orientalis*

Color		**Size of leaves**	**Shape of leaflets**	**Presence on leaflets**			**Form of bush**
Flowers	**Leaves and stems**			**Pubescence**	**Venation**	**Spinelet**	
White	Light-green	Medium	Ovale	–	–	+	Erect
Lilac	Green	Medium	Ovale	–	+	+	Erect
Light-blue	Green	Medium	Elliptical	+	+	–	Semi-erect
	Dark-green	Large	Ovale	–	–	+	Erect
Blue	Green	Large	Lanceolate	–	+	+	Semi-erect
		Medium	Ovale	–	+	+	Semi-erect
	Dark-green	Medium	Ovale	–	–	+	Erect

TABLE 5.2 *(Continued)*

Color		Size of leaves	Shape of leaflets	Presence on leaflets			Form of bush
Flowers	**Leaves and stems**			**Pubes-cence**	**Vena-tion**	**Spinelet**	
	Dark-green with antho-cyanin	Small	Ellipti-cal	–	+	–	Semi-erect
		Medium	Lanceo-late	–	+	+	Erect
Purple	Green	Medium	Ovale	+	+	–	Semi-erect
	Dark-green	Medium	Ovale	–	–	+	Semi-erect
		Large	Ovale	–	+	+	Erect
			Lanceo-late	+	+	+	Semi-erect
	Anthocy-anin	Medium	Ovale	–	+	+	Semi-erect
			Lanceo-late	–	+	+	Semi-erect
		Large	Ovale	–	+	+	Erect
Dark-pur-ple	Dark-green	Small	Lanceo-late	–	+	+	Semi-erect
	Intense an-thocyanin	Small	Lanceo-late	–	+	+	Semi-erect
		Large	Ovale	–	+	+	Semi-erect
		Medium	Lanceo-late	–	–	+	Semi-erect

To assess the qualitative traits visual observations were carried out according to which differences among variety samples were defined. Since the studied gene pool differed in qualitative traits much more compared with the cultivated varieties we have identified all possible variations and a table of characteristics for the development of special methods for testing varieties for distinguishability, uniformity and stability was compiled (Table 5.3).

TABLE 5.3 Traits of *Galega orientalis* For Establishing Distinguishability, Uniformity and Stability of Varieties

Traits	Degree of expression	Index
Plant		
Height before cutting the herbage in late autumn (in the first year of life)	Very short (50–59 cm)	1
	Short (60–69 cm)	3
	Medium (70–79 cm)	5
	High (80–90 cm)	7
	Very high (> 90 cm)	9
Height at the budding phase during spring regrowth (in the second year of life)	Very short (90–104 cm)	1
	Short (105–119 cm)	3
	Medium (120–134 cm)	5
	High (135–150 cm)	7
	Very high (> 150 cm)	9
Height at the beginning of flowering phase before harvesting for hay (in the second year of life)	Very short (90–107 cm)	1
	Short (108–125 cm)	3
	Medium (126–143 cm)	5
	High (144–160 cm)	7
	Very high (> 160 cm)	9
Type of bush at the beginning of flowering phase of the first cut (in the second year of life)	Reclining	5
	Semi-erect	7
	Erect	9
Stem		
Anthocyanin color at the beginning of flowering phase (in the second and third years of life)	No	1
	Weak	3
	Medium	5
	Strong	7
	Very strong	9
The length of the central stem including blooming cluster during the flowering phase of the first cut (in the second and third years of life)	Very short (70–90 cm)	1
	Short (91–120 cm)	3
	Medium (121–150 cm)	5
	Long (151–180 cm)	7
	Very long (> 180 cm)	9

Number of internodes on the stem during the flowering phase, pcs.	Very low (< 7)	1
	Low (9)	3
	Medium (11)	5
	High (13–14)	7
	Very high > 14	9
The thickness of the most developed stem during the phase of full bloom	Very thin (4–6 mm)	1
	Thin (7–9 mm)	3
	Medium (10–12 mm)	5
	Thick (13–15 mm)	7
	Very thick (> 15 mm)	9
Leaf		
Color of sprouts during the phase of 3–5 true leaves in the year of sowing	Light-green	1
	Green	3
	Dark-green	5
Color during flowering phase(in the second and third years of life)	Light- green	1
	Green	3
	Dark -green	5
	Light violet-green	7
	Dark violet-green	9
Leaf length during the flowering phase in the second and third years of life	Very short (14–17 cm)	1
	Short (18–21 cm)	3
	Medium (22–25 cm)	5
	Long (26–29 cm)	7
	Very long (> 29 cm)	9
Number of leaflets on the leaf during the flowering phase in the second and third years of life	Very small	1
	Small	3
	Medium	5
	Large	7
	Very large	9
Form of leaves during the flowering phase in the second and third years of life	Lanceolate	1
	Elliptical	2
	Ovate	3

Width of leaves	Very narrow (1 cm)	1
	Narrow (2 cm)	3
	Medium (3 cm)	5
	Wide (4 cm)	7
	Very wide (> 4 cm)	9
Length of leaflets	Very short (4 cm)	1
	Short (5 cm)	3
	Medium (6 cm)	5
	Long (7 cm)	7
	Very long (> 8 cm)	9
Presence of spinelet on the top leaflet	+	1
	–	2
Flower		
Color	White	1
	Lilac	3
	Light-blue	5
	Blue	7
	Purple	9
Dates of flowering	Very early	1
	Early	3
	Medium	5
	Late	7
	Very late	9
Bean		
Form of bean	Straight	1
	Curved	9
Seeds		
Color of testa	Bright-yellow	1
	Opaque-yellow	2
	Olive	3
Weight of 1000 seeds (conditional), g	Very low (up to 5 g)	1
	Low (5.1–6.0 g)	3
	Medium (6.1–7.0 g)	5
	High (7.1–8.0 g)	7
	Very high (8.1–9.0 g)	9

In addition, we have identified constant variety samples with different combination of morphological traits to use as variety samples-standards for comparison of new varieties in their identification. The selected constant variety samples differed not only in qualitative, but also quantitative traits.

Five phenotypically different variety samples-standards were selected by the color of flowers and vegetative organs and were vegetative propagated. Among them are SEG-1-with white colored flower and light-green leaves and stems; SEG-2-with lilac colored flower and dark-green leaves and stems; SEG-3-with light-blue colored flower and dark-green leaves and stems; SEG-4-with blue colored flower and dark-green leaves and stems; SEG-5-with purple colored flower and anthocyanin leaves and stems. Plant height in variety samples-standards ranged from 130 to 140 cm and the number of stems per plant – from 28 to 40 pieces (Table 5.4).

TABLE 5.4 The evaluation results of variety samples-standards of *Galega orientalis* for seed production

Variety samples-standards	Height of plant, cm	Number of stems per plant, pcs.	Number per one stem				Seeds, per plant, g	Seeds, in a pod, pcs.	Weight 1000 seeds, g
			Clusters, pcs.	Beans, pcs.	Seeds				
					pcs.	g			
SEG –1	130	44	5.7	101.4	170.3	1.2	54.6	1.8	7.3
SEG –2	130	40	6.4	102.8	204.7	1.5	60.0	2.0	7.4
SEG –3	140	28	5.9	92.0	259.3	1.8	50.4	2.8	7.1
SEG –4	135	33	4.9	75.2	165.3	1.3	42.9	2.2	7.7
SEG –5	130	38	6.1	120.5	246.2	2.0	75.0	2.0	8.1

Variety samples also differed in seed production. In one stem there were formed from 4.9 to 6.4 clusters, from 75 to 120 beans, from 165.3 to 259.3 pcs. or 1.2–2.0 g of seeds. Productivity per plant reached 42.9–75.0 g of seeds and their weight of 1,000 ranged from 7.3 to 8.1 g. The number of seeds per pod was 1.8–2.8 pcs. Thus, variety samples-standards are characterized by a certain set of quantitative and qualitative traits making it possible to carry out identification of varieties.

5.4 CONCLUSIONS

1. New source material for breeding *Galega orientalis* created at the Belorusian State Agricultural Academy is characterized by considerable diversity of morphological and economically useful traits.
2. Variability of traits in the new source material (both quantitative and qualitative) is manifested in a wide range, which allowed to compile a table of

traits for the identification of the criteria of distinguishability, uniformity and stability of new varieties of *Galega orientalis*.

3. As a result of the research constant variety samples-standards of *Galega orientalis* with various combinations of morphological traits were identified and propagated; a comprehensive assessment was given to them. The table of traits and the identified constant variety samples-standards were transferred to the state establishment "State Inspection for Testing and Protection of Plant Varieties" of the Republic of Belarus for testing varieties of *Galega orientalis* for patentability.

KEYWORDS

- **copyright**
- **distinguishability**
- **patent**
- **traits**
- **uniformity**

REFERENCES

1. Law of the Republic of Belarus "*On Patents of Plant Varieties*" of (13.04.1995), *3725, XII*, Minsk (1995) 22p (in Russian).
2. Law of the Republic of Belarus "On the Accession of the Republic of Belarus to the International Convention for the Protection of New Varieties of Plants" *(115–123)* of (29.06.2002) Legislation Pravo, R. B. (2004) 14p (in Russian).
3. International Convention for the Protection of New Varieties of Plants, On the Accession of the Republic of Belarus to the International Convention for the Protection of New Varieties of Plants, March 9, Geneva (1991) (in English).
4. International Convention for the Protection of New Varieties of Plants, Geneva International Union for the Protection of New Varieties of Plants (2004) 28p (in English).
5. Methodology for Testing Plant Varieties for Distinguish ability, Uniformity and Stability (2004). State Inspection for Testing and Protection of Plant Varieties, Ministry of Agriculture and Food of the Republic of Belarus, Minsk Information Centre of the Ministry of Finance of the Republic of Belarus, 274p (in Russian).
6. State Register of Varieties, and Trees and Shrubs of the Republic of Belarus (2006). Ministry of Agriculture and Food of the Republic of Belarus, the State Inspection for Testing and Protection of Plant Varieties of the Republic of Belarus, Tankevich, S. S. Ed. Minsk 148p (in Russian).
7. Rubtsov, M. I. (1997). Creation of a Genetic Collection of Perennial Legumes Based on Markers and Donors of Economically Useful Traits, Rubtsov, M. I., Kozlov, N. N., Makarenkov, M. A. et al. Feed Production in Russia, Collection of Scientific Papers on the 75th Anniversary of the All-Russian Research Institute of Feeds named after Williams, V. R., Moscow 291–297 (in Russian).

8. Agofodorova, M. N. (1997). Biotechnology of Fodder Crops, Agofodorova, M. N., Solodkaya, L. A., & Ivashuta, S. I. Feed Production in Russia Collection of Scientific Papers on the 75th Anniversary of the All-Russian Research Institute of Feeds named after Williams, V. R. Moscow, 298–309 (in Russian).
9. Agofodorova, M. N. (2002). Creating Source Material by Biotechnological Methods, Agofodorova, M. N., Solodkaya, L. A., & Solozhentseva, L. F. Adaptive Feed Production, Problems and Solutions, A Collection of Scientific Works of the All-Russian Research Institute of Feeds named after Williams, V. R. Moscow, 375–385 (in Russian).
10. Bushuyeva, V. I. (2008). *Galega Orientalis* Monograph, Bushuyeva, V. I. Minsk Ekoperspektiva, 176p (in Russian)
11. Bushuyeva, V. I. (2009). *Galega Orientalis*, Monograph 2nd Edited Extension Bushuyeva, V. I., & Taranukho, G. I. Minsk Ekoperspektiva 204p (in Russian).

PART III

PROBLEMS OF GRAIN CROPS GROWING ON ACID SOILS OF THE EUROPEAN NORTH

CHAPTER 6

PROBLEMS OF GRAIN CROPS GROWING ON ACID SOILS OF THE EUROPEAN NORTH

EVGENIY M. LISITSYN[1,2]

[1]Behalf of N.V. Rudnicki Zonal Research Institute of Agriculture of North-East of the Russian Agricultural Academy, d. 166a, Lenin St., Kirov, 610007 Russia;
E-mail: niish-sv@mail.ru, edaphic@mail.ru

[2]Vyatka State Agricultural Academy, d. 133, October Av., Kirov, 610017, Russia;
E-mail: info@vgsha.info

CONTENTS

6.1 INTRODUCTION

In our epoch world agriculture tends to occupy moderate and subtropical areas. But in 1931 great Russian scientist N.I. Vavilov noticed that 73.7% of the cultivated land worldwide belonged to a strip located between 55 and 23°30′ of north [1]. To the beginning of XXI century only 1/17 part of all arable land of the Earth located on northern areas above 55°.

Expansion of agriculture to the north does not have great value for the Western Europe and the North America owing to the geographical reasons. Huge spaces of Northern Norway, Sweden as well as Greenland and Iceland make by mountains and rocks, lake and glaciers, boulders – results of a glacial age – inaccessible to agriculture. Only some points of agriculture both in Old World and in a New World rise to a Polar Circle and even above. In Norway barley and oats crops exist between 70 and 69° north. In Sweden barley reaches to 68°30′. In Canada the sowings of bread cereals reaches a Polar Circle.

In a broad sense the North is mean all Non-Chernozem strip having podzolic, leached and boggy soils. The general features of this territory are the short growth season, a significant amount of summer and winter precipitations, rather small quantity of heat (less than 2200° for all growth season). So, the basic preconditions of development of agriculture in the North, according to N.I. Vavilov [1], are liming of acid soils, their draining and input of mineral fertilizers.

Acid soils (having pH of a soil solution lower than 5.5) occupy large areas of the lands potentially suitable for agriculture all over the world [2]. By the estimations resulted in article [3], their share makes about 40%. According to some researchers [4], 56.5% of total area of acid soils in Europe or 146 and 75 million hectares belong to podzolic and sod-podzolic soils. All remained 12 types of acid soils (according to the grouping of soils accepted by FAO) occupy the rest 43.5%.

Losses of yield in acid soils only in the Russian Federation make 15–16 million tons of agricultural production in terms of grain per year. Efficiency of mineral fertilizers on strong- and middle-acid soils goes down on 20–25% [5].

In the north-east of Europe acid soils are located basically in the Non-Chernozem Zone of Russia (Fig. 6.1). This region represents the large area of arable lands of the country, called to provide national economy and the population by various products of agriculture and animal industries. The lands of an agricultural purpose occupy here about 48.9 million hectares, including 30.7 million hectares of an arable land.

Results of agrochemical observation of soils, as of January, 1st, 2011, show that acid soils make 30.2 million hectares or 35.6% of the surveyed arable land of the Russian Federation. For prevention of expansion of acid soils it is necessary to liming annually not less than 5 million hectares. Now this work is spent on the area of 228–300 thousand hectares [5]. In detail dynamics of sowing areas on acid soils of a Non-Chernozem zone of Russia is considered in our earlier article [6].

FIGURE 6.1 A zone of distribution of podzolic soils in European part of Russia.

The Central Federal District stands out especially for other Federal Districts of Russian Federation, because 52.6% of its' arable land are acid soils of different type. In the Central Federal District the greatest area of acid soils are in Oryol (74.4%), Kostroma (64.0%) and Smolensk (63.0%) regions, and in two last a considerable part make strong acid soils – accordingly, 14.3% and 12.3% (across whole Russia – 2.3% only). The optimum acidic regime of soil in the district is in Moscow Region in which acid soils make 29.0% of an arable land. This fact is a consequence of intensive liming in last decades most likely.

In three Federal Districts areas of acid soils exceed an average index of the country and make in North-West District – 40.7%, in Privolzhsky District – 34.8%, and in Ural District – 42.1%. Following areas essentially surpass these average indexes in their own districts: Republic of Komi – 76.0%, Republic of Karelia – 61.0%, and also the Vologda region – 54.0%; Penza region – 85.1%, Kirov region – 73.6%, Permsky Kray – 73.0%, Nizhniy Novgorod region – 66.9%, Republic of Mordovia – 68.4%; and Sverdlovsk region – 62.7%.

The greatest areas of acid soils are in Privolzhsky and in Central Federal Districts of Russian Federation, and the least – in North-West Federal District. Among all subjects of the Russian Federation the greatest areas of acid soils are in the Penza,

Kirov, and Tambov regions. As a whole relative density of acid soils is great (over 70%) in Republic of Komi, Kirov region, and Permsky Kray.

As a whole by 2012 in Non-Chernozem Zone of Russia acid soils make not less than half of farmlands more than 30% of which are soils with average and high degree of acidity.

Therefore the question on plasticity, adaptability and resistance against stressful edaphic factors of species and varieties of agricultural crops has got an exclusive urgency last decades. In the end of XIX century V.V. Dokuchayev wrote that agricultural crops will give the greatest results to the person if only they will be adapted to local soil, waters, and a climate [7]. Ecologization of crop farming becomes the integral part of modern strategy of an adaptive intensification of agriculture providing decrease of dependence of agrosystems from an environment [8].

The main factor defining toxicity of acid sod-podzolic soils in the Non-Chernozem Zone of the European part of Russia is high level of mobile (exchange) ions of trivalent aluminum. Aluminum is the most widespread metal in earth crust, making up to 8% of its weight [9]. Though, according to Kabata-Pendias and Pendias [10], Al is an usual component of plants (its average content is about 200 mg/kg of dry matter), and some researchers shown stimulating of seedlings growth with aluminum [11], nevertheless, toxicity of mobile ions Al^{3+} is the major limiting factor at cultivation of plants on acid soils, a principal factor reducing productivity of plants on 67% of all acid soils [12].

Decrease in grain yields in Russia under the influence of soil acidity according to Klimashevsky [13] can reach 85%. Therefore many Russian scientists were engaged earlier and are occupied now with the decision of a problem of an effective utilization of the acid lands. Activity on selection of resistant varieties of grain crops has begun in our country in the middle of twentieth century [14, 15]. In 60th, Klimashevsky has suggested the concept of active opposition of plants to Al in a root zone (rhyzosphere). More than three decades Klimashevsky with co-workers studied various aspects of plants' acid resistance, including questions of Al effects on growth of root systems and efficiency of mineral nutrition of plants.

Since 1984 researches on estimation and revealing of barley varieties resistant against high soil acidity and high content of Al^{3+} ions are conducted in the North-East Agricultural Research Institute of Russian Academy of Agricultural Sciences [16]. There are some successes in breeding of acid- and Al-resistant varieties of the basic agricultural crops of the North-East of the European part of Russia [17, 18].

Barley is the basic cereal culture used in agricultural and in process industry both as grain and as fodder culture, and also as raw materials for manufacture of alcoholic production and beer [17]. Barley is poor adapted for growth on acid soils; its optimum pH level of soil solution is 6.0. On acid soils good yield of barley can be received only at entering of limy materials. However, breeding of acid- and Al-resistant varieties such as 'Dina' and 'Novichok,' allows to receive enough high yield of culture without additional liming of soil [18].

Oats is a fodder culture first of all. Fodder unit of oats has been accepted for the standard of an estimation of fodder units of other cultures. It basically is growing up in a forest zone of the European part of Russia and in a forest-steppe zone. The wide area of distribution of culture is explained with large amount of oats ecotypes and its good fitness to cultivation conditions. The oats are considered as rather resistant culture to edaphic stresses; it occupies position between a rye and wheat on acid-resistance level [19]. Degree of reaction of oats varieties on soil aluminum toxicity considerably varies depending on their ecological-geographical origin [20]. Distribution of Al-resistant and Al-sensitive hexaploid oats varieties by their place of origin has shown that the European continent share makes 42% of resistant forms. The greatest quantity of resistant forms is selected among samples from Spain, Bulgaria, and Slovakia. Presence of middle-resistant forms (59%) is characteristic for the African continent. Basically it is varieties from Tunis, Morocco, Algeria, and Kenya [21].

However at oats cultivation on acid sod-podzolic soils (pH 4.0–4.5) with high content of exchangeable ions of aluminum (14–23 mg/100 g of soil) productivity of oats decreases by 40–50% [22]. The root system suffers first of all, especially at the first stages of ontogeny, further there is a reduction of development of vegetative and generative organs. Depression is observed (up to 30–40%) in formation of the sizes of a stalk, leaves and a panicle, quantity of ears and grains in a panicle are reduced.

Oats crops in Russia occupy 3rd place after wheat and barley. They are extended practically in all grain-producing regions. The oats volume of output in Russia over the last 10 years fluctuated from 4.5 to 7.1 million tons. In 2001 gross yield of oats has made 7.1 million tons, in 2007 – 5.4 million tons. The oats Record harvest was in 1986 – 15.7 million tons. However, the tendency of reduction of areas under this culture is observed last years in world agriculture. The areas were reduced considerably during the period since 1961 on 1965 and since 1978 on 1999. Total reduction of crops has practically made 26 million hectares – from 38 to 12 million hectares. There was a reduction of areas under crops in Russia too. In 1990 the Russian Federation sowed 9.1 million hectares of oats but in 2005 oats crops in Russia had made 3.34 million hectares only.

In 2010 the total size of sowing areas under barley in Russia has made almost 6.8 million hectares, and under oats – almost 3 million hectares. The gross yield of barley grain has reached the minimum values in 2010 (6.7 million tons), maximum – in 2008 (20.5 million tons), of oats accordingly – 3.2 million tons in 2010 and 5.8 million tons in 2008. Average annual cropage of barley grain in 2006–2010 has made 14.6 million tons, of oats – 4.9 million tons.

As a whole for 2003–2008 in the Russian Federation nearby 12–13% of all arable areas were occupied annually under crops of spring and winter barley, 4.3–4.8%

– under oats crops. In Non-Chernozem Zone of Russia barley as the grain and fodder culture occupies nearby 25–60% of all sowing areas.

Considerable territorial and temporal variability of soil and environmental conditions, especially high level of soil acidity, complicate both breeding of new varieties, and their cultivation in high degree. High stable grain yields of spring cereals are provided only with such varieties, which are as much as possible adapted for contrast environmental conditions and have high adaptive potential against stressful factors of biotic and abiotic nature.

For successful cultivation of grain crops under conditions of northern acid sod-podzolic soils it is necessary to combine following abilities in one genotype: (a) ability of root systems to resist to toxic influence of ions of aluminum (that automatically means resistance to low pH of soil solution); (b) ability of the photosynthetic apparatus of plants to assimilate solar energy and to transform it into plastic substances; (c) ability to redistribute flows of the received plastic substances into economic valuable parts of a plant [23].

These three basic abilities are defined by different ecological-genetic systems, which, however, closely interrelate and cooperate among themselves.

For increase of efficiency of selection process on breeding of ecologically resistant varieties of an oats and barley, it is necessary to possess the general knowledge of the physiological and genetic mechanisms used by plants for maintaining of their metabolism under conditions of acid stress.

The activity on breeding of new varieties resistant against acid soil and high contents of ions of trivalent aluminum could be considerably facilitated, if data about the genetic control of plant resistance to the given stressful factors was available to scientists and breeders. By this time researches of genetic bases of Al-resistance have led to the inconsistent data about number of genes [24–26] supervising it, till now there is not accurate and clear picture of genetics of aluminum resistance even for wheat despite many decades of researches of this problem.

All above-stated defines a particular interest of different scientists (physiologists of plants, geneticists, soil scientists, breeders) to problems of seasonal dynamics of level of acidity and the content of aluminum ions in different genetic horizons of acid podzolic soils, studying of physiology-and-genetic mechanisms as well as search of genetic sources of resistance of plants to acid stress. Offered articles show some results of work of researchers of the North-East Agricultural Institute and Vyatka State Agricultural Academy (Kirov, Russia) on the specified problems.

KEYWORDS

- **agrosystems**
- **barley**
- **Central Federal District**
- **oats**
- **podzolic soil**
- **polar circle**
- **Privolzhsky Federal District**
- **sod-podzolic soil**

REFERENCES

1. Vavilov, N. I. (1931). Problem of Northern Agriculture, materials of the Leningrad Emergency Session of Academy of Sciences of the USSR 25–30 XI1931, Leningrad, Publishing House of Academy of Sciences, 15p (in Russian).
2. Borlaug, N. E., & Dowswell, C. R. (1997). The Acid Lands One of Agriculture's Last Frontiers, Plant-Soil Interaction at Low pH, Brasilian Soil Science Society, 5–15.
3. Delhaize, E., Ryan, P. R., Hebb, D. M., Yamamoto, Y., Sasaki, T., & Matsumoto, H. (2004). Engineering High-level Aluminum Tolerance in Barley with the ALMT1 Gene, Proc National Academy Science USA, 15249–15254.
4. VonUexkull, H. R., & Mutert, E. (1995). Global Extend Development and Economic Impact of Acid Soils, Plant Soil, *171,* 1–15.
5. The Report on a Condition and Use of the Lands of an Agricultural Purpose (2011). Moscow "Rosinfoagroteh" 148p (in Russian).
6. Lisitsyn, E. M., Batalova, G. A., & Shchennikova, I. N. (2012). Dynamics of Sowing Areas and Productivity of Barley and Oats in the Various Regions of the European Russia having Acid Sod-Podzolic Soils, in Breeding of Oats and Barley Varieties for Acid Soils, The Theory and Practice, Palmarium Academic Publishing, Saarbrucken Germany, 11–28 (in Russian).
7. Dokuchayev, V. V. (1899). To the Doctrine About Nature Zones, Saint Petersburg, 28p (in Russian).
8. Zhuchenko, A. A. (1993). Problems of Adaptation in Modern Agricultures, Agricultural Biology, *5,* 3–35 (in Russian).
9. FitzPatrick, E. A. (1986). An Introduction to Soil Science, N. Y. Longman Scientific and Technical, 2–55.
10. Kabata-Pendias, A., & Pendias, H. (1986). Trace elements in Soils and Plants, Boca Raton, Florida CRC Press, Inc., 439p.
11. Schier, G.A., & McQuattie, C. J. (2002). Stimulatory Effects of Aluminium on Growth of Sugar Maple Seedlings, J Plant Nutrition, *25(11),* 2583–2589.
12. Eswaran, H., Reich, P., & Beinroth, F. (1997). Global Distribution of Soils with Acidity, Brazilian Soil Science Society, 159–164.
13. Klimashevsky, E. L. (1966). Essay on Physiology of Variety Moscow "Kolos" Publishing, 116p (in Russian).

14. Butkevich, V. V. (1947). Adaptation of Plants to Chemical Composition of Environment, Selection and Seed-Growing, *6,* 63–76 (in Russian).
15. Kornilov, M. F., Borisov, E. M., & Trunina, Z. V. (1955). Liming of Soil and a Variety (from Researches of Leningrad Branch of VIUAA), Liming of Sod-Podzolic Soils, Bulletin of All-Russia Institute of Fertilizers and Agrochemistry, Moscow, *31,* 202–250 (in Russian).
16. Rodina, N. A. (1986). Estimation of Barley Varieties on Resistance to Acid Soils and Aluminum, Selection of Grain Crops on Resistance to Diseases and Adverse Factors of Environment in Volgo-Vyatka Region, Kirov, North-East Agricultural Institute, 47–55 (in Russian).
17. Golovko, T. K., Rodina, H. A., Kurenkova, S. V., & Tabalenkova, G. N. (2004). Barley in the North Breeding-Genetic and Physiology-Biochemical Bases of Productivity, Yekaterinburg, Publishing House of Ural Branch of the Russian Academy of Sciences, 153p (in Russian).
18. Lisitsyn, E. M., Shchennikova, I. N., & Shupletsova, O. N. (2011). Cultivation of Barley on Acid Sod-Podzolic Soils of North-East of Europe in Barley Production, Cultivation and Uses, New York, Nova Publishing, 49–92.
19. Aniol, A., & Madej, L. (1996). Genetic Variation for Aluminum Tolerance in Rye, Vortr fur Pflanzenzuchtung, *35,* 201–211.
20. Kosareva, I. A., Davydova, G. V., & Semenova, E. V. (1998). Diagnostics of resistance of Oats Plants to High Contents of Aluminum Ions in a Soil Solution, Agricultural Biology, *5,* 73–76 (in Russian).
21. Loskutov, I. G., Kosareva, I. A., & Semenova, E. V. (2001). Features of Aluminum Resistance in Oat Wild Species. Oat Newsletter. Vol. 47.
22. Batalova, G. A. (2000). Oats. Technology of cultivation and selection. Kirov. North-East Agricultural Institute. 206 p. (in Russian).
23. Lisitsyn, E. M., Shikhova, L. N., Tiunova, L. N. (2012). Physiology of aluminum resistance in cereals. In: Breeding of oats and barley varieties for acid soils. The theory and practice. Palmarium Academic Publishing, Saarbrucken, Germany, 89–172. (in Russian)
24. Delhaize, E., Craig, S., Beaton, C. D., Bennet, R. J. V., & Randall, P. J. (1993). Aluminum tolerance in wheat (*Triticum aestivum* L.). 1. Uptake and distribution of aluminum in root apices. Plant Physiol. *103*, 685–693.
25. Bona, L., Carver, B. F., Wright, R. J., Baligar, V. C. (1994). Aluminum tolerance of segregating wheat populations in acidic soil and nutrient solutions. Commun. Soil Sci. Plant Anal. *25*, 327–339.
26. Aniol, A. (1997). The aluminum tolerance in wheat. In: Proceedings of the International Conference: Plant Breeding: Theories, Achievements and Problems. Kedainiai, Lithuania, 4–22.

CHAPTER 7

DYNAMICS OF ACIDITY AND THE ALUMINUM CONTENT IN PODZOLIC SOIL

LYUDMILA N. SHIKHOVA[1], OSTAP A. PALADICH[2], and EUGENE M. LISITSYN[1,2]

[1]Vyatka State Agricultural Academy, 133, Oktyabrsky Prospect, Kirov, 610017, Russia; E-mail: niish-sv@mail.ru, info@vgsha.info

[2]Behalf of N.V. Rudnicki Zonal Research Institute of Agriculture of North-East of the Russian Agricultural Academy, 166-a, Lenin St., Kirov, 610007, Russia; E-mail: shikhova-l@mail.ru

CONTENTS

ABSTRACT

For soils of North Europe regions, the high acidity and high exchangeable aluminum content, associated to low pH, are the main constraints for agricultural production. Estimation of acid-resistance level of different species and varieties of plants can be directly and uniquely determined on measurement of an economic crop under field conditions only. But seasonal and long-term dynamics of acidity and the content of aluminum ions can influence on level of plants field resistance. For studying of this dynamics three sites of the natural and cultivated podzolic soils of a middle taiga zone of the European part of Russia have been chosen. Soil samples were selected by soil drill from three top genetic horizons of soil: organic (O_i for woodland; A_p for arable soil; A_h for fallow soil), eluvial (A_{he}) and illuvium (B) horizons from May till September, in the middle of each month. As researches have shown, content of exchangeable ions of aluminum had considerable influence on seasonal dynamics of pH_{KCl} in podzolic soil of moss-covered spruce forest. Thus seasonal dynamics of the aluminum is significant for horizon B only. In arable soil spatial heterogeneity of pH_{KCl} is insignificant; however it increases a little with depth. So, in horizon A_p coefficient of variation of spatial change of pH_{KCl} fluctuate from 0.7 to 4.2%; in horizon A_{he} – from 0.7 to 12.4%; in horizon B – from 0.4 to 12.0%. The amplitude of seasonal dynamics in plowed horizon makes 0.3–0.7, in eluvial horizon – 1.1, and in illuvium horizon – 0.5–1.1 pH units; in various terms of the growth season the given soil can be referred to different categories on acidity degree – from weakly acidic to close to neutral. Seasonal dynamics of acidity of fallow soil is low also; in humus horizon coefficient of variation fluctuated from 2.5 to 5.2%, amplitude of fluctuations thus made from 0.3 to 0.7 units; in horizon A_{he} coefficients of variation characterizing seasonal variability of a parameter, reached 17.3%, and amplitude – 1.8 pH units. Essential correlation between change of pH_{KCl} and the content of organic matter is revealed; dependence between seasonal dynamics of pH_{KCl} and Al^{3+} is high ($r = -0.76$ at $P = 0.99$). Fallow land differs from arable soil by presence of exchangeable aluminum in horizon A_{he}. Exchangeable Al^{3+} ions in horizon A_{he} of fallow soil are found out not during whole growth season, but in separate terms only. Usually there are beginning of a season and its end.

7.1 INTRODUCTION

Highly acid soils are generally considered those with pH < 5.5 cover about 30% of world's total land area and 70% of the world's potentially arable land [1]. A northern cold temperate belt of acid soils occurring in the humid northern temperate zone is comprised of predominantly organic acid soils. Although in organic acid soils H ions dominate in the soil solution little is known about how plants manage high H ion concentrations. Low pH can directly inhibit root growth [2, 3, 4]. Aluminum (Al) is the most abundant metal and the 3rd most abundant element in the earth crust. Soils contain an average of 7% Al [5]. Plants might adapt to low-pH stress and Al toxicity through the same biological processes [6]. P deficiency is another major soil constraint to agricultural production, limiting crop yield in 30% to 40% of the arable lands in the world [7]. The low P status of acid soils is a particular problem because large amounts of P need to be applied in order to raise concentrations of available soil P to an adequate level [8]. Thus, low P availability, toxic levels of Al, and H^+ rhizotoxicity are considered to be the three major stresses limiting plant growth on acid soils [9].

The characterization of soil acidity components is given by the active acidity, usually expressed as soil pH, and by the potential (total) acidity (at pH 7). The total acidity, by its turn, is the sum of exchangeable acidity and nonexchangeable (hydrolytic) acidity. The exchangeable acidity is given by the aluminum ion (also called exchangeable aluminum = Al^{3+}). The nonexchangeable is related to the content of H covalently bound to colloids, and monomers and polymers of aluminum in soil [10].

Now aluminum resistance in plants is considered as a complex phyto-ecological problem on which decision producing a better crop yield on acid soils depends. The annual production loss due to acidity is estimated to be about 10 percent is in the order of $498 million [11].

Any extreme factor leads to decrease in efficiency of plants – both the general biomass, and an economic-valuable part of a crop. Therefore two kinds of agricultural crops resistance are usually distinguished – biological and agronomical resistance [12]. *Biological resistance* characterizes that limit of stressful loading at which plants are still capable to form viable seeds; quantitatively it is expressed in units of measure of the extreme factor treated on plants. *Agronomical resistance* reflects degree of decrease in a crop under the influence of stressful impact; it is expressed in ratio of change of plants productivity under the influence of the stressor.

There are a lot of methods to evaluate Al tolerance in agricultural crops, include a callus-based assay in tissue culture, hydroponic solution assays, and a soil-based assay. Laboratory- and greenhouse-based techniques are widely employed which are nondestructive, and can be applied in early developmental stages. Even though, screening can be directly done on acid soils under controlled environment, soil based screening is usually preceded by preliminary screening in solution culture. The soil-based assay is time consuming and labor intensive However, the ultimate and most direct method of evaluating for Al tolerance is by measuring economic yield (forage or grain) under field conditions [13]. Typically the plants are grown on soil limed to nontoxic level, and unlimed soil and the data are analyzed and reported as the ratio of grain yield or the trait of interest in unlimed to the lime amended plot [14–17]. The advantage of soil based screening methods compared to nutrient solution culture is that it takes into consideration other soil factors that may influence Al tolerance [18–20]. The two most important problems observed when evaluating for Al tolerance in the field are the presence of fungal pathogens, in which infection is often favored by the application of lime to low pH soils [14], or spatial variability of pH in the surface and subsurface soil layers [15]. However, some authors specify considerable variation of indicators of soil acidity during growth season. It is depended on pattern of moisture regime of soil, temperature, type of phytocenosis, and fertilizer system [21], as well as depth of eluvial soil horizon [22].

Therefore research of seasonal and long-term dynamics of acidity components and content of aluminum ions in the natural and cultivated podzolic soil of a middle taiga zone of the European part of Russia was the objective of the given work.

7.2 MATERIALS AND METHODOLOGY

Three sites of podzolic light-loamy soils of Podosinovets area of Kirov region have been chosen as study objects: an arable land presented by a field of a crop rotation; fallow land of 6–8 years; and soil of middle taiga woodland, under moss-covered spruce forest.

Soils are generated on carbonate-free blanket loam. The given soils are characterized by the brightest demonstration of zone process of soil formation – podzolic one which characteristic signs are high acidity, low content of fulvatic humus, and low content of elements of plant nutrition. Such combination of adverse edaphic factors gives the opportunity to investigate their change in a complex.

For the characteristic of studied soils full profiles have been made on each site in August 2012 and August 2013, the brief physical and chemical characteristic of soils is resulted in Table 7.1.

In moss-covered spruce forest the basic part of a forest stand is presented by a fur-tree, but there is also pine and birch. Underwood is not developed. The grass cover is developed poorly and consists basically of a bilberry, wild strawberry, oxalis, and some gramineous plants. The moss and lichen cover is presented by stains under crones of trees.

The arable soil is occupied by a cereal-grass crop rotation. In 2011, field was occupied with clover of second year of use, in 2012—with spring wheat, in 2013—with oats. Mineral fertilizers in ratio $N_{12}P_{12}K_{12}$ were input for spring wheat and oats. The clover has been cut on hay in July; in August its after-grass has been plowed. Spring wheat and oats had harvested on grain in the beginning of September.

Samples of soil were taken from three upper soil horizons: organic (O_i for woodland; A_p for arable soil; A_h for fallow soil), eluvial (A_{he}) and illuvium (B) horizons. Sampling was carried out annually by five times for the growth season from May till September, in the middle of each month in six-fold replication.

Air dried samples of each soil were homogenized, divided in three subsamples, identified, ground in a porcelain crucible and passed through 0.5 mm mesh sieve and packed. Soil acidity components were characterized as follows: the active acidity was determined in 0.01 mol L^{-1} $CaCl_2$ solution, 1:2.5 (v/v) soil/solution ratio, through pH measurement; the total acidity ($H^+ + Al^{3+}$) directly through the extraction with 1 mol L^{-1} ammonium acetate solution at pH 7.0, followed by titration [23].

TABLE 7.1 Chemical Properties of Investigated Podzolic Soils

Site	Horizon (depth, cm)	pH_{KCl}	H^+	Al^{3+}	Hydro-lytic acidity	$Ca^{2+} + Mg^{2+}$	Base saturation, %	Carbon of humus %	P_2O_5	K_2O
			mg-equivalent 100 g^{-1} of soil						mg kg^{-1}	
moss-covered spruce forest	O_i (0-5)	4.0	0.8	0.6	36.7	10.0	21.4	20.92*	470	680
	A_{he} (5-40)	3.8	0.2	2.8	5.4	4.2	43.8	0.15	75	30
	AB (40-55)	3.7	0.4	3.3	6.2	12.0	65.9	0.18	94	58
	B (55-104)	3.7	0.4	3.4	4.7	21.8	82.3	0.25	127	86
	BC (104-125)	4.1	1.0	1.2	3.1	25.2	89.0	0.29	143	84
	C (125-140)	4.1	1.1	1.4	3.2	26.0	89.0	0.23	165	65
fallow land	O_e (0-3)	5.2	0.7	0.0	-**	-	-	1.58	190	294
	A_h (3-19)	5.3	0.7	0.0	1.5'	10.0	87.0	0.94	133	106
	A_{he} (19-31)	5.2	0.7	0.0	1.1	10.4	90.4	0.20	92	202
	AB (31-44)	4.8	0.7	5.0	6.3	16.6	72.5	0.18	115	210
	B (44-87)	3.6	0.3	3.1	4.2	18.4	81.4	0.14	145	118
	BC (87-119)	4.0	0.8	2.0	3.6	20.0	84.7	0.16	135	84
	C (119-150)	3.9	0.8	1.3	2.2	17.8	89.0	0.23	178	79
arable land	A_p (0-20)	5.2	0.8	0.0	1.0	14.4	84.7	1.39	290	192
	A_{he} (20-34)	4.8	0.6	0.0	1.9	6.6	89.0	0.24	236	242
	AB (34-44)	4.6	0.3	1.0	2.9	13.4	93.5	0.19	182	256
	B (44-108)	3.7	0.3	2.7	4.3	21.0	77.6	0.15	211	128
	BC (108-125)	3.7	0.7	2.0	3.2	18.8	82.2	0.15	220	86
	C (125-136)	3.8	0.3	2.1	2.7	19.0	83.0	0.15	210	84

Note: *for O_i – Total carbon content; ** not determined.

O_i – layer of leaves, pine needles and twigs of organic soil horizon; O_e – partially decomposed organic layer; A – a mineral horizon formed at or near the surface in the zone of leaching or eluviation of materials; A_p – plowed A soil horizon; A_h – a horizon enriched with organic matter, contains less than 17% organic C by weight; A_{he} – an A_h horizon that has undergone eluviation; B – a mineral horizon characterized by enrichment in organic matter, sesquioxides, or clay, or by the development of soil structure; C – a mineral horizon comparatively unaffected by the pedogenic processes operative in A and B; AB – kind of transitional horizon dominated by properties of A master horizon but having subordinate properties of B horizon; BC – kind of transitional horizon dominated by properties of B master horizon but having subordinate properties of C horizon.

The exchangeable acidity (Al^{3+} + H^{+}_{tit}) and the exchangeable aluminum (Al^{3+}) were extracted by 1 mol L^{-1} KCl solution, 1:10 (v/v) soil/solution ratio, and determined by titration of 25 mL KCl extract with 25 mmol L^{-1} NaOH, using 1 g L^{-1} phenolphthalein as indicator, and by back-titration, after acidification with 40 g L^{-1} NaF, with 25 mmol L^{-1} HCl, respectively [24]. The difference between titrable exchangeable acidity and titrable aluminum gave the so-called titrable hydrogen (H^{+}_{tit}) content. For estimation of contents of exchangeable potassium, phosphorus, calcium, and magnesium, methods described in Russian State Standards [24, 25]; content of organic matter by Tjurin method [26]; labile organic matter – using 0.1 M neutral pyrophosphate extract, 1:2 (v/v) soil/solution ratio [27]. Results were submitted to descriptive statistics for estimations of mean, and correlation and regression analysis for selected variables with software packages MS Excel 10.0, AGROS 2.07, STATGRAPHICS Plus for Widows 5.0.

7.3 RESULTS AND DISCUSSION

7.3.1 SEASONAL DYNAMICS OF ACIDITY IN WOODLAND PODZOLIC SOIL

Degree of exchangeable acidity of podzolic soil under moss-covered spruce forest (expressed as pH_{KCl}) has insignificant spatial variability. In the bottom horizons A_{he} and B coefficients of variation of spatial changes of the parameter seldom exceed 2%. In a forest litter spatial heterogeneity is a little above but also is insignificant (Table 7.2). It is possible to consider that the soil of these horizons is homogeneous for the given parameter. Seasonal dynamics of degree of acidity of the given soil is insignificant also. In forest litter and in eluvial horizon pH_{KCl} value varied significantly during growth seasons 2012–2013; in 2011 seasonal changes are not expressed. Amplitude of dynamics of pH_{KCl} can make 0.8–0.7 units in litter, and 0.2–0.1 units only in eluvial horizon. In horizon B seasonal dynamics of acidity was statistically significant in 2012 only.

Decrease in acidity from spring to the beginning of summer is characteristic for a forest litter; variation of pH_{KCl} values are maximum here (0.8 units in 2012 and 0.7 units in 2013) (Fig. 7.1). Probably, it is connected to overdrying of soils after its spring soaking therefore concentration of soil solution has risen. According to [21], it leads to reduction of soil acidity.

TABLE 7.2 Change of pH_{KCl} Values in Podzolic Soil of Moss-Covered Spruce Forest

Soil horizon	Year	Spatial variability						Seasonal variability	
		Parameter*	Month					Parameter	Mean
			May	June	July	August	September		
O_i	2011	X	4.5	4.9	4.7	4.9	4.8	$X_{max}-X_{min}$	n**
		V, %	4.8	6.0	4.8	1.2	2.1	V, %	
	2012	X	4.1	4.4	4.9	4.5	4.5	$X_{max}-X_{min}$	0.8
		V, %	7.0	7.6	5.9	8.1	8.7	V, %	6.4
	2013	X	4.5	5.2	4.7	4.6	5.2	$X_{max}-X_{min}$	0.7
		V, %	9.6	5.7	3.8	2.6	3.8	V, %	6.9
A_{he}	2011	X	4.1	4.1	4.1	4.1	4.2	$X_{max}-X_{min}$	n
		V, %	1.8	2.2	1.0	1.6	1.3	V, %	
	2012	X	4.0	3.9	4.0	3.9	4.0	$X_{max}-X_{min}$	0.1
		V, %	0.7	1.9	0.5	2.9	2.4	V, %	1.4
	2013	X	3.9	3.9	3.9	3.8	3.9	$X_{max}-X_{min}$	0.2
		V, %	2.1	1.7	1.0	0.7	1.4	V, %	2.3
B	2011	X	4.0	4.0	4.0	4.0	4.0	$X_{max}-X_{min}$	n
		V, %	1.3	1.3	0.2	0.1	0.5	V, %	
	2012	X	3.9	3.6	3.7	3.6	3.8	$X_{max}-X_{min}$	0.3
		V, %	2.0	1.0	0.6	1.0	0.6	V, %	3.2
	2013	X	3.7	3.8	3.8	3.7	3.8	$X_{max}-X_{min}$	n
		V, %	1.5	0.5	0.5	0.9	0.5	V, %	

Notes: * – X – arithmetic middling; X_{max}-X_{min} – amplitude of variation; V, % – coefficient of variation; ** n – seasonal dynamics is insignificant statistically.

By the end of summer acidity of soil horizon O_i raises again, and in September can remain at former level (2012), or to decrease (2013). Various patters of changes of pH_{KCl} values in second half of growth seasons of these years, possibly, is caused by distinctions in a mode of humidifying of litter, as well as in dynamics of organic matter and in exchange aluminum content. So, positive functional relation ($r = 0.43$ at $P = 0.99$) is revealed between changes of pH_{KCl} values and content of humus carbon (C_{hum}), whereas negative relation – with content of Al^{3+} ($r = -0.40$ at $P = 0.99$).

Seasonal dynamics of pH_{KCl} in A_{he} soil horizon had similar patterns in 2012 and 2013, thus in August decrease in its value is marked (Fig. 7.1). The seasonal variation of pH_{KCl} of A_{he} horizon has no statistically significant correlation with a seasonal variation of organic matter content. Correlation is also weak negative ($r = -0.36$ at $P = 0.99$) with content of exchange Al^{3+} ions. Degree of acidity in horizon B had significant seasonal changes only in 2012 when it raised a little in summer and then decreased again.

Thus, content of exchangeable Al^{3+} ions has considerable influence on seasonal dynamics of pH_{KCl} in podzolic soil under moss-covered spruce forest. Seasonal dynamics of the aluminum per se is significant only for horizon B in 2011 and 2013. In two top horizons it is not expressed (except for horizon A_{he} in 2012) because of very high spatial variability of the content of this element surpassing seasonal changes. Heterogeneity of spatial distribution of aluminum in a forest litter (Table 7.3) is especially high. Nevertheless, its presence makes some impact on degree of soil acidity of all considered horizons (for example $r = -0.28$ at $P = 0.95$ for horizon B).

Pattern of seasonal dynamics of content of exchangeable Al^{3+}, which is established in soil horizon B only, is distinguished in different years of research. Possibly it is defined by fluctuations in content of total carbon (C_{tot}), weak significant relation ($r = 0.27$ at $P = 0.95$) with which is found out.

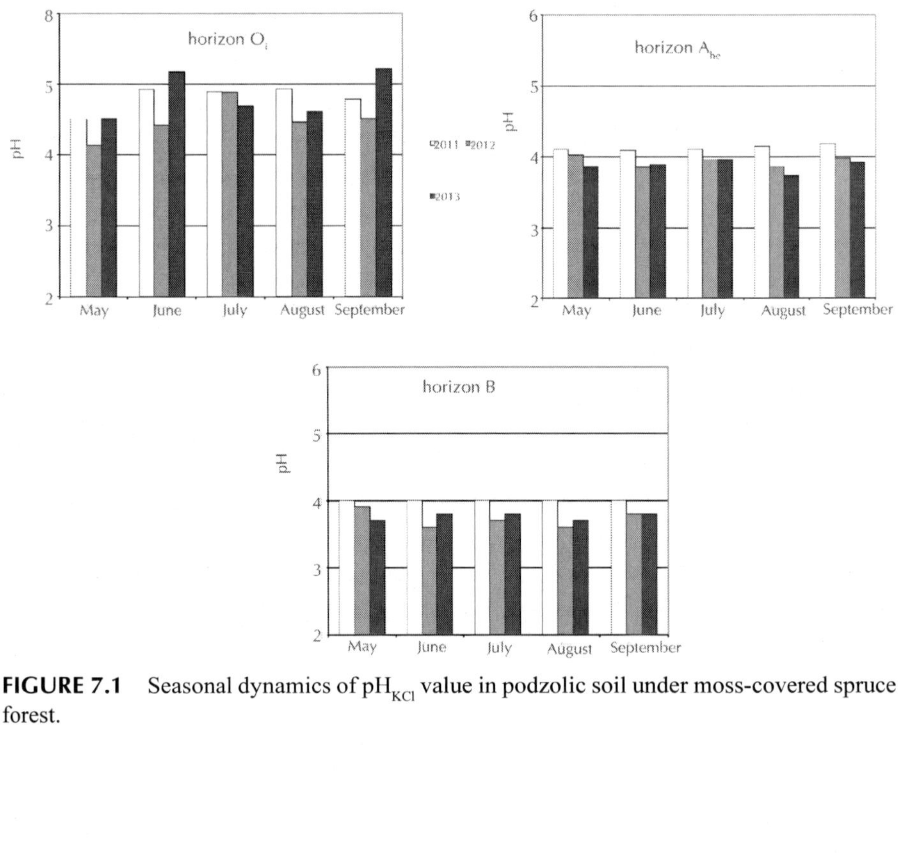

FIGURE 7.1 Seasonal dynamics of pH_{KCl} value in podzolic soil under moss-covered spruce forest.

It is necessary to notice that similar dependence is found out for soil of eluvial horizon too (r = 0.31 at P = 0.99). In horizon O_i, on the contrary, correlation between these two parameters has a negative sign (r = –0.28 at P = 0.95).

It is known that aluminum forms complex salts with organic compounds (including humus acids). Mobility of these complexes depends by nature of organic matter. For example, owing to more complicated chemical structure of humus acids their complexes with aluminum are unstable to sedimentation and almost always drop out in a deposit. Chelate complexes of fulvic acids, which make the basic part of humus of the bottom horizons of podzolic soils remain in soil solution – they are mobile [28].

Hence, positive or negative influence of organic matter on dynamics of the content of exchangeable Al^{3+} ions will be defined, first of all, by its qualitative structure. It is possible to assume that in a forest litter humic acids are extended enough and form insoluble complexes with aluminum; some part of aluminum is in structure of the partially decomposed organic matter. Besides, in this horizon exchangeable acidity is caused by hydrogen rather than aluminum (Table 7.2). In the bottom horizons fulvic acids prevail over humic acids within a complex of organic matter owing to higher migratory ability of the first. Forming complex compounds with aluminum they do not reduce its exchangeable ability.

TABLE 7.3 Content of Exchange Ions Al^{3+} in Podzolic Soil of a Fir Grove (mg-equivalent 100 g^{-1} of Soil)

Soil horizon	Year	Spatial variability						Seasonal variability	
		Parameter*	Month					Parameter	Mean
			May	June	July	August	September		
O_i	2011	X	1.00	0.80	0.43	1.03	0.21	X_{max}–X_{min}	n**
		V, %	52.0	66.1	58.1	34.0	22.9	V, %	
	2012	X	1.20	0.53	0.64	0.71	0.41	X_{max}–X_{min}	n
		V, %	78.4	145.1	114.7	23.8	93.9	V, %	
	2013	X	1.08	0.17	0.65	0.92	0.75	X_{max}–X_{min}	n
		V, %	67.9	122.5	201.0	101.6	109.5	V, %	
A_{he}	2011	X	2.72	3.04	2.54	2.57	2.53	X_{max}–X_{min}	n
		V, %	18.8	40.9	11.8	17.9	29.8	V, %	
	2012	X	2.23	2.41	2.84	2.84	2.14	X_{max}–X_{min}	n
		V, %	18.8	18.8	16.7	24.3	6.1	V, %	
	2013	X	2.47	2.75	2.55	2.60	3.32	X_{max}–X_{min}	0.85
		V, %	28.5	18.0	8.6	5.4	21.4	V, %	12.5

B	2011	X	3.22	3.83	3.62	3.93	3.30	X_{max}–X_{min}	0.71
		V, %	3.8	6.7	2.4	5.2	18.0	V, %	8.8
	2012	X	4.41	4.30	4.37	3.84	4.41	X_{max}–X_{min}	n
		V, %	14.7	12.4	2.5	16.5	2.1	V, %	
	2013	X	3.61	4.43	5.12	3.03	3.86	X_{max}–X_{min}	2.09
		V, %	9.8	11.0	2.8	8.4	4.6	V, %	19.9

Note: See Table 7.2.

7.3.2 SEASONAL DYNAMICS OF ACIDITY IN ARABLE PODZOLIC SOIL

In arable soil the pH_{KCl} as a whole has high enough value in all investigated horizons. Its spatial heterogeneity is thus insignificant, though it increases a little with depth. So, in A_p horizon coefficient of variation of spatial changes of pH_{KCl} fluctuate from 0.7 to 4.2%; in horizon A_{he} – from 0.7 to 12.4%; in horizon B – from 0.4 to 12.0% (Table 7.4). Higher spatial heterogeneity in the bottom horizons can be caused with its nonuniform depth of occurrence.

The amplitude of seasonal dynamics is insignificant also. In plowed horizon (A_p) it is significant in all three years of researches; in the bottom horizons it was insignificant in 2011. But, despite it, amplitude of seasonal dynamics in plowed horizon makes 0.3–0.7, in eluvial horizon – 1.1 and in illuvium horizon – 0.5–1.1 pH units. Thus, in various terms of the growth season the given soil can be classified to different categories on acidity degree – from weak acidic to close to neutral. A sign distinguishing arable soil from natural podzolic forest soil is the high dynamics of a considered indicator in the bottom horizons A_{he} and B, caused by more contrast hydrothermal regime.

Pattern of seasonal dynamics of pH_{KCl} differs on years. In horizon A_p in 2011statistically significant increase in degree of acidity was noted in August, in 2012 – in July and September; in 2013 gradual, from May by September, acidification of soils was observed (Fig. 7.2). Possibly, cultivated crop influences on degree of soil acidity and on character of its change during the growth season. So, in 2011 acidity of soil had the least expressed seasonal dynamics, and the pH_{KCl} value throughout all growth season did not fall lower then 5.6. This year the clover grew in the field, which could smooth fluctuations of soil humidity as has caused absence of pH_{KCl} dynamics during the period from May till July.

As well as in soil under moss-covered spruce forest significant relation between changes of pH_{KCl} and C_{hum} has been revealed ($r = 0.57$ at $P = 0.99$) by means of the correlation analysis. Given dependence is shown in second half of growth season; in the spring and in the beginning of summer it is absent. Possibly, owing to intensive mineralization of C_{hum} during the period from July till August there was a saturation of a soil solution with carbon dioxide.

TABLE 7.4 Change of pH_{KCl} Values in Podzolic Soil of An Arable Land

Soil horizon	Year	Spatial variability						Seasonal variability	
		Parameter*	Month					Parameter	Mean
			May	June	July	August	September		
A_p	2011	X	5.9	5.8	5.8	5.6	5.8	$X_{max}-X_{min}$	0.3
		V, %	0.7	0.9	1.9	1.5	1.7	V, %	1.9
	2012	X	5.5	5.5	5.1	5.4	5.2	$X_{max}-X_{min}$	0.4
		V, %	1.2	4.2	2.6	2.7	1.4	V, %	3.4
	2013	X	5.9	5.7	5.5	5.4	5.2	$X_{max}-X_{min}$	0.7
		V, %	1.4	3.2	2.4	2.0	0.7	V, %	4.9
A_{he}	2011	X	5.7	5.7	5.6	5.5	5.6	$X_{max}-X_{min}$	n**
		V, %	2.7	2.8	6.2	10.4	0.7	V, %	
	2012	X	5.6	5.3	4.9	5.6	4.5	$X_{max}-X_{min}$	1.1
		V, %	12.4	5.8	3.9	10.6	4.6	V, %	9.2
	2013	X	5.6	5.1	5.2	4.9	4.7	$X_{max}-X_{min}$	1.1
		V, %	3.6	5.2	2.9	1.5	1.8	V, %	6.7
B	2011	X	4.3	4.4	4.3	4.3	4.9	$X_{max}-X_{min}$	n
		V, %	2.4	0.4	0.9	8.8	1.2	V, %	
	2012	X	4.5	4.8	4.5	4.4	3.7	$X_{max}-X_{min}$	1.1
		V, %	7.2	8.7	1.5	12.0	0.5	V, %	9.3
	2013	X	4.4	4.5	4.3	4.0	4.5	$X_{max}-X_{min}$	0.5
		V, %	6.1	1.2	1.8	1.2	2.7	V, %	4.5

Notes: See Table 7.2.

At dissociation of formed H_2CO_3 ions H^+, which have appeared in a significant amount could saturate a soil absorbing complex. The increase of pH_{KCl} in September, 2011 is encountered in active mineralization of clover residuals. Decrease in soil acidity at entering of the plant residuals, especially legumes, has been noted by some authors [29, 30].

In 2012, level of acidity and pattern of its seasonal dynamics is absolutely distinct from other years of research. Firstly, acidity degree as a whole has increased; the pH_{KCL} value did not exceed 5.5 units (Table 7.4, Fig. 7.2). Secondly, amplitude of seasonal dynamics of the given parameter has increased, caused considerable acidification of soils in July (pH_{KCl} = 5.1).

Acidifying action of root system of wheat, which excludes H^+ ions in the period of most active uptake of nutritional elements was, possibly, the principal reason, which has caused decrease of pH_{KCl} during the period from June till July. Besides, decrease of pH_{KCl} can be caused with increase in soil humidity and, accordingly, with concentration of H^+ ions as a result of organic acids dissociation. In August when activity of root system has decreased the pH_{KCl} value has increased, and then, in September, has decreased again, the reason for that could be mineralization of the wheat residuals and formation of acidic products.

In 2012, relation between pH_{KCl} and content of C_{hum} was not revealed. Influence of mineral fertilizers should be excluded also because of their low dose.

In 2013, the pH_{KCl} gradually decreased from spring by the autumn. This year the greatest amplitude of seasonal changes of the given parameter was noted which has made 0.7 units. High pH_{KCl} value in the beginning of the growth season (5.9 units in May) could be caused by early overdrying of soils, because in the second decade of May at high daily average air temperature (+13.7 °C) amount of precipitation was insignificant (6.1 mm only).

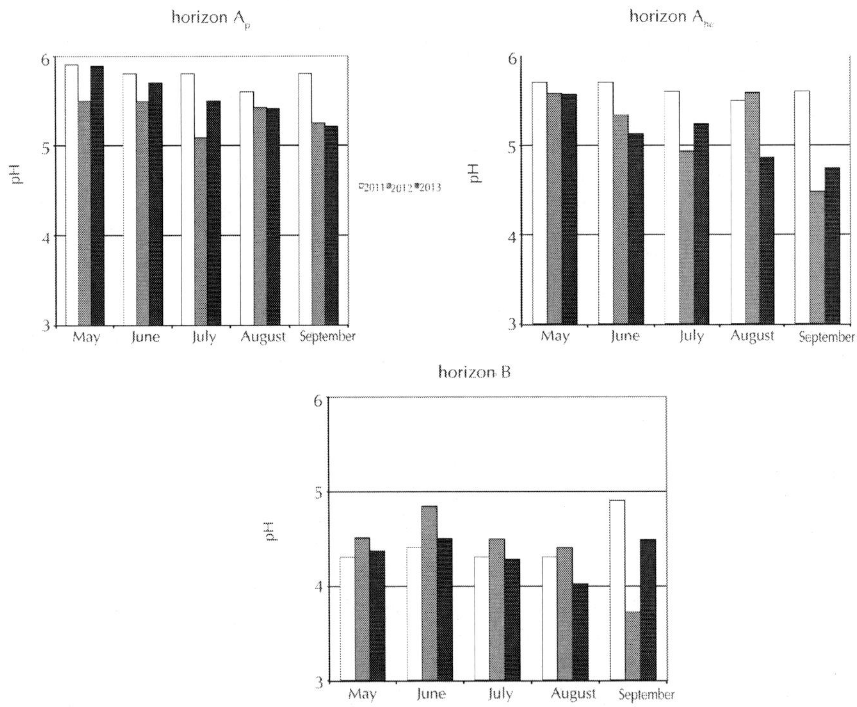

FIGURE 7.2 Seasonal Dynamics of pH_{KCl} in Arable Podzolic Soil.

Further soil acidification could have some reasons. In first half of growth season it could occur owing to increase of soil humidity because of frequent and plentiful precipitations. Since July acidifying action on soil, possibly, was rendered by root system of oats having longer period of active consumption of nutritional elements. Besides, in 2013 significant weak relation is revealed between seasonal dynamics of pH_{KCl} and seasonal dynamics of C_{hum} content (r = 0.35 at P = 0.90).

In eluvial horizon of arable soil seasonal fluctuations of pH_{KCl} in 2011 were insignificant (Table 7.4), and this year the highest level of pH_{KCl} was noted (see Fig. 7.2). The reason of absence of dynamics and low pH_{KCl} value as a whole can be activity of clover root system promoting considerable drying of soil.

In 2012–2013, change of pH_{KCl} in A_{he} horizon were statistically significant and, possibly, are caused by the same reasons, as in A_p horizon, including influence of C_{hum} (r = 0.41 at P = 0.99).

In illuvium horizon the significant seasonal variation of the given parameter was noted in 2012–2013. In 2011, amplitude of spatial variability of pH_{KCl} considerably exceeded its seasonal variability (Table 7.4). As Fig. 7.2 shows, pattern of seasonal dynamics in illuvium horizon differs considerably from that revealed in the top horizons. Most likely, in this horizon other mechanism of regulation of an acid condition of soil starts to play a considerable role. There are exchangeable Al^{3+} ions at a soil-absorbing complex of horizon B, which influence degree of exchangeable acidity. The illuvium horizon of arable land is characterized by very high spatial heterogeneity in content of exchangeable aluminum. Sometimes its coefficient of variation reaches 163.6% (Table 7.5).

Pattern of seasonal changes of aluminum content in 2012 and 2013 was distinguished, and these distinctions were characteristic for the spring and fall period, but within summer change of Al^{3+} content were almost similar. The content of exchangeable Al^{3+} ions correlates closely with degree of soil acidity (r = −0.78 at P = 0.99). Both the first and the second parameter can act as regulating factor.

TABLE 7.5 Content of Exchangeable Al^{3+} Ions in Illuvium Horizon of Arable Soil (mg-equivalent 100 g^{-1} of Soil)

Year	Spatial variability						Seasonal variability	
	Parameter*	Month					Parameter	Mean
		May	June	July	August	September		
2011	X	1.08	1.29	1.60	2.08	0.00	X_{max}–X_{min}	n
	V, %	38.6	106.9	5.8	76.1	0.0	V, %	
2012	X	0.10	0.10	0.04	0.87	2.76	X_{max}–X_{min}	2.72
	V, %	94.0	163.6	49.0	115.6	7.5	V, %	150.1
2013	X	0.89	0.11	0.36	1.24	0.43	X_{max}–X_{min}	1.13
	V, %	68.4	125.5	70.6	9.0	49.4	V, %	74.7

Notes: See Table 7.2.

Moreover, correlation between these two parameters and seasonal dynamics of C_{hus} has been revealed (r = 0.35 for pH_{KCl} and r = −0.51 for Al^{3+} at P = 0.90). More close relation between C_{hum} dynamics and dynamics of Al^{3+} content, possibly, specifies in direct character of this interaction. Influence of C_{hum} content on degree of acidity (pH_{KCl}) has indirect character, and is carried out through regulation of concentration of exchangeable Al^{3+} ions.

Apparently, labile humus components (including newly formed humus acids) form the complex salts with aluminum, which are dropping out in a deposit.

7.3.3 SEASONAL DYNAMICS OF ACIDITY IN FALLOW PODZOLIC SOIL

The pH_{KCl} of fallow soil has high values. In top horizon (humus horizon) it does not differ from pH_{KCl} value of arable soil. The given soil can be determined as weakly acidic or close to the neutral on acidity degree. Acidity increases downwards on a profile.

Spatial heterogeneity of pH_{KCl} value of fallow soil, as well as in two other soils, is expressed slightly. Some increase in coefficients of variation of spatial variability is observed in horizon A_{he} (Table 7.6). As a whole, the pH_{KCl} value concerns the parameters which are not possessing significant spatial variability.

Seasonal dynamics of acidity of fallow soil is low also. In two top horizons A_h and A_{he} seasonal changes of pH_{KCl} were significant in all years of researches; thus pH_{KCl} of soil of horizon A_{he} has appeared more dynamical, as well as in arable soil (Table 7.6). So, in humus horizon coefficient of variation fluctuated from 2.5 to 5.2%, amplitude of fluctuations made from 0.3 to 0.7 units; in horizon A_{he} coefficients of variation characterizing seasonal variability of a parameter reached 17.3%, and amplitude – 1.8 units of pH.

In horizon B dynamics of acidity was significant in 2013 only, and coefficient of variation has made 3.4%, amplitude – 0.3 units of pH.

Pattern of seasonal changes of pH_{KCl} is distinguished in different years of research. In humus horizon A_h maximum and minimum values of pH_{KCl} in different years took place in different months of a growth season (Fig. 7.3). So, in 2011 the greatest acidity has been noted in July, in 2012 – in May and August, in 2013 – in June and September. The reasons of such “shift of phases” on a curve of a seasonal pH_{KCl} variation can be different. Probably, it is caused by distinctions in regimes of soil humidifying and by succession of phytocenosis.

Essential correlation between change of pH_{KCl} and the content of organic matter has been revealed, and unlike arable soil – with content of C_{hum} (r = 0.81 at P = 0.99). Besides, it has appeared that increase of C_{hum} is connected with soil acidification (r = −0.45 at P = 0.99). The increase in content of labile forms of organic matter (C_{lab}) in fallow soil is closely connected with reduction of quantity of total humus (C_{tot}). This fact explains, in this case, existence of negative correlation between changes of pH_{KCl} and a share of C_{lab} in structure of C_{hum}; the intensive mineralization noted in the middle of summer, on the one hand, leads to quantity increase in C_{lab} content, on the other hand – to increase in degree of acidity.

TABLE 7.6 Dynamics of pH_{KCl} in Fallow Soil

Soil horizon	Year	Spatial variability						Seasonal variability	
		Parameter*	Month					Parameter	Mean
			May	June	July	August	September		
A_h	2011	X	-	5.5	5.2	5.5	5.9	$X_{max}-X_{min}$	0.7
		V, %	-	2.7	4.3	1.2	0.3	V, %	5.2
	2012	X	5.1	5.6	5.5	5.4	5.5	$X_{max}-X_{min}$	0.5
		V, %	5.4	3.2	1.6	3.4	4.8	V, %	3.5
	2013	X	5.5	5.3	5.5	5.6	5.3	$X_{max}-X_{min}$	0.3
		V, %	2.4	3.1	1.4	1.1	0.8	V, %	2.5
A_{he}	2011	X	-	4.5	4.2	4.4	6.0	$X_{max}-X_{min}$	1.8
		V, %	-	4.1	3.1	2.4	4.3	V, %	17.3
	2012	X	4.3	5.4	5.5	5.2	5.2	$X_{max}-X_{min}$	1.2
		V, %	9.6	4.2	5.9	8.9	17.4	V, %	9.3
	2013	X	4.7	4.1	5.4	5.8	4.7	$X_{max}-X_{min}$	1.7
		V, %	8.2	4.5	1.5	1.4	1.9	V, %	13.5
B	2011	X	-	4.3	4.2	4.3	4.4	$X_{max}-X_{min}$	n**
		V, %	-	1.3	1.1	0.8	6.1	V, %	
	2012	X	3.8	3.8	3.8	3.8	3.9	$X_{max}-X_{min}$	n
		V, %	3.3	2.3	0.9	2.9	5.6	V, %	
	2013	X	4.0	3.8	4.1	4.0	3.8	$X_{max}-X_{min}$	0.3
		V, %	0.9	0.9	1.8	1.1	1.2	V, %	3.4

Notes: See Table 7.2.

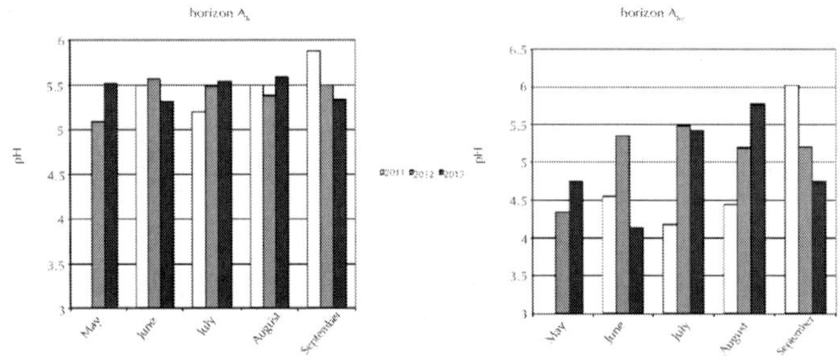

FIGURE 7.3 Seasonal dynamics of pH_{KCl} in fallow podzolic soil.

In eluvial horizon A_{he} seasonal dynamics of pH_{KCl} practically coincides with that of horizon A_h, but has bigger amplitude of fluctuations (Fig. 7.3). Similar pattern of dynamics means the identical reasons. So, its correlation with content of C_{hum} is positive also, but a little less close ($r = 0.62$ at $P = 0.99$).

Unlike the top horizon, here there is one more factor influencing degree of acidity – presence of exchangeable Al^{3+} ions (Table 7.7).

TABLE 7.7 Content of Exchangeable Al^{3+} Ions in Fallow Podzolic Soil

Soil horizon	Year	Spatial variability						Seasonal variability	
		Parameter*	Month					Parameter	Mean
			May	June	July	August	September		
A_{he}	2011	X	–	0.45	1.71	0.88	0.00	$X_{max}-X_{min}$	1.71
		V, %	–	14.2	13.4	33.2	0.0	V, %	95.8
	2012	X	1.28	0.00	0.00	0.00	0.12	$X_{max}-X_{min}$	1.28
		V, %	52.7	0.0	0.0	0.0	138.3	V, %	200.5
	2013	X	0.36	1.26	0.00	0.00	0.13	$X_{max}-X_{min}$	1.26
		V, %	120.9	47.4	0.0	0.0	156.1	V, %	151.3
B	2011	X	–	1.18	1.86	0.96	1.39	$X_{max}-X_{min}$	n**
		V, %	–	17.2	25.1	86.1	76.1	V, %	
	2012	X	3.11	1.97	2.02	3.30	1.94	$X_{max}-X_{min}$	1.36
		V, %	26.3	30.0	10.3	23.9	49.8	V, %	27.4
	2013	X	3.38	3.24	1.19	1.57	2.85	$X_{max}-X_{min}$	2.20
		V, %	14.9	10.1	44.8	21.4	9.3	V, %	40.9

Notes: See Table 7.2.

Dependence between seasonal pH_{KCl} and Al^{3+} dynamics is high ($r = -0.76$ at $P = 0.99$). The fallow soil differs from arable soil by presence of exchange aluminum in horizon A_{he}. Possibly, fallow soil in its agricultural use had low degree of cultivation and influence of fertile plowed horizon did not affect underlying layers of soil. Besides, in neglected fallow soil zone processes of soil formation are restored gradually, amplifies podzol-forming process accompanied by destruction of a mineral part and occurrence of exchangeable aluminum.

Exchangeable Al^{3+} ions in horizon A_{he} of fallow soil are found out not during whole growth season, but in separate terms only. Usually there are the beginning and the end of season though in 2011 it has been found out in the middle of summer too (Table 7.7, Fig. 7.4).

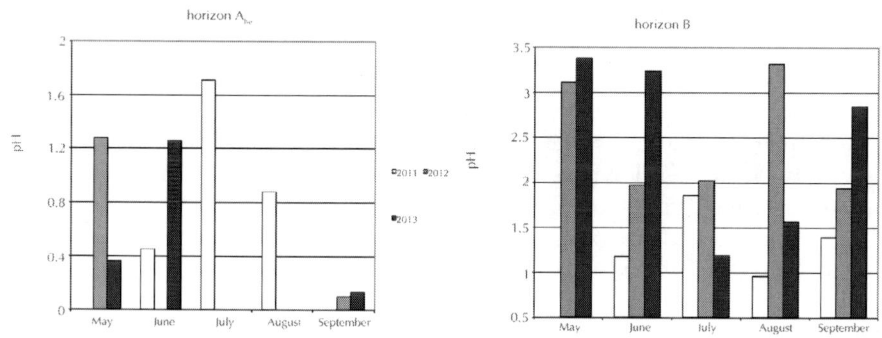

FIGURE 7.4 Seasonal dynamics of the content of exchangeable Al^{3+} ions in fallow podzolic soil.

Such pattern of dynamics can be caused by fluctuations of soil humidity when less acidic solutions from horizon A_h get into eluvial part of a profile and reduce mobility of Al^{3+} ions as well as by dynamics of content of C_{hum}. There was statistically significant correlation between content of aluminum and C_{hum} ($r = -0.40$ at $P = 0.99$). Negative dependence between these parameters, possibly, is caused by the high participation of humus acids in formation of the organic-mineral complexes reducing mobility of Al^{3+}.

The similar pattern is characteristic for soil of horizon B also. Seasonal dynamics of pH_{KCl} has high enough correlation with the content of exchangeable Al^{3+} ions ($r = -0.64$ at $P = 0.99$). Unlike soil of eluvial horizon exchangeable Al^{3+} ions are present at horizon B during whole growth season (Table 7.7). Pattern of seasonal dynamics of the content of exchangeable Al^{3+} ions in horizon B differs on years of research (Fig. 7.4) and, possibly, depends on seasonal dynamics of the content of labile components of organic matter. Correlation is negative also, but weak: $r = -0.21$ at $P = 0.90$.

7.4 CONCLUSION

Thus, seasonal changes of degree of acidity have the different patterns in natural and arable soils, which are differ on horizons too. For a forest litter decrease of acidity to the beginning or the middle of summer and its increase by the end of summer is characteristic. As a rule acidity of eluvial horizon increases in August but decreases in September.

In arable soil at various fluctuations during growth season acidity of plowed and eluvial horizons increases by the fall usually. In illuvium horizon of arable land seasonal changes of pH_{KCl} are differed on years. In fallow soil various patterns of seasonal dynamics of degree of acidity in all investigated horizons is possible.

As a whole, in various terms of growth season arable and fallow soils can be referred to different categories on acidity degree – from weakly acidic to almost neutral. Acidity of the arable soil, which value in the first most responsible period of plants growth and development are close to neutral, does not render stressful influence on cultivated plants.

At all investigated horizons of moss-covered spruce forest soil and in the bottom horizons of arable and fallow soils exchangeable ions of aluminum are present. Its content varies considerably in space, and heterogeneity increases in arable soils. Spatial variability often overlaps the amplitude of seasonal changes and does their insignificant.

Seasonal dynamics of pH_{KCl} and of exchange aluminum is influenced by dynamics of C_{hum} (total organic carbon in O_i) and of C_{lab}. Degree of this influence varies in different horizons and depending on degree of soil cultivation. In a case of exchange aluminum coefficients of correlation can even change a sign: in the bottom horizons of forest soil at increase in content of C_{hum} the content of exchangeable ions of aluminum increases also; in arable and fallow soils – on the contrary. It, possibly, is caused by different qualitative structure of C_{hum} and its labile components. In arable soil, in comparison with others, the great influence on the content of exchangeable aluminum and soil acidity renders the content of labile humus matter.

ACKNOWLEDGEMENT

This work is supported by the Russian Academy of Agricultural Sciences (Project 04.05.02.). We thank the staff of department of plant edaphic resistance (North-East

Agricultural Institute, Kirov, Russia) for their technical help with conducting laboratory and field experiments.

KEYWORDS

- **arable soil**
- **eluvial horizon**
- **fallow**
- **illuvium horizon**
- **moss-covered spruce forest**
- **organic matter**
- **pH**

REFERENCES

1. Borlaug, N. E., & Dowswell, C. R. (1997). The Acid Lands, One of Agriculture's last Frontiers, Plant-Soil Interaction at Low pH, Brasilian Soil Science Society, 5–15.
2. Kinraide, T. B. (2003). Toxicity Factors in Acidic Forest Soils, Attempts to Evaluate Separately the Toxic Effects of Excessive Al^{3+} and H^{+} and Insufficient Ca^{2+} and Mg^{2+} Upon Root Elongation, Eur. J. Soil Science, *54,* 323–333.
3. Koyama, H., Toda, T., & Hara, T. (2001). Brief Exposure to Low-pH Stress Causes Irreversible Damage to the Growing Root in *Arabidopsis Thaliana*, Pectin-Ca Interaction May Play an Important Role in Proton Rhizotoxicity, J. Exp. Bot., *52,* 361–368.
4. Yang, J. L., Zheng, S. J., He, Y. F., & Matsumoto, H. (2005). Aluminum Resistance Requires Resistance to Acid Stress, a Case Study with Spinach that Exudes Oxalate Rapidly When Exposed to Al Stress, J. Exp. Bot., *56,* 1197–1203.
5. FitzPatrick, E. A. (1986). An Introduction to Soil Science, N. Y. Longman Scientific and Technical, 2–55.
6. Sawaki, Y., Iuchi, S., Kobayashi, Y., Ikka, T., Sakurai, N., Fujita, M., Shinozaki, K., Shibata, D., & Kobayashi, M. (2009). *Stop1* Regulates Multiple Genes that Protect *Arabidopsis* from Proton and Aluminum Toxicities, Plant Physiology, *150*, 281–294.
7. Runge-Metzger, A. (1995). Closing the Cycle Obstacles to Efficient P Management for Improved Global Security, Tiessen, H. (Ed) Phosphorus in the Global environment Transfers, Cycles and Management, John Wiley & Sons, New York, 27–42.
8. Sanchez, P. A., & Uehara, G. (1980). Management Considerations for Acid Soils with High Phosphorus Fixation Capacity, Khasawneh, F. E., Sample, E. C., Kamprath, E. J. (Eds) The Role of Phosphorus in Agriculture, American Society of Agronomy, Madison Wisconsin, 263–310.

9. Kochian, L. V., Hoekenga, O. A., & Pineros, M. A. (2004). How do Crop Plants Tolerate Acid Soils? Mechanisms of Aluminum Tolerance and Phosphorous Efficiency, Annual Revision Plant Biology, *55,* 459–493.
10. Thomas, G. W., & Hargrove, W. L. (1984). The Chemistry of Soil Acidity, Adams, F. (Ed) soil Acidity and Liming Madison, ASA, CSSA, SSSA, 3–56.
11. Herbert, A. (2009). Opportunity Costs of Land Degradation Hazards in the South-West Agricultural Region, Department of Agriculture and Food Western Australia, Resource Management ISSN 1039–7205.
12. Udovenko, G. V. (1995). Resistance of Plants to Abiotic Stresses, Physiological Bases of Plant selection Saint Petersburg VIR Publishing, *2*, 293–352 (in Russian).
13. Hede, A. R., Skovmand, B., & Lopez-Cesati, J. (2001). Acid Soils and Aluminum Toxicity, Reynolds, M. P., rtiz-Monasterio, J. I., McNab, O. A. (eds) Application of Physiology in Wheat Breeding, Mexico, D. F. CIMMYT, 172–182.
14. Johnson, J. P., Carver, B. F., & Baligar, V. C. (1997). Productivity in Great Plains Acid Soils of Wheat Genotypes Selected for Aluminum Tolerance, Plant and Soil, *188,* 101–106.
15. Carver, B. F., & Ownby, J. D. (1995). Acid Soil Tolerance in Wheat, Adverse Agron, *54,* 117–173.
16. Deborah, A. S., & Tesfaye, M. (2003). Plant Improvement for Tolerance to Aluminum in Acid Soils a Review, Plant Cell Tissue Organism, *75,* 189–207.
17. Liu, H. (2005). Aluminum Resistance among Seeded Bermudagrasses, Hort Science, *40,* 221–223.
18. Ring, S. M., Fisher, R. P., Poile, G. J., Helyar, K. R., Conyers, M. K., & Morris, S. G. (1993). Screening Species and Cultivars for their Tolerance to Acidic Soil Conditions, Plant Soil, *155–156,* 521–524.
19. Tang, C., Rengel, Z., Diatloff, E., & Gazey, C. (2003). Responses of wheat and Barley to Liming on a Sandy Soil with Subsoil Acidity, Field Crops Research, *80*, 235–244.
20. Silva, I. R., Corrêa, T. F. C., Novais, R. F., Gebrim, F. O., Nunes, F. N., Silva, E. F., & Smyth, T. J. (2008). Protective Effect of Divalent Cation against Aluminum Toxicity in Soybean, R. Bras Ci Solo, *32,* 2061–2071.
21. Pereverzev, V. N. (1996). Seasonal Dynamics of Soil Processes in Woodlands of Kola Peninsula, Abstracts of II Congress of Soil Scientists Society, *2,* 184–185 (in Russian).
22. Liang, C., Piñeros, M. A., Tian, J., Yao, Z., Sun, L., Liu, J., Shaff, J., Coluccio, A., Kochian, L. V., & Liao, H. (2013). Low pH Aluminum and Phosphorus Coordinately regulate Malate Exudation through *GmALMT1* to Improve Soybean Adaptation to Acid Soils, Plant Physiology, *161,* 1347–1361.
23. Agrochemical Methods of Research of Soils (1975). Moscow, Nauka Publishing 656p (in Russian).
24. Arinushkina, E. V. (1970). Hand Books of Reference Methods for Chemical Analysis of Soils, Moscow State University Publishing, 488p (in Russian).
25. State Standard of USSR 26207–26291 Soils, Determination of Mobile Compounds of Phosphorus and Potassium by Kirsanov Method Modified by CINAO (1992). Moscow Standard Publishing, 7p (in Russian).
26. State Standard of USSR 26213–26291, Soils Methods for Determination of Organic Matter (1992) Moscow Standard Publishing, 8p (in Russian).
27. Recommendations for Research of Balance and Transformation of Organic Matter at Agricultural use and Intensive Cultivation of Soils (1984) Moscow Russian Academy of Agricultural Sciences Publishing, 96p (in Russian).
28. Ponomareva, V. V. (1964). Theory of Podzol Forming Process, Leningrad Nauka Publishing, 327p (in Russian).

29. Noble, A. D., & Randall, P. J. (1996). Leaf Litter Ash Alkalinity and Neutralization of Soil Acidity, Plant Soil, *179,* 293–302.
30. Haynes, R. J., & Mokolobate, M. S. (2011). Amelioration of Al Toxicity and P Deficiency in Acid Soils by Additions of Organic Residues a Critical Review of the Phenomenon and the Mechanisms, Nutrient Cycling in Agro Ecosystems, *59,* 47–63.

CHAPTER 8

ROLE OF THE PATERNAL AND MATERNAL COMPONENTS OF CROSSING ON RESISTANCE OF OATS HYBRIDS TO ALUMINUM IONS

GALINA A. BATALOVA[1,2] and EUGENE M. LISITSYN[1,2]

[1]Behalf of N.V. Rudnicki Zonal Research Institute of Agriculture of North-East of the Russian Agricultural Academy, d.166-a, Lenin St., Kirov, 610007, Russia; E-mail: niish-sv@mail.ru, edaphic@mail.ru

[2]Vyatka State Agricultural Academy, d.133, October Av., Kirov, 610010, Russia; E-mail: info@vgsha.info

CONTENTS

ABSTRACT

The toxicity caused by Al^{3+} ions is a major factor limiting productivity of cereals on acid soils all over the world. The trait "aluminum resistance" of plants is considered now physiologically simple, but genetically complex parameter. Researches on genetics of this trait at oats (*Avena sativa* L.) are begun rather recently and till now have not led to unambiguous opinion on number of genes governing aluminum resistance. On an example of oats varieties of different ecological-geographical origin character of influence of paternal and maternal components of crossing on level of F_2 hybrids Al-resistance is studied. It is found out that relative (aluminum in compare with control treatment) level of cation-exchange capacities of roots, their oxidizing activity, as well as activities of acid phosphatase and acid lipase was supervised by a maternal component of crossing (32–66%) and by interaction of components (34–41%), influence of paternal component was not statistically significant (1–8%). More over paternal component had not statistically significant influence on an integrated parameter of resistance – relative root growth (RRG), the part of influence of a maternal component reached 51%, a part of influence of interaction of components of crossing – up to 30%. Depending on what genotype was used as a maternal component of crossing, the direction and force of dominance of parameter RRG varied from negative superdominance up to positive superdominance. The analysis of graphics of orthogonal regression using trait coordinates "weight of a stalk" – "weight of a panicle" has allowed to reveal significant influence of a maternal component of crossing on aluminum resistance of oats F_2 hybrids and action of genetic system of adaptability. The analysis of distinctions between direct and reciprocal hybrids of the second generation has shown that in the studied set of genotypes trait "aluminum resistance" may be governed by at least three nuclear and two cytoplasmic genes.

8.1 INTRODUCTION

Oat is an important cereal grown in different regions of the world. It ranks sixth in world cereal production following wheat, maize, rice, barley and sorghum. Cultivation is most widespread and concentrated in temperate and humid climate zones in North America and Europe [1]. These areas have the cool, moist climate to which oats are best adapted. Russia, Canada, the United States, Finland, and Poland are the leading oat producing countries. Oats are adapted to a wide range of soil types, thus temperature and moisture conditions are the usual limiting factors as to where oats are grown.

Most cultivated varieties of oat belong to *Avena sativa* L., an allohexaploid species, with $2n = 6x = 42$ chromosomes, originated from the aggregation of three ancestral diploid genomes AA, CC, DD [2].

Although world oat production has been stable for the last five years, the demand for oat as a food has risen because of the recognized nutritive value of the oat grain. A better understanding of the genetic control related to the phenotypic variation in morpho-physiological and qualitative-quantitative traits is seen as a fundamental step in breeding programs. On opinion of Nava et al. [3] this allows an improved efficiency in the selection of well-adapted genotypes with high yield potential.

The major limitation to cereal productivity on most acid soils in Russia as well as all over the world is aluminum (Al) toxicity [4]. A great number of aluminum tolerance studies have been carried out in maize, wheat and other grasses, although very few have addressed oats. Thus, first articles were published in 1987 by Foy et al. [5, 6]. Working with several Polish and USA oats varieties they observed a wide range of intraspecific variation in oats Al resistance. But the genetic analysis of Al tolerance in oats have been begun an area of research at the beginning of XXI century only. Genetics studies made in Brazil indicate that aluminum tolerance in oats is controlled by one or two dominant genes [7–9]. Resent results reported in the literature indicate that Al tolerance in cereals is mediated by a genetically complex but physiologically simple mechanism when considered at the whole-plant level [10].

At the analysis of numerous set of oats varieties of Russian and foreign selection it was revealed that genotypes within Avena species vary widely in their Al-resistance level [11, 12]. Reduction of root growth is the most widely recognized symptom of Al toxicity so resistance level were estimated based on Al-inhibited root elongation in nutrient solution or on relative [(with Al/without Al) × 100] seedling root length. The number of genes contributing to a quantitative character is estimated by distribution of average means of the parameters within populations of different generation from crosses of parental varieties having contrast level of Al-resistance. Nava et al. [13] suggested that oats Al-resistance was controlled by one dominant major gene with the tolerant genotypes carrying Al_aAl_a and the sensitive ones al_aal_a alleles.

Twenty-one oat genotypes bred in Brazil were evaluated for their Al-resistance on P_1, P_2, $F_1$and F_2 generations of nine crosses among tolerant × sensitive genotypes within the Universidade Federal do Rio Grande do Sul breeding program [7]. The authors suggested that the tolerant response was conditioned by one gene with dominant effect. Some later they ascertained that this gene has at least two different alleles [14]. Oliveira et al. [15] used four oat genotypes with differential response regarding tolerance to aluminum toxicity (tolerant, sensitive and intermediate) as parents in crossing, and established that the trait aluminum tolerance is controlled by one gene with multiple alleles. Wagner et al. [8] evaluated a larger number of crossings and obtained results indicating the presence of one or two dominant genes involved in aluminum tolerance.

In other researches [16, 17] F_2 populations from crosses of varieties having the same Al-resistance level were used to determine number of genes of tolerance to toxic aluminum concentrations. Genetic analysis indicated that the same gene confers the Al tolerance in each genotype. However, a diverse pattern of Al accumulation and different level of development of oxidative stress was obtained for the genotypes, suggesting that the external detoxification (exudation of ligands which bind Al^{3+}) is not the main mechanism of tolerance.

However, the results of our previous studies on large set of oats varieties and F_1–F_2 hybrids [12] indicated that such estimation of number of genes participating in the control of a trait couldn't be considered as comprehensive. Frequency distribution of the F_2 seedlings shows presence of transgressions, i.e. the plants surpassing parental genotypes by resistant level or having it below parental forms. Possibly initial parental genotypes were not homozygous at the studied trait; each of them should have both dominant and recessive alleles in this case [18, p. 127]. Thus in a case of monogenic inheritance (supposed by authors of above-mentioned articles) parental genotypes were not contrast actually but have the same genotype. Phenotypic distinction between them can be explained by action of multiple alleles, or action of several nonallelic genes. The objectives of this study were to determine the role of parental components of crosses on hybrid level of aluminum resistance.

8.2 MATERIALS AND METHODOLOGY

Researches on studying of genetic control of oats aluminum resistance were carried out in several sets of experiments. Oats genotypes having different ecological-geographic origin were selected: for first set of experiments – varieties Krechet, Faust, E-1643, I-1987 (Kirov, Russia), Minerva, Mernime (France), Waldern (Canada), Freija, 17427 (Sweden), and F_2 hybrids having variety Krechet as paternal or maternal parent; for second set of experiments – varieties Argamak (Russia), Colt (Great Britain), Wilma (The Netherlands) and F_2 hybrids between them; for third set – varieties Argamak (Russia), Freija (Sweden), Petra (Estonia), and direct and reciprocal F_2 hybrids between them.

Inhibition of seminal root growth elicited by Al in test solution was used to quantify Al tolerance, using distilled water at pH 6.0 as control and 1 mM Al sulfate at pH 4.3 as test treatment by method described earlier [19]. The relative root growth index (RRG), calculated as the maximum root length in Al stress culture divided by maximum root length in control [20, 21] was used as an indicator to evaluate Al tolerance. Root cation-exchange capacity was estimated according to [22]; oxidizing activity of roots – according to Ref. [23]; acid phosphatase (EC 3.1.3.2.) and acid lipase (EC 3.1.1.3) activities was estimated using the protocols described in Ref. [24]. A part of influence of each parental component was counted with a method of two-factorial analysis of variance (ANOVA).

8.3. RESULTS AND DISCUSSION

Data of ANOVA of the first set of experiments on growth of oats seedlings roots under control and aluminum treatments has shown (Table 8.1) that variability of root length of hybrids was defined by influence of a maternal variety on 46.0% (control) or on 24.3% (aluminum). There was a statistically significant ($p < 0.05$) influence of maternal variety on variability of trait RRG (14.2%).

TABLE 8.1 Parameters of growth of oat seedling roots.

Variety, Hybrid	Root length, mm		RRG. %
	Control	1 mM Al	
17247	86.4 ± 2.3	81.8 ± 2.3	94.6
E-1643	109.4 ± 2.7	97.2 ± 2.4	88.8
Freija	78.5 ± 2.5	79.4 ± 2.6	101.1
Minerva	88.5 ± 2.1	71.8 ± 1.4	81.3
Waldern	89.8 ± 2.1	75.4 ± 2.1	84.0
I-1987	88.9 ± 2.3	86.1 ± 2.5	96.8
Faust	106.5 ± 2.3	93.9 ± 1.8	88.2
Krechet	82.9 ± 1.5	81.5 ± 2.1	98.3
17247 × Krechet	79.8 ± 1.4	77.4 ± 0.8	96.9
E-1643 × Krechet	105.8 ± 2.2	97.2 ± 2.1	91.9
Freija × Krechet	90.7 ± 2.3	87.7 ± 2.6	96.7
Waldern × Krechet	79.2 ± 2.1	83.9 ± 2.2	106.0
I-1987 × Krechet	89.7 ± 2.1	85.3 ± 2.1	95.1
Krechet × Mernime	91.3 ± 1.8	85.7 ± 2.0	93.8
Krechet × Minerva	92.1 ± 2.4	86.9 ± 1.5	94.4
Krechet × Waldern	89.7 ± 1.9	86.6 ± 1.1	98.5
Krechet × E-1643	107.3 ± 2.5	98.4 ± 1.3	91.7
Krechet × Faust	99.6 ± 2.2	91.9 ± 2.0	92.2

Variety Krechet (having RRG = 99.6%) taking as paternal component at crossing with variety Waldern (RRG = 85,5%) is provided high level of Al-resistance of hybrid equal to Krechet, but taking as maternal component – much lower than both parents (77.7%). On the other hand at crossing of varieties with close levels of resistance Krechet (99.6%) and I-1987 (104.2%) hybrid was much less resistant (87.8%). Crossing of Krechet as maternal form and variety Mernime (97.7%) has led to creation of a hybrid with considerably lowered level of resistance (87.9%).

Previously we established [25, 26] that direction and strength of dominance of RRG index can change considerably (Table 8.2) depending on direction of crossing (i.e. what variety was taken as maternal or paternal form); this direction influences on character of interaction of genes governing oats aluminum resistance.

TABLE 8.2 Inheritance of Root Growth Parameters by Oats F_2 Hybrids

Hybrid	Root length				RRG, %	
	Control		1 mM Al			
	d	H	d	H	d	H
17247 × Krechet	–2.69	0.936	–10.14	0.250	1.35	0.311
I-1987 × Krechet	0.55	0.891	–1.19	0.876	–3.90	0.908
Waldern × Krechet	–2.36	0.989	0.96	0.974	1.92	0.960
Mernime × Krechet	–7.13	0.913	–2.71	0.989	4.52	0.993
E-1643 × Krechet	0.78	0.975	1.80	0.553	0.53	0.432
Freija × Krechet	6.90	0.898	–0.19	0.884	–1.15	0.734
Krechet × Waldern	1.00	0.966	–0.19	0.884	–1.15	0.734
Krechet × Фауст	0.68	0.969	–0.19	0.884	–1.15	0.734
Krechet × Mernime	3.50	0.947	–0.29	0.975	–2.27	0.976
Krechet × E-1643	0.74	0.968	1.15	0.765	–0.26	0.726
Krechet × Minerva	2.63	0.399	0.93	0.986	–0.38	0.941

Note: d – degree of dominance, H – Broad-sense heritability.

Thus, data of Table 8.2 show that use of oats Krechet as maternal parent has led to positive dominance and superdominance on root length under control condition, and to negative dominance and superdominance on RRG index. For a set of hybrids at which oats Krechet was used as paternal component of crosses certain low in pattern of dominance was not revealed for any parameter.

Hybrid analysis is applied usually to research of the genetic control of Al-resistance in which course the parity of resistant and sensitive plants of hybrid generation are calculated, but variability of parent genotypes on investigated parameter is beyond the consideration.

The variability of both hybrids and parents Al-resistance level is taking into account by computer software GENAN [27]. At estimation of number of the genes coding the investigated parameter, it allows to consider variability of RRG index at parental forms of crossing and its influence on character of distribution of hybrid plants on resistance groups. For the hybrids of oats used in the present work the analysis of number of the genes coding Al-resistance level by means of software GENAN has allowed to reveal the following character of control of the parameter:

1. At hybrids 17427 × Krechet, Mernime × Krechet, and Waldern × Krechet – by one gene;
2. At hybrids Freija × Krechet and Krechet × Faust – by two genes;
3. At hybrids Krechet × Waldern, Krechet × E-1643, E-1643 × Krechet, Krechet × Mernime, and Krechet × Minerva – by three genes;
4. At hybrid I-1987 × Krechet – by four genes.

Considering that distinctions between direct and reciprocal hybrids were explained, basically, by action of cytoplasmic genes, the analysis of three pairs of crossings in which there were direct and reciprocal F_2 hybrids has allowed us to come to conclusion (Table 8.3) that in this set of genotypes trait "aluminum resistance" can be governed by up to three nuclear and up to two cytoplasmic genes.

TABLE 8.3 Distinctions in the Genetic Control of Al-Resistance in Direct and Reciprocal Oats F_2 Hybrids

Hybrid	Total number of genes	Number of nuclear genes	Number of cytoplasmic genes
Mernime × Krechet	1	1	2
Krechet × Mernime	3		
Waldern × Krechet	1	1	2
Krechet × Waldern	3		
Krechet × E-1643	3	3	0
E-1643 × Krechet	3		

In the second set of experiments (Table 8.4) the analysis of number of genes governing oats Al-resistance with software GENAN has allowed to reveal differences between varieties Argamak and Wilma in three genes, between varieties Argamak and Colt – in two genes, and between varieties Colt and Wilma – in one gene.

TABLE 8.4 Distribution of Oats Seedlings on Level of Al-Resistance (1 mM Al)

Variety, F_2 hybrid	Groups of seedlings having level of Al-resistance (RRG, %)									
	50	60	70	80	90	100	110	120	130	140
Argamak	3	4	32	48	10	5				
Colt			11	37	41	11				
Wilma		1	1	12	56	29				
Colt *x* Argamak	1	4	15	14	29	20	17	3		
Wilma *x* Colt	6	11	16	9	18	32		6		
Argamak x Wilma		9	24	13		5		23	24	2

Two by two comparison of parental varieties can indicates only number of genes on which the two given varieties differ. But if one organizes data on triad complex (Fig. 8.1) then possibility of mathematical interpretation the data will increase considerably since it will be necessary to analyze three pairs of varieties simultaneously. Fulfilling mathematical requirements by quantity of genes differing in pairs, it is possible to present hypothetically the formula of a studied trait for each of the variety entering into triad complexes, designating by Latin letters these differing genes. Then the structure of the genetic formula of a trait "aluminum resistance" at used oats varieties will look as in Fig. 8.1.

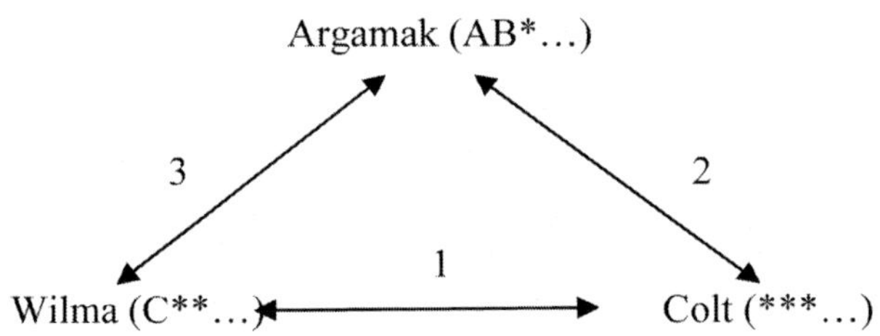

FIGURE 8.1 Structure of the genetic formula of a trait "aluminum resistance" at the investigated oats varieties. Numbers designate quantity of differing genes, * – identical genes.

Data of figure indicates that different genes take part in the structure of genetic formula of the same trait "aluminum resistance" of investigated oats varieties. Considering that oats plants continued growth of root systems under conditions of Al action during all experiment, it is possible to assume that each of them even the most sensitive had at least one such gene. Then the sum of number of various genes governing the studied trait at given set of oats varieties must be not less than four.

Data of the two series of experiments allows to assume the following:

- Considering that plants of all tested oats varieties continued growth in the presence of 1.0 mM of aluminum, it is possible to suggest that each of them harbor at least one gene of Al-resistance;
- At different varieties of the same species (oats) under the same stress condition (1 mM Al, pH 4.3) the trait "aluminum resistance" may be governed with a different quantitative and/or qualitative set of genes;
- Not less than five genes two of which is cytoplasmic take part in the governing of Al-resistance at the studied sample of varieties

The ecologic-genetic model of quantitative trait offered by [28], explains quantitative and qualitative changes in action of plants genetic systems as that the structure of these genetic systems under different growth conditions includes different genes. Thus authors specify that the activity of specific genes per se can keep unchanged, but their products can make or not the contribution to development of a quantitative trait. Considering the high variety of physiological reactions for aluminum, as well as absence of specific reactions for it, one can assume that parental varieties can possess both different genes and the same genes, but different genes (genetic systems) take part in the reaction for stressful influence at different genotypes.

By analogy that [28] have suggested to call "redefinition of the genetic formula of a trait in ontogeny" the phenomenon when the set of the genes governing development of any trait naturally changes under varying conditions of growth, it is possible to offer concept "variability of the genetic formula of a trait" within a plant species when number and/or a set of the genes governing a trait is distinguished at different varieties of the same species or at one variety at different intensity of stressful influence.

Except the analysis of morphological traits (root length or its regrowth) for estimation the level of Al-resistance it is possible to use a set of biochemical parameters though any of them is not specific to aluminum action. Studying of relation between level of Al-resistance of oats genotypes and such biochemical parameters as roots cation-exchange capacity [25]; activities of acid lipase and acid phosphatase [29]; root oxidizing activity [30] has allowed us to conclude that different genotypes of the same species (oats) use the same biochemical mechanisms in different degree in reaction for aluminum stress, and the estimation of only an integrated index (RRG) is insufficient for finding-out of the genetic control of Al-resistance.

The analysis of F_2 generation from crossing of Al-resistant oats varieties Freija and Petra with Al-sensitive variety Argamak has been used for studying of character of inheritance of different indices of aluminum resistance in the third set of experiments. Direct and reciprocal hybrids of the above-named varieties have been investigated. Thus, three parental varieties and six F_2 hybrid combinations between them served as a research material [30]. The two-factorial ANOVA has allowed to reveal following features of influence of parental and maternal components of crossing on development of studied indices of resistance.

8.3.1 ROOT LENGTH

Influence of both parental components is almost identical under control conditions – a part of maternal genotype influence is 19.4%, of paternal genotype – 17.0%; their interaction is considerable also – 58.9%. Action of stressful factor led to strengthening of influence of both parental forms (44.1 and 28.8% accordingly) at the expense of decrease in a part of influence of their interaction (practically by 3 times to 17.7%). Parental component of crossing did not influence on RRG index (influence is insignificant statistically); the part of influence of maternal component on the index has grown up to 51.1%, and a part of interaction of components – up to 30.2%.

8.3.2 ACID LIPASE ACTIVITY

Only influence of interaction of components is statistically significant under control condition – 93.9%, it explains almost all variability in studied population. However, at action of stressful factor the role of maternal component amplifies sharply – up to 38.2%; the part of influence of the paternal component increases to statistically significant value – 8.7%; the part of influence of interaction of components decreases by three times – to 34.5%. Strengthening of a part of influence of maternal component occurs at estimation of relative activity of enzyme – up to 68.8%. The part of influence of the paternal component is low (5.8%), the part of interaction of maternal and paternal components is insignificant too (10.1%).

8.3.3 ACID PHOSPHATASE ACTIVITY

Parts of influence of both parents are almost equal under the control: a maternal component – 32.8%, paternal – 34.3%, interaction of components is weak enough – 16.3%. The contribution of a paternal component on activity of enzyme has increased up to 59.2% at aluminum impact, the part of influence of maternal became statistically insignificant (4.3%), and interaction of components has a little amplified (19.3%). However, level of relative activity of enzyme (aluminum treatment in comparison with control) was governed by maternal component only (37.0%) and interaction of components (36.6%); influence of a paternal component is not significant statistically (1.2%).

8.3.4 ROOT OXIDIZING ACTIVITY

Action of paternal and maternal components are almost equal in the control – 13.8 and 14.2% accordingly, their interaction is very high – 61.9%. At action of stressful factor only an influence of interaction of components of crossing (79.3%) is significant statistically; contributions of paternal and maternal components are not significant (5.2 and 0.4%). But, as well as in a case of acid phosphatase, level of relative root oxidizing activity is under governing of maternal component – it part of influence is 32.5%, as well as interactions of components – 41.1%. Influence of paternal component is not significant statistically (7.7%).

8.3.5 ROOT CATION-EXCHANGE CAPACITY

Parts of influence of a maternal component and interaction of components are almost equal (33.6 and 35.1%) under control conditions at insignificant influence of a paternal component (6.2%). Approximately the same pattern is observed in case of action of the stressful factor, as well as at estimation of degree of change of the index under influence of aluminum. Parts of influence of paternal, maternal components and their interaction are equal accordingly 8.6, 37.4, and 42.9% (Al treatment) and 8.7, 34.2, and 51.4% (relative level of an index).

Study of the given selection material has been continued under field conditions on acid and neutral soil backgrounds. As data obtained earlier testify to participation of several different genes / genetic systems in governing of aluminum resistance then estimation of plants has been spent on complex indicators – on size and direction of shifts in action of genetic systems of adaptability and attraction of plastic substances by a technique offered by V.A. Dragavtsev [28].

The author [28] offers the existing of some total genetic-physiological systems by mean of which breeders empirically improve species by complex quantitative properties of productivity. In our study we use two of them only – system of attraction, which provides transfer of plastic matter from the straw and leaves into the ear; and system of adaptivity to environment as a whole. When plotting in the diagram (two-dimensional system of character coordinates "the mass of an ear of the main stem" – "the mass of a straw of the main stem") the mean values of the characters of the variety one marks only two genetic reason of variability: polymorphism by genes of adaptivity extends the correlation ellipse along the positive regression line (long axis in diagram), and polymorphism by genes of attraction – along the negative regression line (short axis in diagram). If there is no genetic variability by genes of attraction then all the points of the main values will spread on the positive regression line; if there is polymorphism by genes of attraction but is not by genes of adaptivity, then all the points of the mean values will spread on the negative regression line. If polymorphism by adaptivity is greater than that by attraction then we have an ellipse stretched along the positive regression line and conversely.

Plants of hybrid Petra × Freija demonstrate high adaptability on both soil backgrounds (Figure 8.2) where positive shifts in action of these systems is revealed at parental forms too. Reciprocal hybrids of this combination of crossing on neutral and acid soil backgrounds show accordingly weak and strong negative shifts on adaptability.

Positive and negative shifts on genetic system of attraction of plastic substances are revealed at both hybrids (direct and reciprocal accordingly) on a neutral soil background. Hence, direct hybrid (Petra × Freija) had advantage over reciprocal hybrid (Freija × Petra) on action of both genetic systems.

Negative shifts in action of genetic systems of attraction and adaptability were observed at plants of hybrid Petra *x* Argamak on both soil backgrounds. Plants of reciprocal hybrid (Argamak × Petra) demonstrate negative shifts on adaptability, but positive one – on genetic system of attraction at acid soil background.

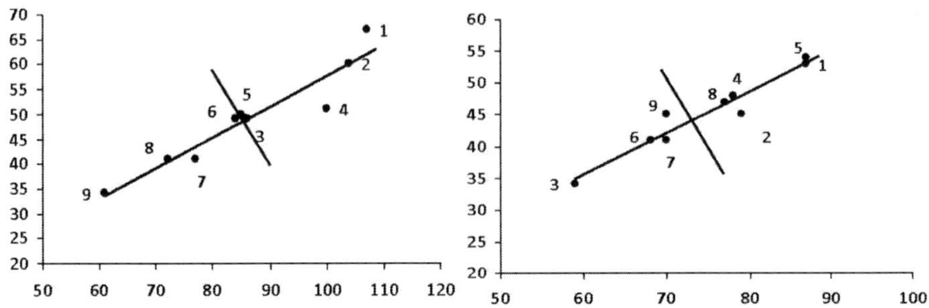

FIGURE 8.2 Graphic analysis of action of genetic systems of adaptability and attraction of plastic substances of oats varieties and F_2 hybrids on neutral (a) and acid (b) soil backgrounds. Axis of abscissa – weight of a stalk (g), axis of ordinates- weight of a panicle (g). Genotypes: 1. – Argamak; 2 – Freija; 3 – Petra; 4 – Petra × Freija; 5 – Freija × Argamak; 6 – Freija × Petra; 7 – Petra × Argamak; 8 – Argamak × Freija; 9 – Argamak × Petra.

Plants of hybrid Argamak × Freija had negative and positive shifts on action of genetic system of adaptability on acid and neutral soil backgrounds accordingly. Reciprocal hybrid Freija × Argamak was shown positive shift in action of genetic system of adaptability on both soil backgrounds; this shift on acid soil background was the greatest for all studied.

Hence, hybrids Freija × Argamak and Argamak × Freija had positive shifts on action of genetic system of adaptability at cultivation on acid soil background. Hybrid Petra × Freija took position in a positive zone of adaptability, however, value of the shift at this combination was lower than at other hybrids. As a whole variety Freija being a paternal component of crossing caused positive shifts in action of genetic system of adaptability of its hybrids to the environment.

Investigated genotypes showed little degree of polymorphism on genes of attraction (which determine distribution of genotypes along short axis of orthogonal regression on Fig. 8.2) at cultivation on acid soil background; only hybrid Argamak × Petra demonstrated positive shifts in action of this genetic system and variety Freija – negative one. All the other genotypes located on the long axis of regression.

Variety Argamak had the greatest adaptability among parental forms, whereas variety Petra – the least. Variety Freija was intermediate. Both hybrids between varieties Argamak and Petra under conditions of a neutral soil background showed negative heterosis on genetic systems of adaptability, under conditions of acid soil background – intermediate inheritance on this trait with deviation towards the worst parental form. The deviation on activity of genetic system of adaptability went towards Petra parental form.

Direct and reciprocal hybrids between varieties Petra and Freija revealed intermediate inheritance of system of adaptability and weak negative heterosis at cultivation on a neutral soil background, but complete dominance and intermediate inheritance with a small deviation towards the worst parental form under conditions of acid soil background. Both hybrids demonstrated deviation on action of genetic system of adaptability towards a paternal component of crossing. Hybrids between varieties Argamak and Freija showed strong negative heterosis on genes of adaptability on a neutral soil background; on acid soil background both hybrids were almost identical on action of genetic system of adaptability with their paternal components.

8.4 CONCLUSION

Thus, researches conducted in laboratory and field conditions give us the reasons to assume that aluminum resistance of oats F_2 hybrids is influenced significantly by maternal component of crossing; the part of influence of a paternal component remains insignificant statistically. Considering a plant as a whole, it is possible to notice that the maternal form has great influences on development of root system of hybrids, whereas paternal form – on parameters of action of genetic system of adaptability to environment conditions estimated by degree of development of aboveground parts of plants. Change of a direction of crossing is reflected on the characteristic of interaction of the genes governing physiology-biochemical parameters of aluminum resistance.

ACKNOWLEDGEMENT

The Russian Academy of Agricultural Sciences supports this work. We thank the staff on laboratory of oat breeding and department of plant edaphic resistance (North-East Agricultural Institute, Kirov, Russia) for their technical help with conducting laboratory and field experiments, and academician V.A. Dragavtsev (Agrophysical Institute, St. Petersburg, Russia) for consultations and discussion.

KEYWORDS

- **aluminum**
- **dominance**
- **genetic systems**
- **hybrids**
- **inheritance**
- **oats**

REFERENCES

1. Murphy, J. P., & Hoffman, L. A. (1992). Origin History and Production of Oat, in Marshall, H. G., & Sorrels, M. E. (Eds) Oat Science and Technology, Crop Science Society of American, Madison, 1–28.
2. Rines, H. W., Molnar, S. J., Tinker, N. A., & Phillips, R. L. (2006). Oat in Chittaranjan Kole (Ed.) Cereals and Millets Genome Mapping and Molecular Breeding in Plants, Springer Berlin, Heidelberg, New York, 211–242.
3. Nava, I. C., Duarte, I. T. L., Pacheco, M. T., & Federizzi, L. C. (2010). Genetic Control of Agronomic Traits in an Oat Population of Recombinant Lines, Crop Breeding and Applied Biotechnology, *10*, 305–311.
4. Floss, E. L. (2004). Breeding Oat for Aluminum Tolerance, Proc 7th International Oat Conference Agrifood research Reports, *51*, 196.
5. Bilski, J. J., & Foy, C. D. (1987). Differential Tolerance of Oat Cultivars to Aluminum in Nutrient Solutions and in Acid Soils of Poland, J. Plant Nutrition, *10(2),* 129–141.
6. Foy, C. D., Smith, D. M., Jr., & Briggle, L. W. (1987). Tolerances of Oat Cultivars to an Acid Soil High in Exchangeable Aluminum, J. Plant Nutrition, *10,* 1163–1174.
7. Sánchez-Chacón, C. D., Federizzi, L. C., Milach, S. C. K., & Pacheco, M. T. (2000). Variabilidade Genética e Heranĉa da Tolerância á Toxicidade do Alumínio Em Aveia, Pesquisa Agropecuária Brasileira, *35,* 1797–1808 (in Portuguese).
8. Wagner, C. W., Milach, S. C. K., & Federizzi, L. C. (2001). Genetic Inheritance of Aluminum Tolerance in Oat, Crop Breeding and Applied Biotechnology, *1*, 22–26.
9. Oliveira, P. H. (2002). Heranĉa Genética E Mapeamento Molecular da Tolerância Átoxicidade Do Alumínio Em Aveia Tese (Doutorado–Agronomia–Fitotecnia), Programa Depós-Graduaĉão Em Agronomia, UFRGS, Porto Alegre, 101f (in Portuguese).
10. Hoekenga, O. A., Vision, T. J., Shaff, J. E., Monforte, A. J., Gung Pyo Lee, Howell, S. H., & Kochian, L. V. (2003). Identification and Characterization of Aluminum Tolerance Loci in Arabidopsis (Landsberg Erecta x Columbia) by Quantitative Trait Locus Mapping, A Physiologically Simple But Genetically Complex Trait, Plant Physiology, *132*, 936–948.
11. Kosareva, I. A., Davidova, G. V., & Semenova, E. V. (1998). Diagnostics of resistance of Oats Plants to the High Content of Aluminum Ions in a Soil Solution, Agricultural Biology, *5*, 73–76 (in Russian).
12. Lisitsyn, E. M. (2005). Potential Aluminum Resistance of Agricultural Plants and its Realization under Conditions of the European North-East of Russia, Thesis Dr Science Biology, Moscow, Timirjazev Agricultural Academy, 361p (in Russian).
13. Nava, I. C., Delatorre, C. A., De Lima Duarte, I. T., Pacheco, M. T., & Federezzi, L. C. (2006). Inheritance of Aluminum Tolerance and its Effects on Grain Yield and Grain Quality in Oats (Avena Sativa L.), Euphytica, *148*, 353–358.
14. Federizzi, L. C., Oliveira, P. H., & Nava, I. C. (2004). Genetics of Aluminum Tolerance in Hexaploid Oats, Process 7th International Oat conference Agrifood Research Reports, *51*, 195.
15. Oliveira, P. H., Federizzi, L. C., Milach, S. C. K., Gotuzzo, C., & Sawasato, J. T. (2005). Inheritance in Oat (Avena Sativa, L.) of Tolerance to Soil Aluminum Toxicity, Crop Breeding and Applied Biotechnology, *5*, 302–309.
16. Castilhos, G. (2010). Estresse Oxidative Em Resposta Ao Aluminio Em Avena Branca, Disscuss Mestrado Em Fitotecnia, Porto Alegre, R. S., Brasil, 81p (in Portuguese).
17. Castilhos, G., Farias, J. G., Schneider, A. B., De Oliveira, P. H., Nicoloso, F. T., Schetinger, M. R. C., & Delatorre, C. A. (2011). Aluminum-Stress Response in Oat Genotypes with Monogenic Tolerance Environment Exp. Botany, *74*, 114–121.
18. Dubinin, N. P. (1987). Genetic Kishinev Shtiintsa, 536p (in Russian).

19. Lisitsyn, E. M. (2000). Intra Varietal Level of Aluminum Resistance in Cereal Crops, J Plant Nutrition, *23(6)*, 793–804.
20. Furukawa, J., Yamaji, N., Wang, H., Mitani, N., Murata, Y., Sato, K., Katsuhara, M., Takeda, K., & Ma, J. F. (2007). An Aluminum-Activated Citrate Transporter in Barley, Plant Cell Physiology, *48*, 1081–1091.
21. Wu, P., Liao, C. Y., Hu, B., Yi, K. K., Jin, W. Z. et al. (2000). QTLs and Epistat is for Aluminum Tolerance in Rice (Oryza Sativa L.) at Different Seedling Stages, Theory Application Genetics, *100*, 1295–1303.
22. Peterburgsky, A. V. (1975). Practical Studies on Agronomical Chemistry, Moscow Kolos Publishing, 496p (in Russian).
23. Klimashevsky, E. L., & Chumakovsky, N. N. (1986). Identifications of Forms of Plants, Sympathetic on Ferrglizers, Herald of Agricultural Science, *3*, 71–77 (in Russian).
24. Methods of Biochemical Research of Plants (1972). Leningrad Kolos Publishing 455p (in Russian).
25. Lisitsyn, E. M., & Lisitsyna, I. I. (2005). Use of Index of Root Cation-Exchange Capacities in Oats Selection on Aluminum Resistance, Herald of Russian Academy of Agrarian Sciences, *3*, 46–48 (in Russian).
26. Lisitsyn, E. M., & Batalova, G. A. (2007). Aluminum Resistance of Direct and Reciprocal F_2 Hybrids of Oats, Herald of Russian Academy of Agrarian Sciences, *1*, 47–49 (in Russian).
27. Merezhko, A. F. (2005). Use of Mendelian Principles in the Computer Analysis of Inheritance of Varying Traits, Ecological Genetics of Cultural Plants, Krasnodar All Russian Rice Research Institute, 107–117 (in Russian).
28. Dragavtsev, V. A. (2002). Algorithms of an Ecological-Genetic Survey of the Gene Pool and Methods of Creating the Varieties of Crop Plants for Yield, Resistance and Quality (Methodical Recommendations, New Approaches) St. Petersburg, VIR 41p (in Russian).
29. Lisitsyn, E. M., & Kropotov, A. V. (2002). Dynamics of Enzymatic Activity of Oat Seedling Roots under the Effect of Acid Stress, Russian Agricultural Sciences, and Allert on Press, Inc., *3*, 6–8.
30. Lisitsyn, E. M., & Batalova, G. A. (2005). About Dependence of Adaptation of Oats F_1 Hybrids to the High content of Aluminum Ions in Growth Media on Crossing Components, Agricultural Biology, *3*, 60–64 (in Russian).

CHAPTER 9

EXPERIENCE OF STUDYING OF GENETIC CONTROL OF BARLEY ALUMINUM RESISTANCE

IRINA N. SHCHENNIKOVA[1] and EUGENE M. LISITSYN[1,2]

[1]Behalf of N.V. Rudnicki Zonal Research Institute of Agriculture of North-East of the Russian Agricultural Academy, d. 166-a, Lenin St., Kirov, 610007, Russia; E-mail: niish-sv@mail.ru, edaphic@mail.ru

[2]Vyatka State Agricultural Academy, d. 133, Oktyabrsky Prospect, Kirov, 610010, Russia; E-mail: info@vgsha.info

CONTENTS

ABSTRACT

Although progress is being made in understanding the mechanisms of Al tolerance, there are conflicting reports of its genetics basis. Till now there is not a common opinion about number of the genes governing barley (*Hordeum vulgare* L.) aluminum resistance. These inconsistencies may be due in part to the different techniques, which have been used to investigate the genetic control of the trait. For example, validity of approaches to segregation of plants of hybrid populations on resistant and sensitive and accordingly to choice of hypothesis about number of genes coding the trait causes a question. On a basis of studies of direct and reciprocal barley F_2 hybrids it is shown as the estimation of number of coding genes changes at acceptance of different levels of resistance (index of relative root growth – RRG) as a segregating line. Only in five cases of the considered 12 hybrid combinations it is possible theoretically to explain received division of whole sample on Al-resistant and Al-sensitive plants by action of one gene with a complete dominance (three – at segregating line equal to RRG = 50%, and two – at RRG = 75%); action of two genes is possible in 25 cases; and in all cases without an exception control of the trait is possible by three genes. It points out genetic complexity of barley Al-resistance and its quantitative character. Having spent comparison of parental forms by number of different Al-resistance genes, we have found out that in the studied samples of barley the structure of the genetic formula of a trait "aluminum resistance" includes various genes at different varieties. All studied plants of barley continue growth of roots during all 5 days of stressful influence, so it is possible to assume that each of them possessed at least one of such genes. Then in total for the given set of samples of barley the number of the various genes participating in the control of Al-resistance was estimated as not less than five. Besides, significant influence of direction of crossing on level of Al-resistance is shown, i.e. the strong parental effect specifying participation of cytoplasmic genes has take place.

9.1 INTRODUCTION

Among the small grains, barley (*Hordeum vulgare*) is the most Al-sensitive species, while rye (*Secale cereale*), wheat (*Triticum aestivum*), maize (*Zea mays*), rice (*Oryza sativa*) and oats (*Avena sativa*) appears to possess the higher degree of Al tolerance [1, 2]. Despite a large amount of research the genetics of both aluminum toxicity and tolerance in cereals are far from being completely understood. Thus, Rigin, Yakovleva [3] proposes that Al resistance in barley is controlled by two or more genetic loci. Other studies have proposed that a single major gene can control Al resistance in a range of barley genotypes [4, 5]: Stølen and Andersen [6] found a single gene (*Pht*) explaining most of the differences for low soil pH tolerance in spring barleys, whereas Reid [7] reported a single gene (*Alp*) controlling aluminum tolerance in winter cultivars. Theoretically, *Pht* and *Alp* could be the same gene, located on chromosome 4 [4, 8]. Main gene (*Alp*) located on the long arm of chromosome 4H [9, 10] is essential for normal Al tolerance expression in barley and other cereals as well [11–13]; Minella and Sorrells [10] suggested that different alleles of the gene conferring different degrees of tolerance to aluminum. Gene for *Alp* (4H locus) was identified as belonging to the multidrug and toxic compound extrusion (MATE) family [14].

These inconsistencies may be due in part to the different methods, which have been used to investigate the genetic control of the resistance: Tang et al. [4] crossed Dayton, highly Al-tolerant six-rowed winter barley that harbors the Al tolerance gene *Alp*, and Harlan Hybrid, a moderately Al-sensitive six-rowed spring barley, and the hematoxylin method was used to evaluate Al tolerance; Raman et al. [15] used middle-resistant (Yambla) and resistant (WB229) varieties and estimated resistance level on regrowth of roots in hydroponic conditions; Ma et al. [16] used resistant and sensitive varieties (Murasakimochi and Morex), estimating level of hybrids resistance by the ratio of roots to shoot mass.

Effects of minor gene located on chromosomes 3H, 4H, 5H, and 6P also have been postulated to explain discrepancies between observed and expected ratios for a single gene hypothesis [7, 12, 17, 18]. Oram [19] noted that some progeny from the cross of two Al-sensitive barley varieties (C17115 and Weeah) were relatively tolerant, and took this as evidence for the transgressive segregation of more than one Al tolerance gene. QTL analysis of 100 plants of second generation from crossing of resistant variety Murasakimochi and sensitive variety Morex has revealed a simple locus of Al-resistance on a chromosome 4H which explained about 50% of phenotypic variation of an investigated trait in the given population [16]. Locus *Alp* has been marked on a chromosome 4H on the high-resolution map created with participation of varieties Dayton and Zhepi 2 [14]. Minor gene effects for Al tolerance have also been reported [7, 12, 18], and have been mapped to chromosomes 2H, 3H, 4H, 5H and 6H [12]. According to Navakode et al. [20] chromosomes 2H and 4H harbor QTL loci responsible for resistance against low Al concentration (10 μM) whereas chromosome 3H – against high concentration (20 μM). On the other hand, QTL analysis has the limitation that only the alleles segregated between the parents of DH or RIL population could be identified [21]. Hence the genetic diversity of tolerant genotypes used in QTL analysis was limited.

The main mechanism of resistance offered earlier for other cereals has now been attributed to the Al-dependent release of malate anions from roots (citrate is released from maize [22], sorghum [23] and barley [24], and malate is released from wheat [25]). A major locus controlling aluminum tolerance in rye is located in long arm of the chromosome 4, the same chromosome location found in wheat and barley [26–29]. But it is necessary to note that the mechanisms leading to increase of secretion of organic acid anions differ considerably at wheat, rye and barley. So, at barley the secretion was characterized by a rapid, temperature-dependent pattern, but does not depend on a dose of aluminum. At wheat activation occurs quickly also and depends not on temperature, but on aluminum concentration. At a rye, unlike these two crops, activation of anion channels is delayed for several hours after exposure to Al, and depends both on temperature, and on concentration of the stressful agent. The amount of excreted anions of organic acids decreases follows the order: rye > wheat > barley. These results suggest that the pathways leading to the secretion of organic acid anions differ between barley and other plant species. Perhaps the genetic factors governing Al-resistance at different cereals have similar chromosomal location but the architecture and expression of these genes can differ considerably.

Furukawa et al. [30] and Wang et al. [14] identified candidate gene *HvMATE* (*HvAACT = Hordeum vulgare A*luminum *A*ctivated *C*itrate *T*ransporter 1) in cv. Murasakimochi and Dayton, respectively, which is responsible for Al-induced citrate secretion and located on the long arm of chromosome 4H. Gene *HvAACT1* appears to affect constitutive in plant roots but presence of Al ions is necessary for activation of citrate efflux. However, Gruber et al. [31, 32] reported that overexpression of *HvALMT1*, which encodes an anion channel protein to facilitate organic anion transport, confers enhanced Al tolerance.

At the same time researches of the genetic control of Al-resistance in wheat [33, 34] and rice [35] have led to conclusion of multiple mechanisms for Al tolerance in plants that, in turn, considerably complicates breeding work. Such differences in estimation can be explained partially by distinctions of the techniques used for studying of the genetic control of a trait [36]. But we offer one more possible reason namely that researchers use improper level of the hierarchical organization – specific or varietal. There are some publications in which authors have shown high intravarietal variability of some parameters used for analysis of plant growth under aluminum treatment [37, 38]. Certainly, varieties of cross-pollinated plant species should consist of a mix of various genotypes. But for self-pollinated species there is much less information about intravarietal variability on physiological and morphological traits in the scientific literature.

Long-term data of our field and vegetative experiments indicates that the basic differences between resistant and sensitive barley plants concern generative organs, but very seldom – vegetative one [39]. Analysis of these data has shown that coefficient of correlation between such indices as level of agronomical resistance (estimated on depression of plant productivity) at cultivation on acid soil and degree of decrease in "number of grain in the ear" or "1000 grains mass" for 50 barley varieties did not exceed 0.46 in all years of researches [40, 41]. In other words it is rather difficult to distinguish Al-resistant and Al-sensitive plants by these last two indices visually.

Therefore, field selections of plants "the best on a phenotype" will lead to that in the selected population there will be both dominant and recessive allele of Al-resistance gene. As it is known, to receive the maximum number of hybrid seeds during breeding work, crossings were spending under nonstressful conditions where differences of plants of parental varieties on stress-resistance are not demonstrated.

Pollen can be collected from the several plants differ on Al-resistance alleles; more over maternal plants pollinated with this mix of pollen can also differ on presence of this or that allele. Then at crossing, under Mendel's laws, mix of genetically different hybrids will be obtained in the first generation already – depending on what plants were taken for crossing following genotypes – aa, Aa, AA – will appearance in a casual proportion.

At self-pollination of received plants the number of heterozygotes will decrease with each following generation, and, finally balance of two homozygotes aa and AA will be established in the population. In case of the polygenic control of Al-resistance dominant and recessive alleles of different genes will be combined in various ratio at different plants of the same variety.

So at individual plants of population the different genes governing a quantitative trait "aluminum resistance" will be presented both in a different ratio, and by different allele variants (dominant and recessive) of individual genes. Therefore in offered article we try to analyze the data about segregation of barley hybrid populations on level of Al-resistance and on this basis to calculate the minimum number of the genes governing the trait.

9.2 MATERIALS AND METHODOLOGY

Direct and reciprocal F_2 hybrids of four breeding lines of barley (565–98, 889–93, 999–93, and 1030–93) from the North-East Agricultural Research Institute breeding program were evaluated for their reaction to aluminum toxic levels in nutrient solutions. The Al-resistance level of all genotypes was assayed by the procedure developed earlier [42]. The seeds of each genotype were planted in filter paper towels (16 cm × 100 cm^2, 1 cm from the top). Each towel contained 95–105 seeds. Then towels were placed in glass beakers with distilled water (control treatment) and 0.5 or 1.0 mM solution of $Al_2(SO_4)_3$, pH 4.3 (Al treatment). Beakers were placed into thermostat under 21–23 °C for 5–7 days. The relative root growth index (RRG), calculated as the maximum root length in Al stress culture divided by maximum root length in control [30, 43] was used as an indicator to evaluate Al tolerance. Higher value of RRG corresponds to higher Al-resistance level. Each estimation ware evaluated with three replicates. The number of genes governing aluminum tolerance was determined with packages of computer software AGROS 2.07 [44] and GENAN [45]. The deviations between the observed and expected frequencies for the trait were compared by the Chi-square (χ^2) test.

9.3 RESULTS AND DISCUSSION

By results of preliminary laboratory analyzes parental forms differed significantly on level of Al-resistance that corresponded to the aims of researches. Pattern of segregation of individual plants in F_2 hybrid population is used by many authors for finding-out of number of the genes governing that index of aluminum resistance on which population is estimated. The main difficulty which researchers face at the analysis of segregations is what value of an index should be accepted for segregating line between resistant and sensitive plants. Mather and Jinks [46] wrote that theoretically at division of split population into three or more number of classes (to six classes inclusive) it is impossible to estimate number of genes more than two (i.e., in such cases the estimation will always give number of genes equal to one or two). However, this position is fair only under following conditions: genes have equal additive effects, without dominance and in the absence of phenotypic variability. In case of division into two groups it is possible to estimate number of the genes coding a trait as 1 up to 4 if one accepts the following conditions: there is an interaction between genes and action of genes is characterized by different degree of dominance.

Table 9.1 presents frequencies of segregation of genotypes of barley F_2 hybrid populations at acceptance of three different values of RRG (50, 75 and 100%) as segregating line.

TABLE 9.1 Segregation of Plants of Barley F_2 Hybrid Population on Al-Resistant and Al-Sensitive at Acceptance of Different Segregating Lines (Values of RRG)

Hybrid population	RRG = 50%	RRG = 75%	RRG = 100%
565–98 × 889–93	86:25	1:110	0:111
565–98 × 999–93	25:86	2:109	0:111
565–98 x1030–93	45:58	1:102	0:103
889–93 × 565–98	99:1	49:51	9:91
889–93 × 999–93	80:13	15:78	2:91
889–93 × 1030–93	93:4	29:68	3:94
999–93 × 565–98	70:33	13:90	4:99
999–93 × 889–93	99:16	14:103	0:117
999–93 × 1030–93	48:41	12:78	1:89
1030–93 × 565–98	111:3	7:107	0:114
1030–93 × 889–93	91:3	19:75	0:94
1030–93 × 999–93	48:61	7:102	0:109

As data of Table 9.1 testify at segregation of plants by segregating line RRG = 50% direct and reciprocal hybrids of genotype 565–98 differed considerably on a ratio of Al-resistant and Al-sensitive seedlings. In a case of crossing of a genotype 565–98 with genotypes 889–93 and 999–93 similar tendency remained at all variants of segregating line (RRG = 50, 75 and 100%). Direct and reciprocal hybrids between genotypes 999–93 and 1030–93 differed among themselves at all variants of division also. As a whole differences between direct and reciprocal hybrids are noted in 72% of cases.

Calculations with AGROS software have shown that the number of genes and possible intergene interactions at different RRG levels (as segregating line) differ a little for different hybrid combinations. Theoretically possible numbers of genes significantly explaining the obtained data are resulted in Table 9.2.

TABLE 9.2 Theoretical Number of the Genes Significantly Explaining Actually Obtained Data of Segregation of Barley F_2 Hybrid Populations

Hybrid population	RRG = 50%	RRG = 75%	RRG = 100%
565–98 × 889–93	1, 2, 3 (6)*	3 (2)	3 (1)
565–98 × 999–93	1, 2, 3 (6)	2, 3 (4)	3 (1)
565–98 × 1030–93	2, 3 (5)	3 (2)	3 (1)
889–93 × 565–98	3 (2)	2, 3 (3)	2, 3 (5)
889–93 × 999–93	2, 3 (4)	2, 3 (5)	2, 3 (4)
889–93 × 1030–93	2, 3 (3)	1, 2, 3 (8)	2, 3 (4)
999–93 × 565–98	1, 2, 3 (7)	2, 3 (4)	2, 3 (4)
999–93 × 889–93	2, 3 (4)	2, 3 (4)	3 (1)
999–93 × 1030–93	2, 3 (5)	2, 3 (4)	3 (2)
1030–93 × 565–98	2, 3 (4)	2, 3 (4)	3 (1)
1030–93 × 889–93	2, 3 (4)	1, 2, 3 (7)	3 (1)
1030–93 × 999–93	2, 3 (5)	2, 3 (4)	3 (1)

* – in brackets the number of the hypotheses significantly describing different types of intergene interactions is resulted.

As data of Table 9.2 show it is possible to explain received segregation of plants of hybrid combinations on Al-resistant and Al-sensitive by action of one gene with a complete dominance in five cases only (three – at RRG = 50% and two – at RRG = 75%); action of two genes is possible in 25 cases; and in all cases without an exception control of the trait is possible by three genes.

From our point of view that fact is especially interesting what even at acceptance as a segregating line of such low value of RRG as 50% the amount of sensitive plants exceeded amount of resistant one at three hybrids: 565–98 × 999–93; 565–98 × 1030–93, and 1030–93 × 999–93. The only hybrid at which there was not excess of amount of sensitive plants over resistant at a segregating line RRG = 75% was 889–93 × 565–98. In seven cases it was not observed resistant plants at all at segregating line RRG = 100%.

Let's stop especially on a problem of definition of segregating line between resistant and sensitive plants. It is possible to offer following two approaches to its choice.

First, it is known that aluminum causes fast (within several hours) inhibition of root growth due to death of the meristematic root tissue [47]. In our experiments duration of aluminum treatment was five days. Thus it is possible to assume that all plants having length of roots in aluminum media more than length of the grain (by analogy to estimation of seed germination) will possess certain level of resistance to the given stressful factor. Hence, the difference in length of roots of single plants indicates not presence or absence of resistance, but its different level. Distinctions of level of resistance can be explained by effect of a dose of genes: the higher the resistance level, the more number of the genes governing the investigated trait represent as a dominant allele. Therefore to estimate a ratio of resistant and sensitive plants is possible having counted number of seeds germinated in the presence of aluminum and compared it with control germination ability. As consequence of such point of view is acceptance of polygene mode of governing of barley Al-resistance.

Secondly, it is possible to spend division of plants on RRG level of 75–100%, i.e. the plants, which have not shown depression of root growth under influence of aluminum will be considered resistant and plants with RRG level below 75% – sensitive. Thus, of course, there will be a question – why some (or many) individual plants have RRG above 100%, i.e. show stimulating effect of aluminum? As the assumption it is possible to put forward the same thought, as in the first case: the control of level of resistance by several genes. At "super-resistant" plants these genes will be in a dominant position and to possess ability to strengthen action of each other mutually. Other possible explanation is that products of the genes providing Al-resistance of such plants have some depressive effect on root growth in the absence of aluminum.

Thus, obtained results specify in genetic complexity of barley aluminum resistance under laboratory conditions and its quantitative character. Within investigated set of barley F_2 hybrids received segregation on Al-resistant and Al-sensitive plants is explained theoretically by action of one gene with a complete dominance in 14% of cases only (three – at RRG = 50% and two at RRG = 75%); in 69% cases action of two genes is possible, and in all cases without an exception the trait is governed by three genes.

The traditional approach in studying of the genetic control of Al-resistance consists in application of crossing in which course ratio of resistant and sensitive plants of hybrid population is counted [48], but thus variability of parental forms on the investigated trait do not take into consideration. Therefore for our calculations [49] we used the computer software GENAN [45]. At an estimation of number of the genes governing the investigated trait, the given program has allowed to consider intravarietal variability of parental components of crossing and its influence on pattern of segregation of hybrid plants on resistance groups.

Distribution of plants of the investigated parental forms and F_2 hybrids of barley on level of resistance to 1.0 mM aluminum are presented in Table 9.3.

TABLE 9.3 Distribution of Individual Plants of Barley on Level of Resistance to 1 mM Aluminum (RRG, %)

Variety, F_2 hybrid	Groups of seedlings with following resistant level (RRG, %)									
	20	30	40	50	60	70	80	90	100	110
565–98				14	45	25	7	4	3	1
889–93			4	6	43	32	10	1	2	1
999–93	1	2	12	29	31	15	7	1	2	
1030–93	1	1	43	33	14	5	3			
565–98 × 889–93	1	2	5	17	46	39	1			
565–98 × 999–93	1	11	25	49	19	4	1	1		
889–93 × 999–93	2			11	37	28	13			2
889–93 × 1030–93	1	1		1	38	26	24	2	3	
1030–93 × 999–93	3	1		57	33	8	3	4		

For the genotypes used in given study the analysis of number of the genes governing Al-resistance by means of software GENAN has allowed to elicit following interesting facts.

The number of genes on which barley parental genotypes differed, changed from 2 to 4 depending on comparison pair: genotypes 565–98 and 999–93 differed by two genes; genotypes 1030–93 and 999–93 – by four genes; distinctions in pairs 1030–93 and 889–93, 565–98 and 889–93, 889–93 and 999–93 were explained by action of three genes.

Paired comparison of parental forms specifies that number of genes only which distinguishes the two genotypes. If we organize the data in the form of a triad complex, then possibilities of mathematical interpretation of the received information will increase considerably as it will be necessary to analyze three pairs of intervarietal differences simultaneously. Carrying out of mathematical requirements on number of genes differing in pairs, it is probably to present hypothetically formulas of the investigated trait for each studied genotype included in a triad complex, having designated these different genes with different Latin letters. Then the structure of the genetic formula of a trait "Al-resistance" for barley genotypes used in our studies will look as follows (Fig. 9.1).

FIGURE 9.1 The scheme of genetic formulas of a trait "Al-resistance" in the investigated barley genotypes under action of 1.0 mM aluminum. Numbers 2–4 designate number of the various genes which are taking part in the control of a trait, * – the same genes

As Fig. 9.1 shows the architecture of the genetic formula of the same trait for the studied barley genotypes estimated on RRG will include various genes. Taking into account that all studied plants of barley continued growth of roots during all time of experiments; it is possible to conclude that each of them possessed at least one of such genes. Then in total for the given set of barley genotypes the number of the various genes participating in the control of a trait "Al-resistance" should be not less than five.

According to many authors [50, 51] the basic biochemical mechanism of plant resistance to action of the high Al concentrations is chelating of ions of trivalent aluminum by anions of organic acids secreted by roots into the rhizosphere. However, there is a lot of the actual data contradicting opinion on the predominating role of organic acids in resistance to aluminum [52, 53].

In our opinion, the dual nature of the genetic control of the mentioned mechanism has been often overlooked. If the composition of a cell membrane (and accordingly presence or absence of cation and anion channels) is coded by nuclear genes then synthesis of organic acids occurs basically in mitochondria of root cells. Therefore it is quite logical to assume that final level of Al-resistance should be under the control with not nuclear only, but cytoplasmic genes also.

The simplest proof of participation of cytoplasmac genes in Al-resistance of plants is analysis of the direct and reciprocal hybrids from crossing of barley genotypes having contrast as well as close level of resistance. Therefore we [54, 55] have studied a role of a direction of crossing in transfer of the trait on an example of barley F_2 hybrids.

The pattern of dominance for index of growth of hybrids F_2 plants has been calculated under the formula [56]:

$$d = \frac{\bar{x}F_2 - MP}{\bar{x}HP - MP}$$

where d – degree of dominance; $\bar{x}\,F_2$ – average values for F_2; MP – average value for both parents; HP – average value for more resistant parent.

The analysis of root growth of barley F_2 hybrids under control (without aluminum) and stressful (0.5 and 1.0 mM of aluminum) conditions has shown (Table 9.4) that variability of root length at F_2 hybrids in the control is defined on 46.0% by influence of a maternal genotype, on 5.5% – by paternal genotype, and on 43.7% – by interaction of parents [57]. At correct selection of pairs for crossing the total coefficient of heritability (h^2) will make 95.2%.

TABLE 9.4 The Analysis of Seedling Growth of Parental Genotypes and F_2 Hybrids of Barley

Genotype, F_2 hybrid	Root length, mm			RRG, %	
	Control	0.5 mM Al	1 mM Al	0.5 mM Al	1 mM Al
565–98	112.4±2.2	92.6±1.5	70.1±1.2	82.4±0.8	62.4±0.6
889–93	102.7±2.2	68.6±1.4	60.6±1.2	66.8±0.8	58.9±0.7
999–93	99.7±1.7	75.4±1.0	52.6±1.2	75.6±0.6	52.8±0.7
1030–93	115.6±1.2	67.2±1.4	52.6±1.1	58.1±0.7	45.5±0.5
565–98 × 889–93	109.2±1.5	71.2±1.2	56.3±1.0	65.2±0.6	51.6±0.5
889–93 × 565–93	103.0±1.1	84.7±1.3	71.3±1.2	82.3±0.7	69.2±0.7
565–98 × 999–93	113.6±1.6	71.8±1.8	48.3±1.2	63.2±0.9	42.6±0.6
999–93 × 565–98	103.3±2.2	65.8±0.9	55.3±1.2	63.7±0.5	53.5±0.7
565–98 × 1030–93	112.7±1.1	74.2±1.1	57.0±1.3	65.8±0.6	50.6±0.7
1030–93 × 565–98	110.8±1.4	79.3±1.3	65.2±1.0	71.6±0.7	58.8±0.5
889–93 × 999–93	107.7±2.2	76.0±1.7	61.0±1.5	70.5±0.7	56.6±0.8
999–93 × 889–93	106.9±1.1	70.7±1.5	58.2±0.8	66.1±0.8	54.5±0.4
889–93 × 1030–93	107.6±1.4	75.0±2.0	64.5±1.5	69.7±1.1	59.9±0.8
1030–93 × 889–93	102.1±1.6	79.0±1.0	62.9±0.8	77.4±0.6	61.6±0.5
999–93 × 1030–93	104.9±1.6	75.6±1.6	52.2±1.7	72.0±0.9	49.8±0.9
1030–93 × 999–93	111.2±1.2	77.4±1.0	59.6±1.2	69.5±0.5	53.6±0.6

Under 0.5 mM aluminum conditions influence of the maternal genotype on parameter "root length" has considerably decreased and has made 11.3%; at 1.0 mM of aluminum it has increased twice – to 24.3%. The paternal genotype influenced at a high concentration of the stressful factor more strongly (3.2% for 0.5 mM and 31.8% for 1 mM Al). The share of interaction of parental forms remained high – 84.6 and 43.8% for 0.5 mM and 1.0 mM Al accordingly. Thus, at correct selection of pairs for crossing the total coefficient of heritability of root length (h^2) will make 99.1% (control), 99.9% (0.5 mM Al), and 81.5% (1.0 mM Al).

Variability of RRG index at action of 0.5 mM Al depended on interaction of parental forms only (81.5%), but at increasing of stressor impact influence of maternal and paternal genotypes has amplified considerably (14.2% and 20.5%, accordingly). In both used stressful conditions (0.5 and 1.0 mM Al) coefficient of heritability of RRG was high enough that allows to conduct the directed selection on it in the given populations of F_2 hybrids at laboratory conditions.

Data of Table 9.4 also demonstrates that in most cases the crossing direction has made significant impact on seedling root growth and their degree of aluminum resistance. Thus, genotype 565–98, acting as a maternal component of crossing, leads to considerable decrease in Al-resistance level of a hybrid. Genotype 1030–93 – on the contrary, being a maternal component, raises this level. Genotypes 889–93 and 999–93 do not show accurately expressed maternal effect in crossing with other genotypes.

The data presented to Table 9.5 testifies that depending on exact combination of crossing dominance of root length in the control and aluminum treatments, as well as RRG index can have positive or negative value, changing from negative superdominance to positive superdominance. Pattern and direction of dominance may coincide for root length both in the control and in aluminum conditions, and may have an opposite direction.

TABLE 9.5 Influence of a Parent Component of Crossing on Character of Dominance of Growth Parameters of Barley F_2 Hybrids

Genotype, F_2 hybrid	Degree of dominance of the parameter				
	Root length, mm			RRG, %	
	Control	0.5 mM Al	1.0 mM Al	0.5 mM Al	1.0 mM Al
565–98 × 889–93	0.33	–0.78	–1.94	–1.21	–5.35
889–93 × 565–93	–0.94	0.34	1.26	0.99	5.00
565–98 × 999–93	1.19	–1.42	–1.51	–4.65	–3.13
999–93 × 565–98	–0.44	–2.12	–0.70	–4.50	–0.85
565–98 × 1030–93	–0.81	–0.45	–0.51	–0.37	–0.40
1030–93 × 565–98	–2.00	–0.05	0.44	0.11	0.13
889–93 × 999–93	4.33	1.18	1.10	–0.16	0.23
999–93 × 889–93	3.80	–0.38	0.40	–1.16	–0.47
889–93 × 1030–93	–0.25	10.14	1.98	1.67	1.15
1030–93 × 889–93	–1.11	15.96	1.58	3.47	1.40
999–93 × 1030–93	–0.35	1.05	–4.00	0.59	0.17
1030–93 × 999–93	0.44	1.49	7.00	0.30	1.22

Direction of crossing caused opposite character of dominance of investigated parameters of resistance for hybrids 565–98 × 889–93 and 889–93 × 565–98. At the hybrids received from crossing of genotypes 565–98 and 1030–93 this tendency is hardly less expressed. At the same time direct and reciprocal hybrids of genotypes 565–98 and 999–93 have shown only different degree of dominance but not its different direction. Direct and reciprocal hybrids of genotypes 889–93 and 1030–93 have lowest differences in direction and character of dominance.

The similar phenomena have been noted by a number of authors who investigated aluminum resistance in other cereals. Thus, Camargo [58] specified that aluminum resistance in wheat F_2 populations was governed by dominant genes at 3 mg/L aluminum, but it became recessive at increase of concentration of the stressful factor up to 10 mg/L. Similar results were obtained in the work of Bona et al. [59]. Frost resistance of wheat also changed a dominance direction at change of temperature of freezing in experiments of Sutka and Viesz [60].

At the same time it is noticed that the expression of genes of Al-resistance depends appreciably on concentration of aluminum in growth media; degree of resistance of all hybrids without an exception decreased with increase in Al concentration [55]. It can testify that Al-resistance in the given set of barley hybrids is possible to be explained by monogenic type of inheritance not always. Moreover, coefficient of heritability of RRG index of some hybrids were higher at 0.5 mM aluminum, whereas of others – at 1.0 mM Al. Thus, it is necessary to select the resistant plants from the given populations at different level of Al treatment. As it is known, high coefficients of heritability at a wide variation of the index can be used in breeding work at selection of the most phenotypic outstanding plants while at low level of coefficient of heritability selection will not lead to desirable results. On the average coefficients of heritability were higher for absolute values of root length, than for RRG index, and for 0.5 mM Al they were higher than for 1.0 m Al.

9.4 CONCLUSION

Thus, the data received in our researches, specifies that in many cases influence of a maternal component of crossing on potential resistance of hybrid progeny to action of high aluminum concentration is statistically significant. Probably, it is depended on what exact physiological-biochemical mechanisms of Al-resistance prevail in the integrated reaction of plant to the given type of edaphic stress. This, in turn, specifies in necessity of an estimation of a complex of parameters of Al-resistance of a genotype but not just degrees of depression of root growth (RRG) for decision making on use of the given genotype in breeding practice.

After numerous researches there was paradoxical enough situation: aluminum resistance is usually inherited as a simple trait in the developed models of crossing; but if it is estimated in a wide set of genotypes the wide and continuous variation from extremely sensitive to the most resistant genotypes is observed. That specifies in action of complex polygene system with numerous intergene effects. Therefore it is difficult enough to combine this data with genetic model of the control of a trait by one or several loci.

ACKNOWLEDGEMENT

This work was supported by the Russian Academy of Agricultural Sciences (Project 04.05.02.). We thank the staff at the laboratory of Barley Breeding and Department of Plant Edaphic Resistance (North-East Agricultural Research Institute, Kirov, Russia) for their technical help with conducting laboratory and field experiments.

KEYWORDS

- **aluminum**
- **barley**
- **dominance**
- **genetic systems**
- **hybrids**
- **inheritance**

REFERENCES

1. Bona, L., Wright, R. J., Baligar, V. C., & Matuz, J. (1993). Screening Wheat and Other Small Grains for Acid Soil Tolerance, Landscape and Urban Planning, *27(2–4),* 175–178.
2. Ishikawa, S., Wagamatsu, T., Sasaki, R., & Manu, P. O. (2000). Comparison of the Amounts of Citric and Malic Acids in Al Media of Seven Plant Species and Two Cultivars Each in Five Plant Species, Soil Science Plant Nutrition, *46*, 751–758.
3. Rigin, B. V., & Jakovleva, O. V. (2001). Genetics Aspects of Barley Tolerance to Toxic Aluminum Ions, Genetic Resources of Cultural Plants, Saint Petersburg All Russian Institute of Plant Growing, *397* (in Russian).
4. Tang, Y., Sorrells, M. E., Kochian, L. V., & Garvin, D. F. (2000). Identification of RFLP Markers Linked to the Barley Aluminum Tolerance Gene *Alp*, Crop Science, *40*, 778–782.
5. Raman, H., Karakousis, A., Moroni, J. S., Raman, R., Read, B., Garvin, D. F., Kochian, L. V., & Sorrells, M. E. (2003) Development and Allele Diversity of Microsatellite Markers Linked to the Aluminum Tolerance Gene *Alp* in Barley, Aust. J. Agriculture Res., *54(12),* 1315–1321.
6. Stølen, O., & Andersen, S. (1978). Inheritance of Tolerance to Low Soil pH in Barley, Hereditas, *88,* 101–105.
7. Reid, D. A. (1971). Genetic Control of Reaction to Aluminum in Winter Barley, Barley Genetics II proc. 2nd Int. Barley Genetics Symp, Pullman, Wash., 409–413.
8. Minella, E., & Sorrells, M. E. (2002). Genetic Analysis of Aluminum Tolerance in Brazilian Barleys, Pesq Agropec, Brasília, Brasília, *37(8),* 1099–1103 (in Portuguese)
9. Minella, E., & Sorrells, M. E. (1992). Aluminum Tolerance in Barley Genetic Relationships among Genotypes of Diverse Origin, Crop Science, *32*, 593–598.

10. Minella, E., & Sorrells, M. E. (1997). Inheritance and Chromosome Location of *Alp* a Gene Controlling Aluminum Tolerance in 'Dayton' Barley, Plant Breed, *116*, 465–469.
11. Ma, J. F., Shen, R., Nagao, S., & Tanimoto, E. (2004). Aluminum Targets Elongating Cells by Reducing Cell Wall Extensibility in Wheat Roots, Plant Cell Physiology, *45(5),* 583–589.
12. Raman, H., Wang, J. P., Read, B., Zhou, M. X., Vengatanagappa, S., Moroni, J. S., O'Bree, B., & Mendham, N. (2005). Molecular Mapping of Resistance to Aluminum Toxicity in Barley, in proceedings of Plant and Animal Genome XIII Conference, January 15–19, San Diego, 154.
13. Wang, J. P., Raman, H., Read, B., Zhou, M. X., Mendham, N., & Vengatanagappa, S. (2006). Validation of an *Alt* Locus for Aluminum Tolerance Scored with Eriochrome Cyanine R Staining Method in Barley Cultivar Honen (*Hordeum Vulgare* L), Australia J. Agriculture Res., 57, 113–118.
14. Wang, J. P., Raman, H., Zhou, M. X., Ryan, P., Delhaize, E., Hebb, D. M., Coombes, N., & Mendham, N. (2007). High-Resolution Mapping of the *Alp* Locus and Identification of a Candidate Gene *HvMATE* Controlling Aluminum Tolerance in Barley (*Hordeum Vulgare* L.), Theoretical Applied Genetics, *115(2),* 265–276.
15. Raman, H., Moroni, J. S., Sato, K., Read, B. J., & Scott, B. J. (2002). Identifications of AFLP and Micro-Satellites Markers Linked with an Aluminum Tolerance Gene in Barley (*Hordeum Vulgare* L), Theoretical Applied Genetics, *105(2, 3)* 458–464.
16. Ma, J. F., Nagao, S., Sato, K., Ito, H., Furukawa, J., & Takeda, K. (2004). Molecular Mapping of a Gene Responsible for Al-Activated Secretion of Citrate in Barley, J. Exp. Botany, *55(401)*, 1335–1341.
17. Minella, E. (1989). Aluminum Tolerance in Barley In heritance, Chromosome Location, Genetic Relationships of Sources of Diverse Origins and Breeding Implications, Dissertation (Ph D) Cornell University, Ithaca, 70p.
18. Echart, C. L., Barbosa-Neto, F. J., Garvin, D. F., & Cavalli-Molina, S. (2002). Aluminum Tolerance in Barley Methods for Screening and Genetic Analysis, Euphytica, *126*, 309–313.
19. Oram, R. N. (1983). Breedings Barley Tolerant to High Soil Acidity and Water Logging, in Driscoll, C. J. (Ed.), Proc. Aust. Plant Breeding Conference Adelaide, South Australia, 71–73.
20. Navakode, S., Weidner, A., Varshney, R. K., Lohwasser, U., Scholz, U., & Borner, A. A. (2009). QTL Analysis of Aluminum Tolerance in Barley, using Gene-Based Markers, Cereal Resolution, Communication, *37(4)*, 531–540.
21. Cai, S., Wu, D., Jabeen, Z., Huang, Y., Huang, Y., & Zhang, G. (2013). Genome-Wide Association Analysis of Aluminum Tolerance in Cultivated and Tibetan Wild Barley, Plus One, *8(7),* e69776.
22. Pellet, D. M., Grunes, D. L., & Kochian, L. V. (1996). Multiple Aluminum-Resistance Mechanisms in Wheat, Roles for Root Apical Phosphate and Malate Exudation, Plant Physiology, *112*, 591–597.
23. Magalhaes, J. V., Liu, J., Guimaraes, C. T., Lana, U. G. P., Alves, V. M. C., Wang, Y. H., Schaffert, R. E., Hoekenga, O. A., Pineros, M. A., Shaff, J. E., et al. (2007). A Gene in the Multidrug and Toxic Compound Extrusion (MATE) Family Confers Aluminum Tolerance in Sorghum, Nat Genetics, *39,* 1156–1161.
24. Zhao, Z., Ma, J. F., Sato, K., & Takeda, K. (2003). Differential Al Resistance and Citrate Secretion in Barley (*Hordeum vulgare* L.), Planta, *217*, 794–800.
25. Sasaki, T., Yamamoto, Y., Ezaki, B., Katsuhara, M., Ahn, S. J., Ryan, P. R., Delhaize, E., & Matsumoto H. A. (2004). Wheat Gene Encoding an Aluminum-Activated Malate Transporter, Plant J., *37(5),* 645–653.
26. Aniol, A., & Gustafson, J. P. (1984). Chromosome Location of Genes Controlling Aluminum Tolerance in Wheat, Rye and Triticale, Can. J. Genetics Cytology, *26*, 701–705.

27. Lagos, M. B., Fernandes, M. I., Camargo, C. E. O., Federizzi, L. C., & Carvalho, F. I. F. (1991). Genetics and Monossonic Analysis of Aluminum Tolerance in Wheat (*Triticum aestivum* L) Revista Brasileira De Genetica, *14,* 1011–1020.
28. Miftahudin, G. J., & Gustafson, J. P. (2002). AFLP Markers Tightly Linked to the Aluminum-Tolerance Gene *Alt3* in Rye (*Secale Cereale* L.), Theor. Appl. Genetics, *104,* 626–631.
29. Ryan, P. R., Raman, H., Gupta, S., Horst, W. J., & Delhaize, E. A. (2009). Second Mechanism for Aluminum Tolerance in Wheat Relies on the Constitutive Efflux of Citrate from Roots, Plant Physiology, *149*, 340–351.
30. Furukawa, J., Yamaji, N., Wang, H., Mitani, N., Murata, Y., Sato, K., Katsuhara, M., Takeda, K., & Ma, J. F. (2007). An Aluminum-Activated Citrate Transporter in Barley, Plant Cell Physiology, *48,* 1081–1091.
31. Gruber, B. D., Ryan, P. R., Richardson, A. E., Tyerman, S. D., Ramesh, S., et al. (2010). *HvALMT1* from Barley is involved in the Transport of Organic Anions, J. Exp. Botany, *61(5),* 1455–1467.
32. Gruber, B. D., Delhaize, E., Richardson, A. E., Roessner, U., James, R. A. et al. (2011). Characterization of *HvALMT1* Function in Transgenic Barley Plants, Functional Plant Biology, *38(2),* 163–175.
33. Cai, S., Bai, G. H., & Zhang, D. (2008). Quantitative Trait Loci for Aluminum Resistance in Chinese Wheat Landrace FSW, Theoretical Applied Genetics, *117*, 49–56.
34. Ma, H. X., Bai, G. H., & Lu, W. Z. (2006). Quantitative Trait Loci for Aluminum Resistance in Wheat Chinese Spring, Plant Soil, *283*, 239–249.
35. Nguyen, B. D., Brar, D. S., Bui, B. C., Nguyen, T. V., Pham, L. N., & Nguyen, H. T. (2003). Identification and Mapping of the QTL for Aluminum Tolerance Introgressed from the New Source, *Oryza Rufipogon* Griff, into Indica Rice (*Oryza sativa* L.), Theoretical Applied Genetics, *106*, 583–593.
36. Wheeler, D. M., Edmeades, D. C., Christie, R. A., & Gardner, R. (1992). Comparison of Techniques for Determining the effect of Aluminum on the Growth of, and the Inheritance of Aluminium Tolerance in Wheat, Plant Soil, *146*, 1–8.
37. Devine, T. E., Foy, C. D., Fleming, A. L., Hanson, C. H., Campbell, T. A., Mc Murtrey, J. E. III, & Schwartz, J. W. (1976). Development of Alfalfa Strains with Differential Tolerance to Aluminum Toxicity, Plant Soil, *44,* 73–79.
38. Blamey, F. P. C., Edmeades, D. C. & Wheeler, D. M. (1990). Role of Root Cation-Exchange Capacity in Differential Aluminum Tolerance of Lotus Species, J. Plant Nutrition, *13*, 729–744.
39. Kropotov, A. V. & Gerasimova, S. P. (1997). Physiological-Biochemical Specificity of Oats Varieties in Relation to Toxic Action of Aluminum, Reports of Russian Academy of Agrarian Science, *6,* 1–3 (in Russian).
40. Johnson, J. P., Carver, B. F., & Baligar, V. C. (1997). Productivity in Great Plains Acid Soils of Wheat Genotypes Selected for Aluminum Tolerance, Plant Soil, *188*, 101–106.
41. Ruiz-Torres, N. A., Carver, B. F., & Westerman, R. L. (1992). Agronomic Performance in Acid Soils of Wheat Lines Selected for Hematoxylin Staining Pattern, Crop Science, *32,* 104–107.
42. Lisitsyn, E. M. (2000). Intravarietal Level of Aluminum Resistance in Cereal Crops, J. Plant Nutrition, *23(6),* 793–804.
43. Hoekenga, O. A., Vision, T. J., Shaff, J. E., Monforte, A. J., Lee, G. P., Howell, S. H., & Kochian, L. V. (2003). Identification and Characterization of Aluminum Tolerance Loci in Arabidopsis (Landsberg Erecta x Columbia) by Quantitative Trait Locus Mapping, A Physiologically Simple But Genetically Complex Trait, Plant Physiology, 132, 936–948.
44. The Package of Breeding-Focused and Biometrical-Genetic Software AGROS, Version 2.07 Tver, (1997) (in Russian).

45. Merezhko, A. F. (2005). Use of Mendelian Principles in the Computer Analysis of Inheritance of Varying Traits, Ecological Genetics of Cultural Plants, Krasnodar All Russia Rice Scientific Institute, 107–117 (in Russian).
46. Mather, K., & Jinks, J. L. (1982). Biometrical Genetics, the Study of Continuous variation, Third ed. Chapman and Hall, London, New York, 403p.
47. Taylor, G. J. (1988). The Physiology of Aluminum Phytotoxicity, Metal Ions in Biological Systems, *24*, Aluminum and its Role in Biology, New York, USA Marsel-Dekker, 123–163.
48. Rigin, B. V., & Jakovleva, O. V. (2006). Genetic Analysis of Barley Resistance to toxic Aluminum Ions, Genetics (Moscow), *42*(3), 385 390 (in Russian).
49. Shchennikova, I. N., & Lisitsyn, E. M. (2009). Intensity of Stressful Influence and the Genetic Control of Barley Al-Resistance, Congress of Geneticists and Breeders to the 200th Ch Darwin Anniversary V Congress of the Vavilov Society of Geneticists and Breeders, Moscow, *1*, 369 (in Russian).
50. Ma, J. F., & Furukawa, J. (2003). Recent Progress in the Research of External Al Detoxification in Higher Plants a Mini Review, J Inorg Biochemistry, *97(1),* 46–51.
51. Gonçalves, J. F. C., Cambraia, J., Mosquim, P. R., & Araújo, E. F. (2005). Aluminum Effect on Organic Acid Production and Accumulation in Sorghum, J. Plant Nutrition, *28(3)*, 507–520.
52. Parker, D. R., & Pedler, J. F. (1998). Probing the "Malate Hypothesis" of Differential Aluminums Tolerance in Wheat by Using Other Rhizotoxic ions as Proxies for Al, Plantation, *205*, 389–396.
53. Wenzl, P., Patiño, G. M., Chaves, A. L., Mayer, J. E., & Rao, I. M. (2001). The High Level of Aluminum Resistance in Signal Grass is not associated with Known Mechanisms of External Aluminum Detoxification in Root Apices, Plant Physiology, *125*, 1473–1484.
54. Lisitsyn, E. M., & Shchennikova, I. N. (2009). Change in Aluminum Resistance Potential of Barley Hybrids under the Effect of the Maternal Cultivar, Russian Agricultural Sciences, *35(6),* 71–373.
55. Lisitsyn, E. M., & Lisitsyna, I. I. (2007). Dominance and Recessiveness of Parameters of Al-Resistance of Barley F_2 Hybrids at Different Concentrations of Stress Factor, Barley Genetics Newsletter, *37,* 1–4.
56. Petr, F. C., & Frey, K. J. (1966). Genotype Correlations, Dominance and Heritability of Quantitative Characters in Oats, Crop Science, *6*, 259–262.
57. Lisitsyn, E. M., Shchennikova, I. N., & Tiunova, L. N. (2006). Pattern of Inheritance of Al-Resistance in Barley F_2 Hybrids, Bulletin of Russian Academy of Agrarian Sciences, *5*, 36–38 (in Russian).
58. Camargo, C. E. O. (1984). Wheat Improvement IV Heritability Studies on Aluminum Tolerance Using Three Concentrations of Aluminum in Nutrient Solutions, Bragantia, *44*, 49–64.
59. Bona, L., Carver, B. F., Wright, R. J., & Baligar, V. C. (1994). Aluminum Tolerance of Segregating wheat Populations in Acidic Soil and Nutrient Solutions, Communication Soil Science Plant Analogy, *25*, 327–339.
60. Sutka, J., & Viesz, D. (1988). Reversal of Dominance in a Gene on Chromosome 5A Controlling Frost Resistance in Wheat, Genome, *30*, 313–317.

CHAPTER 10

ACTION OF SOME GENETIC SYSTEMS OF PLANTS OF OATS AND BARLEY DIFFERING ON RESISTANCE TO ENVIRONMENTAL ABIOTIC STRESSORS

EUGENE M. LISITSYN[1,2], GALINA A. BATALOVA[1,2], IRINA N. SHCHENNIKOVA[1], and LYUDMILA N. SHIKHOVA[2]

[1]Behalf of N.V. Rudnicki Zonal Research Institute of Agriculture of North-East of the Russian Agricultural Academy, 166-a Lenin St., Kirov, 610007, Russia; E-mail: niish-sv@mail.ru, edaphic@mail.ru

[2]Vyatka State Agricultural Academy, 133 Oktyabrsky Prospect, Kirov, 610010, Russia; E-mail: info@vgsha.info

CONTENTS

ABSTRACT

The ecological-genetic model of organization of a complex quantitative trait of plants starts with a postulate that the genetic formula represents a system of interrelated polygenes and environmental limiting factors, and at presence of environmental stressors the trait is coding by a various set of genes. The phenomenon of multidirectional displacements of genetic and environmental deviations established by Dragavtsev et al. [21, 22] has allowed to create a new method of the genetic analysis of an initial and breeding material. The analysis of two-dimensional orthogonal system of coordinates "the weight of an ear of the main stem" and "weight of a straw of the main stem" gives the chance to divide studied varieties into 4 groups with a various combination of positive and negative deviations on genetic systems of attraction and adaptivity, and the analysis of system of traits "weight of grain in an ear" and "the weight of chaff in an ear" allows to estimate genetic systems of adaptivity + attraction and system of microdistribution of photosynthates within an ear. In our experiments on an example of several oats varieties (Krechet, Dens, Ulov, Selma, and Faust) and barley varieties (Dyna, Elf, Ecolog, and Novichok) it is shown existence of intravarietal heterogeneity of plants on indices of resistance to abiotic factors (such as soil drought and presence of aluminum ions) and its influence on action of above-mentioned genetic systems. Significant differences in action of the specified genetic systems at the main and lateral shoots of the studied grain crops are shown. So, at calculation only on the main ear genetic system of microdistribution of plastic substances of Novichok barley works four times more effectively, than for Ecolog barley. However, at calculation on lateral ears this system works in one and a half time more effectively for variety Ecolog, therefore at calculation on all ears the specified genetic system works for both varieties equally effective. In the main ear plastic substances pass into grain from chaff more effective energetically, than at lateral ears. At the tested oats varieties, in contrast, plastic substances flow from chaff into grain with identical energetic efficiency both in the main panicle, and in lateral panicles.

10.1 INTRODUCTION

Aluminum toxicity represents a most limiting factor for yield in cereals in acid soils [1] because aluminum contained in soil and available to plant promotes reduction in root growth [2] and inhibits simulation and transport of nutrients, influencing negatively on yield [3]. Now the exudation of anions of some organic acids by roots is considered basic physiological mechanism of resistance of grain crops plants to toxic influence of aluminum [4–7]. The genetic control of this mechanism is offered [8]. Based on the timing of secretion, two patterns of Al-induced organic anion secretion have been proposed [9, 10]. Both oats [7] and barley [11] belong to Pattern I plants, such as wheat [12] in which roots Al may simply activate a transporter in the plasma membrane to initiate organic acids anion secretion, and the induction of genes is not required [10, 13]. In Pattern II plants, such as rye [14] and triticale [15] Al may induce the expression of genes and the synthesis of proteins involved in organic acids metabolism or in the transport of organic acid anions [16]. As organic acids anions at plants of the first type are secreted into rhizosphere irrespective of aluminum impact then it is possible to study a role of this physiological mechanism in action of some genetic systems of plants of oats and barley in absence of stressful influence using plants with different level of Al-resistance.

It is also known that plant involves numerous physiological mechanisms in its reaction to stressful influence of aluminum ions accompanied by change of activity of a great number of genes (genetic systems) [17, 18]. The assumption on pyramiding of actions of the numerous genetic systems involved in aluminum resistance is proposed [19]. Moreover, some researchers specify that external chelating of aluminum by organic anions is not the most important mechanism in oats Al-resistance reaction [20].

The ecological-genetic model of organization of a complex quantitative trait offered by Dragavtsev and co-workers [21, 22], starts with a postulate that at various stressful conditions of environment the trait is supervised by a various set of genes. This model consists not only of genetic characteristics of an investigated material, but limiting factors of environment and different stages of plant ontogenesis also. The authors of a model found out that environmental deviations of a quantitative trait and deviations caused by genetic effects may have different direction and values that, in turn, has allowed to create a new method of the genetic analysis of an initial and breeding material [22]. Using graphic data presentation in orthogonal two-dimensional systems of some coordinates it is possible to identify a plant genotype on its phenotype.

According to authors of model, the success of breeding work is defined by empirical use of six genetic-physiological systems [22]: 1) System of attraction (which provides the attraction of plastic matters from the straw into the ears in the period of grain forming); 2) System of microdistributions of attracted matter between the grain and chaff in the cereals; 3) System of adaptivity (frost resistance, cold hardiness, drought resistance, etc.); 4) System of "feed paying" under a limiting factor of soil nutrition (efficiency of applying nitrogen, phosphorus, potassium); 5) System of tolerance to density; 6) System of variability of periods of ontogenesis.

The analysis of two-dimensional orthogonal system of trait coordinates "the weight of an ear of the main stem" and "weight of a straw of the main stem" gives the opportunity to divide studied varieties into four groups with a various combination of positive and negative deviations on genetic systems of attraction and adaptivity, and the analysis of system of traits "weight of grain in an ear" and "the weight of chaff in an ear" allows to estimate genetic systems of "adaptivity + attraction" and system of microdistribution of products of photosynthesis within ear simultaneously. It allows also to made division of varieties on a combination of positive and negative deviations of the given systems.

10.2 MATERIAL AND TECHNIQUE

The first series of researches: For decrease influence of intravarietal variability seedlings of oats (*Avena sariva* L.) and barley (*Hordeum vulgare* L.) were used having equal length of germinal roots. For this purpose about 400 seeds of each oats varieties Krechet and Ulov, and barley varieties Dyna and Elf were germinated for 4 days in rolls of a filter paper in distilled water, pH 6.0. After this one hundred seedlings per genotype, chosen on the basis of their having similar root lengths average for the genotype, have been selected. This initial length of a root has been noted for each individual plant. Further these plants have been placed for 3-day additional growth in 1 mM aluminum solution at pH 4.3. Upon termination of stressful influence the length of longest root of each individual plant has been measured again for calculation of root regrowth and, accordingly, level of Al-resistance of the given plant by relative root elongation (RRE = (final root length – initial root length) / initial root length). All plants have been sowing under conditions of greenhouse experiment on neutral soil for further growth and development. Upon termination of growing season parameters of crop structure [23] have been estimated at each individual plant.

For the second series of researches following varieties have been used: oats – Dens, Ulov, Selma, Faust; barley – Ecolog and Novichok. Under laboratory conditions selection of four-day seedlings on level of resistance against two stressful factors was done: high content of aluminum (the basic stressful factor of acid podzolic soils) – 1 mM of aluminum sulfate pH 4.3; high osmotic pressure of a solution – 3 atmospheres (drought imitation).

Distilled water pH 6.0 was used as a control. Two groups of seedlings were selected from each variety – most and the least resistant, estimating length of the longest root (resistant seedlings had longer roots). The average length of short-root fractions (SR) of seedlings made 1.5 cm, of long-root fractions (LR) – 6 cm. Seedlings of control variants did not be divided on fractions. For carrying out of laboratory selections about 1000 grains of each variety were used; the selected seedlings (both fractions together) made nearby 10–12% of initial sample. Further the selected seedlings were sown under field conditions and were grown up to full maturity of grain according to [23]. The estimation of action of genetic systems was spent by a technique offered by [22].

Statistical processing of the received data and construction of graphs of orthogonal regression are spent with use of a software package Microsoft Excel 2007.

10.3 RESULTS AND DISCUSSION

10.3.1 ACTION OF GENETIC SYSTEMS OF ADAPTIVITY AND PLASTICITY OF THE MAIN EARS/PANICLES

The average root length of initial plants has made on varieties: Ulov – 70.8 ± 0.8 mm; Krechet – 64.9 ± 0.7 mm; Dyna – 71.0 ± 0.7 mm; Elf – 81.3 ± 0.8 mm. The average regrowth of root systems for three days of aluminum treatment on varieties has made: Ulov – 3.6 ± 0.3 mm (5% to initial length); Krechet – 6.2 ± 0.4 mm (10% to initial length); Elf – 3.7 ± 0.3 mm (5% to initial length), and Dyna – 1.7 ± 0.3 mm (2% to initial length). However, it is necessary to note high intravarietal variability of degree of this regrowth (Table 10.1): at oats Ulov 18% of tested seedlings had zero regrowth, at oats Krechet – 9%, at barley Elf – 16%, and at barley Dyna – 53%.

Thus, data of Table 10.1 show that in each investigated variety there are plants with different level of Al-resistance.

TABLE 10.1 Intra-Varietal Variability of Oats and Barley on Regrowth of Seedling Roots (1 mM Al, 3 days of Treatment), % of Total Sample

Variety	Re-growth, %			
	0	1–5	5–10	>10
Oats Ulov	18	39	28	15
Oats Krechet	9	21	29	41
Barley Elf	16	42	26	16
Barley Dyna	53	31	11	5

However, any variety of self-pollinated plants is genetically homogeneous, that is all plants of a variety should have the same set of genes. The possible explanation of various reactions on aluminum of plants within variety can be found in the theory of the ecological-genetic organization of quantitative trait [21, 22]. This theory specifies that distinctions in the genetic control of a trait under various ecological conditions can indicate a difference in a quantitative and/or qualitative set of the genes influencing development of this trait. In other words, under some conditions development of the trait is controlled by one set of genes, but under other conditions – by a little different set. In doing so the genotype of a plant does not change.

It is known that transport of assimilates in the whole plant is under the genetic control. However, environment factors (light, temperature, moisture, elements of a mineral nutrition, physiologically active substances, etc.) make essential impact on a direction of distribution of assimilates. It is noticed that studying of such displacement approaches us to possibility of regulation of its use for growth or for their deposition in storage [24].

It is possible to assume that different plants within variety use different genes (genetic systems) for adaptation to stressor (such as aluminum). Distinctions between plants in action of a system of sink-source relations should serve as confirmation of the idea that will be reflected in an arrangement of points of these plants on the diagram of orthogonal regression in a system of coordinates "the weight of an ear/panicle – weight of straw " [22].

The diagrams of orthogonal regression constructed for studied varieties are presented on Fig. 10.1 (a-d). As one can see, distinctions of individual plants within variety on level of Al-resistance revealed at early stages of development (4–7 days after beginning of grain swelling), at adult plants (at a stage of full maturity) are not reflected in changes of action of genetic systems of the general adaptivity and plasticity of these plants.

Near all segregated groups of plants had samples both with positive and with negative deviations in action of studied genetic systems. Polymorphism on genetic system of the general adaptivity is higher, than polymorphism on genetic system of plasticity.

As an explanation of such distribution of genotypes it is possible to assume the following. As it has been pointed out earlier, such complex trait as "Al-resistance" of plants at different genotypes can be provided with functioning of both various physiological-biochemical mechanisms, and the various relative contribution of each of these mechanisms to final value of the trait.

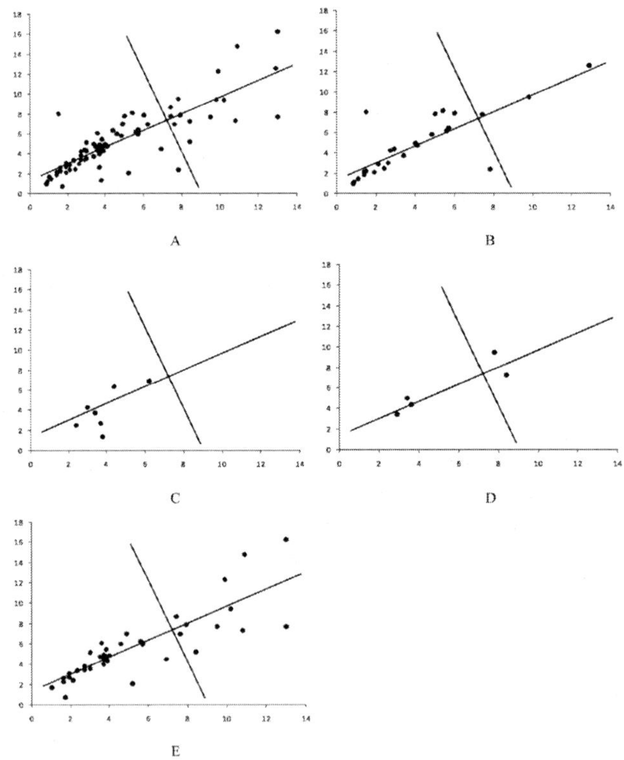

FIGURE 10.1 (A) Distribution of individual plants of barley Dyna in a system of coordinates "weight of an ear – weight of straw": A – all plants; B – plants with 1–5% root regrowth; C – plants with 5–10% root regrowth; D – plants with more than 10% root regrowth; E – plants without root regrowth. An axis of abscises – weight of straw (g), an axis of ordinates – weight of an ear (g).

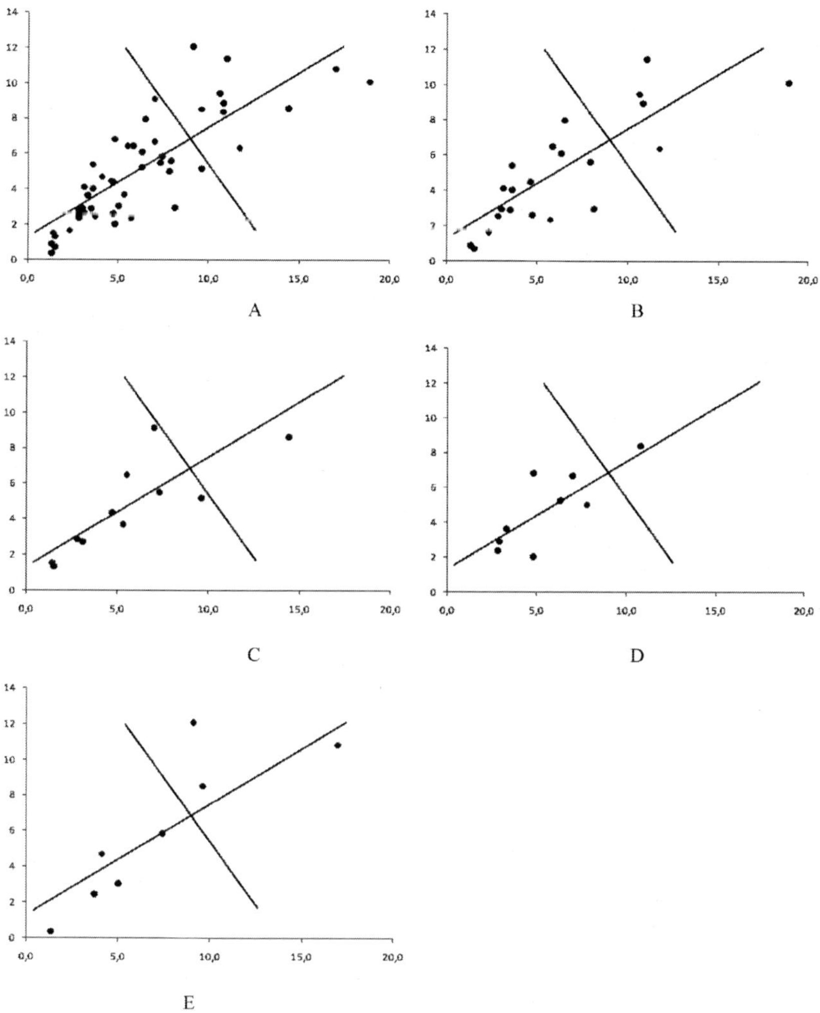

FIGURE 10.1 (B) Distribution of individual plants of barley Elf in a system of coordinates "weight of an ear – weight of straw": A – all plants; B – plants with 1–5% root regrowth; C – plants with 5–10% root regrowth; D – plants with more than 10% root regrowth; E – plants without root regrowth. An axis of abscises – weight of straw (g), an axis of ordinates – weight of an ear (g).

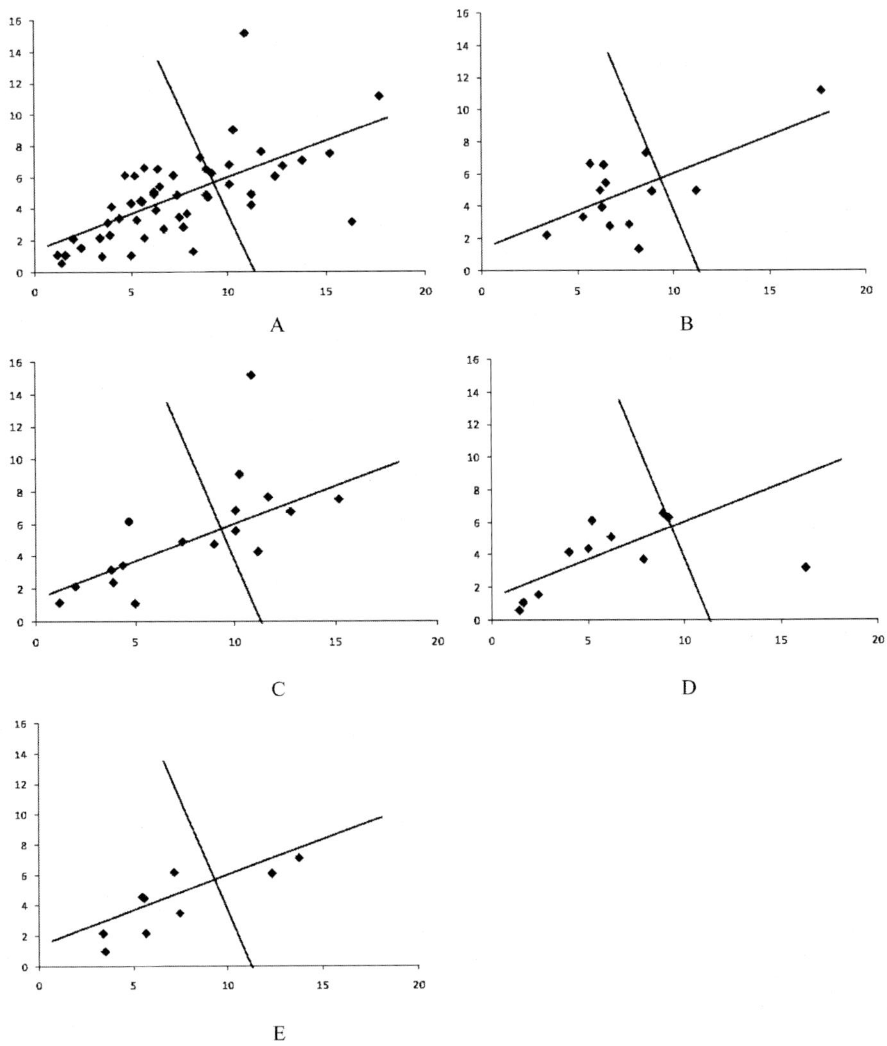

FIGURE 10.1 (C) Distribution of individual plants of oats Ulov in a system of coordinates "weight of an ear – weight of straw": A – all plants; B – plants with 1–5% root regrowth; C – plants with 5–10% root regrowth; D – plants with more than 10% root regrowth; E – plants without root regrowth. An axis of abscises – weight of straw (g), an axis of ordinates – weight of a panicle (g).

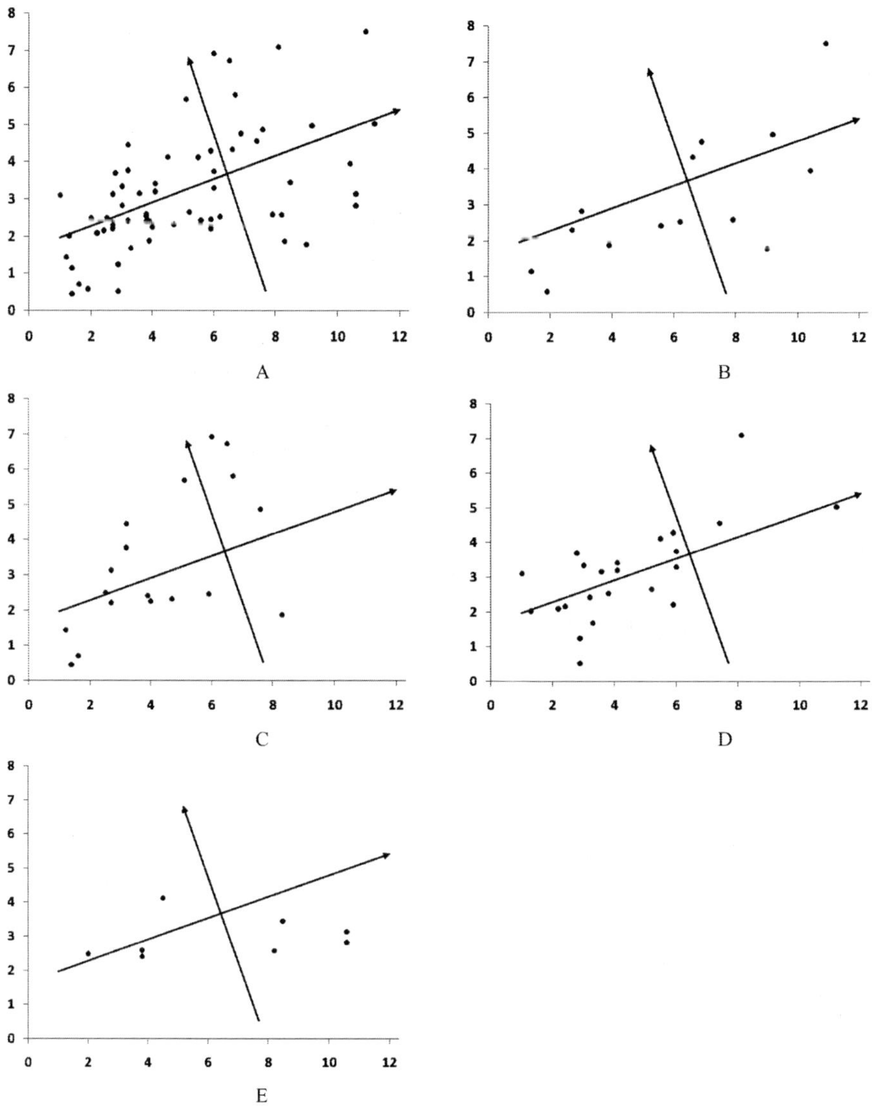

FIGURE 10.1 (D) Distribution of individual plants of oats Krechet in a system of coordinates "weight of an ear – weight of straw": A – all plants; B – plants with 1–5% root regrowth; C – plants with 5–10% root regrowth; D – plants with more than 10% root regrowth; E – plants without root regrowth. An axis of abscises – weight of straw (g), an axis of ordinates – weight of a panicle (g).

Distribution of plants of the same variety in a system of trait coordinates "weight of an ear/panicle" – "the weight of straw" shows distinctions in the genetic control of these traits (intravarietal genetic polymorphism) even under control (not-stressful) conditions. These distinctions are explained by distinctions in a quantitative and qualitative set of the genes governing a trait (by different "genetic formula" of the trait at individual plants).

Therefore it is possible to assume that, on the one hand, aluminum leads to different level of depression/strengthening of action of the various genes, which are responsible for reaction on stressor, and on the other hand, these genes participate in the genetic control of development of plants under normal conditions in various degrees.

In other words, the genes providing Al-resistance of plants, being the genes which activity is shown in normal growth conditions (absorption of nutrients; secretion of organic acids by roots; activity of enzymes of the general metabolism etc.) can bring different contribution to plant growth and development.

Let's admit that aluminum has led to change of expression of any gene, for example, responsible for secretion of organic acids from root cell. Strengthening of this secretion will lead to relative strengthening of growth of that genotype at which the given mechanism is the most important in development of resistance, and will not affect relative growth of a genotype, which uses other mechanism of resistance. Further, if this gene brings the essential contribution to growth of plants under normal conditions, one will find out a close correlation between growth both under norm and under stressful conditions; otherwise this correlation will not be shown.

So it become clearly why plants differ in their level of Al-resistance can show deviations in action of genetic systems of adaptivity and plasticity identical on value and on direction; and on the contrary, plants identical on Al-resistance levels will settle down in different parts of the presented diagrams.

Possibly, plants with high level of Al-resistance which points are located in the top right square of the diagrams will combine this resistance with ability to provide high crops under normal conditions. The plants of the same group located in the bottom left square of the diagram, being highly resistant under the conditions of aluminum impact, will not be high-yielding in the normal environment.

10.3.2 COMPARISON OF ACTION OF THE GENETIC SYSTEMS AT LATERAL AND MAIN STEMS OF GRAIN CROPS

As lateral stems of grain crops (an oats and barley) appear on plants in little bit other environmental conditions than the main stems, it is interesting to consider feature of action of genetic systems "adaptivity + attraction" and "microdistributions of products of photosynthesis" at the main and lateral stems of the given agricultural crops.

On Fig. 10.2 (barley) the designation of variants corresponds to the following scheme:

1 – Ecolog – control   6 – Novichok – control
2 – Ecolog – a drought – SR   7 – Novichok – a drought – SR

3 – Ecolog – a drought – LR	8 – Novichok – a drought – LR
4 – Ecolog – aluminum – SR	9 – Novichok – aluminum – SR
5 – Ecolog – aluminum – LR	10 – Novichok – aluminum – LR

On Fig. 10.3 (oats) the designation of variants corresponds to the following scheme:

1 – Faust – control	11 – Ulov – control
2 – Faust – a drought – SR	12 – Ulov – a drought – SR
3 – Faust – a drought – LR	13 – Ulov – a drought – LR
4 – Faust – aluminum – SR	14 – Ulov – aluminum – SR
5 – Faust – aluminum – LR	15 – Ulov – aluminum – LR
6 – Selma – control	16 – Dens – control
7 – Selma – a drought – SR	17 – Dens – a drought – SR
8 – Selma – a drought – LR	18 – Dens – a drought – LR
9 – Selma – aluminum – SR	19 – Dens – aluminum – LR
10 – Selma – aluminum – LR	20 – Dens – aluminum – LR

Figure 10.2 shows how the arrangement of variants changes relative each other at an estimation of main and lateral stems separately and of all stems of a plant in the sum.

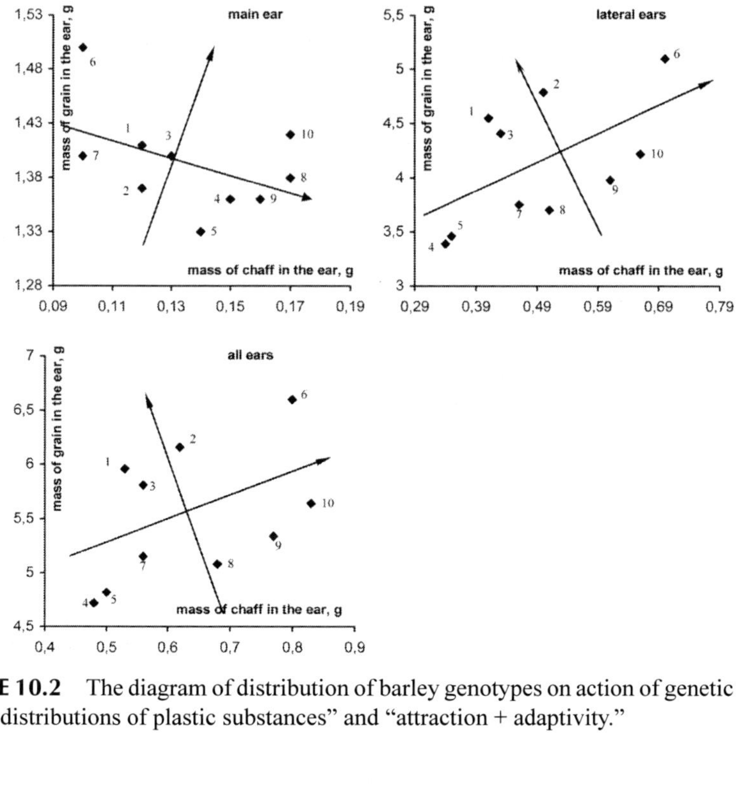

FIGURE 10.2 The diagram of distribution of barley genotypes on action of genetic systems: "micro-distributions of plastic substances" and "attraction + adaptivity."

Figure 10.2 visually shows that the overall picture of distribution of variants coincides appreciably at calculations for all plant and for lateral ears only, and considerably differs from the estimations made on the main ears: the angle of slope of regression lines, and relative positioning of the points corresponding to variants of experiment changes essentially. So, at calculation on the main ear only genetic system of micro-distribution of plastic substances (MDP) of Novichok barley (control) works four times more effectively (the point is located in 4 times further from the beginning of coordinates in a direction to the left-upwards), than for Ecolog barley (control). However, at calculation on lateral ears this system works more effectively for Ecolog (in one and a half time); at calculation on all ears MDP genetic systems work equally effectively for both varieties.

In work of MDP genetic system following distinctions are visible: variant N7 (selection on drought resistance, SR group of Novichok) at estimation on the main ear gives strong positive deviation; by consideration of lateral ears and total ears – negative deviation in action of the given genetic system. There were not qualitative changes in action of MDP genetic system by variants at variety Ecolog – at all three methods of calculation variants of experiment settle down equally relative to each other, quantitative distinctions are noted only.

At an estimation on the main ear both SR variants of variety Novichok (NN 7 and 9) show negative deviations in action of system of attraction; both variants of LR (NN 8 and 10) – positive deviations. By consideration of lateral ears for this variety, selections on drought resistance give negative and selections on Al-resistance – positive deviations in action of system of attraction. For variety Ecolog moving of a variant N2 of group with negative deviations in work of this genetic system (main ear only) into group with positive deviations (lateral and total ears) was observed.

Comparison of values of an angle of slope of a positive line of regression shows that in the main ear plastic substances pass more effectively energetically from chaff into grain, than at lateral ears.

As a whole, both selections on Al-resistance from Ecolog barley have shown the worst results of action of all studied genetic systems. Probably, influence by aluminum causes so considerable changes in a metabolism of the plants which have tested short-term influence to the stressor that they cannot further restore its normal level even in absence of aluminum in growth media, i.e. those systems which allow plant to adapt to acid growth media and aluminum, simultaneously cause decrease of productive abilities in plants of variety Ecolog.

For Novichok barley such accurate laws was not revealed. It is possible to notice only that selection of LR plants on reaction to aluminum at all ways of estimation shows positive deviations in action of genetic system of adaptivity.

By consideration of Figure 10.3 exact difference of barley and oats is visible first of all: on diagrams both for the main panicles, and for lateral and total panicles angles of slope of a positive regression line are almost equal, i.e. plastic substances flow from chaff into grain with identical energetic efficiency both in the main panicles, and in lateral panicles for all analyzed samples as a whole.

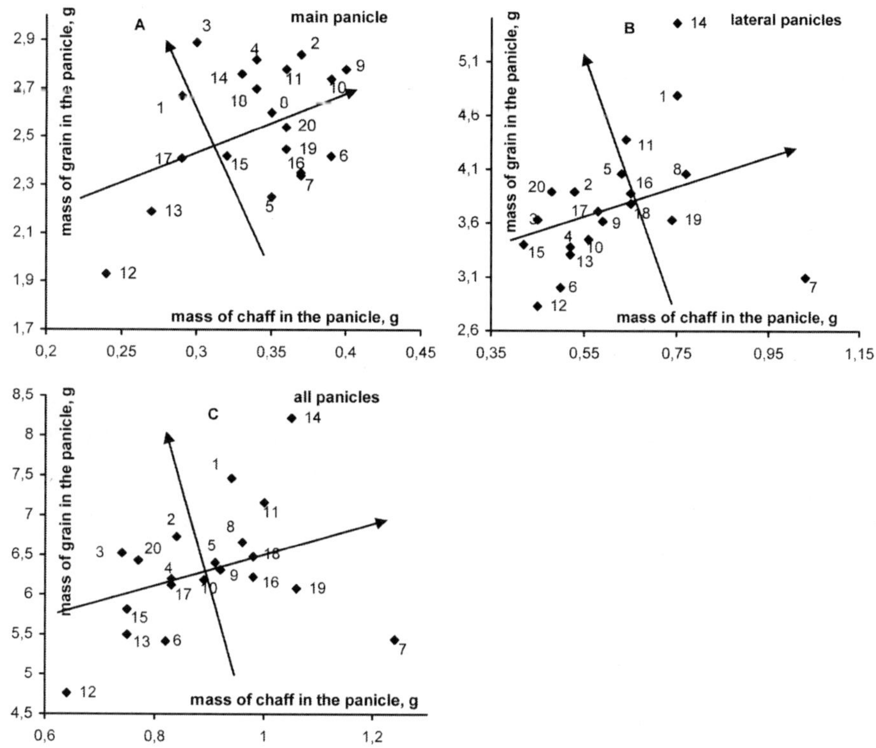

FIGURE 10.3 The diagram of distribution of oats genotypes on action of genetic systems "micro-distributions of plastic substances" + "attraction and adaptability."

As for individual variants of experiment distinctions between the data received for lateral and main panicles exist nevertheless. So if to compare control variants among themselves by consideration of the main panicles the greatest attraction and adaptivity had variety Selma; by consideration of lateral panicles the same variety becomes the worst on action of the given genetic systems. Also it is possible to note a difference in an assessment of action of MDP genetic system of variety Dens – for the main panicle its action has negative deviation concerning a regression line, but by consideration of lateral panicle – positive one.

The distinctions exist in action of genetic systems ("attraction" + "adaptivity") by consideration of lateral and main panicles for following variants of experiment: all variants of variety Faust (NN 1–5 on Fig. 10.3) change signs on deviations in action of these systems for the opposite; for variety Selma change of signs on opposite is marked in variants of the control and both selections on Al-resistance (NN 6, 9, 10); at variety Dens both LR variants and control change signs (NN 16, 18, 20); for variety Ulov qualitative differences are characteristic only for LR variant which have been selected on reaction to aluminum.

As a whole for oats there were twice less qualitative changes in action of MDP genetic system at different ways of an estimation: these were both selections on Al-resistance of variety Faust (NN 4, 5), of variety Selma (NN 9, 10); and the same variants of variety Dens, as in the previous case (NN 16, 18, 20).

Thus, at estimation on the main panicle SR variants of variety Faust on reactions to a drought, and both groups (SR and LR) of variety Selma on reaction to aluminum have appeared the most adaptive to the given conditions of cultivation and the most favorable on action of genetic system of attraction. Both selections on drought resistance of variety Ulov and SR group of variety Dens on reaction to a drought have appeared the least valuable on action of these genetic systems.

Based on estimation on lateral panicles (and on sum of panicles) the SR variant of variety Ulov selected on reaction to aluminum has appeared the most perspective on action of all three genetic systems, considered in experiment; the least valuable – SR selection of the same variety on reaction to a drought.

All selections from variety Faust were characterized by deterioration of action of all investigated genetic systems. For variety Selma LR group of plants on reaction to both stressors and both groups on Al-resistance improve action of all three genetic systems (attraction, adaptivity, and microdistribution of plastic matters). Only SR group of plants of variety Ulov selected on Al-resistance surpassed the initial variety (without selection) on action of the given genetic systems.

Selections from variety Dens had the greatest deviations in action of genetic systems: LR variant of selection on reaction to drought has surpassed initial variety by all systems; SR selection on reaction to aluminum was better than an initial variety on action of systems "attraction + adaptivity," but conceded on action of MDP system. Two other selections from this variety were less adaptive and with the worst attraction, but surpassed an initial variety in action of MDP system.

10.4 CONCLUSION

Thus, the data obtained in researches shows that short-term influence of abiotic stressful factor on seedlings of grain crops leads to considerable changes in action of genetic systems of attraction, adaptivity, and micro-distributions of plastic substances of the adult plants which have grown further in the absence of action of the given factors. Changes in action of genetic systems of micro-distribution of plastic substances and mutual action of systems of attraction and of adaptivity differ considerably for the main ear/panicle and lateral ears/panicle. Outflow of plastic substances from chaff into grain in the main ear of barley goes with higher energetic efficiency, than in lateral ears. Outflow of plastic substances from chaff into grain in lateral and main panicles of oats goes with quite identical energetic efficiency.

ACKNOWLEDGEMENT

This work is supported by the Russian Academy of Agricultural Sciences (Project 04.05.02.). We thank the staff on laboratories of oat breeding, barley breeding, and department of plant edaphic resistance (North-East Agricultural Research Institute, Kirov, Russia) for their technical help with conducting laboratory and field experiments, and academician V.A. Dragavtsev (Agro-physical Institute, St. Petersburg, Russia) for consultations and discussion.

KEYWORDS

- **aluminum**
- **genetic systems**
- **intravarietal heterogeneity**
- **resistance**
- **soil drought**

REFERENCES

1. Valle, S. R., Carrasco, J., Pinochet, D., & Calderini, D. F. (2009). Grain Yield, Above-Ground and Root Biomass of Al-tolerant and Al-sensitive Wheat Cultivars under Different Soil Aluminum Concentrations at Field Conditions, Plant Soil, *318,* 299–310.
2. Tang, C., Diatloff, E., Rengel, Z., & Mc Gann, B. (2001). Growth Response to Sub surface Soil Acidity of Wheat Genotypes Differing in Aluminum Tolerance, Plant Soil, *236,* 1–10.
3. Kochian, L. V., Pineros, M. A. & Hoekenga, O. A. (2005). The Physiology, Genetics and Molecular Biology of Plant Aluminum Resistance and Toxicity, Plant Soil, *274,* 175–195.
4. Boscolo, P. R, Menossi, M. & Jorge, R. A. (2003). Aluminum-Induced Oxidative Stress in Maize Phytochemistry, *62(2),* 181–189.
5. Cai, S., Bai G. H., & Zhang, D. (2008). Quantitative Trait Loci for Aluminum Resistance in Chinese Wheat Landrace FSW, Theoretical Applied Genetics, *117,* 49–56.
6. Ryan, P. R., Raman, H., Gupta, S., Horst, W. J., & Delhaize, E. (2009). A Second Mechanism for Aluminum Tolerance in Wheat Relies on the Constitutive Efflux of Citrate from Roots, Plant Physiology, *149,* 340–351.
7. Radmer, L., Tesfaye, M., Somers, D. A., Temple, S. J., Vance, C. P., & Samac, D. A. (2012). Aluminum Resistance Mechanisms in Oat (*Avena sativa* L.), Plant Soil, *351(1)*, 121–134.
8. Delhaize, E., Ma, J. F., & Ryan, P. R. (2012). Transcriptional Regulation of Aluminum Tolerance Genes, Trends in Plant Science, *17(6),* 341–348.
9. Ma, J. F. (2000). Role of Organic Acids in Detoxification of Aluminum in Higher Plants, Plant Cell Physiology, *41(4)*, 383–390.

10. Ma, J. F. (2005). Physiological Mechanisms of Al Resistance in Higher Plants, Soil Science Plant Nutrition, *51(5)*, 609–612.
11. Zhao, Z., Ma, J. F., Sato, K., & Takeda, K. (2003). Differential Al Resistance and Citrate Secretion in Barley (*Hordeumvulgare* L), Plantation, *217*, 794–800.
12. Ryan, P. R., Delhaize, E., & Randall, P. J. (1995). Characterization of Al-Stimulated Efflux of Malate from the Apices of Al-Tolerant Wheat Roots, Plantation, *196(1)*, 103–110.
13. Ma, J. F. (2007). Syndrome of Aluminum Toxicity and Diversity of Aluminum Resistance in Higher Plants, International Review of Cytology, *264,* 225–252.
14. Yang, Z. M., Sivaguru, M., Horst, W. J., & Matsumoto, H. (2000). Aluminum Tolerance is achieved by Exudation of Citric Acid from Roots of Soybean (*Glycine Max*), Physiology Plantar, *110(1),* 72–77.
15. Ma, J. F., Taketa, S., & Yang, Z. M. (2000). Aluminum Tolerance Genes on the Short Arm of Chromosome 3R are linked to Organic Acid Release in Triticale, Plant Physiology, *122(3),* 687–694.
16. Ma, J. F., Ryan, P. R., & Delhaize, E. (2001). Aluminum Tolerance in Plants and the Complexing Role of Organic Acids, Trends in Plant Science, *6(6),* 273–278.
17. Drummond, R. D., Guimarães, C. T., Felix, J., Ninamango-Cárdenas, F. E., Carneiro, N. P., Paiva, E., & Menossi M. (2001). Prospecting Sugarcane Genes Involved in Aluminum Tolerance, Genetics Molecular Biology, *24(1–4),* 221–230.
18. Houde, M., & Diallo, A. O. (2008). Identification of Genes and Pathways Associated with Aluminum Stress and Tolerance using Transcript me Profiling of Wheat Near-Isogenic Lines, BMC Genomics, *9*.
19. Hervé, C. B., Calai, F. A., Nava, I. C., & Delatorre, C. A. (2013). Tolerância Ao Alumínio Tóxico Em Germoplasma Brasileiro Elite De Aveia, Ciência Rural, *43(8),* 1364–1370 (In Portuguese).
20. Castilhos, G., Farias, J. G., de Bernardi Schneider, A., de Oliveira, P. H., Nicoloso, F. T., Schetinger, M. R. C., & Delatorre, C. A. (2011). Aluminum-Stress Response in Oat Genotypes with Monogenic Tolerance, Environment Exp. Botany, *74*, 114–121.
21. Dragavtsev, V. A., Litun, N. P., Shkel, I. M. & Nechiporenko, N. N. (1984). Model of Ekological-Genetic Control of Quantitative Traits of Plants, Reports of Academy of Sciences of the USSR, *274(3),* 720–723 (in Russian).
22. Dragavtsev, V. A. (2002). Algorithms of an Ecological-Genetic Survey of the Gene Pool and Methods of Creating the Varieties of Crop Plants for Yield, Resistance and Quality (Methodical Recommendations New Approaches St. Petersburg VIR, 41p.
23. Technique of State Variety Testing of Agricultural Crops (1989). Moscow, Kolos Publishing 250p (in Russian).
24. Kursanov, A. L. (1976). Transport of Assimilates in a Plant Moscow Science Publishing 646p (in Russian).

PART IV

SCIENTIFICALLY SUBSTANTIATED SOIL AND CLIMATICAL REGIONS FOR THE INDUSTRIAL CULTIVATION OF CROPS

CHAPTER 11

SCIENTIFICALLY SUBSTANTIATED SOIL AND CLIMATICAL REGIONS FOR THE INDUSTRIAL CULTIVATION OF FRUIT CROPS IN UKRAINE

MYKOLA O. BUBLYK, LIUDMYLA A. FRYZIUK, and
LIUDMYLA M. LEVCHUK

Institute of Horticulture of NAAS of Ukraine, 23, Sadova St., Kyiv, Ukraine 03027;
E-mail: mbublyk@mail.ru

CONTENTS

ABSTRACT

The authors have determined the soil and climatic regions in Ukraine that are favorable for the industrial cultivation of apple, pear, plum, cherry, sweet cherry, apricot, peach and black currant where the plants are able to use their biological potential in maximum.

The methods of those regions determination have been proposed based on the scientific substantiation of the crop (cultivar) requirements to weather and soil factors and the integral evaluation of the degree of their biological potential realization under concrete conditions, the potential being determined by the weighted sum of the weather factors normed deviations from the given ideal.

11.1 INTRODUCTION

The fruit crops distribution in the regions of a country according to their requirements to the weather and soil growing conditions makes it possible to achieve the optimum productivity and fruits quality by means of realizing the biotic factors and crops characteristics valuable for economy that is to use to the fullest degree their biological potential, which means the total spectrum of the characteristics variability that manifest themselves as a result of the genotype and environment interaction. It is just the factors, which ensure the optimum productivity and fruits quality that are of the greatest importance in this variability [1–4].

The methods of the fruit crops distribution existing at the present time [2, 5] are not always applicable under the conditions of Ukraine. Therefore the methods were elaborated based on the crops capability to display their biological potential in the concrete soil and climatic conditions. On the basis of those methods the regions have been chosen in Ukraine which are the most favorable for the industrial cultivation of the main fruit crops – apple, pear, plum, cherry, sweet cherry and apricot.

11.2 MATERIALS AND METHODOLOGY

The objects of the researches were production and experimental orchards, which are in the major soil and climatic regions of Ukraine. For the analysis the 20-year data about the yield of those crops main cultivars were used as well as the everyday information from the meteorological stations adjoining to the investigated orchards. The weather conditions computer base was created to process the received data, which enables to analyze them with the use of the existing methods [6–8].

The methods of the region integral criterion as concerns weather conditions were formed on the basis of obtained equations and formula for the calculation of the weighted sum of standardized deviations from the predetermined ideal [9].

The degree of the soil favorability for each crop was determined on the basis of the generalized field data experiment data and corresponding scientific literature.

11.3 RESULTS AND DISCUSSION

The proposed methods are based on the selection of regions favorable for the fullest concrete crop biological potential realization and include a number of stages. Let us consider the realization of those methods using the example of sweet cherry.

11.3.1 FORMULATION OF THE FRUIT CROP REQUIREMENTS TO THE WEATHER CONDITIONS AND DETERMINATION OF THE COUNTRY'S REGIONS, WHICH MEET THOSE REQUIREMENTS

As a results of generalizing the data of studying the sweet cherry reaction to the main environmental factors we determined [10], that for the fullest realization of this crop biological potential the necessary amount of temperatures above 10 °C is 2600 °C, the number of days with the temperature above 15°C for the vegetation period must be not less than 110 and minimum temperatures during the sweet cherry dormancy period not below –23 °C. Besides, those regions should be considered favorable for the sweet cherry cultivation where the temperatures below –23 °C occur not oftener than during 20% of winters, limitly favorable are those where such temperatures are in 21–39% of winters and in unfavorable regions such temperature reduction takes place in more than 40% of winters. Short-term reductions of the temperature in the phase of 'white bud' may not exceed –4 °C and in the flowering period –0.6—2 °C. The risk of such temperature occurring exists to a certain degree in all the regions but their probability corresponds on the total to that of the temperature reduction during the dormancy period and does not demand the correction of the above-mentioned zones.

So favorable as to the weather conditions for the sweet cherry cultivation are larger part of the Steppe and partly the Forest-steppe of Ukraine.

11.3.2 THE SOILS CLASSIFICATION CONCERNING THEIR FAVORABILITY FOR A CONCRETE FRUIT CROP

Sweet cherry is more exacting to the soil conditions as compared to other fruit crops. This crop total state and productivity correlate directly with the humus horizon power, especially on the skeletal, high-carbonate, sandy and other low-productive soils. Sweet cherry is one of the most sensitive crops to the clay content [11, 12]. The sweet cherry requirements to soil are determined to an important degree by a rootstock. For the sweet cherry on the mahaleb cherry the light-textured soils are favorable even with the high carbonate content.

On the sandy loam chernozems sweet cherry on this rootstock is more productive than on the cherry. On the heavy-textured soils, it on the mahaleb cherry, on the contrary, develops badly and this rootstock is unfavorable here [11].

The following soils are favorable for sweet cherry: leached chernozems, podzolized chernozems, regarded chernozems, typical chernozems, typical moist chernozems, common chernozems, common meadow chernozems, southern chernozems, dark-residually alkaline soils, light gray forest, chestnut gray forest soils, dark gray podzolized soils, soddy-podzolic loamy and sandy loam soils, dark gray podzolized regarded soils, soddy brown warm soils.

We refer to the limitly favorable soils as follows: common chernozems, southern alkaline chernozems, chernozems on fine clay, dark chestnut alkaline soils, chestnut alkaline soils, typical meadow chernozems, chernozems on the eluvium of the carbonate rocks, soddy carbonate soils on the eluvium of the carbonate rocks, chernozems on the eluvium of the thick rocks, soddy soils on the eluvium of the thick rocks, soddy podzolized loamy gley soils, podzolic brown soils, podzolic gleyed soils, brown mountain soils, medium eroded chernozems, podzolized soils.

Unfavorable for sweet cherry are: soddy chernozems, sandy soils, weakly humus sands, soddy podzolic sandy soils, soddy sandy soils, meadow soils, meadow alkaline soils, meadow bog and bog soils, peat-boggy soils, turf-peats, meadow chernozem soils, meadow chernozem alkaline soils, solonetzs, meadow chernozems, soddy solodized gley soils, degraded solonetzs, strongly eroded chernozems, podzolized soils, brunizems, dark brown soils, mountain meadow acidic soils, alkaline chernozems on fine clay.

The soils, which are favorable for sweet cherry, lie compact in the Fores-steppe and Steppe. It should be noted that among the favorable soils there are many heavy ones. They are unfavorable for sweet cherry on the mahaleb cherry seedlings, however it may be grown successfully here on other rootstocks.

The limitly favorable soils occur most frequently in the Donbas, on the left bank of the Dnieper, in Crimea, Southern Steppe, Western Foresr-steppe and Transcarpathia. Most of the unfavorable soils are in the Polissya, Carpatians, Mountaninous Crimea and in the valleys of rivers.

11.3.3 INTEGRAL CRITERION OF THE FRUIT CROP (CULTIVAR) BIOLOGICAL POTENTIAL REALIZATION DEGREE IN THE CONCRETE SOIL AND CLIMATIC CONDITIONS

We consider that the crop distribution to regions on the basis of taking into consideration main soil and climatic factors can be merely approximate because it does not reflect in full the peculiarities of certain cultivars and the genotype interaction with the total spectrum of the environment conditions. That's why we suggest that the most favorable for a certain crop (cultivar) regions should be specified by means of the integral criterion of this crop (cultivar) biological potential realization degree under the concrete cultivation conditions. This process includes the following stages.

1. The collection of the data about the crop (cultivar) productivity and weather conditions of regions for a certain period. This collection can ensure the authenticity on the 95% level. The calculation of regression equations of the productivity dependence on the weather conditions.
2. The calculation of the weather factors possible contribution into the crop (cultivar) productivity realized by means of the multiplication of the difference between minimum and maximum value of each factor to corresponding coefficients under variables in regression equations.
3. The calculation of the integral criterion (*IC*) of the regions favorability for crops (cultivars):

$$IC = \sum_{a=1}^{N} Y_a \left(X_a - \bar{X}_a \right) / C_a,$$

where Y_a is weight coefficient; C_a is the standard deviation; X_a and X_a is a desirable level of the weather factor a and its virtual value; N is the amount of factors.

4. The analysis of the integral criterion values: better conditions for the crop cultivation correspond to the lower value.

Taking into consideration the data of the integral criterion (IC) of the sweet cherry biological potential realization in the concrete soil and climatic conditions it is the Donbas (IC = 0.41), Podillya (IC = 1.1), Prydnistrovya (IC = 4.4), Southern Steppe (IC = 4.9) that are the most favorable regions for this crop. Much more favorable are the conditions in the northern and eastern parts of the Forest-Steppe (IC = 10.1–12.9).

11.3.4 PREPARATION OF THE RECOMMENDATIONS FOR THE CROP (CULTIVAR) DISTRIBUTION ON THE TERRITORY OF UKRAINE

The important result of our researches is the determination of the regions favorable, limitedly favorable and unfavorable for the main fruit crops cultivation by means of the complex estimation of the soil and climatic conditions and integral criterion of the degree of realizing the crop (cultivar) biological potential. Besides, we have analyzed the correspondence of the actual orchards distribution to the scientifically grounded requirements. The obtained data have been used for the preparation of the Program of the Development of the Horticulture in Ukraine for the years 2014–2020.

11.3.4.1 SWEET CHERRY

The total surface of the orchards will be 15.5 thousands of hectares, the surface in the Steppe (Dnipropetrovsk, Donetsk, Kherson, Mykolaiv, Odessa, and Zaporizhzhya regions) being about 70% (Fig. 11.1). The rest of the orchards are to be in the Podillya and Prydnistrovya [13].

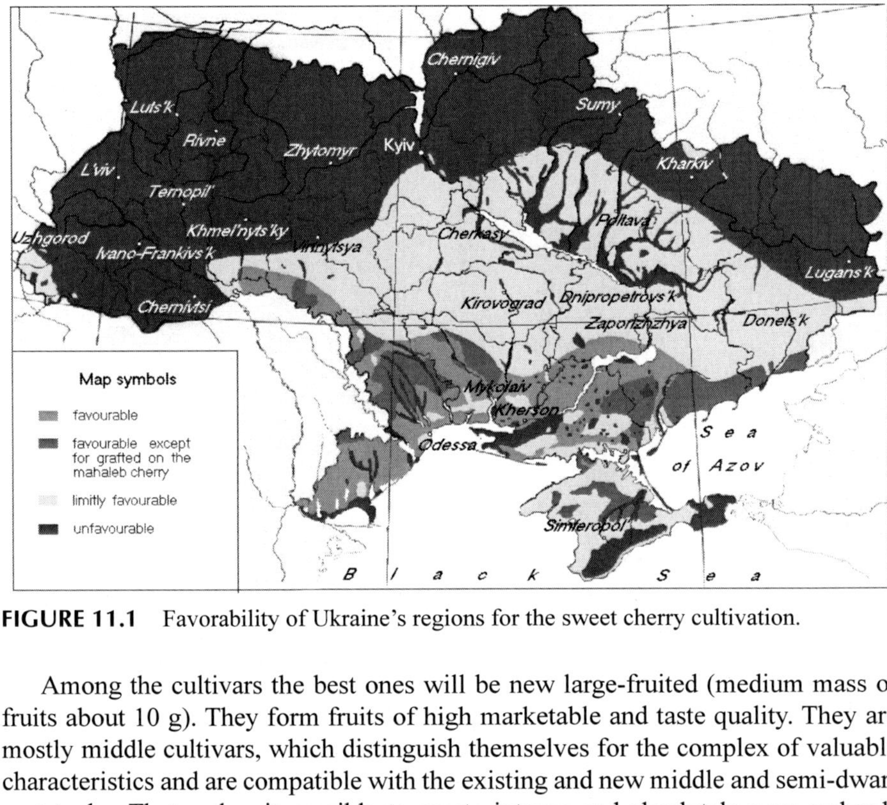

FIGURE 11.1 Favorability of Ukraine's regions for the sweet cherry cultivation.

Among the cultivars the best ones will be new large-fruited (medium mass of fruits about 10 g). They form fruits of high marketable and taste quality. They are mostly middle cultivars, which distinguish themselves for the complex of valuable characteristics and are compatible with the existing and new middle and semi-dwarf rootstocks. That makes it possible to create intense and absolutely new orchards which ensure the beginning of fruit-bearing in the 3 or 4 years after planting and in the period of the full fruit-bearing the yield will be not less than 20–25 t/ha. In the Steppe we shall prefer the inland cultivars 'Annushka,' 'Donets'ka, 'Jerelo,' Krasavytsya,' 'Krupnoplidna,' 'Melitopol's'ka Chorna,' 'Proshchal'na Taranenko,' 'Valeriya,' and 'Vasilisa Prekrasna.' In the other regions 'Amazonka' 'Annushka,' 'Donchanka' 'Donets'ky ugolyok,' 'Etyka' 'Lyubava,' 'Nizhnist,' 'Otrada,' 'Vasilisa Prekrasna,' 'Yaroslavna,' and others will be grown with the main role of their fruit is to be consumed as fresh.

11.3.4.2 APPLE

Apple is the main crop for fresh fruit production in Ukraine. The entire surface of the fruit-bearing orchards will be 132 thousands of hectare. Approximately 40% of them should be located in the Western Forest-Steppe, 40% in the Central Steppe and Crimea, the rest in other regions of the country. For apple it is optimal to grow cultivars of different ripening terms in the following correlation: summer cultivars 5%, autumn ones 15% and winter cultivars 80%.

In the most favorable regions of the Steppe and Forest-Steppe the leaders (about 60%) will be the introduced cultivars 'Champion,' 'Delicia,' 'Golden Delicious,' and clones, 'Jonagold' and clones as well as the inland cultivar 'Ranette Symyrenko' and its clones with a production level of 25–35 t/ha and in the regions with the less amount of warmth which lie farther to the north the Ukrainian cultivars (about 80%) 'Amulet, "Askol'da,' 'Papirovka,' and 'Slava Peremozhtsyam.' In order to augment the raw material base for processing (juices, purée, jams) and obtain fruit products with the pesticide load in the least possible degree and a yield of 25–30 t/ha in the Forest-Steppe and Western Polissya the high resistant and immune to the main fungous diseases cultivars will be preferable, namely: the inland ones 'Amulet,' 'Antonivka,' 'Askol'da,' 'Caleville Donets'ky' 'Edera,' 'Mavka' 'Radogost,' 'Sapfir,' and 'Skifs'ke zoloto' (about 70–80%) and the foreign cultivars 'Ligol,' 'Topaz,' and others (up to 20–30%).

11.3.4.3 PEAR

The total fruit-bearing orchards surface will be 16 thousands of hectares, 48% of them being proposed to be placed in the Steppe and Crimea, 40.6% – in the Forest-Steppe, 8.7% in the Polissya and 2.7% in the Carpathian region. In comparison to the existing distribution the amount of the orchards in the Carpathians is suggested to be decreased and that in the Steppe and Forest-Steppe be to increased considerably (Fig. 11.2).

The surfaces have been differentiated depending on the region and ripening terms. For instance, in the Pollisya summer and autumn cultivars only are proposed to be grown in the correlation of 46% and 54%, respectively. In the Forest-Steppe autumn cultivars should be preferred (46.7%), then winter cultivars (35.2%) and at last summer ones (18%). In the Steppe more than a half of surfaces have been assigned for winter cultivars, one third for autumn cultivars and the rest for summer ones. In the Peredcarpathians and Transcarpathians 37.5% of surfaces will be for winter and autumn cultivars each and a quarter for summer ones. The greatest part of the surfaces for summer cultivars is suggested for Crimea, southern regions and Carpathian region in order to provide with fruits tourists and those who rest in sanatoria.

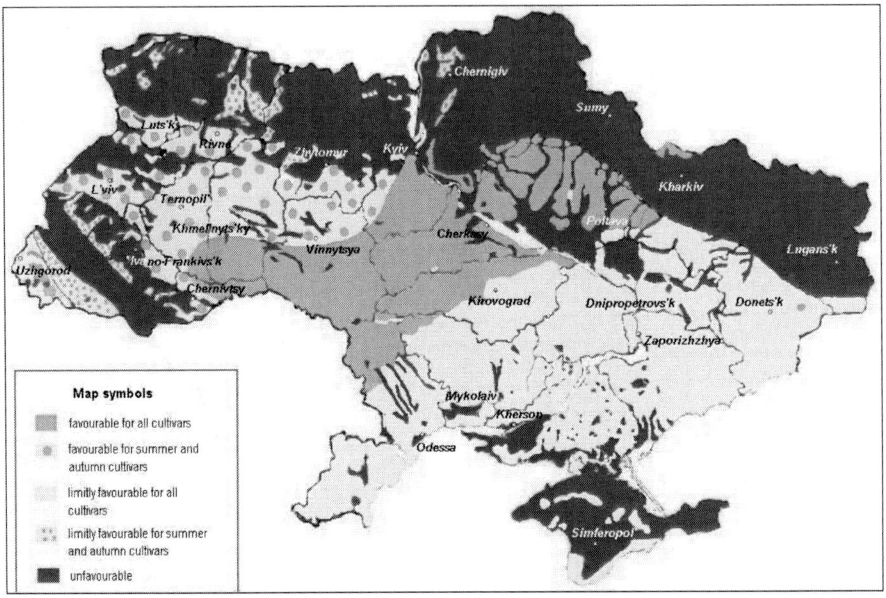

FIGURE 11.2 Favorability of Ukraine's regions for the pear cultivation.

Taking into consideration the main role of autumn which ensure a yield of 20–25 t/ha of high quality fruits the surfaces will be increased with the use of cultivars cultivar that are resistant to bacteriosis and fungous diseases of leaves (the inland cultivars 'Bukovynka,' 'Cheremshyna,' 'Khotynchanka,' 'Krupnoplidna,' 'Malivchanka,' 'Osin' Bukovyny,' 'Stryis'ka,' 'Yablunivs'ka,' and some introduced ones. The importance of the large-fruited and high quality cultivars 'Noyabrs'ka Moldavii,' 'Veresneve Devo,' 'Vyzhnytsya' will increase as well. The cultivars like 'Conference' will be important for the storage of fruits as frozen.

11.3.4.4 CHERRY

The entire surface of the orchards will be 22.1 thousands of hectares. Almost 60% of the industrial orchards are to be located in the Steppe (mainly the western and southern parts), approximately 30% in the Forest-Steppe (mostly the eastern part) and Prydnistrovya, about 10% in the other regions (Fig. 11.3).

The principal requirements to the cultivars must be their ecological tolerance, early ripening, high fruits quality and favorability for the mechanized harvesting. Approximately 25% of them are dessert cultivars of different ripening items. Other cultivars will be used for different types of processing.

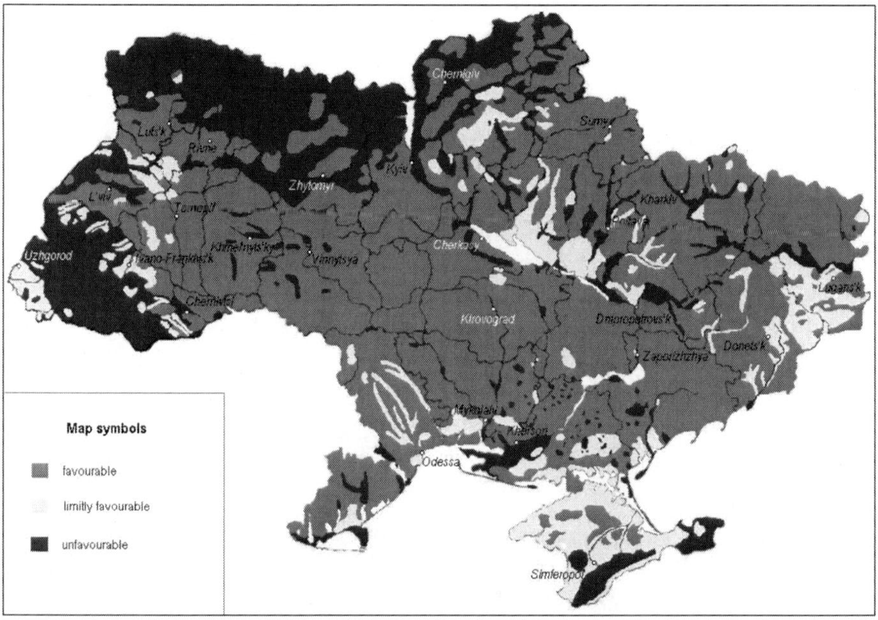

FIGURE 11.3 Favorability of Ukraine's regions for the cherry cultivation.

In the southern regions the cultivars 'Chudo-Vyshnya,' 'Igrushka,' 'Kseniya,' 'Melitopol's'ka Radist,' 'Shalunya, "Slavyanka' 'Vstryecha,' and others will be preferable that will ensure in the intense orchards a yield of about 20 t/ha.

The cultivars 'Kseniya,' 'Nochka,' 'Podbyel's'ka,' 'Shalunya, "Slavyanka,' 'Turgenyevka,' 'Vstryecha' with a yield of 15–20 t/ha will be more purposeful in the Forest-Steppe and Southern Polissye. The correction of the pesticide load on their fruits will be carried out taking into consideration the conditions of the year and *Coccomyces hiemalis* Higg. resistance.

The creation of the cherry orchards for the fruits processing exclusively will have an important role in the horticultural processing industry.

It is self-fertile and partly self-fertile cultivars with high quality fruits, resistant to fungous diseases (*Coccomyces hiemalis* Higg. and monilia) ('Molodizhna,' 'Nord Star') and favorable for the mechanized harvesting ('Lotovka' and others) that will be competitive ones.

11.3.4.5 PLUM

The total orchards surface will be 22.0 thousands of hectares. Approximately 50% of the industrial orchards are to be concentrated in the Steppe (Donet'sk, Zaporizhzhya, Dnipropetrovs'k, Kirovograd, Mykolaiv and Odessa regions) and about

40% in the Forest-Steppe (Kharkiv, Poltava, Vinnytsya, and Khmelnyts'ky regions), the rest in the other regions, which are favorable for plum, those orchards, as a rule, being nonmarketable on the individual patches. The surfaces of the classical but not Plum pox virus resistant cultivars 'Ugorka Italiis'ka,' 'Veresneva,' 'President' etc. will be decreased while for the ukrainian cultivars 'Nen'ka,' 'Lagidna,' 'Reineclaude Karbysheva,' 'Reineclaude Rannii,' 'Ugorka Donets'ka Rannya' and introduced ones 'Hanna Spät,' 'Stenley' with a yield of over 20 t/ha will be increased. It is the existing and new cultivars, Plum pox virus resistant with high fruit marketable qualities and a yield level of 20–30 t/ha ('Khanita,' 'Chachakska Lepotica' and others that will enjoy the greatest demand. Concerning the ripening terms it is just the universal cultivars (for freezing, drying, processing) that will occupy up to 75–80% of surfaces, that is late cultivars from the group of 'ugorkis' ('Stenley,' 'Donets'ka Konservna,' 'Ugorka Donets'ka' and others). At the same time taking into consideration the severe affection of ugorkis by plum fruit moth and chalcid wasps it is impossible to reduce their fruits pesticide load.

11.3.4.6 APRICOT

The entire surface of the orchards will be 10.6 ha. Almost 35% of all the industrial orchards should be located on the favorable soils of the Southern Steppe, approximately 40% in the Central Steppe, Podillya, Prydnistrovya and partly in the Volyn.' In this region it is best to use patches with the most favorable conditions protected from the northern winds.

About 25% of the orchards may be placed on the favorable soils in the regions with risky cultivation – the L'viv, Volyn,' Zhytomyr, Kyiv, Poltava, Kharkiv, and Donets'k regions.

On the condition of growing in the most favorable regions the existing ('Chervonoshchoky,' 'Melitopol's'ky Piznii,' 'Kyivs'ky Krasen,' ' 'Olimp,' 'Parnas' and others) and new inland and foreign ('Early Gold') large-fruited cultivars will be preferable. At the same time it is impossible to avoid necessary sprinkling with fungicides against monilia at different stages of its manifestation. Early cultivars ('Melitopol's'ky Rannii and others) will not occupy large surfaces. Those for the cultivars with the high organic oil content in the kernel will increase for the usage in the perfumery and cosmetic industry.

11.3.4.7 PEACH

From the total orchards surface 7.7 thousands of hectares approximately 85% of the peach industrial orchards should be placed in the Southern Steppe and about 15% in the Transcarpathians, Podillya, Prydnistrovya and Central Steppe of Ukraine. The dynamics of the surfaces increase will take place for the cultivars with yellow flesh of fruit ('Redhaven,' 'Lyubimyets 2,' 'Potomok,' 'Donetsky Zhyolty') of the universal usage (preservation, making juice with flesh, mixing with other fruits as well as for cultivars with white flesh which are resistant to the hibernation condition and leaf curl.

11.3.4.8 BLACK CURRANT

The entire orchards surface is 5.4 thousands of hectares. The favorable regions for the industrial cultivation are in the Ukraine's Forest-Steppe where approximately 85% of all the orchards will be placed and the rest in the Northern Steppe (Fig. 11.4).

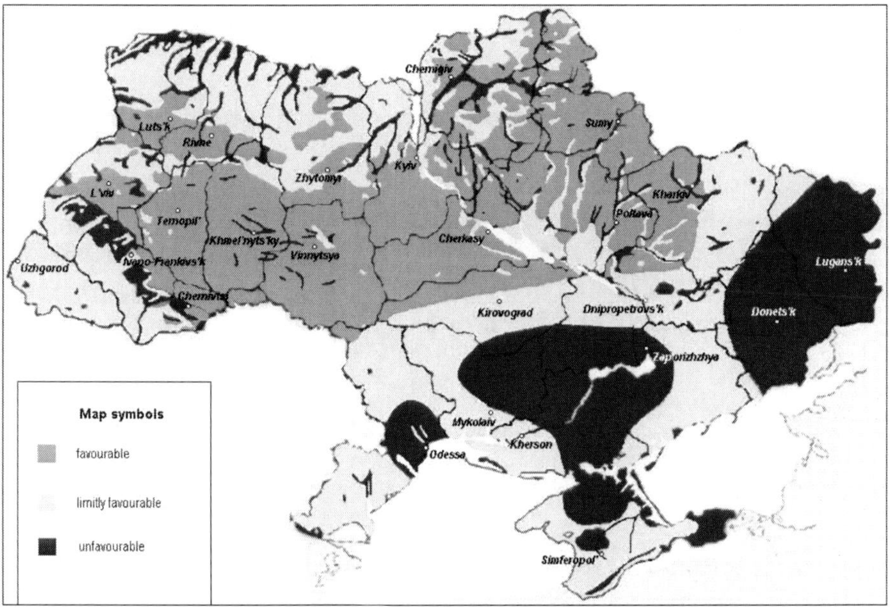

FIGURE 11.4 Favorability of Ukraine's regions for the black currant cultivation.

The policy concerning cultivars will be based on the inland assortment. Cultivars of Ukrainian breeding will constitute no more than 90% of the whole industrial assortment. 'Konsul,' 'Kopanya,' 'Krasa Lvova,' 'Sophiivs'ka,' 'Sanyuta,' 'Vernisazh,' 'Volodymyrs'ka,' 'Yuvileina' will remain of the especial importance. However, they will be gradually substituted for those of the new generation – 'Kazkova,' 'Melodiya,' 'Muza,' 'Nadbuzhans'ka,' 'Nimfa,' 'Oriana,' 'Raduzhna,' and 'Verbna.'

11.4 CONCLUSIONS

The regions of Ukraine that are the most favorable for the industrial cultivation of apple, pear, plum, cherry, sweet cherry, apricot, peach and black currant have been determined on the base of the proposed methods of the fruit crops distribution which are grounded on the integral evaluation of displaying their biological potential in concrete soil and climatic conditions.

KEYWORDS

- **biological potential**
- **fruit trees and shrubs**
- **soil**

REFERENCES

1. Dragavtseva, I. A. (1990). Ecological Resources of the Apricot Productivity in Southern Russia, Krasnodar, 94p (in Russian).
2. Kashin, V. I. (2000). Manifestation of the Orchard Plants Biological Potential, Orchard Plants Biological Potential and ways of its Realization Papers of the International Conference (July, 19–22, 1999) VSTISP, Moscow, 3–15 (in Russian).
3. Lebedev, V. M. (1999). Climate Getting Warmer Influence on Fruit Plants, Scientific Fundamentals of the Resistant Horticulture in Russia, Reports of the Conference, (March, 11–12 1999), Russian Academy of Agricultural Sciences Michur in Research Institute of Horticulture, Michurinsk, 50–53 (in Russian).
4. Methodological Approaches to the Southern Horticulture Development (2000) Krasnodar, 49p.
5. Dragavtseva, I. A. (1999). Ecological Method of the Fruit Crops Optimal Distribution, Papers of International Scientific and Practical Conference "Horticulture and Viticulture of the 21st Century", Part 2 Horticulture, Krasnodar, 38–41 (in Russian).
6. Alekseev, R. P. (1998). Methods of the Plants Development Analysis Taking into Consideration Meteorological Factors, Methods of the Researches and Analysis of Variance in the Scientific Horticulture, *1*, Michurinsk, 106–107 (in Russian).
7. Bublyk, M. O. (1985). Mathematical Modeling in the Horticultural Experiments, Fruit and Vegetables Farmings, *6*, 17–21 (in Russian).
8. Moroz, V. N. (1988). Regression Analysis of the Agricultural Crops Cultivars Resistance, Measuring and Computer Engineering in the Management of the Production Processes in the Agro Industrial Complex, Part 2 Leningrad, 314–317 (in Russian).
9. Bublyk, M. O. (2002). Integral Criterion of the Region Favorability for the Fruit Crops Cultivation as Concerns the Weather Factors, Herald of the Agrarian Science, *6,* 31–33 (In Ukeainian).
10. Bublyk, M. (2002). Influence of Weather Factors on the Stone Crops Productivity in Ukrain, Fruit, Nut and Vegetable Production Engineering, Proceeding of the 6th International Symposium Held in Potsdam (2001) Potsdam-Bornim, 117–121.

11. Ivanov, V. F. (1998). Ecology of the Fruit Crops, Ivanov, V. F., Ivanova, A. S., Opanasyenko, N. Y. et al. Kyiv Agrarna Nauka, 410p (in Russian).
12. Negovyelov, S. F., & Val'kov, V. F. (1985). Soils and Orchards Rostov, 192p (in Russian).
13. Bublyk, M. O., Barabash, L. O., Fryziuk, L. A., & Chorna, G. A. (2010). Rational Distribution of the Farm Orchards on the Main Fruit Crops in Ukraine Sweet Cherry and Apricot, Vegetables and Fruits, February, 32–35.

CHAPTER 12

SUGAR BEET PRODUCTIVITY FORMATION DEPENDING ON FOLIAR APPLICATION OF MICROELEMENTS

VOLODYMYR A. DORONIN[1] and LESYA M. KARPUK[2]

[1]Institute of Bioenergy crops and Sugar Beet of NAAS of Ukraine, 25 Klinichna Str., Kyiv, 03141, Ukraine; E-mail: doronin@tdn.kiev.ua

[2]Bila Tserkva National Agrarian University, 8/1, Soborna Square, Bila Tserkva, Kyiv region, 09117, Ukraine; E-mail: zuikes@ukr.net

CONTENTS

ABSTRACT

It was analyzed the research results of sugar beet foliar application influence, at stated periods of vegetation, by various types and norms of microelements on the yield and root crops quality formation in the conditions of unstable moistening in the right bank of the Central Forest-Steppe Zone of Ukraine. It was revealed that the sugar beet yield formation, sugar content and sugar yield are significantly depends on plants foliar application in their respective phases of vegetation forms of microelements and norms of their application. Foliar application is conducting in closing leaves in a row phase and by the 30 days before harvesting by different microfertilizers at the norms of application from 3.0 to 7.0 L/ha was promote the macronutrients better absorption from the soil, which in turn influences on the final sugar beet productivity. The most effective is foliar application, which provides roots growth productivity and sugar yield. Foliar application is provides a significant increase not only roots productivity, but their sugar yield from hectare at conducting in a month before harvesting.

12.1 INTRODUCTION

The questions about plant organism relationship patterns and environment are open the unlimited possibilities for human on the growth and plants development influence in order to obtain the high and stable yields [1, 2]. The one of the ways to address on the production process is the using sugar beet foliar application of microelements. The foliar application effectiveness is depends on the types of micronutrients, norms of their consumption and timing of feeding. Micronutrients are accelerate the plants development and seed ripening, increase the plant resistance to adverse environmental conditions, as well as make them resistant against several bacterial and fungal diseases. It was established [3] that for the plants are most effectiveness the biologically activeness microelements in the form of chelates microfertilizers – kompleksons (chelated metal compounds). Foliar feeding can improve the rates of assimilation of nutrients from fertilizers; significantly reduce their doses without crop productivity reducing, to achieve a uniform distribution of micronutrients within a field area, to minimize the stress of pesticide treatments, to provide all the necessary elements of plant nutrition at critical periods of their development [4]. Through the leaves and stems is allows to optimize the norm and ratio between nutrients during the growing season. A lot of attention was paid in the Zarishnyak [4], Bulygin [3], etc. works for foliar application issues by the chelate forms of fertilizers. Earlier research works were established [3], that foliar feeding of sugar beet hybrids that created on cytoplasmic male sterility (CMS) basis within Reakom-R-beet microfertilizer is positively effected on plant growth and development, the leaf surface formation, mass of roots increase, dry matter accumulation, which ultimately leads to in yield of roots and sugar increases.

The goal of our researches was influence of new microfertilizers Reastim-humus-beet and Reakom-plus-beet study, in which the ratio of microelements is balanced, with taking into account their needs for sugar beet productivity and photosynthesis, respectively – on sugar beet yield. Microelements that are part of these micronutrients are in a biologically active form, so easily and quickly by plants absorbed. In the control variant micronutrients were not used. As the standard was used domestic microfertilizer Reakom-R-beet.

12.2 MATERIALS AND METHODOLOGY

Field experiments were performed during 2010–2012. It was directed on the effect of foliar application of sugar beet in the growing period by the different types of microelements and norms on the yield formation and quality of roots in the field conditions of Bila Tserkva National Agrarian University, which is located in the zone of unstable moistening.

Under programmed sugar beet yields of 70 t/ha was created the common background with the application of organic fertilizers and mineral fertilizers. At the same time was taken into account the nutrient reserves in the soil. For the problem of increasing the productivity of photosynthesis of sugar beet, respectively, – the yield of roots solving was conducted three-factor experiment which investigated the influence of the timing of foliar feeding application (factor A), types of micronutrients (factor B) and norms of their application (Factor C). In the experiments were use microfertilizers of Ukrainian production research and production center "Reakom": Reakom-R-beet (standard), Reastim-humus-beet and Reakom-plus-beet. Microelements were added in two terms: closing leaves in a row phase and one month before harvesting by different norms from 3.0 to 7.0 L/ha.

Determination of the pure productivity of photosynthesis, which is measured in grams of dry matter per m^2 of leaf area per one day (g dry matter/m^2 leaf surface per day) was determined by Nichiporovich method [5]. The components of the net productivity of photosynthesis are leaf surface area, which is measured in thousands of square meters of leaf area per hectare (thousands m^2/ha) and photosynthetic potential (mln $m^2 \times$ day/ha). This experience gives the opportunity to comprehensively assess the effectiveness of this agrotechnological technique.

A plant analyzes and other observations were performed according to existing methods that have been used in domestic practice [6]. Experiment was laid by the method of split plots location repetitions systematically, consistently. Experiments repeated were fourfold. For research it was use hybrid seeds of domestic selection Ukrainian ChS 72.

The statistical data processing was performed on a personal computer by R. Fisher method [7].

12.3 RESULTS AND DISCUSSION

Sugar beet plants productivity is the result of all metabolic process in which was balanced the steps of organic matter forming and its expenditure on growth, development, breathing and other vital processes. Influence of individual agricultural methods research including foliar application and its impact on this important biological process is theoretical and practical significance.

Our researches have shown that during all sugar beet vegetation period, raw mass of roots and leaves growth were passed irregularly. In the first half of the growing season was happened quite intense growth of the assimilation apparatus that due to plant genetic features. From the biological point of view is warranted, as it leaves apparatus is synthesizes dry matter, which in the second half of vegetation more intensively accumulated in the sugar beet roots. At the end of vegetation the leaf surface area and leaf mass in the terms of a balanced fertilizer system are naturally decreases. Root mass was increased throughout the growing season.

On September 1st it is established that on the root mass and leaf mass increase have essentially influenced the types and norms of microelements using in the both terms of their application (Table 12.1).

On the I date of accounting – September 1, in the phase of leaves closing in a row, at the application norm of 3 L/ha of microfertilizer Reakom-plus-beet the leaf mass was 120.8 g, at the norm of 5 L/ha –163.2 g and the highest leaf mass was noted in the variant with microfertilizer application in the norm of 7 L/ha – 170.7. In all variants, the growth of leaves was significantly higher than on the control. At microelements application a month before harvesting were obtained similar results. The greatest value of leaves mass was also obtained in the variant with the microelements application of Reakom-plus-beet fertilizer at the norm of 7 L/ha – 185.4 g. Analogous results were obtained with the foliar application of Reastim-humus-beet microfertilizer. With the increase of the microelements norms using is increases the mass of leaves. Mass of leaves in the closing leaves in rows phase in the norm of 5 L/ha at the foliar feeding by Reakom-R-beet (standard) microfertilizer was 140.6 g, by Reastim-humus-beet – 156.5 g, Reakom-plus-beet microfertilizer – 163.2 g (SSD_{05} = 12.7 g) on September 1. Significant difference in a leaves mass gain is depending on types of micronutrients is not found.

On the second date of accounting – before harvesting (October 30) was set to decrease leaves mass gain in all variants, compared with the first date of accounting, which is related with the sugar beet biological features, i.e. a decrease in the assimilation apparatus functionality and a significant nutrients outflow from leaves to the roots. If at the first term of Reastim-humus-beet microelements application in the norms of 5 and 7 L/ha on September 1 the leaf mass gain was 156.5 and 163.1 g, then before harvesting it was 101.4 and 104.6 g, respectively. At Reakom-plus-beet and Reakom-R-beet (standard) microfertilizers application were obtained the similar results. It is worth to noting that at the foliar application in the second term (30 days before harvest) most effective was Reakom-plus-beet microfertilizer, which provided the roots mass gain at the all application norms compared with Reastim-humus-beet t micronutrient.

TABLE 12.1 The Dynamics of Sugar Beet Root and Leave Mass Increases Depending on the Timing, Types and Application Norms of Microelements in Feeding (Average of 2010–2012)

Type of microfertilizers (factor B)	Application norm, L/ha (factor C)	Mass of leaves, g		Mass of roots, g	
		September 1	before harvesting	September 1	before harvesting
Phase of closing leaves in a row (factor A)					
Without feeding (control)	-	126.3	57.8	301.9	342.5
Reakom-R-beet (standard)	5.0	140.6	73.7	340.9	382.3
Reastim-humus-beet	3.0	137.3	91.9	324.8	386.8
	5.0	156.5	101.4	357.0	414.7
	7.0	163.1	104.6	396.3	448.9
Reakom-plus-beet	3.0	129.8	85.8	339.6	421.9
	5.0	163.2	90.1	380.1	461.5
	7.0	170.7	131.1	418.9	482.6
A month before harvesting (factor A)					
Without feeding (control)	-	117.7	51.6	304.3	337.9
Reakom-R-beet (standard)	5.0	133.6	69.4	328.6	364.3
Reastim-humus-beet	3.0	124.0	88.4	330.6	409.9
	5.0	150.1	90.5	351.5	457.1
	7.0	164.7	125.3	409.0	477.1
Reakom-plus-beet	3.0	137.6	98.1	361.1	432.5
	5.0	164.4	103.2	387.4	470.2
	7.0	185.4	144.1	438.1	503.3
SSD_{05} factor A (term of application)		12.9	19.8	10.9	26.1
SSD_{05} factor B (type of micro-fertilizer)		12.7	18.2	15.1	21.2
SSD_{05} factor C (application norm)		3.3	6.7	5.8	8.4

Root mass was also increased significantly with increasing consumption norms depending on microelements and their application norms. At the first term of Reakom-plus-beet microfertilizer application in the leaves closing in a row phase, at the norm of 3 L/ha, at the first of September, the root mass was 339.6 g, in the norms 5 and 7 L/ha – 380.1 and 418.9 g, respectively. At Reastim-humus-beet microfertilizer foliar feeding in norms of 3, 5 and 7 L/ha root mass were 324.8, 357.0 and 396.3, respectively. According to the root masses depending on the type of micronutrients, it was established that under the same application norm of preparations – 5 L/ha the higher mass of root – 380.1 g was obtained by Reakom-plus-beet microfertilizer foliar feeding or above on 23.1 g compared with Reastim-humus-beet microfertilizer and 39.2 g, compared with standard Reakom-R-beet (SSD_{05} = 21.2 g).

At the second term of micronutrient fertilizer application was observed a similar dependence on the root mass growth, depending on the term, type and application norms of micronutrients (SSD_{05} factor A = 26.1 g, factor B = 21.2 g factor C = 8.4 g). Roots mass in the variant with the standard was higher by 26.4 g, on the variants with the Reastim-humus-beet application by the different norms above 72–139.2 g, and on the variants with Reakom-plus-beet application in different norms above 94.6–165.4 compared to the control (without fertilization).

Thus, the raw mass of roots and leaves growth during of sugar beet growing season is undergo more intensive in variants with foliar feeding application of new Reakom-plus-beet t and Reastim-humus-beet microfertilizers. The growth trend of root raw mass in variants with foliar fertilizer with microelements was persisted during the harvest season. Significant influence on plant growth and development, both in the closing plants in a row phase, and before the harvest, had types of micronutrients, their norms and terms of application. With the application norms of micronutrients increasing from 3 to 7 L/ha was increased leaves and root wet mass. Under foliar application conducting during the second term (30 days before harvest) effectively was Reakom-plus-beet microfertilizer.

Dynamics of roots and leaf mass growth is inextricably linked with the intensity of photosynthetic processes transmission.

Photosynthesis process regulation i.e. its productivity increasing – is one of the most effective methods of influence on the sugar beet productivity, and for them – an important means of yield level increasing.

Researches have established that on the efficiency of sugar beet photosynthesis is significantly influence the timing of micronutrients foliar feeding, types and norms of their application. Observation of the growth and development of sugar beet plants assimilation surface on the first date of registration (September 1) when microelements is making in the closing leaves in a row phase was showed that leaf surface area depending on the term, types and norms of microelements application on average was ranged between 34.5 to 46.6 thousand m^2/ha (Table 12.2).

TABLE 12.2 Leaf Surface Area and Photosynthetic Productivity Depending on the Types, Norms and Terms of Microelements Applying in Feeding on September 1 (Average of 2010–2012)

Type of micro-fertilizers (factor B)	Application norm, L/ha (factor C)	Leaf surface area, thousand m^2/ha	Photosynthetic potential, million m^2 × days/ha	Pure photosynthesis productivity, g dry matter/m^2 of leaf area per day
Phase of closing leaves in a row (factor A)				
Without feeding (control)	-	34.5	1.03	5.48
Reakom-R-beet (standard)	5.0	38.4	1.15	5.67
Reastim-humus-beet	3.0	37.5	1.12	5.61
	5.0	42.7	1.28	5.96
	7.0	44.5	1.34	6.26
Reakom-plus-beet	3.0	35.5	1.06	5.88
	5.0	44.6	1.34	6.50
	7.0	46.6	1.40	6.61
A month before harvesting (factor A)				
Without feeding (control)	-	32.1	0.96	4.65
Reakom-R-beet (standard)	5.0	36.5	1.09	5.29
Reastim-humus-beet	3.0	33.9	1.02	5.54
	5.0	41.0	1.23	5.67
	7.0	45.0	1.35	5.73
Reakom-plus-beet	3.0	37.6	1.13	5.37
	5.0	44.9	1.35	5.97
	7.0	50.6	1.52	6.31

On the control variant (without feeding), average for research years, the leaf surface area was 34.5 thousand m^2/ha and its higher value was obtained in the variant of Reakom-plus-beet application in the norm of 7 L/ha – 46.6 thousand m^2/ha, that is caused by optimal area of plant nutrition and the best leaf surface assimilation apparatus formation. The smallest value of the leaf surface index, relative to the control, was obtained in the areas with Reastim-humus-beet and Reakom-plus-beet micronutrients application in the norm of 3 L/ha, respectively, 37.5 and 35.5 thousand m^2/ha. That is, on leaf surface increasing a significant influence had as a form of micronutrients and the norm of application. On a variant with using for foliar feeding Reastim-humus-beet at the application norms of microelements increasing from 3 to 7 L/ha the leaf surface area was increased on 7.0 thousand m^2/ha, and on the plots with Reakom-plus-beet application the leaf area was increased on 11.1 thousand m^2/ha.

According to the research results on the variants with the application of various types of micronutrients with different norms, on the first of September the photosynthetic potential was average and was within 1.03–1.40 million m^2 days/ha. So, on the variant with the application of Reakom-R-beet (standard) micronutrient in the norm of 5 L/ha the index of photosynthetic potential was 1.15 million m^2 days/ha, which is on 0.12 million m^2 days/ha higher than on the control variant (without application).

High indicators of photosynthetic potential were obtained in variants with Reastim-humus-beet and Reakom-plus-beet microfertilizers application in the norms of 5 and 7 L/ha. In comparison with the application norm of 3 L/ha these indicators were increased on 0.06–0.34 million m^2 days/ha.

Reakom-R-beet, Reastim-humus-beet and Reakom-plus-beet microfertilizers foliar feeding application amid a general background of fertilizer is establishing appropriate conditions for the photosynthetic process intensity increasing, especially pure photosynthetic productivity. The most favorable were areas in which the feeding was carried out in closing leaves in a row phase with the norm of micronutrients application of 5 and 7 L/ha. At Reakom-R-beet application in the recommended norm of 5 L/ha for the production the pure photosynthetic productivity was 5.67 g of dry matter/m^2 leaf area per day, at Reastim-humus-beet application in the norms of 5–5.96 and 7 L/ha – 6.26 g dry matter/m^2 leaf area per day, respectively. After Reakom-plus-beet application in the norms of 5 and 7 L/ha, the pure photosynthetic productivity was 6.50 and 6.61 g dry matter/m^2 leaf area per day, respectively. Given that in these variants there was a high photosynthetic potential (1.15–1.40 million m^2 days/ha) and has created a favorable physiological background for productive work of each plant cell by the expense of micronutrients application, it were created the necessary conditions for a high level of photosynthesis process passing.

On September 1, at the second period of micronutrients application were also determined the indicators of photosynthesis productivity. It should be noted that the indicators of leaf surface area in all variants were almost in a par, as in the variants after the first term of micronutrients foliar feeding were in the range of 32.1 to 50.6 thousand m^2/ha. Photosynthetic potential value (0.96–1.52 million m^2 days/ha) and pure photosynthetic productivity (4.65–6.31 g dry matter/m^2 leaf area per day) were lower in comparison with the first period of micronutrient application.

Summing up it should be noted that foliar feeding application in closing leaves in a row phase is delivers the productivity of photosynthesis increasing, particularly in variants where used Reakom-R-beet microfertilizer at application norm of 5 L/ha,

Reastim-humus-beet at application norms of 5 and 7 L/ha and the Reakom-plus-beet at the same norms, which resulted the high indicators of leaf area from 38.4 to 46.6 thousand m^2/ha, the photosynthetic potential of 1.15 to 1.40 million m^2 days/ha and pure photosynthetic productivity from 5.67 to 6.61 g dry matter/m^2 leaf area per day, and this in turn is impact on sugar beet final productivity. On pure photosynthetic indicators productivity it is possible to predict the sugar beet productivity depending on the norms and types micronutrient application in feeding.

Assimilation surface of sugar beet plants growth and development observation in the second registration date (October 30) at the microelements application in closing leaves in a row phase was showed that leaf surface area depending on the term, types and application norms of microelements, on average, ranged from 15.8 to 35.8 thousand m^2/ha that at 10.8–18.7 thousand m^2/ha less in comparison with the similar period on the first date of registration – the first of September (Table 12.3). This is related to the sugar beet biological features because in the autumn at the harvest time, leaf mass does not develop, but on the contrary, begins to die and most share of nutrients goes from the leaves to the roots.

On the control variant (without feeding) the leaf surface area, in average of research years, was 15.8 thousand m^2/ha and its higher value was obtained in the variant with Reakom-plus-beet micronutrient in norm of 7 L/ha application – 35.8 thousand m^2/ha, that due to the optimal area of plant nutrition and the best assimilation apparatus of leaf surface formation. The smallest value of leaf surface area, concerning the control was in the areas with Reastim-humus-beet and Reakom-plus-beet micronutrients application in norm of 3 L/ha was obtained of 25.1 and 23.4 thousand m^2/ha, respectively. That is, on the leaf surface area increase was affected as the timing of foliar feeding type of micronutrients and their norms of application. In an variant with using Reastim-humus-beet micronutrient for foliar feeding with increasing application norms of microelements from 3 to 7 L/ha the leaf surface area was increased to 3.5 thousand m^2/ha, and in the plots with Reakom-plus-beet micronutrient application leaf surface area was increased to 12.4 thousand m^2/ha.

According to the research results in variants with the application of various types of micronutrients with different norms of application, at 30 October the photosynthetic potential was unsatisfactory and average, and was in the range from 0.47 to 1.07 million m^2 days/ha. On the variant with the Reakom-R-beet micronutrient (standard) application in the norm of 5 L/ha the photosynthetic potential indicator was 0.60 million m^2 days/ha, which is on 0.13 million m^2 days/ha more, than on control variant (without feeding).

TABLE 12.3 Leaf Surface Area and Photosynthetic Productivity Depending on the Types, Norms and Terms of Microelements Applying in Feeding on October 30 (Average of 2010–2012)

Type of microelements (factor B)	Application norm, L/ha (factor C)	Leaf surface area, thousand m^2/ha	Photosynthetic potential, million m^2 days/ha	Pure photosynthesis productivity, g dry matter/m^2 of leaf area per day
Phase of closing leaves in a row (factor A)				
Without feeding (control)	-	15.8	0.47	2.78
Reakom-R-beet (standard)	5.0	20.1	0.60	4.40
Reastim-humus-beet	3.0	25.1	0.75	4.34
	5.0	27.7	0.83	4.46
	7.0	28.6	0.86	4.57
Reakom-plus-beet	3.0	23.4	0.70	4.35
	5.0	24.6	0.74	4.48
	7.0	35.8	1.07	4.82
A month before harvesting (factor A)				
Without feeding (control)	-	14.1	0.42	2.56
Reakom-R-beet (standard)	5.0	19.0	0.57	4.36
Reastim-humus-beet	3.0	24.1	0.72	4.19
	5.0	24.7	0.74	4.36
	7.0	34.2	1.03	4.81
Reakom-plus-beet	3.0	26.8	0.80	4.57
	5.0	28.2	0.85	4.70
	7.0	39.3	1.18	4.95

The high indicators of photosynthetic potential were obtained in variants with Reastim-humus-beet and Reakom-plus-beet microfertilizers application in norms of 5 and 7 L/ha. In comparison with the norm of 3 L/ha application these indicators were increased on 0.03–0.37 million m^2 days/ha.

Application of Reakom-R-beet, Reastim-humus-beet and Reakom-plus-beet microfertilizers in foliar feeding amid a general background of fertilizer, at application the microelements in both terms is establishing the appropriate conditions of photosynthetic process intensity increasing, especially its pure productivity. However, in comparison with the first registration date these indicators were slightly lower that caused by physiological and biological features of sugar beet. The most favorable were plots on which the feeding is carried out in a clamping leaf in row phase with the flow norm of microelements of 5 and 7 L/ha. After Reakom-R-beet application in the recommended norm for the production of 5 L/ha the pure photosynthetic productivity was 4.40 g of dry leaf surface matter/m^2 leaf surface per day, Reastim-humus-beet in the norms of 5 and 7 L/ha – 4.34 and 4.57 g dry matter/m^2 leaf surface per day, and for Reakom-plus-beet application in this norm, – 4.35 and 4.82 g dry leaf surface matter/m^2 per day, respectively.

On October 30th, at the second term of microelements entering were also determined the photosynthetic productivity indicators and studying the influence of microelements on this process. It should be noted that the indicators of leaf surface area in all variants with Reakom-R-beet in the norm of 5 L/ha, Reastim-humus-beet in the norms of 5 and 7 L/ha and Reakom-plus-beet in the same norms application were almost at the level as in the variants for the first period make of microelements application in feeding and ranges from 19.0 to 39.3 thousand m^2/ha. The highest value of photosynthetic potential was obtained in the variant with Reakom-plus-beet in norm 7 L/ha application – 1.18 million m^2 per day/ha, which is on 0.11 million m^2 per day/ha, more than by the first term of micronutrients application. At the second period of Reakom-plus-beet microfertilizer making at application norms of 3 and 7 L/ha was also received high pure photosynthetic productivity (4.57 and 4.95 g dry matter/m^2 leaf surface per day), compared with the first period of microelements application in feeding.

The data is summarizing, it should be noted that the foliar feeding use a month before harvest was ensured the photosynthetic productivity growth, particularly in variants where was the studied the effectiveness of Reakom-R-beet micronutrients at application norm of 5 L/ha, Reastim-humus-beet at application norms of 5 and 7 L/ha and the Reakom-plus-beet at the same norms of application. Sugar beet foliar feeding by micronutrients of various types, and norms in different periods is enhances the intensity of the photosynthesis process passage.

The higher norms of micronutrients use for sugar beet foliar feeding in closing leaves in a row phase is provide a photosynthetic productivity increase. On the first date of registration the leaf surface area in closing leaves in a row phase was increased on 3–16%, a month before harvest – on 5–37%: on the second date of registration in in closing leaves in a row phase at 21–56% and per month before harvest at 26–64%. On the first date of the registration photosynthetic potential in closing leaves in a row phase was increased on 3–26%, a month before harvest on 6–37%, on the second date of registration in closing leaves in a row phase – on 20–55%, a month before harvest on 25–63%. On the first date of the registration pure photosynthetic productivity in closing leaves in a row phase was increased on 2–17%, a month before harvest – on 12–26%, in the second date of registration p in closing leaves in a row phase on 36–42%, a month before harvest on 39–48%.

Before sugar beet sowing and after harvesting was performed the determination of macronutrients in soil in variants, where spent the foliar feeding by various types of microfertilizers. The results were showed that hydrolyzable nitrogen (N), movable phosphorus compounds, and potassium (K_2O) amount was decreased (Table 12.4), which earlier results is confirms that microelements are facilitate the absorption of nutrients of plants from the soil [5].

TABLE 12.4 Content of Nutrition Elements in Soil (mg/kg of soil), Depending on the Timing, Types and Application Norms of Microelements in Sugar Beet Feeding (Average of 2010–2012)

Type of microelements (factor B)	Application norm, L/ha (factor C)	In the period of sowing			In the period of harvest		
		N	P_2O_5	K_2O	N	P_2O_5	K_2O
Phase of closing leaves in a row (factor A)							
Without feeding (control)	-	335.3	305.0	282.7	234.7	197.9	76.3
Reakom-R-beet (standard)	5.0	335.3	305.0	282.7	201.2	195.0	70.2
	3.0	335.3	305.0	282.7	227.7	194.5	74.9
Reastim-humus-beet	5.0	335.3	305.0	282.7	200.0	191.9	72.6
	7.0	335.3	305.0	282.7	191.2	188.6	72.2
	3.0	335.3	305.0	282.7	229.4	194.8	74.6
Reakom-plus-beet	5.0	335.3	305.0	282.7	201.4	193.0	73.3
	7.0	335.3	305.0	282.7	192.5	189.1	70.8
A month before harvesting (factor A)							
Without feeding (control)	-	335.3	305.0	282.7	234.7	197.9	76.3
Reakom-R-beet (standard)	5.0	335.3	305.0	282.7	202.2	196.6	72.0
	3.0	335.3	305.0	282.7	229.0	196.8	74.6
Reastim-humus-beet	5.0	335.3	305.0	282.7	201.2	193.4	72.4
	7.0	335.3	305.0	282.7	194.2	190.3	71.5
	3.0	335.3	305.0	282.7	229.6	197.5	74.4
Reakom-plus-beet	5.0	335.3	305.0	282.7	200.7	194.1	71.5
	7.0	335.3	305.0	282.7	192.7	188.6	70.6

Decrease in the content of macroelements in soil was observed both in foliar feeding in the closing leaves in a row phase, and a one-month before harvest. At the same time, there was a not significant difference on this indicator, depending on the timing of foliar application.

At application norms of microelements from 5.0 to 7.0 L/ha as in the closing leaves in a row phase, and 30 days before harvest in all variants were significantly decreased the content of hydrolyzable nitrogen, movable phosphorus compounds and potassium in the soil after sugar beet harvesting, compared with variants where foliar fertilizer was performed with application norm of 3.0 L/ha. The effect of microelements various types on the macroelements assimilation by plants, the efficiency of them was almost the same. After the first term of micronutrients application in foliar feeding, at the Reakom-R-beet microfertilizer in closing leaves in a row phase at a norm of 5 L/ha the hydrolyzable nitrogen in the soil was left 201.2 mg/kg, movable phosphorus compounds – 195.0 mg/kg soil and potassium – 70.2 mg/kg. At Reastim-humus-beet microfertilizer application in the same norm after sugar beet harvesting, the content of hydrolyzable nitrogen was 200 mg/kg, movable phosphorus compounds – 191.9 mg/kg and potassium – 72.6 mg/kg in the soil, at Reakom-plus-beet microfertilizer in the soil was hydrolyzable nitrogen of 201.4 mg/kg, movable phosphorus compounds – 193.0 mg/kg and potassium – 73.3 mg/kg, respectively.

In variants with Reastim-humus-beet and Reakom-plus-beet microelements at application norm of 7.0 L/ha was also observed the content decrease of microelements in the soil compared with the norm of application 5.0 L/ha. In the variant, with Reastim-humus-beet application in feeding the hydrolyzable nitrogen content decreased on 8.8 mg/kg, movable phosphorus compounds – 3.3 mg/kg and potassium – on 0.4 mg/kg. In the variant with Reakom-plus-beet application the hydrolyzable nitrogen content was decreased on 8.9 mg/kg, movable phosphorus compounds – on 3.9 mg/kg and potassium – 2.5 mg/kg. The similar results were obtained by the balance of hydrolyzable nitrogen, movable phosphorus compounds and potassium and on the second term of application (one month before harvest) micronutrients in feeding.

An important indicator is characterizing the effectiveness of microelements and the level of productivity in the technology of sugar beet growing. The sugar beet yields, sugar content and sugar yield formation is significantly depends on the foliar feeding plants application in their respective phases of vegetation, forms of microelements and their application norm (Table 12.5).

TABLE 12.5 Sugar Beet Productivity Depending on the Timing, Types and Application Norms of Microelements in Feeding (Average of 2010–2012)

Type of microelements (factor B)	Application norm, L/ha (factor C)	Plants density before harvest, thousand/ha	Yield, t/ha	Sugar content, %	Sugar yield, t/ha
Phase of closing leaves in a row (factor A)					
Without feeding (control)	-	97.8	46.3	15.3	7.1
Reakom-R-beet (standard)	5.0	102.5	51.8	14.8	7.7
	3.0	102.4	49.5	14.4	7.1
Reastim-humus-beet	5.0	102.4	52.3	14.5	7.6
	7.0	102.5	56.5	14.5	8.2
	3.0	101.9	48.5	14.2	6.9
Reakom-plus-beet	5.0	102.2	53.2	14.4	7.7
	7.0	103.4	58.0	14.3	8.3
A month before harvesting (factor A)					
Without feeding (control)	-	98.7	47.4	15.4	7.3
Reakom-R-beet (standard)	5.0	102.9	52.6	14.7	7.7
	3.0	102.5	49.7	14.3	7.1
Reastim-humus-beet	5.0	103.4	56.3	14.5	8.2
	7.0	104.4	64.6	14.2	9.2
	3.0	104.9	54.4	14.7	8.0
Reakom-plus-beet	5.0	105.6	63.9	14.4	9.2
	7.0	107.3	71.2	14.4	10.3
SSD_{05} factor A (term of application)		1.1	1.9	0.1	0.3
SSD_{05} factor B (type of microfertilizer)		0.9	2.7	2.2	2.5
SSD_{05} factor C (application norm)		3.3	3.4	0.7	0.3

It is proved that sugar beet productivity was influenced on the terms of microelements application in fertilizer, micronutrients types and application norm. At in the second term of foliar feeding (one month before harvest) of Reastim-humus-beet microfertilizer in the application norm of 5 L/ha, the yield of roots was 56.3 t/ha or was higher on 4.0 t/ha, compared the first period. The similar results were obtained according to increasing norms up to 7 L/ha and Reakom-plus-beet microfertilizer foliar feeding, regardless of the application norm of micronutrient. According to the types of micronutrients, at the first term of foliar feeding was observed a tendency the yield of root crops to increase or decrease, and by the second term was established a significant increase at Reakom-plus-beet microfertilizer application, compared with other microfertilizers. On the sugar beet yield was more significantly affected the application norms of micronutrients. With application norms increasing of Reastim-humus-beet microfertilizer in the first period with the application from 3 to 7 L/ha the yield of root was increased from 49.5 to 56.5 t/ha (SSD_{05} = 3.4 t/ha). Similar results were obtained with Reakom-plus-beet micronutrient using.

The highest sugar beet productivity was received by Reakom-plus-beet microelement foliar feeding a month before harvest at the application norm of 7 L/ha. At almost the same density of evenly spaced plants in a row (coefficient of variation was 3.6%) on the high agricultural background, foliar application was obtained the high yields of root – 71.2 t/ha. High yield of roots has secured the yield of sugar more than 10 tons/ha at their sugar content of 14.4%.

Foliar feeding conducting by Reakom-plus-beet microfertilizer in closing leaves in a row phase also was ensured the high productivity of sugar beet, but it was slightly lower than in micronutrients application a month before harvest at the all norms of application. Foliar application conducting in time a month before harvest by various types of microelements, with application norms of 5 and 7 L/ha was increase the yield of roots. If in the application norm of 5 L/ha in the closing leaves in a row phase of Reakom-plus-beet microfertilizer the root yield was 53.2 t/ha, their sugar content – 14.4%, at the feeding of this microfertilizer in the same norm a month before harvest, the yield was increased – on 10.7 t/ha, and the sugar content was remained at the level of 14.4%, a significant impact on the sugar content had the type of microelements (SSD_{05} factor B = 2.2%).

The similar relationship was established and with foliar feeding of sugar beet complex of Reastim – Humus – beet chelated micronutrients in the norms of 5 and 7 L/ha. But the productivity of sugar beet was lower than when using Reakom-plus-beet micronutrients. On the control – without foliar feeding at the same high agricultural background at the microelements application in closing leaves in a row phase the yield of roots was significantly lower and amounted of 46.3 t/ha, but the sugar content was higher – 15.3%, sugar yield in this case was – 7.1 t/ha.

After the first term of micronutrient application the high sugar yield was obtained in the variant with Reakom-plus-beet micronutrient using in norm of 7 L/ha – 8.3 t/ha, while in the variant with Reastim-humus-beet using on 0.1 t/ha less. After the second term of microfertilizers application the high sugar yield was received in the variant with Reakom-plus-beet micronutrient using in norm of 7 L/ha and was 10.3 t/ha. In the variant with Reastim-humus-beet using it was on 1.1 t/ha below. This suggests that a significant influence on the yield of sugar per 1 ha index has the form of micronutrient (SSD_{05} factor B = 2.5 t/ha). Other factors influence that has been studied – was negligible.

New chelate micronutrients Reakom-plus-beet and Reastim-humus-beet using on the high agricultural background was provided a significant increase in sugar beet productivity, not only in comparison with the control, where foliar feeding was not performed, and in the variant with Reakom-R-beet (standard) chelated micronutrient using. This is explained by the composition of microelements that positively influence on the growth and development of plants and the accumulation of organic mass. For example Reakom-plus-beet microfertilizer is a liquid concentrated solution based on ultramicro and micronutrients in chelate form, which also contains two different natures of chelating agent. Due to this microelements are part of fertilizer more stable and biologically active and Reastim-humus-beet microfertilizer is a balanced composition of humic substances and chelates microelements and considering of sugar beet requirement [8].

12.4 CONCLUSIONS

1. During the growing season raw mass of roots and leaves of sugar beet is intensive increase in variants with using of Reakom-plus-beet and Reastim-humus-beet new microfertilizers in foliar feeding. The growth trend of root raw mass was in variants with foliar feeding of microelements and persisted during the harvest season.
2. Significant influence on the growth and development of plants in closing leaves in a row phase, and a month before harvest had the types of micronutrients, their norms and terms of application. With norms of micronutrients from 3 to 7 L/ha increasing is increases raw mass of leaves and roots. At foliar feeding holding during the second term (30 days before harvest) was effective Reakom-plus-beet microfertilizer.
3. Foliar feeding using, a month before harvest, is delivers productivity of photosynthesis increased, particularly in variants where was studied the effectiveness of Reakom-R-beet at application norm of 5 L/ha, Reastim-humus-beet and Reakom-plus-beet micronutrients at application norms of 5 and 7 L/ha.
4. Sugar beet foliar feeding of various types and norms in different periods is enhances of the passage of the process of photosynthesis intensity.
5. Foliar feeding in the closing leaves in a row phase and 30 days before harvest by different microfertilizers at application norms of 3.0 to 7.0 L/ha is promotes better absorption of macronutrients from the soil, which in turn influence on the final productivity of sugar beet.
6. New chelate micronutrients Reakom-plus-beet and Reastim-humus-beet using in the high agricultural background was provided a significant increase in sugar beet productivity, not only in comparison with the control, where foliar feeding was not performed, and in the variant with chelated micronutrient Reakom-R-beet (standard) using.

7. The most effective is foliar feeding by Reakom-plus-beet chelated micronutrient during the second term in the norm of 7 L/ha, which provides productivity gains of roots – 18.6 t/ha and yield of sugar 2.6 t/ha compared with foliar feeding of Reakom-R-beet (standard) microfertilizer.
8. Foliar feeding conducting a month before harvest is provides a significant increase not only of roots productivity, but their yield of sugar per hectare.

KEYWORDS

- **microfertilizers**
- **photosynthetic potential**
- **sugar**

REFERENCES

1. Sabluk, V. T., Gryschenko, O. N., Polovynchuk, O. Y., & Nikitin, M. N. (2011). Sugar Beet Productivity Increase, Sugar Beet, *1*, 11–12 (in Ukrainian)
2. Lebedev, S. I. (1967). Plant Physiology K, Urogay, 384p (in Ukrainian).
3. Buligin, S. Y., Demishev, L. F., Doronin, V. A. et al. (2007). Micro Elements in Agriculture, Dnipropetrovs'k Sich, 104p (in Russian).
4. Zaryshnyak, A. S. (2006). Foliar Application of Fertilizers in Sugar Beet Growing, Sugar Beet, *4*, 17–19 (in Ukrainian)
5. Nichiporovich, A. A., Strogonov, L. E., & Chmora, S. N. (1961). Photosynthetic Activity of Plants in Fields (Methods and Accounting Task in Connection with the Formation of Yields) MoscowAcademy of Science of the USSR, 133p (in Russian).
6. Zubenko, V. F., Borisiuk, V. A., Balkov, I. J. et al. (1986). Research Methods of Sugar Beet, Kyiv Union Scientific Research Institute of Sugar Beet, 292p (in Russian).
7. Fisher, R. A. (2006). Statistical Methods for Research Workers, New Delhi Cosmo Publications, 354p.
8. Butov, V. M., & Porudyeyev, V. A. (2007). Influence of Basic Fertilizer on Yield of Sugar Beet under Irrigation 8, Conditions Proceedings of the Institute of Agriculture of the Southern Region of the Ukrainian Academy of Agrarian Sciences, *51*, 61–65 (in Ukrainian).

CHAPTER 13

MARKET DEVELOPMENT TRENDS OF CROP PRODUCE IN UKRAINE

ANDRIY V. DORONIN

National Academy of Agrarian Sciences of Ukraine, 37 Vasylkivska Str, Kyiv, Ukraine 03022; E-mail: andredor@meta.ua

CONTENTS

ABSTRACT

Market development trends of crop produce in Ukraine, which provides both food safety and food produce export of the country, were analyzed in this paper. High profitability of the cultivation of corn, sunflower, soybean and rape encourages producers to increase their sown areas. There is a tendency of increasing crop produce. It has been defined that raw crop produce is the major item of Ukrainian export. Thus, the latter depends on the fluctuations of market conditions very much. The polynomial model, which describes the level of grain and leguminous crops productivity in Ukraine for the period from 1913 to 2012 and envisages further yield increase of these crops was developed. The ways of enhancing the adaptability of crop production under current conditions were identified.

13.1 INTRODUCTION

Crop production has always been and remains the main branch of the agrarian sector of Ukraine, which provides both food safety and food produce export of the country. The provision of the population with crop and livestock produce depends on its development. There are all the required conditions: fertile soils, favorable climatic conditions, large domestic and foreign markets, a beneficial geopolitical location of Ukraine in the middle of Europe, the availability of seaports, etc. And yet, the dynamic development of crop industry does not provide the population of the country with animal products, which in turn leads to the necessity to balance consumer market through a large part of pork import.

At the same time the efficiency of crop production is not sufficient. Therefore, foreign economic activity for Ukrainian producers plays a primary role in the context of crop produce export and its processed products and import regulations in accordance with the requirements of the World Trade Organization.

The purpose of the research is to determine the level of the market development of crop produce in Ukraine and the enhancement ways of production efficiency in current conditions.

It is important to state that the achievements associated with the domestic market development of crop produce in the conditions of constant transformations require further research. The search of external segments of the world food markets as well as the availability of the domestic agricultural products there is of great significance.

13.2 MATERIALS AND METHODS

The data of the State Statistics Committee of Ukraine and the author's own calculations were used in the process of writing this article. Prices and production costs were recalculated/reduced according to the official exchange rate of Hryvnia to U.S. dollar set by the National Bank of Ukraine for a due period. Prices are given without taking into account value added tax (VAT – 20%), budget subsidies and supplements. Methods of system analysis and logic generalization to learn the experience of bioethanol production from plant growing produce were used: a comparative analysis – to analyze the process of statistic information, economic and mathematical modeling – to develop a polynomial model which describes the level of grain and leguminous crop productivity in Ukraine, a settlement and constructive method – to determine economic efficiency indicators in crop production, induction and deduction – to summarize research results, an abstract and logic method – to make conclusions and proposals.

13.3 RESULTS AND DISCUSSION

In Ukraine in the early 90s of the XX century during the transition to market relations, the conditions for free pricing were created for all branches of the economy, and approximate prices were introduced for agriculture; the prices of farm products grew much slower than those of goods and services consumed in agriculture which resulted in price disparity in favor of farmers. This, in turn, affected the structure of the sown areas, those under industrial crops (as they are more profitable) being increased.

Thus, in the years of 2001–2012 a noticeable increase of industrial crop share in the total sown area was observed in the structure of the sown area under agricultural crops – 13.5% in 2001 to 26.9% in 2012 (Table 13.1) [1–7]. The increase of the industrial crop share in the total sown area was due to the expansion of the sown areas under sunflower – 9.0% in 2001 to 17.1% in 2012, and beginning from 2008 the emergence of soybeans and rapeseed crops – 4.1% and 3.1% in 2012, respectively.

TABLE 13.1 Structure of the Sown Areas in Ukraine in 2001–2012 (Farms of All Categories)

Agricultural crop	Structure of the sown areas by the years, %						
	2001	2005	2008	2009	2010	2011	2012
Grain and leguminous crops	55.8	57.6	57.6	59.1	56.0	56.6	56.8
including:							
food crops	32.8	30.3	29.8	28.9	26.4	24.1	27.4
wheat (winter and spring)	25.5	25.6	26.2	25.5	23.9	21.1	24.5
rye (winter and spring)	3.3	2.4	1.7	1.8	1.1	1.1	1.0
rice	0.1	0.1	0.1	0.1	0.1	0.1	0.1
millet	1.3	0.5	0.6	0.4	0.4	0.7	0.6
buckwheat	2.6	1.6	1.1	1.0	0.8	1.0	1.1

TABLE 13.1 *(Continued)*

Agricultural crop	Structure of the sown areas by the years, %						
	2001	**2005**	**2008**	**2009**	**2010**	**2011**	**2012**
haricot	0.1	0.1	0.1	0.1	0.1	0.1	0.1
grain fodder crops	23.0	27.3	27.8	30.2	29.6	32.5	29.4
barley (winter and spring)	14.7	17.3	15.5	19.1	16.7	12.5	13.7
corn	4.6	6.6	9.3	8.1	10.1	17.2	13.1
oat	2.1	1.8	1.7	1.6	1.2	1.1	1.0
leguminous crops (without haricot)	1.4	1.5	0.9	1.3	1.5	1.1	1.3
sorghum	0.1	0.1	0.4	0.1	0.1	0.6	0.3
Industrial crops	13.5	20.2	25.0	24.1	27.1	27.5	26.9
including:							
sunflower	9.0	14.4	15.9	15.5	17.0	17.8	17.1
sugar beet (factory)	3.5	2.5	1.4	1.2	1.9	1.7	1.9
soybean	–	–	2.1	2.4	4.0	5.4	4.1
rapeseed (winter and spring)	–	–	5.2	4.0	3.4	2.1	3.1
Potatoes and vegetable and melon crops including:	7.9	7.8	7.3	7.2	7.3	7.3	7.3
potato	5.7	5.8	5.2	5.2	5.2	5.2	5.2
vegetables	1.8	1.8	1.7	1.7	1.7	1.8	1.8
Fodder crops	22.8	14.4	10.1	9.6	9.6	8.6	9.0

Due to the expansion of corn sown area (4.6% in 2001 to 13.1% in 2012), we have the increase of the share of grain and leguminous crops – 55.8% in 2001 to 56.8% in 2012. Besides, the share of food crops decreased – 32.8% in 2001 to 27.4% in 2012, while the share of grain forage crops increased – 23.0% in 2001 to 29.4% in 2012.

A significant decrease of forage crop share – 22.8% in 2001 to 9.0% in 2012 was observed, it had a negative effect on the development of livestock production in Ukraine. Potatoes and vegetable and melon crops remain almost the same in the structure of the sown area – 7.9% in 2001 to 7.3% in 2012 [1–7].

Farm producers work out the structure of sown areas taking into considerations many factors. The main ones include the structure of agricultural lands, their quality, specialization, product demand, the availability of means of production and labor resources, and climatic conditions.

The rational structure of the sown areas ensures enough grain, industrial and forage crops, potatoes, vegetables in the required range for successful production, and all crops have good forecrops; it facilitates the creation of appropriate agronomic and economic conditions, and on this basis it enhances yielding capacity. This structure makes it possible to better use farmland, to introduce proper crop rotations as each crop requires a good predecessor. In recent years the share of the crops which are in demand, namely sunflower, corn, soybean, has been increased in the structure of crop rotation, this resulted in the damage of scientifically grounded crop rotations. Failure to follow crop rotation, a simplified system of tillage and plant care led to a deterioration of phytosanitary state of the fields, and in turn to the increase of pesticide load, the latter had a negative effect on both environment and people's safety.

In Ukraine the socioeconomic crisis had a negative impact both on the development of agriculture in general and on crop production (Table 13.2) [8, 9].

TABLE 13.2 Production of the Main Crops in Ukraine in the Years of 2001–2012 (Farms of All Categories)

Agricultural crop	Production by the years, thousand tones					2012 in % to	
	2001	2005	2010	2011	2012	2001	2011
Grain and leguminous crops	39706.1	38015.5	39270.9	56746.8	46216.2	116.4	81.4
including:							
wheat (winter and spring)	21348.5	18699.2	16851.3	22323.6	15762.6	73.8	70.6
rye (winter and spring)	1822.5	1054.2	464.9	578.9	676.8	37.1	116.9
rice	68.9	93.0	148.0	169.9	159.8	231.9	94.1
millet	266.5	140.6	117.1	278.8	157.4	59.1	56.5
buckwheat	387.6	274.7	133.7	281.6	238.7	61.6	84.8
barley (winter and spring)	10185.7	8975.1	8484.9	9097.7	6936.4	68.1	76.2
corn	3640.7	7166.6	11953.0	22837.8	20961.3	575.7	91.8
oat	1115.7	790.7	458.5	505.6	629.7	56.4	124.5
leguminous crops	827.3	757.5	592.3	491.0	473.4	57.2	96.4
Industrial crops	–	–	–	–	–	–	–
including:							
sunflower	2250.6	4706.1	6771.5	8670.5	8387.1	372.7	96.7

TABLE 13.2 *(Continued)*

Agricultural crop	Production by the years, thousand tones					2012 in % to	
	2001	2005	2010	2011	2012	2001	2011
sugar beet (factory)	15574.6	15467.8	13749.2	18740.5	18438.9	118.4	98.4
soybean	73.9	612.6	1680.2	2264.4	2410.2	3261.4	106.4
rapeseed (winter and spring)	134.6	284.8	1469.7	1437.4	1204.4	894.8	83.8
Potatoes and vegetable and melon crops	–	–	–	–	–	–	–
including:							
potato	17343.5	19462.4	18704.8	24247.7	23250.2	134.1	95.9
vegetables	5906.8	7295.0	8122.4	9832.9	10016.7	169.6	101.9
Fodder roots	7712.7	8015.1	6770.8	7428.9	6993.0	90.7	94.1

In Ukraine in 1990 the production of grain and leguminous crops amounted to 51.0 million tons and in 2000 it was only 24.4 million tons;, respectively, sugar beet production reduced from 44.2 million tons in 1990 to 13.2 million tons in 2000, but sunflower production increased from 2.5 million tons in 1990 to 3.4 million tons in 2000 [8]. And in the following years, grain production grew rapidly: from 39.7 million tons in 2001 to 46.2 million tons in 2012, that is, 16.4%. It should be noted that corn production (among other grain crops) increased by 5.7 times – up to 20.9 million tons in 2012.

The same year, 2012, wheat production reduced by 26.2% – up to 15762.6 thousand tons, that of rye by 62.9% – to 676.8 thousand tons, millet production by 40.9% – to 157.4 thousand tons, that of buckwheat by 38.4% – to 238.7 thousand tons, barley production by 31.9% – up to 6936.4 thousand tons, that of oat by 43.6% – to 629.7 thousand tons, leguminous crop production by 42.8% – to 473.4 thousand tons.

A significant increase in sunflower production is observed – from 2.2 million tons in 2001 to 8.3 million tons in 2012, or by 3.7 times, soybean by 32.6 times – to 2.4 million tons, rapeseed by 8.9 times – to 1.2 million tons, and also a slight increase of sugar beet production by 18.4% – to 18.4 million tons. Potato production is growing dynamically by 34.1% – to 23.2 million tons in 2012 as well as vegetables by 69.6% – to 10.0 million tons. Over the years production of fodder roots decreases gradually – to 6.9 million tons in 2012 [8, 9].

Following the major elements of crop growing technology, namely, plant nutrition, plant protection from pests, diseases and weeds along with soil and climatic conditions of Ukraine during this period – all this favored the productivity increase of such crops as grain and leguminous crops by 15.1% – to 3.12 t/ha in 2012, in particular that of rye by 9.7% – to 2.27 t/ha, rice by 69.7% – to 6.21 t/ha, buckwheat by 29.9% to 1.65 t/ha, sugar beet by 2.25 times – to 41.08 t/ha, soybean by 69.3% – to 1.71 t/ha, rape by 77.4% – to 2.2 t/ha, potato by 48.9% – to 16.1 t/ha, vegetables by 61.7% – to 19.92 t/ha, fodder root crops by 25.1% – to 30.18 t/ha (Table 13.3) [8, 9].

TABLE 13.3 Yield of the Main Crops in Ukraine in 2001–2012 (Farms of All Categories)

Agricultural crop	Yield by the years, t/ha					2012 in % to	
	2001	2005	2010	2011	2012	2001	2011
Grain and leguminous crops	2.71	2.60	2.69	3.70	3.12	115.1	84.3
including:							
wheat (winter and spring)	3.10	2.85	2.68	3.35	2.80	90.3	83.6
rye (winter and spring)	2.07	1.73	1.67	2.07	2.27	109.7	109.7
rice	3.66	4.34	5.05	5.73	6.21	169.7	108.4
millet	1.06	1.17	1.37	1.78	1.03	97.2	57.9
buckwheat	0.67	0.69	0.67	0.99	0.87	129.9	87.9
barley (winter and spring)	2.60	2.06	1.97	2.47	2.11	81.2	85.4
corn	3.24	4.32	4.51	6.44	4.79	147.8	74.4
oat	2.00	1.76	1.48	1.81	2.09	104.5	115.5
leguminous	2.01	1.89	1.51	1.46	1.56	77.6	106.8
Industrial crops	–	–	–	–	–	–	–
including:							
sunflower	0.94	1.28	1.50	1.84	1.65	175.5	89.7
sugar beet (factory)	18.26	24.82	27.95	36.33	41.08	225.0	113.1
soybean	1.01	1.45	1.62	2.04	1.71	169.3	83.8
rapeseed (winter and spring)	1.24	1.46	1.70	1.73	2.20	177.4	127.2
Potatoes and vegetable and melon crops	–	–	–	–	–	–	–
including:							
potato	10.81	12.84	13.25	16.80	16.10	148.9	95.8
vegetables	12.32	15.71	17.36	19.50	19.92	161.7	102.2
Fodder roots	24.12	27.31	27.58	31.29	30.18	125.1	96.5

Thus in the years of 2001–2012 the growth in production of some agricultural crops is recorded due to the increase of their share in the total sown area and their yield increase, they are grain and leguminous crops, sunflower, soybean, rapeseed. Due to the yield growth alone the production of rice, sugar beet, potatoes and vegetables was increased. The production of rye, buckwheat, oat and fodder root decreased because their sown areas were reduced. The decrease of wheat, barley and legumes production occurred due to both the reduction of their share in the total sown area and that of the productivity.

It should be noted that in recent years grain and leguminous crops have occupied a significant share in the structure of the sown areas (2012–56.8%), including wheat (2012–24.5%), barley (2012–13.7%), corn (2012–13.1%) [8, 9].

Based on the polynomial model, which describes the level of grain and leguminous crop productivity in Ukraine in 1913–2012, the yield forecast of these crops, which envisaged its increase by 15%, was made (Fig. 13.1).

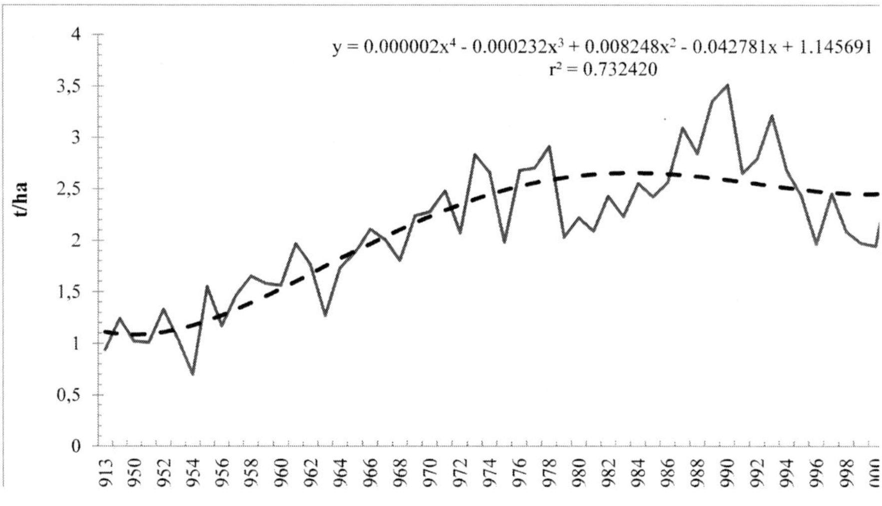

FIGURE 13.1 The dynamics of the grain and leguminous crop yields in Ukraine by the years.

Thus, in Ukraine there are all preconditions for the production increase of grain crops, which provide not only food safety but also product export.

It is important to emphasize the extensive development of Ukraine's trade relations in the grain product market. In 2005 the geography of exports comprised 75 countries, in the year of 2012–101 countries, or 26 countries more. In 2005–2012, the main importers of Ukrainian wheat were Spain, Egypt, Israel and Tunisia, which imported 7.6, 6.0, 4.2 and 3.0 million tons over the years, respectively. The main corn importers were Belarus, Spain, Iran, Tunisia and Egypt, having purchased 1.1, 5.5, 3.9, 1.8 and 8.2 million tons, respectively. As to the third major grain crop – barley, the main importers were Middle East countries: Saudi Arabia, Iran, Syria and Jordan, which, in 2005–2012, carried 18.3, 2.5, 2.0 and 1.9 million tons from Ukraine, respectively [10].

The main agro-food products of Ukrainian exports to European Union countries are raw output of crop production (seed of grain and oil-bearing crops), crude sunflower and other oils. In the year of 2012, grain crops constituted 40.2%, seed and fruits of oil-bearing crops – 24.9%, respectively, fats and oils – 16.6%, waste of food industry – 10.6% in the export structure of agro products from Ukraine to the EU. Along with this, 60% of Ukrainian exports cover five countries: Spain, the Netherlands, Poland, Italy and France. Over 50% of grain exports go to Spain, over 20% of oil crops and almost a quarter of the oil is shipped to the Netherlands [10].

In Ukraine farm producers increase the sown areas under industrial crops taking into consideration favorable conditions of external markets and price advance. In particular, the high profitability level of sunflower, soybean and rapeseed production stimulates producers to increase their sown areas, compared with those under other crops (Table 13.4) [11–15].

TABLE 13.4 Economic Production Efficiency of the Main Agricultural Crops in Ukraine in 2008–2012 (Agricultural Enterprises)

Index	Year					2012 in % to	
	2008	2009	2010	2011	2012	2008	2011
Wheat							
Total cost of 1 ton, USD	122.1	96.6	125.5	141.6	174.1	142.6	123.0
Average selling price of 1 ton, USD	143.6	102.2	137.5	166.5	194.6	135.5	116.9
Level of profitability, %	17.6	5.8	9.6	17.6	11.8	-	-
Barley							
Total cost of 1 ton, USD	133.2	97.6	121.0	145.5	177.6	133.3	122.1
Average selling price of 1 ton, USD	159.5	92.8	120.5	168.7	197.9	124.1	117.3
Level of profitability, %	19.8	-4.9	-0.4	16.0	11.4	-	-
Corn							
Total cost of 1 ton, USD	125.9	92.0	120.5	122.6	158.9	126.2	129.6
Average selling price of 1 ton, USD	139.2	111.7	156.5	170.0	190.3	136.7	112.0
Level of profitability, %	10.6	21.5	29.9	38.6	19.8	-	-
Seed of sunflower							
Total cost of 1 ton, USD	222.8	172.2	231.7	257.8	308.2	138.3	119.5
Average selling price of 1 ton, USD	263.8	243.5	381.6	404.8	449.4	170.3	111.0

TABLE 13.4 *(Continued)*

Index	Year					2012 in % to	
	2008	2009	2010	2011	2012	2008	2011
Level of profitability, %	18.4	41.4	64.7	57.0	45.8	-	-
Sugar beet							
Total cost of 1 ton, USD	39.7	39.2	52.6	47.7	46.5	117.1	97.5
Average selling price of 1 ton, USD	42.5	53.7	61.4	65.2	53.8	126.6	82.5
Level of profitability, %	7.1	37.0	16.7	36.5	15.7	-	-
Soybean							
Total cost of 1 ton, USD	334.9	260.9	282.0	295.2	348.4	104.0	118.0
Average selling price of 1 ton, USD	339.2	349.8	328.3	366.3	429.8	126.7	117.3
Level of profitability, %	1.3	34.1	16.4	24.1	23.4	-	-
Rapeseed							
Total cost of 1 ton, USD	271.5	245.3	291.8	395.4	404.5	149.0	102.3
Average selling price of 1 ton, USD	410.7	303.6	369.3	522.4	491.0	119.5	94.0
Level of profitability, %	51.3	23.8	26.6	32.1	21.4	-	-
Potatoes							
Total cost of 1 ton, USD	199.1	145.3	175.1	218.7	184.3	92.6	84.2
Average selling price of 1 ton, USD	214.8	164.1	283.9	257.3	144.6	67.3	56.2
Level of profitability, %	7.9	12.9	62.1	17.7	-21.5	-	-

It can be seen from Table 13.4, the level of production profitability of sunflower seed increased from 18.4% in 2008 to 45.8% in 2012. During the same period the profitability level increased, respectively – soybean – from 1.3% to 23.4%, but that of rapeseed, conversely, decreased from 51.3% to 21.4%. In 2012 efficiency reduction of rapeseed production was due to the production cost increase of 1 ton of rapeseed by 49%, while the average selling price of 1 ton of rapeseed increased only by 19.5%. Accordingly, the production cost of 1 ton of sunflower seed, during this period, increased by 38.3% and the average selling price of 1 ton increased by 70.3%, the cost of producing of 1 ton of soybeans increased only by 4.0%, while the average selling price of 1 ton increased by 26.7%. Efficiency of sugar beet production increased from 7.1% in 2008 to 15.7% in 2012.

During the period of 2008–2012 the profitability level of corn production increased from 10.6% in 2008 to 19.8% in 2012, but during the same period the wheat production effectiveness reduced from 17.6% to 11.8% and that of barley – from 19.8% to 11.4%. In 2012 the decrease of wheat production efficiency was due to the increase in production cost of 1 ton of wheat by 42.6%, respectively, the average selling price of 1 ton of wheat rose only by 35.5%, the reduction of the efficiency of barley production was due to the increase in the cost of production of 1 ton of barley by 33.3%, accordingly the average price of 1 ton grew only by 24.1%.

In recent years the reason of the loss of grain industry efficiency has been the lack of the development of agrarian market and infrastructure of grain marketing, the lack of an effective mechanism of state pricing regulation of grain products in the conditions of both overproduction of grain and harvest failure, high level of credit rates and inadequate budgetary support. In current conditions, government regulation of the grain market in Ukraine does not play an incentive role in relation to grain production and fails to effectively respond to the challenges of the global economy.

In 2012, potato production became unprofitable (–21.5%). It should be noted that the year of 2012, comparing with the previous year, is characterized by the production efficiency decrease of the major agricultural crops.

It is advisable to create and involve the effective mechanisms for the regulation of crop produce market on a national level. To stabilize prices of crop products the government must carry out purchasing and commodity interventions coordinate and adjust pricing policy in the national and international markets.

The European Union is very important among the external economic partners of Ukraine concerning the produce of the agrarian sector of the national economy. Trade balance of agro-food products between Ukraine and the EU is positive, except for the year of 2010 (Fig. 13.2) [10].

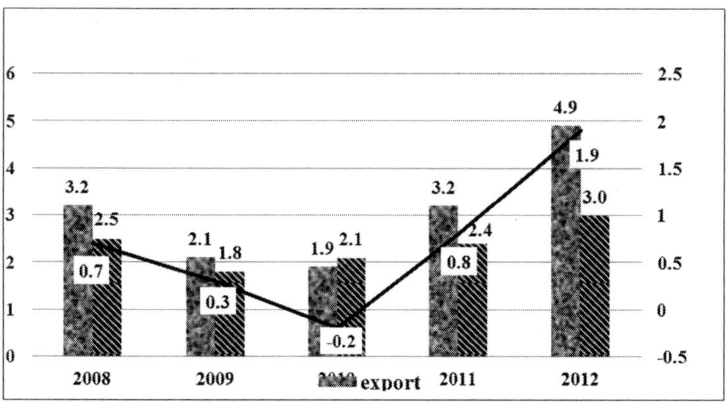

FIGURE 13.2 Dynamics of foreign trade of agro-food products of Ukraine with the European Union, billions U.S. dollars, 2008–2012.

In 2008, Ukraine exported agro-food products to the EU countries to the amount of 3.2 billion U.S. dollars, imported products to the amount of 2.5 billion U.S. dollars, trade balance was positive (0.7 billion U.S. dollars); this could be explained by Ukraine's entry to the World Trade Organization. In 2012, the increase of agro product export from Ukraine to the EU countries up to 4.9 billion U.S. dollars was recorded; import amounted to 3.0 billion U.S. dollars, trade balance was positive (1.9 billion U.S. dollars), which gives hope of favorable perspectives for Ukraine to create a free trade area with the EU.

Grain crops and seed and fruit of oil plants constitute the basis of Ukrainian exports to the EU, in 2008 their share was 30.2 and 37.4%, respectively, while in 2012 the share of grain crops increased to 40.2%, but the share of seed and fruit of oil-bearing crops decreased up to 24.9% (Table 13.5) [10].

TABLE 13.5 The Structure of Foreign Trade of the Main Agro-Food Products of Ukraine with the European Union in the Years of 2008–2012

Products	2008		2009		2010		2011		2012	
	$ million	%	$ million	%	$ million	%	$ million	%	$ million	%
Export										
Grain crops	962	30.2	454	21.8	158	8.1	1032	32.3	1981	40.2
Seed and fruits of oil crops	1190	37.4	757	36.3	704	36.2	920	28.8	1228	24.9
Fats and oils	659	20.7	469	22.5	632	32.5	666	20.8	818	16.6
Residues and waste of food industry	119	3.7	170	8.2	194	10.0	285	8.9	520	10.6
Processing products of vegetables, fruits	17	0.5	17	0.8	31	1.6	20	0.6	88	1.8
Edible fruits and nuts	82	2.6	57	2.7	75	3.9	99	3.1	79	1.6
Sugar and sugar confectionery	9	0.3	15	0.7	17	0.9	38	1.2	44	0.9
Vegetables and roots	19	0.6	23	1.1	23	1.2	18	0.6	34	0.7
Others	129	4.0	123	5.9	113	5.8	119	3.7	132	2.7
Total	3186	100	2085	100	1947	100	3197	100	4924	100
Import										
Meat and edible by-products	465	18.5	261	14.2	239	11.3	199	8.1	397	13.4

TABLE 13.5 *(Continued)*

Products	2008		2009		2010		2011		2012	
	$ million	%	$ million	%	$ million	%	$ million	%	$ million	%
Different food products	344	13.7	253	13.8	262	12.3	325	13.3	319	10.7
Edible fruits and nuts	133	5.3	179	9.8	199	9.4	167	6.8	307	10.3
Grain crops	69	2.7	28	1.5	76	3.6	155	6.3	188	6.3
Residues and waste of food industry	167	6.6	156	8.5	160	7.5	161	6.6	180	6.1
Seed and fruits of oil crops	148	5.9	66	3.6	82	3.9	140	5.7	150	5.1
Fats and oils	121	4.8	101	5.5	101	4.8	120	4.9	125	4.2
Processing products of vegetables, fruits	116	4.6	81	4.4	107	5.0	123	5.0	121	4.1
Others	949	37.8	707	38.6	896	42.2	1059	43.2	1181	39.8
Total	2512	100	1832	100	2122	100	2449	100	2968	100

Raw material orientation of Ukrainian exports makes Ukraine's position in foreign markets vulnerable, as the demand for raw goods is unstable and characterized by significant price variability. Therefore it is necessary to ensure an export increase of Ukrainian goods with high added value. In particular, the export growth of processed vegetables and fruits from 17 million U.S. dollars in 2008 to 88 million U.S. dollars in 2012 was recorded, in turn their share in the export structure increased from 0.5% in 2008 to 1.8% in 2012. There is also the export increase of a product group "sugar and sugar confectionery" from 9 million U.S. dollars in 2008 to 44 million U.S. dollars in 2012, accordingly their share in export structure grew from 0.3% in 2008 to 0.9% in 2012. The export of the group "fats and oils" grows from 659 million U.S. dollars in 2008 to 818 million U.S. dollars in 2012, but its share in the export structure decreased from 20.7% in 2008 to 16.6% in 2012. Along with this, the export of residues and waste of the food industry increases from 119 million U.S. dollars in 2008 to 520 million U.S. dollars in 2012, and their share in the export structure grew from 3.7% in 2008 to 10.6% in 2012.

Goods with a high added value constitute the basis of Ukrainian import from the EU, namely, meat and meat by-products, ready-to-eat food products. However, import of meat and meat by-products decreased from 465 million U.S. dollars in 2008 to 397 million U.S. dollars in 2012, and their share in the structure of import was 18.5% in 2008 and 13.4% in 2012. The second by specific weight in Ukrainian import from EU is the produce of a product group "various food items" (e.g., tea, coffee, sauces, ketchup, mustard, and various semifinished products of industrial use, food flavoring, etc.). Import of various food products decreased from 344 million U.S. dollars in 2008 to 319 million U.S. dollars in 2012 and their share in the import structure was 13.7% in 2008 and 10.7% in 2012.

It is necessary to mention that analyzing some commodity groups one can observe the import increase which can restrict the growth of domestic production. Thus, the import of the group "edible fruits and nuts" increased from 133 million U.S. dollars in 2008 to 307 million U.S. dollars in 2012, respectively their share in the structure of import increased from 5.3% in 2008 to 10.3% in 2012. The increase in grain crop import from 69 million U.S. dollars in 2008 to 188 million U.S. dollars in 2012 was observed, and their share in the structure of import was 2.7% in 2008 and 6.3% in 2012. The import of residues and waste of food industry, seed and fruits of oil-bearing crops, fats and oils and preserved vegetables and fruits increased as well.

It should be noted that in Ukraine grain crop export increases, and meat and food by-products have the largest share of import from the EU; in future this tendency can have a negative influence on the development of the domestic livestock industry.

Hence, the main directions of enhancing the adaptability of crop production in current conditions are: to cultivate agricultural crops in favorable soil and climatic conditions, to develop cultivars and hybrids, well adapted to natural conditions of Ukraine, to transfer crop production to postindustrial development models, to provide raw material base for bio-energy, to create forage supplies for livestock, to develop organic production, to structure the market crop produce.

To achieve these goals it is necessary to solve the issue of soil fertility improvement through the introduction of the complex of organizational and technological measures such as: to use scientifically grounded crop rotation while growing crops, to focus on organic fertilization systems, to increase the efficiency of fertilizer application by optimizing the doses, timing and methods of application, to use soil-protecting technologies of tillage, to reduce a negative impact of plant protection chemicals on a plant itself and useful microflora of the soil.

To provide arable farming with adapted cultivars and hybrids, which have higher productivity and better environmental resistance requires the organization of seed production system in Ukraine. Thus, even the cultivars and hybrids adapted to weather and environmental conditions cannot always guarantee high productivity of the crops because of the processes of global warming on our planet.

The application of postindustrial development models of crop production industry envisages the development and implementation of resource saving bio-adaptive farm practices, integrated pest management, optimization methods of the process of plant growth and development, taking into consideration climate changes and the principles of precise farming.

When the problem of supplying our country with affordable by price energy carriers becomes urgent, it is appropriate to speed up the production of alternative fuels, including those manufactured from crop produce. Production and use of biofuels will accelerate the solution of such strategic objectives for the development of Ukraine, in particular its agriculture, as the decrease in producers' dependence on imported fuel and satisfaction of demand for these products at a lower possible price. Therefore it is necessary to create appropriate raw base for bio energy development – to optimize rapeseed sown areas for biodiesel production and those of sugar beet – for bio-ethanol manufacture. Production and use of bio ethanol and biodiesel will make it possible to arrange the production of environmentally clean alternative fuel, to create new jobs, to increase profits, and to reduce Ukraine's dependence on imported fuel.

To meet the needs of livestock production in fodder supplies, it is necessary to expand fodder crop fields, including corn for silage and perennial grasses, and to reduce the share of corn grain and soybean exports, instead to use them for the development of domestic livestock production.

It is necessary to implement the development of organic production through the prohibition of the use of chemically synthesized fertilizers and plant protection products, hormones, antibiotics, genetically modified organisms, growth regulators, etc., and to appropriately label products which meet the requirements of organic ones. Unfortunately, the lack of relevant legislation in Ukraine makes it impossible to organize the system of accreditation and normalization of the certification agency activity and it also prevents the country to be an exporter of organic products in the international market. In 2012 the number of countries where the organic standards were introduced was 86, another 26 countries are in the process of relevant legislature development [16], including Ukraine, Russia, Bosnia and Herzegovina – in Europe; Bangladesh, Hong Kong, Kyrgyzstan, Laos, Nepal, Pakistan, Sri Lanka, Syria, Vietnam – in Asia; Jamaica and St. Lucia – in the Caribbean; Egypt, Kenya, Senegal, South Africa, Tanzania, Zambia, Zimbabwe – in Africa [17].

To improve the structure of plant growing produce it is necessary to manufacture the products which are in great demand in the world market – wheat, corn, barley, and to optimize the sown areas under sunflower (not more than 15% in rotation) – to produce oil for domestic consumption and for export, those under rapeseed – to produce biodiesel, the sown areas under sugar beet – to produce bio ethanol, those under corn for silage, soybean and perennial grasses – to satisfy the needs of livestock production in required fodder supplies.

13.4 CONCLUSIONS

In Ukraine there are all preconditions for the increase of plant growing produce, which provides not only food safety of the country but also product export. The production increase of some crops occurs due to both their share increase in the total acreage and their productivity increase; they are grain and leguminous crops, sunflower, soybean, rapeseed. The production increase of rice, sugar beet, potatoes and vegetables was accomplished due to the crop yield increase alone. The yield increase of grain and leguminous crops by 15% is envisaged.

Taking into consideration favorable conditions of external markets and price advance, farm producers increase sown areas under industrial crops in Ukraine. In recent years the reason of the loss of grain industry efficiency has been the lack of the development of agrarian market and infrastructure of grain marketing, the lack of an effective mechanism of state pricing regulation of grain products in the conditions of both overproduction and harvest failure, high level of credit rates and inadequate budgetary support of the industry.

The main function of the state in regulating the market crop produce must be the prediction of the market situation. The process of the creation of the national market of plant growing products in the conditions of globalization is still in the infancy, so this market neither functions properly nor implements its tasks, namely, supply and demand balance, profitability growth of agricultural producers, the solution of social and environmental problems of the industry. The use of efficient mechanisms of government regulation is to ensure the market stability; they consist in ensuring the balance of quantitative and qualitative parameters of supply and demand for the output.

The European Union is very important among the external economic partners of Ukraine concerning the produce of the agrarian sector of the national economy. Trade balance of agro-food products between Ukraine and the EU is positive, except for the year of 2010, which gives hope for favorable perspectives of Ukraine to create a free trade zone with the EU. Raw material orientation of Ukrainian exports makes Ukraine's position in foreign markets vulnerable. Goods with a high added value constitute the basis of Ukrainian import from the EU, namely, meat and meat by-products, ready-to-eat food products. In Ukraine grain crop export grows, and meat and meat by-products constitute the largest import share from the EU; in future this tendency may have a negative influence on the development of the domestic livestock industry.

The main directions of enhancing the adaptability of crop production in current conditions are: to cultivate agricultural crops in favorable soil and climatic conditions, to develop cultivars and hybrids, well adapted to natural conditions of Ukraine, to transfer crop production to postindustrial development models, to provide raw material base for bio-energy, to create forage supplies for livestock, to develop organic production, to structure the market crop produce.

ACKNOWLEDGEMENT

The author expresses special thanks to Sergey Kvasha, Doctor of Economic Sciences, professor, academician of the National Academy of Agrarian Sciences of Ukraine, for his assistance in writing this paper.

KEYWORDS

- **export**
- **food safety**
- **government regulation**
- **grain market**
- **import**
- **market condition**

REFERENCES

1. The Results of the Final Accounting of Agricultural Sown Areas for the Harvest of 2001. Statistical Bulletin Kyiv State Statistics Committee of Ukraine, (2001) 53p (in Ukrainian).
2. The Results of the Final Accounting of Agricultural Sown Areas for the Harvest of 2005. Statistical Bulletin Kyiv State Statistics Committee of Ukraine (2005) 53p (in Ukrainian).
3. Sown Areas of Agricultural Crops under the Harvest of 2008. Statistical Bulletin Kyiv State Statistics Committee of Ukraine (2008) 53p (in Ukrainian).
4. Sown Areas of Agricultural Crops under the Harvest of 2009. Statistical Bulletin Kyiv State Statistics Committee of Ukraine (2009) 53p (in Ukrainian).
5. Sown Areas of Agricultural Crops under the Harvest of 2010. Statistical Bulletin, Kyiv State Statistics Committee of Ukraine (2010) 53p (in Ukrainian).
6. Sown Areas of Agricultural Crops under the Harvest of 2011. Statistical Bulletin, Kyiv State Statistics Committee of Ukraine (2011) 53p (in Ukrainian).
7. Sown Areas of Agricultural Crops under the Harvest of 2012. Statistical Bulletin, Kyiv State Statistics Committee of Ukraine (2012) 53p (in Ukrainian).
8. Plant Growing of Ukraine, Statistical Digest of the Year of 2005. Kyiv State Statistics Committee of Ukraine (2006) 111p (in Ukrainian).
9. Plant Growing of Ukraine, Statistical Digest of the Year of 2012. Kyiv State Statistics Committee of Ukraine (2013) 110p (in Ukrainian).
10. Kvasha, S. N., Vlasov, V.I., Kryvenko, N. V. et al. (2013). Export and Import of Production of Agricultural Sector in Ukraine, Status and Tendencies, Kyiv Institute of Agricultural Economics, 80p (in Ukrainian).
11. Basic Economic Indicators of Agricultural Production in Agricultural Enterprises, Statistical Bulletin of the Year of 2008. Kyiv State Statistics Committee of Ukraine (2009) 76p (in Ukrainian).
12. Basic Economic Indicators of Agricultural Production in Agricultural Enterprises, Statistical Bulletin of the Year of 2009. Kyiv State Statistics Committee of Ukraine (2010) 80p (in Ukrainian).
13. Basic Economic Indicators of Agricultural Production in Agricultural Enterprises, Statistical Bulletin of the year of 2010. Kyiv State Statistics Committee of Ukraine (2011) 88p (in Ukrainian).
14. Basic Economic Indicators of Agricultural Production in Agricultural Enterprises, Statistical Bulletin of the Year of 2011. Kyiv State Statistics Committee of Ukraine (2012) 88p (in Ukrainian).

15. Basic Economic Indicators of Agricultural Production in Agricultural Enterprises, Statistical Bulletin of the Year of 2012. Kyiv State Statistics Committee of Ukraine (2013) 88p (in Ukrainian).
16. The World of Organic Agriculture (2013). Key Indicators and Leading Countries, FiBL & IFOAM (2013) URLhttp://www.organic-world.net/fileadmin/documents/yearbook/2013/web-fibl-ifoam-2013-25-34.pdf.
17. Beate Huber (2013). The World of Organic Agriculture, Regulations and Certification Emerging Trends, Research Institute of Organic Agriculture, URL: http://orgprints.org/22324/7/huber-2013-standards.pdf.

CHAPTER 14

EFFECT OF NUTRIENT MEDIA CONTAINING NATURAL AND CHEMICALLY MODIFIED STARCHES ON HAPLOID PRODUCTION IN SPRING BARLEY ANTHER CULTURE *IN VITRO*

OLENA V. BILYNSKA

V.Ya. Yurjev Plant Production Institute of National Academy of Agricultural Sciences of Ukraine, d. 142, Moskovsky av., Kharkiv, 61060, Ukraine; E-mail: bilinska@ukr.net

CONTENTS

ABSTRACT

The efficiency of agar replacement in the inductive media for spring barley anther culture *in vitro* by chemically modified and natural starches from seeds of corn lines and pea cultivars possessing mutant alleles of endosperm structure genes – *wx*, su_2, *ae* and *r* controlling the ratio of amylose and amylopectin has been investigated. Gels produced with natural starches, which had high amylose content (corn mutations ae, su_2 and pea that one *r*) were characterized by better structural and mechanical properties than those ones produced with starch of normal type. Data concerning positive effect of all types of starch on the process of a direct somatic embryo formation have been obtained. There have been shown genotypic differences in the reaction to gelatinized substances of medium for androgenetic structure induction and plant regeneration. Perspective preparations are recommended for application in plant biotechnology.

14.1 INTRODUCTION

Anther and isolated microspore culture *in vitro* are the most effective methods of the fast homozygous line production in plant breeding. Based on the induction of haploid microspores to form multicellular structures which later develop into callus, embryos and plantlets, these methods allow to decrease duration of breeding process up to 5 years and increase the efficiency of outstanding genotypes evaluation and selection [1–3].

Since it has been proved that auxins and cytokinins added to the medium in a certain ratio ensured the regulation of morphogenesis via the dedifferentiation and differentiation processes, the improvement of nutrient media for plant cell, tissue and organ culture *in vitro* in the most cases aimed to optimize growth regulator composition [4–6]. Salts of macro, microelements, vitamins, amino acids etc. were also showed to be very impotent supplementations and their quantity were the object of medium modification [7, 8].

A gelatinizing component of solid media as a rule is presented by agar, which is assigned to be an inert supporter determining the structural, mechanical and osmotic properties of medium. That is why much less attention was paid to investigation on gelling substances. However, it is known to be that the completion of morphogenesis program from cell to plant also depends on the chemical nature and structural properties of gelling base of medium [9].

The most outstanding result was obtained when barley starch was used as agar substitute in medium for barley anther culture *in vitro* [10, 11]. This replacement allowed drastically increase plant regeneration efficiency.

We for the first time demonstrated positive effect of several preparations of chemically modified starches applied as agar substitution in the media for spring barley anther culture *in vitro* [12]. The most interesting results were connected with the ability of starch containing media to promote a direct embryo formation and to suppress callus growth. So long as somatic embryos (embryoids) are known to be structures with the highest regeneration ability [13], stimulation of direct embryoid formation in different *in vitro* systems including anther culture *in vitro* is the fastest and the most effective way of plant regeneration. Besides it has been shown in our investigations that among plants regenerated from embryoids inducted in the starch solidified media vitreous plants (with high water content in leaves) were absent. The latter observation is in agreement with that one revealed in the micropropagation of vegetable plants [14].

Our experiment was also continued in the direction of evaluation of gelling and morphogenetic properties of natural starches of different origin. Taking in consideration that barley starch is an exotic chemical produced only in Finland and lacking in the catalogs of the prominent firms specialized on the production of chemicals for biochemistry and biotechnology, we decided to use corn and pea starches with a different amylose and amylopectin content.

Choose of these crops as a starch sources is explained by their diversity concerning endosperm structure gene mutations, which regulate a content of starch copolymers amylose and amylopectin. It has been described 20 mutant genes of endosperm structure in corn [15, 16]. These mutations are well studied and many lines possessing different types of starch have been created [17]. The most visible phenotypic effect have such mutations as *wx* (more than 90% amylopectin in the starch of seeds), su_2 with amylose content up to 40% and *ae* possessing 50–80% of amylose in seed starch. Among pea forms, which are considered to be a source of gelling component of medium much attention should be paid to carriers of mutation *r* (*rugosus*), which blockades the activity of starch branching enzyme and increases amylose content from 30 to 75% [18].

Since the results of our investigations have already shown a possibility to replace agar by more effective for morphogenesis and less coast starch gelling components the aim of the present study was to test availability of new preparations of natural and chemically modified starches in nutrient medium composition for barley haploid production in anther culture *in vitro*.

14.2 MATERIALS AND METHODOLOGY

14.2.1 PLANT MATERIAL, SPIKE COLLECTING, STERILIZATION

Three spring barley (*Hordeum vulgare* L.) genotypes possessing a contrast androgenic capacity were used as an initial material. Doubled haploid line DH00-126 is a high responsive genotype of androgenetic origin produced from F_1 hybrid Kharkovsky 74×Exotik. It is characterized by the ability to produce a high percentage of morphogetetic structures via a direct embryogenesis and green plants. Fenix is a cultivar with low androgenic responsibility. Cultivar Exotik possesses a high capacity of callus induction and a low that one to plant regeneration [19, 20].

Plants were grown under field conditions. Spikes with microspores on the late uninucleate stage were cut and were pretreated at 4 °C for 5–6 days in the dark. A stage of microspore development was determined under the light microscope using acetic carmine squashing preparations. Collected spikes in leaf sheaths were surface sterilized for 10–15 min in a thin layer of a cotton wood wetting with 70% ethanol.

14.2.2 CULTURE MEDIA

A 10 mm diameter tubes or 5 cm diameter Petri dishes were used as cultural vessels. Anthers were excised from spikes in a laminar flow cabinet and transferred onto inductive medium NMSmod.2 [21]. This medium contained macroelements N6 [22] and microelements MS [23] and other components (mg/L): vitamins (thiamine HCl – 10.0, B6–1.0, PP –1.0; myo-inositol – 100. 0), growth regulators (2,4-D – 2.0, BAP – 0.5); aminoacids (glutamine – 200.0, proline – 100.0; alanine – 100.0), maltose (Merck, Germany) – 90,000; agar (Difco, USA) – 8000; pH 5.7–5.8. All organics were of Serva (Germany) production.

In experimental variants of inductive medium agar was substituted for natural starches with a high amylose content obtained from seeds of corn lines AE-392A and AC-11, which were the carriers of natural mutations of endosperm structure genes ae and su_2. Besides starches obtained from seeds of pea cultivars Banan and Alfa which were carriers of dominant (R) and recessive (r) alleles of gene controlling amylose content were used for agar substitution. Starch concentration varied from 4.5% for pea starch of normal type *(RR)* to 6.5% for *ae-*, su_2-, *rr*-type of starch with high amylose content. All natural starches were produced via standard procedure with some modifications [24] and kindly supplied by Tymchuk and Derebizova (Yurjev Plant Production Institute of National Academy of Agricultural Sciences of Ukraine).

Preparations of chemically modified starches D-5a, D-5b, D-5 g, D-6, D-7 [14] were kindly supplied by Dulnyev (Institute of bioorganic and oil chemistry of National Academy of Sciences of Ukraine) and used for agar substitution in medium for callus and embryo induction. Concentration of these gelling substances was 6.0 and 12.0%.

Regeneration medium contained macro and microelements MS [23] and such components (mg/L): vitamins (thiamine HCl – 10.0, B6–1.0, PP –1.0; myo-inositol – 100. 0), growth regulators (NAA – 0.2, BAP – 0.2); aminoacids (glutamine – 200.0, proline – 100.0; alanine – 100.0), sucrose (Merck, Germany) – 30,000; agar (Difco, USA) – 8,000; pH 5.7–5.8. Media were sterilized in autoclave at 121 °C for 20 min.

14.2.3 ANTHER CULTURE AND PLANT REGENERATION

Anther culture was incubated at 24±2 °C in dark for 20–30 days. Callus and embryoids (in the most cases responding anthers with callus and embryos) visible to naked eye were transferred onto regeneration medium. They were cultivated for 3 days in dark and then were transferred to the climate chamber with 16 h photoperiod, 100 $\mu E\ m^{-2}\ s^{-1}$ light intensity and temperature 24±2 °C. After 2–4 weeks green plantlets were transferred into 20 mm diameter tubes containing medium differed from regenerative one by omitting growth regulators and were grown in the climate chamber for 4–6 weeks at the above mentioned light and temperature conditions. For root induction medium supplemented with 2.0 mg/L NAA was applied.

14.2.4 EXPERIMENTAL DESIGN AND STATISTICAL ANALYSIS

Anthers isolated from a spike were inoculated onto a medium in one culture vessel. 300–600 anthers from 7–10 spikes were planted per every experimental variant. A number of responding anthers (%) and a number of regenerated green plantlets per cultivated anthers (%) were determined in order to evaluate effect of gelling agent nature on the efficiency of haploid production in anther culture *in vitro.* Statistical analysis was carried out by t-test with Microsoft Office (Excel 2003).

14.3 RESULTS AND DISCUSSION

In order to be applied as a gelling component of nutrient medium both natural and chemically modified starches would correspond to some demands. First of all gels made of different types of starch must have good structural and mechanical properties which remain stable during all period of anther cultivation lasting at least 30–35 days. Their density would be sufficient for explants inoculation. Besides, paste after boiling would be suitable for easy transfer to culture vessels without remainders.

Our previous investigations on starch gelling capacities included test of 18 preparations of chemically modified and 7 preparations of natural starches (5 from corn seeds and 2 pea seeds) [25–28]. Results showed (Figs. 14.1 and 14.2) that preparations differed by structure, color and gelling capacities.

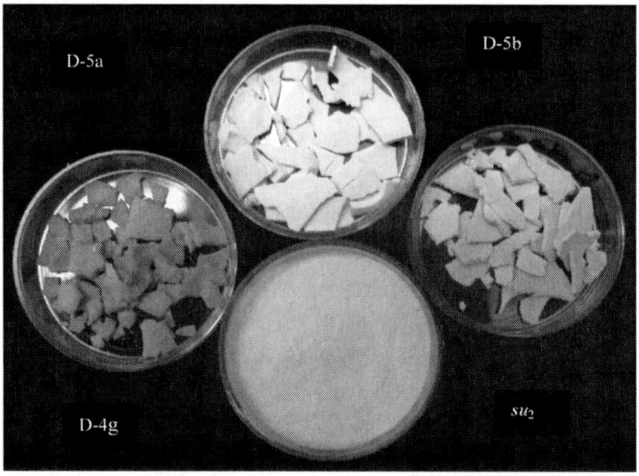

FIGURE 14.1 Natural corn starch type su_2, chemically modified starches D-5a, D-5b, D-4 g.

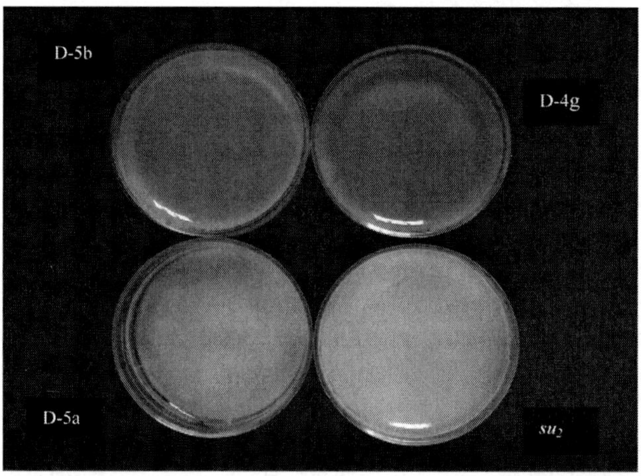

FIGURE 14.2 Gels produced with chemically modified starches chemically modified starches D-5a, D-5b, D-4 g and natural corn starch type su_2.

Twelve preparations with good gelling properties were selected among testing chemically modified starches. Another formed very dense paste when were added in concentration 12.0% and did not solidify in lower concentration. Chemically modified starch D-2 had the best paste characteristics. It turned into transparent solution after boiling and could be replaced in a tube or Petri dish without any traces on the walls of glass in which it was boiled. Other preparations formed transparent paste after sterilization in autoclave. They solidified in 24 h. Gels had good structure and smooth surface. However, all preparations except of D-5a, D-6 and D-7 had such negative trait as crack formation (Fig. 14.3) in two weeks after transfer into culture vessels.

Gels made of preparations D-6 and D-7 were the most stable. Chemically modified starch D-5a formed gel, which maintained its structure during 25–30 days. Nevertheless crack appearance had not visible effect on androgenic structure formation and plant regeneration (Fig. 14.4) it could decrease a commercial value of starch preparations.

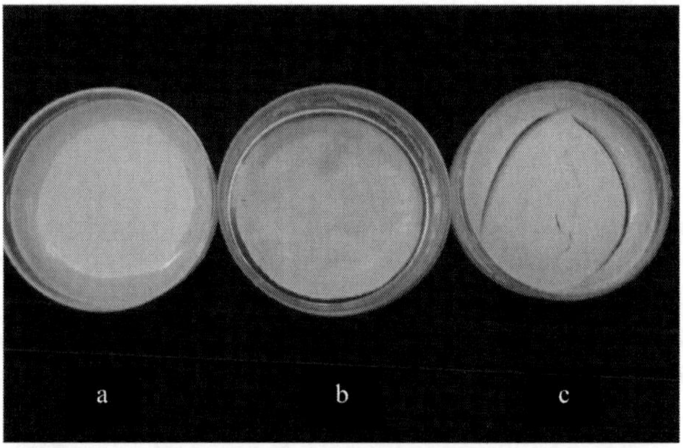

FIGURE 14.3 Media with different gelling component: (a) corn starch *su*$_2$, (b) chemically modified starch D-7, (c) chemically modified starch DKKmod. Crack formation in media containing chemically modified starch DKKmod in 12 days after media production.

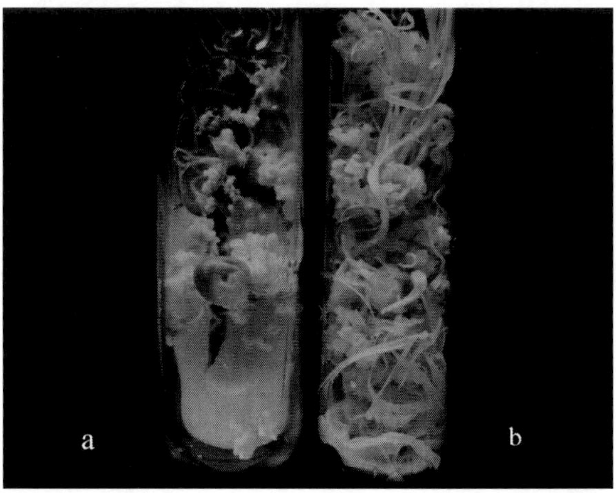

FIGURE 14.4 Morphogenetic structure formation and plant regeneration in anther culture *in vitro* of spring barley line DH00–126 on the media containing chemically modified starch D-5A (a) and agar after (b) 40 days of anther inoculation.

It should be noted that cracks were absent in the media solidified with natural starches. The results of our investigations were in agreement with previously known fact concerning a dependence of natural starch gelling properties on amylose and amylopectin content [29]. Corn starch obtained from seeds of line which was a carrier of mutant gene *wx* (90% of amylopectin) was appeared to be unsuitable for use as a component of medium. At the same time gels of good quality were formed by starches with high amylase content obtained from seeds of corn (mutations *ae*, su_2) and pea (mutation *rugosus*).

There were differences in gelling capacities of corn and pea starches of normal type obtained from seeds of lines with dominant alleles of endosperm structure genes. Gel produced with corn starch of this type lost its structure through very fast deformation coursed by drying in a day after its formation while normal pea starch formed stable gel in concentration 6.5% [28] and 4.5% (unpublished data).

The best starch preparations were selected with the aim to determine their effect on morphogenesis in spring barley anther culture *in vitro*. Several experiments were carried out in order to evaluate the possibility of replacement of agar by natural starches. The most interesting results were obtained when corn starches *ae* and su_2-type were used as a gelling agent of the inductive medium.

Analysis of the data presented in Figs. 14.5 and 14.6 showed that regardless of the medium variant barley genotypes remained their peculiarities of morphogenesis. DH00–126 had the highest yield of morphogenetic anthers and regenerated plan. Cultivar Fenix was much less responsive. In Exotic sporophyte development of microspores were promoted in significant number of anthers but it did not result in corresponding number of green plants.

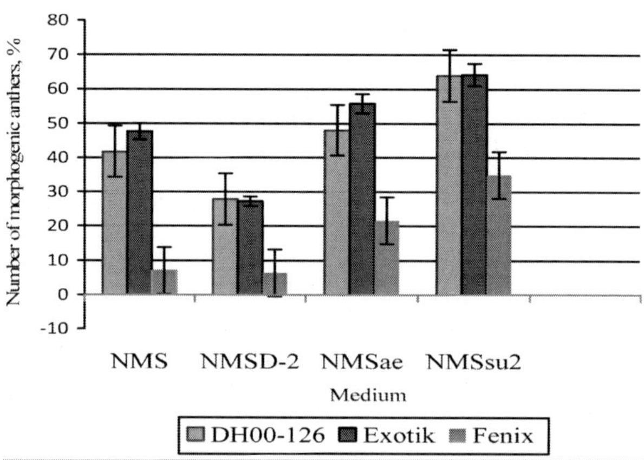

FIGURE 14.5 Formation of morphogenetic structures in spring barley anther culture *in vitro* on the media containing different gelling agents: NMS – agar (0.8%, Difco, USA); NMSD-2 – chemically modified starch D-2 (12.0%), NMSae, NMSsu_2 – corn starches (6.5%) obtained from seeds of lines which were carriers of mutant gens *ae* and su_2, respectively.

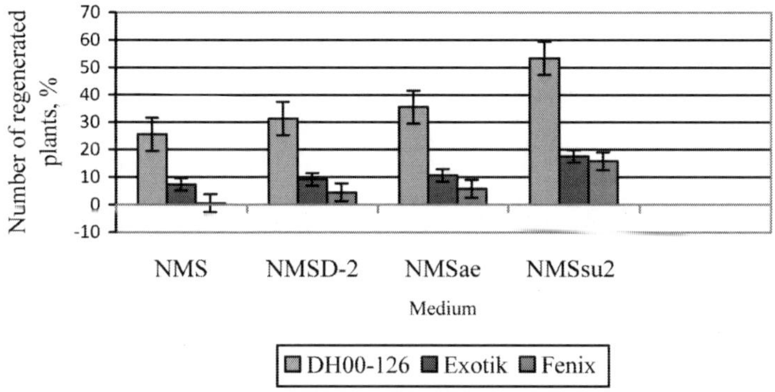

FIGURE 14.6 Plant regeneration in anther culture *in vitro* of spring barley line ДГ00–126 on the media containing different gelling agents: NMS – agar (0,8%, Difco, USA); NMSD-2 – chemically modified starch D-2, NMSae, NMSsu$_2$ – corn starches (6.5%) obtained from seeds of lines which were carriers of mutant gens *ae* and *su*$_2$, respectively.

Chemically modified starch D-2 depressed the process of androgenic structures induction. At the same time it hadn't negative effect on plant regeneration. Moreover this preparation increased plant regeneration rate from 0.56 to 4.43% in cultivar Fenix (LSD_{05}=3.46). As in previously conducted experiments, use of natural starches stimulated both embryogenesis and plant regeneration. Plants were obtained directly from embryoids on the inductive medium (Fig. 14.7). Moreover, androgenic structures formed in starch containing media have a higher morphogenetic capacity than those ones formed on agar containing medium.

FIGURE 14.7 Induction of callus, embryoids and plant regeneration in anther culture *in vitro* of spring barley line DH00–126 on media containing agar (a) and corn starch of su$_2$-type (b).

Positive effect of starches on plant regeneration efficiency was apparent in all genotypes. The highest plant yield was obtained in line DH00–126 (53.42%). Two-fold increase of this trait was revealed in Exotik. But the most contrast results were obtained in less responsive cultivar Fenix. In particular, the regeneration rate in medium containing starch of su_2-type was reached 15.81%. Agar substitution for ae-type starch was less effective in comparison to that one for su_2-type starch. However, this preparation was better solidifying component of the media than agar.

When corn starches are used in medium composition it should be taken into account some technical problems connected with difficulties of starch separation from proteins which are presented by fraction soluble in ethanol [24]. Pea starches can be obtained via more simple procedure because their grain proteins consist of fractions soluble in water and in salt solutions [18]. Since gens regulating high amylose content in starch are recessive, and corn is cross-pollinating crop, corn plants carrying these genes demand space isolation. Therefore, a very impotent advantage of pea as a source of media solidifying substance lies in its biological peculiarity as self-pollinating crop.

Keeping in view a crucial role of amylose content in determination of corn starch gelling properties, quality evaluation of gels forming with pea starches of normal type (genotype RR, 30% amylose) and high amylose type (mutant homozygous genotype rr, 70% amylose) were conducted. The results of this investigation [28] showed that both types of starch at concentration 6.5% formed gels suitable for anther inoculation and remained their structure during 30 days. But gel produced with starch of normal type had high density and negative effect on androgenic structure induction. Solidifying properties of pea high amylose starch was close to these ones of corn high amylose starch of su_2-type. Effect of these starch preparations on embryo induction and plant regeneration was also similar.

In order to improve structural properties of media supplemented with pea starch of normal type, study on lower concentration (4.5 instead of 6.5%) effect was carried out. Data showed (Table 14.1) that when concentration of normal type starch was 4.5% differences between embryo formation and plant regeneration on agar and normal type containing media were not significant. Positive effect of high amylose starch media on androgenesis *in vitro* were conformed in this experiment. Both media containing pea starch had good structural properties and exceeded those ones of chemically modified starch D-5a, because cracks didn't appear (Fig. 14.8). Plant regeneration was observed in the inductive media without transfer of embryoids to special medium with low growth regulator concentration. Nevertheless, a majority of plants were obtained via transfer of embryos and callus to regeneration medium supplemented with agar.

TABLE 14.1 Effect of Gelling Component of Inductive Media on Androgenesis *in vitro* in Spring Barley (*H. vulgare* L.)

Medium	No. of inoculated anthers	No. of responding anthers	Morphogenic anther frequency (%)	No. of green plants	Green plant regeneration frequency (%)
NMSmod.2	528	161	30.49±2.00	111	21.02±1.77
NMSRR	608	187	30.75±1.87	148	24.34±1.74
NMSrr	552	245	44.38±2.11*	284	51.96±2.13**

* **P** ≥ 0.05;
**P ≥ 0.01.
NMSmod.2 contained 0.8% agar Difco (USA); NMSRR contained 4.5% pea starch of normal type; NMSrr contained 6.5% pea starch of high amylose type.

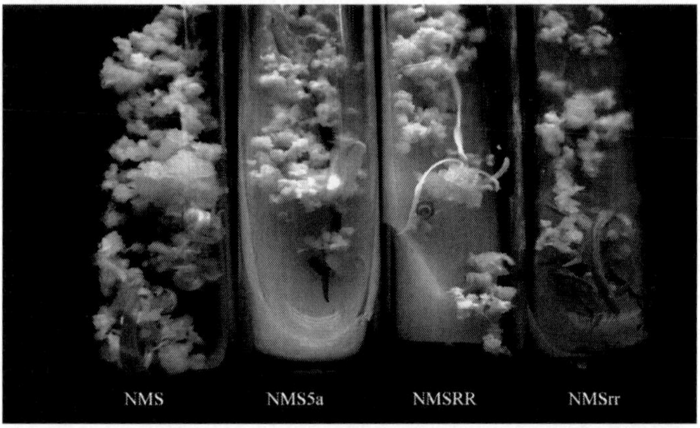

FIGURE 14.8 Embryo formation and plant regeneration in spring barley (*Hordeum vulgare* L.) anther culture *in vitro* on media differed in solidified component: NMS contained 0.8% agar Difco (USA), NMS5a contained 12.0% chemically modified starch D-5a, NMSRR contained 4.5% pea starch of normal type; NMSrr contained 6.5% pea starch of high amylose type.

Natural starches undoubtedly are ecological pure source of nutrient media gelling component for plant cells, tissue and organ culture. They can be obtained in laboratory using simple techniques and corresponding plant raw material. But preparations of natural starch can be improved by chemical [14, 30, 31], and physical modification [32], which allow enhance their technological properties in the direction of solubility of paste.

The latter is very impotent for easy pouring of media into culture vessels. On the other hand, it is apparently that high technological properties of gelling component of media will be meaningless without the ability to promote morphogenesis *in vitro*.

Our first investigation aimed to study the effect of replacement of agar to chemically modified starch DKKmod. In the medium for cultivation of spring barley anthers *in vitro* revealed both positive and negative peculiarities of starch gelling component [12]. Promotion of direct embryo formation, i. e. without callus phase, and enhancement of the morphogenetic ability of callus through depression of growth nonmorphogenic one were the most impotent positive effect of starch.

The other very useful consequence of starch medium application was a lack of vitreous plants among those ones regenerated both in the inductive medium and after transfer embryoids or embryogenic calli to regeneration medium supplemented with agar. Plants regenerated from androgenic structures formed in the starch containing media has increased viability and better survival rate in *ex vitro* conditions after transfer into soil pots. This fact was in agreement with the results of early observations concerning lower percentage of vitreous plants of some vegetable crops in micropropagation [14]. It should be noted that natural starches didn't decrease a number of vitreous plants. Moreover, when concentration of pea normal type starch was 4.5% instead of 6.5% a number of vitreous plants were enhanced.

Germination of embryoids in the media supplemented with chemically modified starches was less effective than in the media solidified with natural starches and agar. Obviously, high density of the first type of media inhibited a process of embryo germination and plant regeneration. In order to study whether chemically modified starch would stimulate plant regeneration when it would be used in regeneration media, special experiment has been carried out. Data obtained have conformed negative effect of chemically modified starch D-2 on plant regeneration when this preparation was added to regeneration medium [33]. Only in cultivar Fenix possessing a low androgenic capacity a higher efficiency of plant regeneration was obtained in starch containing regeneration media that could be explained by low speed of the androgenic structure development in recalcitrant genotypes and preferential regeneration through indirect embryogenesis.

Preparations of chemically modified starches differed in their morphogenetic effect in spring barley anther culture *in vitro*. The first evaluated preparations (DKKmod, D-2) reduced a number of embryogenic anthers in comparison with agar. Further investigations allowed find preparation D-5b which hadn't negative effect on androgenic structure induction (Fig. 14.8). Starch D-5a inhibited androgenic structure formation but in much less extent than D-2 and D-4 g.

Plant regeneration efficiency was higher in media solidified with chemically modified starches D-2, D-5a and D-5b than in agar containing media (Fig. 14.9).

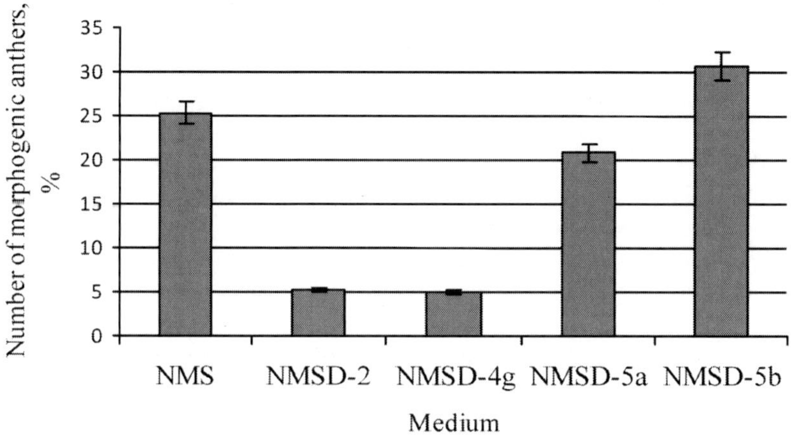

FIGURE 14.9 Formation of morphogenic structures in anther culture *in vitro* of spring barley line ДГ00–126 in the media containing different gelling agents: NMS – agar (0,8%, Difco, USA); NMSD-2, NMSD-4 g, NMSD-5a, NMSD-5b – chemically modified starches D-2, D-4 g; D-5a, D-5b (12.0%).

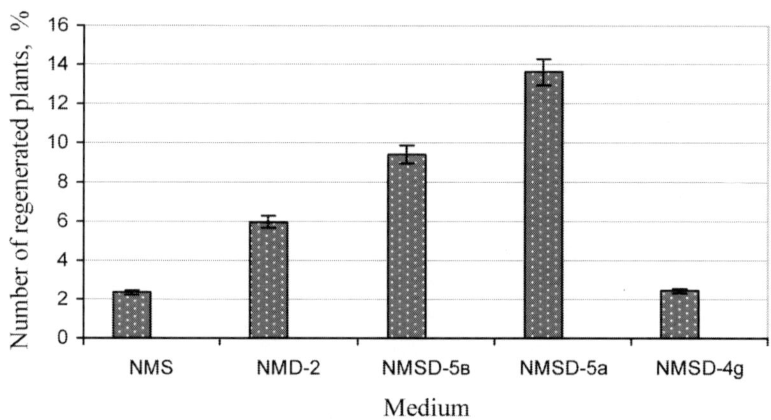

FIGURE 14.10 Green plant regeneration in anther culture *in vitro* of spring barley line ДГ00–126 in the media containing different gelling agents: NMS – agar (0.8%, Difco, USA); NMSD-2, NMSD-4 g, NMSD-5a, NMSD-5b – chemically modified starches D-2, D-4 g; D-5a, D-5b (12.0%).

The best result of plant regeneration was obtained when D-5a was used as agar substitution. Taking in consideration data of this experiment starch D-5a was considered to be the most perspective preparation for plant biotechnology. This starch produced paste, which after boiling could very easy pouring into tubes before autoclave sterilization. If nutrient media or only organics (vitamins, hormones, amino acids, etc.) are filter sterilized and added to paste which sterilized in autoclave, this procedure don't differ from that one in the case of use agar as a gelling component. The matter is that, chemically modified starch D-5a after autoclave sterilization produced paste looked as absolute transparent solution. It was impossible to distinguish high qualified (and high expensive) agar "Difco" and D-5a after autoclave sterilization. Naturally, after mixing both components of media (filter and autoclave sterilized) the media can be pouring into culture vessels without any difficulties. Since chemically starch containing media demand longer period of time for gel formation in comparison to agar containing media, organics and salts can be added to paste in an hour or much later after autoclave sterilization. Solution temperature could reach 20–25 °C that was very impotent for addition of special thermo unstable components of the media.

Our study on morphogenetic effect of chemically modified starch D-5a in anther culture of spring barley showed that this preparation enhanced the effectiveness of plant regeneration without decrease androgenic structure production (Table 14.2).

TABLE 14.2 Effect of Genotype and Gelling Component of Inductive Media on Androgenesis *in vitro* in Spring Barley (*H. vulgare* L.)

Genotype	Gelling component	No. of inoculated anthers	No. of responding anthers	Morphogenic anther frequency (%)	No. of green plants	Green plant regeneration frequency (%)
DH00–126	agar	384	134	34.90±2.43	117	30.47±2.33
	D-5a	383	203	53.00±2.55**	176	45.95±2.55**
Exotik	agar	503	127	25.25±1.94	12	2.38±0.68
	D-5a	336	139	41.36±2.67**	30	8.93±1.56*
Fenix	agar	656	63	9.60±1.15	17	2.59±0.62
	D-5a	400	63	15.75±1.70*	31	7.75±1.33*

* $P \geq 0.95$;

** $P \geq 0.99$.

Media contained 0.8% agar Difco (USA) and 12.0% chemically modified starch D-5a.

Moreover, percentage of morphogenetic anthers demonstrated sufficient growth in all genotypes. The dynamics of the embryoid and callus development was the same regardless of a nature of gelling component of media. The first visible structures were appeared on the anthers in 20 days after anther inoculation both on agar and chemically modified containing media. However, there were significant differences in further development of these structures. In agar solidified medium embryoids formed via direct embryogenesis if they didn't transfer to regeneration medium began to germinate. Other globular embryo like structures gradually lost their morphogenetic capacity in inductive medium. Active growth of friable callus possessing a very low regeneration potential took place in culture vessels. Results of our observations has conformed a necessity to transfer androgenic structures to regeneration medium when they reached 1–2 mm as quickly as possible, in order to prevent the loss of their ability to produce plantlets [11].

Structures formed in medium solidified with starch D-5a had smaller size and slower growth. Callus growth was also depressed. Globular structures (undifferentiated embryoids and compact callus) after replacement to regeneration medium (as a rule together with anther) completed their development and regenerated into plantlets. Differentiated embryoids had also smaller size and germinated with lower frequency than in agar containing inductive medium. However, they had higher ability to develop into plantlets in regeneration agar containing medium.

Possible explanation of starch based media useful properties, including promotion of direct embryo formation was based on the idea that anther walls should be able to hydrolyze starch with α- and β-amylase This idea was confirmed when the largest number of anthers forming callus and green plantlets was obtained in sucrose-free barley starch gelatinized medium in barley anther culture *in vitro* [10]. Interestingly, use starch instead of agar allowed replacement of expensive carbon source maltose by less coast sucrose with less negative effect on androgenesis *in vitro* in spring barley [35]. A very useful feature of starch gelatinized media was a possibility to increase duration between transferring of cultivated explants onto fresh medium, for example, in micropropagated rapeseed plants [36] that could be very impotent if plant material maintained *in vitro* in gen banks.

14.4 CONCLUSIONS

Thus having based on the results of estimation of natural corn and pea starches obtained from lines which were carries of endosperm structure mutant genes and series of chemically modified starches preparations suitable for application as solidifying agents of nutrient media have been selected. Promotion of direct embryoid formation and sufficient decrease number of vitreous plants (in the case of chemically modified starches) are the most impotent consequences of agar replacement by starches in nutrient media for spring barley haploid production via anther culture *in vitro*. Improvement of morphogenesis and very low price of starches as compared to those of agar should be taken into account if mass production of DH-lines for plant breeding purposes is aimed to carry out.

Further studies are needed to perfect technological properties of the chemically modified starch preparations. New series of experiments should be done to evaluate the efficiency of this innovative part of haploid production technology in a broad genetically diverse plant material including F_1 hybrids from breeding programs.

Nevertheless, isolated microspore culture is consider to be more contemporary and effective method of barley haploid production, anther culture *in vitro* as much more simple method which needn't special equipment can be used in plant breeding.

KEYWORDS

- **amylopectin**
- **amylose**
- **androgenesis**
- **anther**
- **auxin**
- **cytokinin**
- **plant regeneration**
- **starch**

REFERENCES

1. Thomas, W. T. B., Forster, B. P., & Gertsson, B. (2003). Doubled Haploids in Breeding, Doubled Haploid Production in Crop Plants [Eds. Maluszynski, M. et al.] Dordrecht Kluwer Academic Publishers, 337–349.
2. Kasha, K. J., Simion, E., Oro, R. & Shim, Y. S. (2003). Barley Isolated Microspore Culture Protocol, Doubled Haploid Production in Crop Plants [Eds. Maluszynski, M. et al.] Dordrecht Kluwer Academic Publishers, 43–48.
3. Devaux, P., & Kasha, K. J. (2009). Overview of Barley Doubled Haploid Production, Advances in Haploid Production in High Plants [Eds. Touraev, A. et al.] Springer Science+Business Media, 47–64.
4. Mitrofanova, I. V. (2009). Somatic Embryogenesis as an *in vitro* System of Cultivated Plants, Physiology and Biochemistry of Cultivated Plants, *41(6)*, 496–508 (in Russian).
5. Kasha, K. J., Simion, E., Oro, R., Yao, Q. A., Hu, T. C., & Carlson, A. R. (2000). An Improved *in vitro* Technique for Microspore Culture of Barley, Euphytica, *120(3),* 319–385.
6. Bugaenko, L. A., & Zalivchaya, L. S. (2007). Callus Culture in *Echinacea* L. Achievements and Problems of Genetics, Breeding and Biotechnology, Kyiv, Logos, *2*, 452–456 (in Russian).
7. Finnie, S., Powell, W., & Dyer, A. F. (1989). The Effect of Carbohydrate Composition and Concentration on Anther Culture Response in Barley (*Hordeum Vulgare* L.), Plant Breeding, *103,* 110–118.
8. Manninen, O. (1998). Optimizing Anther Culture for Barley Breeding, Agricultural and Food Science in Finland, *6,* 389–398.
9. Kohlenbach, H. W., & Wernike, W. (1978). Investigation on the Inhibitory Effect of Agar and the Function of Active Carbon in Anther Culture, Ztsch, Pflanzenphysiol, *86(5)*, 463–472.
10. Sorvari, S. (1986). The Effect of Starch Gelatinized Nutrient Media in Barley Anther Culture, Annales Agriculture Finnie, *25,* 127–133.
11. Kuhlmann, U. & Foroughi-Wehr, B. (1989). Production Haploid in Frequencies Sufficient for Barley Breeding Programs, Plant Cell Rept., *8(2)*, 110–118.

12. Kruglova, N. N. (2001). Morphogenesis in Wheat Anther Culture, Embryological Approach, Ufa Gilem, 203p (in Russian).
13. Patent 52031 Ukraine, MPK 7 C08B31/02, A01N57/00, A01N59/00, A01C1/06, Method of Polymer Material Production, Dulnyev, P. G., Kondrqatenko, S. I., Chrernyshenko, T. V,. Miroshnichenko, V. P., Ivchenko, N. V., & Goncharova, S. A. (17.01.2005) Bul., *12*, (in Ukrainian).
14. Belynskaya, E. V., & Dulnyev, P. G. (2007). Modified Starch DKK mod as a Component of Nutrient Medium for Barley Haploid Production in Anther Culture *in vitro*, Physiology and Biochemistry of Cultivated Plants, *39(2),* 136–143 (in Russian).
15. Coe, E., & Polacco, M. (1994). Maize Gene List and Working Maps, Maize Genet News Letters, *68,* 156–191.
16. Whitt, S. R., Wilson, L. M., Tenaillon, M. I., Gaut, B. S., & Buckler, E. S. (2002). Genetic Diversity and Selection in the Maize Pathway, Proc. National Academy Science USA, *99*, 12959–12962.
17. Tetlow, I. J., Morell, M. K., & Emes, M. J. (2004). Recent Developments in understanding the Regulation of Starch Metabolism in Higher Plants, J. Exp. Botany, *55*, 2131–2145.
18. Hedley, C. L, Bogracheva, Ya T., Lloyd, J. R., & Wang, T. L. (1996). Manipulation of Starch Composition and Quality in Pea Seeds, Agri-Food Quality, an Inter Disciplinary Approach [Fenwick, G. R., Hedley, C., Richards, R. L., & Khokhar, S. (Ed)] Cambrige, the Royal Society Chemistry, 138−148.
19. Belinskaya, E. V. Production of Sign Collection Concerning Androgens is *in vitro* Capability and its Application in Genetics and Biotechnology Investigations, Bulletin of the Ukrainian Society of Genetics and Selections,*5(1, 2),* 11–20 (in Russian).
20. Belinskaya, E. V. (2008). Inheritance of Potential for *in vitro* Androgens is in Spring Barley, Cytology and Genetics, *42(4),* 237−245.
21. Bilynska, O. V., Vesna, S. V., & Manzyuk, V. T. (2002). Application of Anther Culture *in vitro* for Production of Initial Material in Hulless Barley Breeding, Breeding and Seed Production, *86,* 164–172 (in Ukrainian)
22. Chu, C. C. (1978). The N6 Medium and its Application to Anther Culture of Cereal Crops, Plant Tissue Culture Procedure Symp Peking Science Press, 43–45.
23. Murashige, T., & Skoog, F. (1962). A Revised Medium for Growth and Bioassays with Tobacco Tissue Cultures, Physiology Plant, *15,* 473–497.
24. Rikhter, M., Augustat, Z., & Shirbaum, K. (1975). Selected Methods of Starch Investigation, Moscow Food Industry, 183p (in Russian).
25. Bilynska, O. V., Dulyev, P. G. (2011). Chemically Modified Starches as Factor of Morphogenesis in Spring Barley Anther Culture *in vitro*. Regulation of Plant Growth and Development Physiological, Biochemical and Genetic Aspects II Abstracts of International Conference, Kharkiv, 55–56 (in Ukrainian).
26. Patent 34859 Ukraine MKB A 01 G7/00, C12N 5/00, Method of Barley Haploid Production in Another Culture *in vitro*, Bilynska, O. V., Tymchuk, S. M., Dulnyev, P. G., & Derebizova, Yu O. (28.08. 2008) Bul, *16* (in Ukrainian).
27. Patent 42192 Ukraine. MB A 01 G7/00, C 12N 5/00, Method of Increasing of Green Plant Yield in Barley Anther Culture *in vitro*, Bilynska, O. V., Tymchuk, S. M., Dulnyev, P. G., & Derebizova, Yu O. (02.02. 2009) Bul, *12*, (in Ukrainian).
28. Patent 103426 Ukraine A01H 4/00, C12N 5/02 2006.1, Artificial Nutrient Medium for Barley Haploid Production in Anther Culture *in vitro,* Bilynska, O. V., Tymchuk, S. M., & Derebizova, Yu O. (10.10.13.) Bul, *19* (in Ukrainian).
29. Klucinec, J. & Thompson, D. B. (1999). Amylose and Amylopectin Interact in the Gelation of Dispersed High-Amylose Starches, Cereal Chemistry, *76,* 282–291.

30. Patent 3975206 US, C08B31/18, Methods of Peroxide Thinning Granular Starch, Lotzqesell, J. A., Moser, K. B. & Hurst, T. L. (17.08.1976) http://www.patentstorm.us.
31. Patent 2159252 RF, MPK6, C08B30/12, Method of Modified Starch Production, Zhushman, A. I., Veksler, R. I., Karpov, V. G., Zavada, E. N., & Grachova, E. V. (20.11.2000) http://www.ru-patent.info (in Russian).
32. Patent 2063402 RF, MPK6, C08B30/12, Method of Modified Starch Production, Tyrsin, Yu A., Korotchenko, N. V., Pashalov, E. R., & Sharpatyi, V. A. (10.07.1996) http://www.ru-patent.info (in Russian).
33. Bilynska, O. V., & Dulnyev, P. G. (2008). Use of Modified Starch as a Gelatinized Component of Nutrient Media for New Formation Induction and Plant Regeneration in Barley Anther Culture *in vitro*, Bulletin of Kharkivsky National Agricultural University Named after Dokuchaev, V. V. Series Biology, *2(14)*, 83–89 (in Ukrainian).
34. Sorvari, S., & Schider, O. (1987). Influence of Sucrose and Melibiose on Barley Anther Culture in Starch Media, Plant Breeding, *99(2)*, 164−171
35. Bilynska, O. V. (2009). Effect of Gelatinized and Nutrient Carbohydrate Components of Media on Haploid Induction in Barley Anther Culture *in vitro*, Bulletin of National University of Life and Environmental Sciences of Ukraine, *132*, 17–24 (in Ukrainian)
36. Bilinska, O. V. & Sokolnikova, Ya N. (2012). Genotypic Peculiarities of Microclonal Propagation in Oil Rapeseed *Brassica napus* L. Achievements and Problems of Genetics, Breeding and Biotechnology, Kyiv, Logos, *4*, 432–437 (in Ukrainian).

PART V

BIO-ORGANIC FARMING

CHAPTER 15

EFFICIENT STRAIN OF SLOW-GROWING NODULATING BACTERIA *BRADYRHIZOBIUM JAPONICUM* 84KL AS A BASIS OF BIOFERTILIZER SOYARHIZ

LILIYA E. KARTYZHOVA, IRINA V. SEMYONOVA, NATALIA V. KOROLYONOK, ZINAIDA M. ALESCHENKOVA, and LYUDMILA V. ROMANOVA

Institute of Microbiology, National Academy of Sciences, d 2, Kuprevich St., Minsk, 220141, Belarus; E-mail: Liliya_Kartyzhova@mail.ru

CONTENTS

ABSTRACT

Most efficient competitive strain of slow-growing nodulating bacteria *Bradyrhizhobium japonicum* 84KL possessing growth-promoting and nitrogen-fixing activities was isolated and selected from nodules of soybean cultivated in Belarus. Technological parameters and conditions for submerged fermentation of a new bacterial strain were defined, and survival rate of the strain during storage in liquid culture and upon immobilization on substrate-carrier was examined. It was found that presowing inoculation of soya seeds with biofertilizer SoyaRhiz promoted rise in soya seed productivity stimulated by mineral $P_{50}K_{120}$ supply by 7.8 c/ha, 5.4% increase in seed protein content, accumulation of ammonium nitrogen in soil in amount 4.3–6.3 mg/kg with concomitant fall of nitrate nitrogen level by 60–70%.

15.1 INTRODUCTION

The relevance of protein problem for fodder production in Belarus requires to include into agricultural practice nontraditional legume crops, like soybean (*Glycine max*). Soya protein is known to approximate to animal proteins in nutritive value. Varied chemical composition of soya seeds enables to use them for alimentary, feed and technical purposes.

Seed inoculation with specific nodulating bacteria *B. japonicum* is a vital agrotechnological procedure for productive soya cultivation in Belarus. Lack of corresponding bacteria in local soils urges to focus research efforts on isolation and selection of symbiotic *B. japonicum* strains from nodules of soybean varieties cultivated in Belarus and elaboration of derived energy-saving methods of producing efficient biofertilizers.

Aim of this investigation – development of technology for manufacturing biofertilizer SoyaRhiz designed for seed treatment and effective soybean cultivation by researchers of Institute of Microbiology, National Academy of Sciences, Belarus.

15.2 MATERIALS AND METHODOLOGY

Microbiological materials: strain of slow-growing nodulating bacteria *B. japonicum* 84KL was isolated from nodules of *Glycine max* tilled at experimental plots of Institute of Genetics and Cytology, National Academy of Sciences, Belarus. The strain was selected as most effective, displaying growth-promoting and nitrogen-fixing activities and mitiating fruitful symbiosis with soya varieties adapted in Belarus.

Field trials were conducted on sod-podzol sandy loam 30–50 cm deep lying on sand bed. The test field located at Uzda-region belonged to Suvorov experimental station of Institute of Agronomy and Agrochemistry. Agrochemical parameters of tilled soil horizon: humus content – 2.0–2.2%, mobile P forms (0.2 n HCl) 180–200 mg/l kg soil, metabolic K (0.2 n HCl) 150–200 мg/l kg soil, pH_{KCl} 5.8–6.0.

Plants: soybean variety "Polesskaya-201."

Submerged fermentation of slow-growing nodulating bacteria *B. japonicum* was carried out on laboratory shaker (180–200 rpm) in 250 mL Erlenmeyer flasks containing 100 mL of liquid nutrient medium.

Composition of nutrient media, g/l: mannitol-lupine – K_2HPO_4 – 0.5; $MgSO_4$ – 0.2; NaCl – 0.2; $CaSO_4$ – 0.1; mannitol – 10.0; $(NH_4)_6Mo_7O_{24}\cdot 4H_2O$ – 0.05 (traces); lupine extract – 100 mm prepared as follows: 5 g of lupine germinated seeds are sterilized in 100 mL of water in autoclave at ¾ atm. pressure during 20 min [1]; molasses-lupine: K_2HPO_4 – 0.5; $MgSO_4$ – 0.2; NaCl – 0.2; $CaSO_4$ – 0.1; $(NH_4)_6Mo_7O_{24}\cdot 4H_2O$ – 0.05 (traces); molasses – 20.0; lupine extract – 100 mL; molasses-peas –K_2HPO_4 – 0.5; $MgSO_4$ – 0.2; NaCl – 0.2; $CaSO_4$ – 0.1; $(NH_4)_6Mo_7O_{24}\cdot 4H_2O$ – 0.05 (traces); molasse – 20.0 r; peas extract – 100 mL (5 g of shelled peas seeds are prepared like lupine extract); molasses-corn – K_2HPO_4 – 0.5; $MgSO_4$ 0.2; NaCl – 0.2; $CaSO_4$ 0.1; $(NH_4)_6Mo_7O_{24}\cdot 4H_2O$ – 0.05 (traces); molasses – 20.0; corn steep liquor – 6.0; molasses-yeast – K_2HPO_4 – 0.5; $MgSO_4$ 0.2; NaCl – 0.2; $CaSO_4$ – 0.1; $(NH_4)_6Mo_7O_{24}\cdot 4H_2O$ – 0.05 (traces); molasses – 20.0; yeast extract – 2.0.

Optimization of nutrient media was performed according to the following criteria: maximal number of *B. japonicum* 84KL cells for minimal time during submerged fermentation; symbiotic and biometric soya characteristics in artificial light culture upon seed inoculation with nodulating bacteria *B. japonicum*. Test runs to evaluate conditions and parameters of submerged fermentation of strain *B. japonicum* 84KL and laboratory technology of manufacturing biofertilizer were conducted in 10 L bioreactor ANKUM-10. 5–7-day seed material in ratio 10 and 20% of media volume was used to inoculate the fermentor. The titer of *B. japonicum* 84KL cells in inocula and biofertilizer SoyaRhiz was determined by plating sequential dilutions on agar mannitol-lupine medium.

Effect of soya seed inoculation was estimated in artificial light culture: germinated sterile seeds were inoculated with cultural liquid of nodulating bacteria and suspension derived from biofertilizer. In field experiments soyabean seeds were inoculated with working solution of biofertilizer. Efficiency of seed inoculation with cultural liquid and biofertilizer was assessed via biometric(height and weight of plants) and symbiotic parameters(virulence, nodulating capacity, nitrogen –fixing activity); qualitative and quantitative productivity data (crude protein content, weight of 1000 grains, grain harvest per 1 hectare) [2–4].

15.3 RESULTS AND DISCUSSION

SoyaRhiz fertilizer manufacturing technology was elaborated using isolated symbiotically active slow growing nodulating bacterial isolates *B. japonicum* 84K. Symbiotic parameters of *B. japonicum* 84KL recorded in artificial light culture with soybean of Yaselda variety exceeded characteristics of reference strain *B. japonicum* 604K: the number of formed nodules equaled 24 per plant, that is, 26% more than that of reference strain; growth- stimulating effect reached 45%, while fixation of atmosphere nitrogen increased 8.5-times to attain the value 1.7 μg N_2/l plant by 30 min in artificial light culture versus 5.04 μg N_2/l plant by 30 min in microvegetation experiment. The strain was deposited at collection of nonpathogenic microorganisms, Institute of Microbiology, National Academy of Sciences, Belarus under the registration number BIM – B-501.

Elaboration of biofertilizer production technology based on efficient strain of nodulating bacteria *B. japonicum* stimulating growth and development of soya envisaged defining optimal nutrient medium, inocula volume, testing stability of symbiotic properties and efficiency of soybean seed treatment in light culture and field trials. It is known that mannitol-lupine medium where mannitol serves as the carbon source is the optimal nutrient substrate for active development of slow-growing nodulating bacteria. Possibility of replacing mannitol by alternative less expensive carbon sources was regarded to optimize composition of nutrients for specific *B. japonicum* strain and to raise profitability of biofertilizer process and product. To check this option effect of new media components on cell reproduction of strain *B. japonicum* 84K was studied during submerged fermentation of soya nodulating bacteria on laboratory shaker 200 (rpm) at 28°C for 120 h. Molasses containing 40–55% sucrose and ash substances – waste product of sugar refining was used as alternative carbon sources. Lupine, peas, yeast extracts and corn steep liquor were chosen as growth factors in the experiment.

Table 15.1 presents growth dynamics of *B. japonicum* cells in the course of 120 h submerged culture on diverse nutrient media. It was found that maximal cell titer of nodulating bacteria *B. japonicum* was registered by 120 h of fermentation on three nutrient media – mannitol-lupine, molasses-yeast, molasses-lupine – 1.5×10^{10} CFU/ml, 1.02 $\times10^9$ CFU/ml and 1.8×10^9 CFU/ml, respectively, whereas minimal cell titer was observed on molasses-peas media– 3.8×10^8 CFU/ml. Cells of nodulating bacteria were not detected in the molasses-corn media after 12 h of submerged culture. Cultural liquid resulting from 120 h of fermentation was applied for soybean seed inoculation.

TABLE 15.1 Cell Titer of Bacterial Strain *B. japonicum* 84KL in the Course of Shaker Flask Culture

Fermentation time, h	CFU/ml of the media				
	mannitol-lupine	molasses-lupine	molasses-peas	molasses-corn	molasses-yeast
0	$(2.0\pm0.06)\times10^7$	$(3.7\pm 0.2)\times10^7$	$2.7\pm 0.3\times10^7$	$(1.2\pm0.3)\times10^7$	$(1.8\pm 0.02)\times10^7$
12	$(6.6\pm0.2)\times10^6$	$(3.5\pm 0.3)\times10^7$	$3.4\pm 0.6\times10^7$	-	$(0.8\pm 0.1)\times10^7$

TABLE 15.1 *(Continued)*

Fermentation time, h	CFU/ml of the media				
	mannitol-lupine	molasses-lupine	molasses-peas	molasses-corn	molasses-yeast
24	$(8.1\pm0.2)\times10^{6}$	$(6.0\pm0.03)\times10^{7}$	$6.5\pm1.2\times10^{7}$	-	$(4.0\pm0.2)\times10^{6}$
72	$(2.7\pm0.1)\times10^{7}$	$(7.9\pm1.04)\times10^{7}$	$1.0\pm0.02\times10^{8}$	-	$(5.6+0.4)\times10^{7}$
96	$(3.3\pm0.08)\times10^{8}$	$(1.2\pm0.01)\times10^{8}$	$1.2\pm0.06\times10^{8}$	-	$(8.9\pm1.2)\times10^{8}$
120	$(1.5\pm0.04)\times10^{10}$	$(1.8\pm0.03)\times10^{9}$	$3.8\pm0.1\times10^{8}$	-	$(1.02\pm0.8)\times10^{9}$

Monitoring of soybean dynamics in artificial light culture demonstrated that inocula derived from molasses-lupine, molasses-peas, and molasses-yeast media promoted growth of seedlings, biomass accumulation and root development. Yet, the obtained data did not exceed the control values (mannitol-lupine medium). It was found that soya seed treatment with inoculum originating from molasses-yeast medium increased strain nodulating ability by 113% in comparison with the control variant. Generation of nodules was also recorded in case of inocula collected from molasses-peas medium but their amount on soybean roots was 50% lower and strain virulent activity was 34% inferior.

It was established that application of cheaper sources of carbon and energy for submerged fermentation of slow-growing nodulating bacterial strain *B. japonicum* 84KL does not secure stability of nitrogen-fixing activity (Fig. 15.1). The final choice of optimal nutrient medium providing for active proliferation of nodulating bacteria *B. japonicum* 84KL and stability of physiological-biochemical characteristics was grounded on parameters of nitrogen-fixing activity.

Ability of soya-colonizing bacteria to bind atmosphere nitrogen was lost during fermentation on molasses-corn medium. 95% reduction of nitrogen-fixing potential was observed in experiments utilizing molasses-peas and molasses-yeast media. Submerged fermentation of soya nodulating bacteria in molasses-lupine medium caused less inhibitory effect than variants with molasses-peas and molasses yeast media. It was proved experimentally that mannitol-lupine medium is the best for cell propagation of nodulating bacteria *B. japonicum* 84KL. The inocula derived from this medium accelerate seed germination and stimulate growth of seedlings. The top nitrogen–fixing activity of strain *B. japonicum* 84KL was revealed following soybean seed treatment with inoculum produced on mannitol-lupine medium.

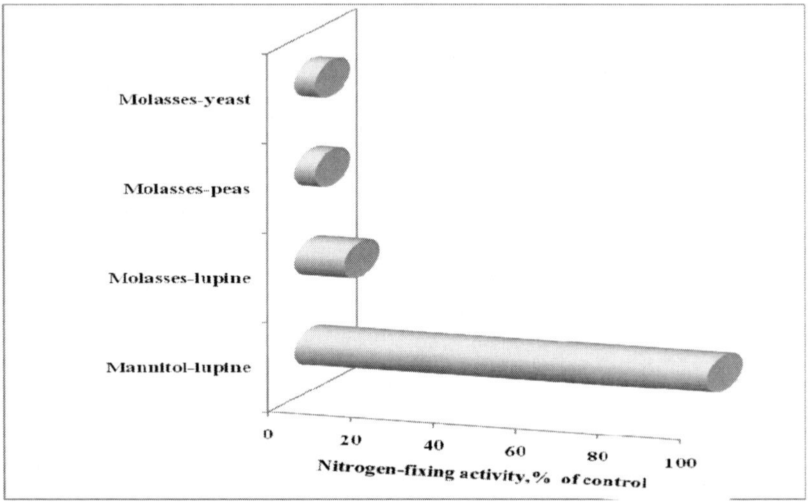

FIGURE 15.1 Effect of different nutrient media on level of nitrogen–fixing activity in strain *B. japonicum* 84KL.

Application of different (10 or 20%) of seed material to generate inocula on mannitol-lupine medium in shake flask culture during 168 h resulted in different saturation of cultural liquid with Rhizobial cells. Maximum microbial titer achieved by 120 h of fermentation with 10% inoculum equaled $6.5\pm0.32\times10^{9}$ CFU/ml – the value smaller by an order if compared with 20% inoculum (2.8×10^{10} CFU/ml), which is essential for strain immobilization on supports (Fig. 15.2).

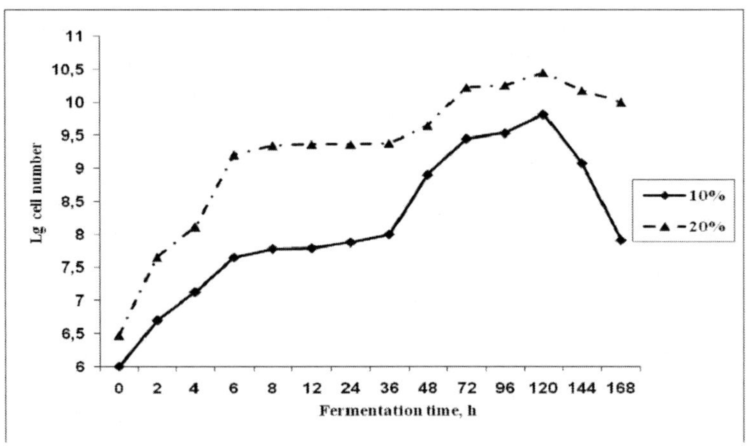

FIGURE 15.2 Growth dynamics of strain *B. japonicum* 84K as a function of different volume of seeded bacterial suspension.

Significant distinctions in impact of inocula varying in bacterial cell density on host plant and formation of symbiotic relationship in artificial light culture were revealed. Cell concentration in inoculum based on initial 20% seed material facilitates generation of solid root system and surface vegetative part, promotes virulent activity of bacteria 4- fold and enhances nodulating capacity 9-times as compared to the inoculum derived from 10% seed material (Fig. 15.3).

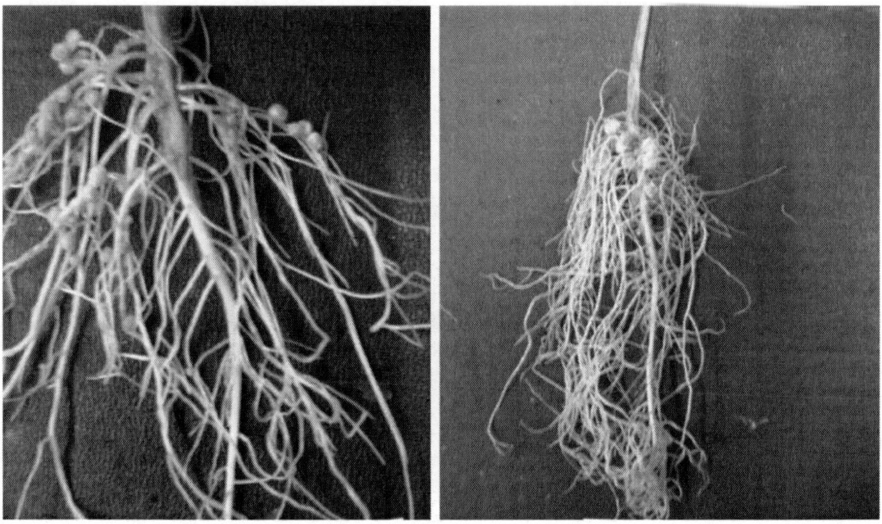

A Б

FIGURE 15.3 Effect of inocula derived from various volumes of seed material on nodulating capacity of strain *B. japonicum* 84KL in symbiotic association with soybean cultivars
Note: the ratio of inoculation material in total volume of nutrient medium: A–20%, B–10%.

It is clear that 20% volume of inoculum contributing to stability of symbiotic properties of nodulating bacteria *B. japonicum* 84KL in association with host plant was approved as optimal for further investigations. The optimal acidity of the nutrient medium favoring high activity and elevated cell titer of nodulating bacteria, stability of symbiotic potential, productive association with host plant was determined. It was found that at pH of the medium 6.8 intensive propagation of soya nodulating bacteria occurred so that after 5 days of submerged culture on mannitol-lupine medium in flasks on the shaker the maximal cell titer of strain *B. japonicum* 84KL was achieved—2.8×10^{10} CFU/ml in contrast to microbial titer at pH 6.05–1.0×10^{8} /ml. Submerged culture in laboratory fermentor during 120 h on mannitol-lupine medium at temperature 28°C, aeration 1 liter of air/1l medium/min, agitation rate 200 rpm, pH 6.8–7.0 produced the maximal cell titer of strain *B. japonicum* 84KL 1.44×10^{10} CFU/mL.

Biofertilizer SoyaRhiz was manufactured by immobilization of soya nodulating bacteria on beat carrier in proportion 90 mL of cultural liquid per 200 g of sterile peat substrate. The end product is fully damp dark homogeneous mass with the following characteristics: degree of peat decomposition—30–35%, pH 6.8, moisture content—42%, cell titer of strain *B. japonicum* 84KL—at least 1.02×10^8 CFU/g. One-hectare portion of biofertilizer weighs about 200 g.

Completed field trials of SoyaRhiz in combination with mineral fertilizer confirmed its application efficiency (Table 15.2).

TABLE 15.2 Symbiotic Soybean Parameters after Seed Treatment with Nodulating Bacteria (Field Tests)

Experimental variants	Nodulating ability, number of nodules per plant	Nitrogen-fixing activity	
		μg N_2/plant/h	μg N_2/m^2
$P_{50}K_{120}$	–	–	–
$P_{50}K_{120}$ + *B. japonicum* 84KL (cultural liquid)	10	0.11	6.1
$P_{50}K_{120}$ + *B. japonicum* 84KL (SoyaRhiz)	8	0.19	10.5

Note: N_2 – atmospheric nitrogen bound by nodulating bacteria and calculated on the basis of content of ethylene generated by acetylene method in analyzed plant samples (roots with nodules).

Performed field trials demonstrated that application of biofertilizer resulted in increment of seed yield expressed as crude protein equivalent to 345 kg/ha (61%), increase of protein ratio in seeds by 5.4%, and facilitated accumulation of ammonium nitrogen in soil (4.3–6.3 mg/kg) coupled to 60−70% decrease of nitrate nitrogen level.

15.4 CONCLUSIONS

Studies were conducted to define and optimize composition of nutrient medium and its capacity for active propagation of soybean nodulating bacteria. The appropriate volume of seed material from strain *B. japonicum* 84KL was chosen to produce inocula with high cell titer of slow–growing nodulating bacteria ensuring effective symbiosis with soy cultivars. Maximum cell titer of strain *B. japonicum* 84KL (1.44×10^{10} CFU/mL) was achieved by 120 h of submerged fermentation in bioreactor ANKUM-10 on mannitol-lupine medium at pH 6.8, temperature +28°C, aeration – 1 L air/L/min, agitation rate 200 rpm.

Biofertilizer SoyaRhiz was produced by immobilization of soya nodulating bacteria on sterile peat carrier in amount 90 mL of inoculum per 200 g of peat substrate. Cell titer of strain *B. japonicum* 84 KL in 1 g of substrate carrier is at least 1.02×10^8 CFU/g. The shelf life of liquid fertilizer is 3 months; the storage term of peat-based SoyaRhiz is extended to 7 months. Pre-sowing inoculation of soya seeds with biofertilizer SoyaRhiz promoted rise in seed productivity backed by mineral $P_{50}K_{120}$ supply up to 7.8 centners per hectare providing for additional protein yields of 345 kg/ha, extra accumulation of ammonium nitrogen in amount 4.3–6.3 mg/kg soil, 60–70% fall of nitrate nitrogen level.

KEYWORDS

- **bacterial culture**
- **inoculum**
- **light culture**
- **nitrogen fixation**
- **soya**
- **symbiosis**

REFERENCES

1. Method to Study Nodulating Bacteria (1981). Methodological Recommendations for Courses to Upgrade Qualifications of Researchers Specializing in Agricultural Microbiology, Leningrad, 48p (in Russian).
2. Zvyagintsev, D. G., Aseeva, I. V., Babyeva, I. P. & Mirchik, T. G. (1980). Methods of Soil Microbiology and Biochemistry, Moscow, 63–130 (in Russian).
3. Gerhardt, F. (1983). Methods of General Bacteriology, Moscow, Mir Publishers, *1*, 536p (in Russian).
4. Acetylene Method Application Guidelines during Selection of Legume Crops for Increased Symbiotic Nitrogen Fixation (1982) Leningrad, 12p (in Russian).

CHAPTER 16

SELECTION OF COMPETITIVE AND EFFICIENT STRAINS OF *RHIZOBIUM GALEGAE*

LILIYA E. KARTYZHOVA

Institute of Microbiology, National Academy of Sciences, Kuprevich str., 2, 220141, Minsk, Belarus

CONTENTS

ABSTRACT

This chapter presents results of selecting effective competitive local strains of *Rhizobium galegae*. Using qualitative parameters (nodulating ability and nitrogen –fixing activity) evaluating rhizobial symbiotic relationship with host plant, the most efficient competitive *Rhizobium galegae* variants were selected. Quantitative analysis of isolates was performed to define the types of competitive ability (saprophytic, rhizospheric, nodulating). The selected cultures of nodulating bacteria were identified as *Rhizobium galegae* strains infecting exclusively seeds of *Galega orientalis* Lam. and capable to set up efficient symbiosis with host plant.

16.1 INTRODUCTION

Agrocenoses based on nonconventional promising perennial crops, like *Galega orientalis* Lam. are a vital reserve of productive, sustainable and long functioning agroecosystems aimed at raising soil fertility, improving ecological situation, reducing adverse technogenic effects, fostering forage basis, increasing vegetable and animal protein yields.

The attractiveness of *Galega orientalis* – a legume cultivar not habitual for Belarus agronomy is determined by high biological flexibility, durability and biomass/seed productivity. Its efficient cultivation in the republic requires optimization of agrotechnological procedures, including presowing seed inoculation ensuring efficient symbiotic relationship with *Rhizobium galegae*.

Constructing stable plant-microbial association between *Galega orientalis* and *Rhizobium galegae* is intended to promote productivity of *Galega orientalis* cultivated in local conditions and to elaborate fertilizing compositions based on competitive specific strains of nodulating bacteria to be used for seed inoculation.

One of key parameters defining legume crop productivity is nitrogen–fixing activity secured by 100-year-odd record of seed inoculation with various species of nodulating bacteria. Despite ubiquitous distribution of rhizobia, there are some geographic and soil type variations in spread of specific microbial groups. Nodulating bacteria-symbionts of clover, peas, beans (*R. leguminosarum trifolii, R.L. viciae, R.L. phaseoli*) and alfalfa (R. meliloti) are typical for moderate latitudes [1–3].

Analysis of population and frequency rate of *Aerbaspirillum* genus capable to carry out nitrogen fixation in rhizosphere of 17 forage grass species demonstrated that its microbial titer ranged from 10^2 to 10^4 cells per 1 g soil [4].

Most numerous species are represented by *R. trifolii* reaching the titer 1.7×10^5 cells/g soil. *R. meliloti* population may be referred to low-density, accounting for 17 cells per 1 g soil [5]. *R. galegae* was not revealed in the examined soils.

Specificity of *R. galegae* was confirmed using phagotyping technique, polysaccharide analysis, polymerase chain reaction, deciphering plasmid composition and structure of genome loci. Two biotypes of *R. galegae* species isolated from two host plant varieties (*Galega orientalis* and *Galega officinalis*) showed significant differences and ability to form efficient symbiosis only with cultivar serving as their isolation source [6].

Galega orientalis culture in Belarus lacking presowing seed inoculation and native rhizobial microbiota in soil results by 2–3 years in crop degeneration and weed replacement. Inoculation of *Galega orientalis* seeds by specific nodulating bacteria led to significant rise of galega green mass yields (40–60%), whereas nitrogen content in the over-ground part of the plant rose by 30–80% [7].

Application of a mixture composed of minced plant roots and bacterial nodules increased biomass productivity of *Galega orientalis* in the first year to 171.4c/ha, or 14.6% up the control [8].

Using a binary inoculum (*R. galegae* +*R. meliloti*) coupled to root molybdenum supply facilitates shaping of more active nodules, higher biomass yields with elevated content of solids and protein in leaves [9].

An effective technique allowing to double or triple *Galega orientalis* yields, to raise productivity of nitrogen fixation 2–4-fold is introduction of strain *R. galegae* 914 and *Arthrobacter myzorens* (myzorin) separately or in mixture [10].

The afore-mentioned data support the essential need to isolate from natural sources local competitive and active *R. galegae* strains capable to initiate efficient symbiosis with *Galega orientalis* in Belarus soils.

Main criteria for screening of nodulating bacterial strains are qualitative manifestations of rhizobia symbiotic potential expressed as nodule-generating capacity and nitrogen-fixing activity [11].

It was found that major determinants of efficient symbiosis are quantitative characteristics of highest value: competitive ability–capacity of tested rhizobial strain to form nodules on host plant roots in the presence of other nodulating bacteria; symbiotic efficiency–ability to set up efficient symbiosis and raise productivity of legume cultivars [12]. Availability of such properties in bacterial isolates lays the basis for their selection as potential constituents of fertilizing biopreparations.

Aim of this study was isolation and selection of local competitive and efficient *Rhizobium galegae* strains.

16.2 MATERIALS AND METHODS

Various microbiological objects were used for isolation of local *Rhizobium galegae* strains:

- natural enrichment culture of nodulating bacteria (suspension);
- soil (nonsterile);
- nodules (sterile, not sterile);
- nodules + soil (not sterile);
- natural isolates *Rhizobium galegae* №№ 1–11;
- natural isolates *Rhizobium galegae* №№ 1, 5, 8 (rifampicin – resistant).

Plant objects: *Galega orientalis* Lam. varieties: "Gale" and "Nesterka."

Soil objects: sod-podzol soil from selection nursery of Agricultural Academy, Gorki.

Bacteria *R. galegae* were isolated from nodules of old plants *Galega orientalis* Lam.

The screening of most efficient strains was based on qualitative (nodulating capacity and nitrogen –fixing activity) and quantitative (competitive ability) parameters [11, 13–15]. Drug–resistant natural isolates of rhizobia № 1, 5, 8 were derived by gradual adaptation to increased rifampicin doses (150 μg/mL).

Efficiency of natural *R. galegae* isolates was evaluated by resistance technique via competitive ability: saprophytic (SCA) [13], rhizospheric (RCA) [14, 15], nodulating (NCA), wherein the nodules formed on crop roots inoculated with a mixture of rhizobial strains were counted within 30 days following inoculation. Each nodule was tested for drug resistance. Indirect method was also applied to assess NCA [16–18].

16.3 RESULTS AND DISCUSSION

16.3.1 RECOVERY OF NATURAL R. GALEGAE ISOLATES

A 1–3 and 15 year-old summer specimens of *Galega orientalis* Lam. were used for isolation of efficient strains of nodulating bacteria representing local populations of *Rhizobium galegae* bv. *galegae*. The nodules were not detected on roots of 1–3 year-old Galega orientalis plants. Specimens of *Galega* plants aged 15 years possessed a well-developed vegetative mass and mighty root system with attached few rosy nodules engaged with rhizospheric sod-podzol tilled soil horizon of selection nursery in isolation of local *R. galegae* strains specific for *Galega orientalis* Lam. Natural *R. galegae* isolates were recovered and screened by analytical selection method.

The number of nodules generated in artificial light culture varied depending on the type of inoculation material (sterile and nonsterile nodules, nodules+soil). The maximum amount of nodules following seed inoculation with suspension based on nonsterile nodules +soil averaged 12 per plant. Large-size nodules having rosy pigmentation were chosen for isolation of *R. galegae* variants. Table 16.1 presents symbiotic properties of 11 *R. galegae* natural isolates and their effect on biometric parameters of host plant.

TABLE 16.1 Qualitative Characteristics of *Rhizobium Galegae* Isolates

R. galegae strains	Virulence, the moment of nodule appearance, days	Nodulating capacity, number of nodules per 1 plant	Phytomass/ root weight, g per 1 plant		Plant height, cm	Nitrogen-fixing activity, µg N_2/plant in 30 min
			crude	dry		
1	10	23	164/82	23/11	8,7	4,6
2	17	7	74/29	16/3	5,2	0,6
3	11	13	63/25	15/4	5,0	-
4	-	-	-	-	-	-
5	10	13	142/51	23/6	8,3	2,2
6	18	5	70/30	16/3	4,8	0,5
7	13	8	76/30	18/5	7,0	0,8
8	10	10	142/46	25/7	8,4	2,5
9	10	5	78/35	16/4	5,2	0,6
10	17	7	84/32	18/3	6,8	0,8
11	15	7	88/28	17/4	7,6	0,9

Note: *R. galegae* isolates were selected according to qualitative parameters of symbiotic activity: maximum values are given in semibold type and minimum values in *semibold italics.*

16.3.2 SELECTION OF MOST EFFICIENT AND LESS EFFICIENT BUT COMPETITIVE NATURAL R. GALEGAE ISOLATES

According to qualitative characteristics of symbiotic potential (nodulating capacity, nitrogen – fixing activity) four most efficient variants were sorted out of 11 isolated cultures of nodulating bacteria colonizing *Galega orientalis*: *R. galegae* № 1, 5, 8 and one isolate competitive in nodulating capacity and not displaying nitrogen –fixing activity—*R. galegae* № 3.

It was found that natural isolate № 1 was distinguished by the highest nodulating capacity (23 nodules per plant) and nitrogen –fixing activity (4.6 µg N/plant in 30 min). Seed inoculation with *R. galegae* № 3 and №5 produced the same number of nodules, yet virulent activity of strain № 3 was 10% lower than that of *R. galegae* № 1.

The number of nodules formed as a result of seed inoculation with *R. galegae* №3, №5, №8 equaled 56, 56 and 43% of corresponding *R. galegae* № 1 values. In *R. galegae* № 3 case reduction of biometric parameters was recorded (Table 16.1): crude/dry biomass weight and plant height as compared to *R. galegae* № 1, 5, 8, probably determined by lack of nitrogen-fixing activity in isolate № 3.

The nodules formed on roots of *Galega orientalis* due to *R. galegae* №3 infection were not pink-colored which provides an auxiliary evidence of not effective symbiosis with host plant. Analysis of qualitative parameters of symbiotic activity demonstrated that natural isolate *R. galegae* № 3 was quite competitive among local rhizobial species but it was unable to maintain symbiotic relationship. It appears logical therefore that *R. galegae* №3 was chosen as test culture for comparative examination of competitiveness and efficiency of selected natural isolates № 1, 5, 8 displaying nitrogen-fixing activity.

Resistance method based on mixed seed inoculation with natural isolates № 1, 5, 8 and test culture enabled to carry out comparative quantitative analysis and to assess competitive ability of each studied natural isolate (Table 16.2).

TABLE 16.2 Quantitative Parameters of *Rhizobium Galegae* Isolates (Resistance Method, %)

Isolate number	SCA		RCA		NCA	
	Control	Experiment 4 days	Control	Experiment 4 days	Control	Experiment 36 days
1	45	57	47	73	53	57
5	52	62	54	82	58	73
8	42	48	40	70	50	60

Note: control – initial percentage of natural isolate at the onset of mixed fermentation with test culture; experiment – percentage of natural isolate during mixed fermentation with test culture after definite time period.

Laboratory trials to estimate symbiotic activity of natural isolates № 1, 5, 8 revealed superiority of variant № 5 in SCA, RCA and NCA values (Table 16.2) securing its better adaptability upon introduction into ecosystems with different soil-climatic conditions. It was found that isolate №5 was capable to assimilate C, N, P and showed tolerance to extreme environmental factors (temperature, pH, humidity, pesticide pollutants in soil). Elevated rhizospheric and nodulating ability of *R. galegae* 5 is determined by high mobility, root adsorption, fast propagation rate promoting enhanced colonization and infection of plant roots. Maximal compatibility with species *Galega orientalis* Lam. enables this isolate to generate active nodules on root surface and set up fruitful symbiosis with host plant. Isolate № 1 distinguished by promising rhizospheric competitive ability (RCA 73%) displayed good motility and reproduction in soil, crop root adsorption and colonization.

It was established that nodulating competitive ability (NCA) of *R. galegae* 1 constituted 57% – less than respective indices of *R. galegae* 5 and 8. It is evident that root penetration rate and compatibility with host plant were also lower. It should be noted, however, that survival rate of *R. galegae* 1 introduced into ecosystem is 9% higher as compared to isolate 8 but 5% inferior to that of isolate 5. Increased phizospheric competitive ability (RCA 70%) of isolate № 8 secures rapid soil mobility and intensive propagation rate in rhizosphere of legume crop. According to Onishchuk (1995), genetically controlled capacity of host plant to sort out definite strains of nodulating bacteria from microbial cenoses or from mixed inocula is realized in combination with rhisopheric competitive ability of isolate *R. galegae* 8, contributing to its nodulating potential.

Adsorption of a large number of *R. galegae* 8 cells on root surface of host plant and acceleration of their penetration into the roots facilitates adequate compatibility with *Galega orientalis*. As a result high nodulating capacity of isolate 8 and specificity of host plant genotype govern creation of effective symbiotic relations. Application of indirect method for evaluation of nodulating competitive ability (NCA) indicating more massive production of nodules by efficient strains vs. test culture № 3 during mixed fermentation, allowed to confirm the data on competitive ability of each natural isolate (1, 5, 8) determined earlier by resistance method.

Quantitative assessment of NCA in examined isolates reflected by coefficient K demonstrated excellent competitive ability of *R. galegae* № 1, 5, 8 as compared to test culture – 108%, 155% and 100%, respectively. Isolate № 5 is characterized by outstanding competitive edge proved by top value of coefficient K – 155%.

The largest amount of nodules was produced by *R. galegae* № 1, 5, 8 mixed with *R. galegae* № 3 in 1:10 ratio.

Summing up, analysis of quantitative characteristics inherent to isolates 1–5, 8 by resistance and indirect methods corroborated high competitive potential and efficiency of *Rhizobium galegae*.

Morphology of *R. galegae* colonies and cells was studied (Fig. 16.1) and their physiological– biochemical properties were characterized using Bergey's taxonomic keys.

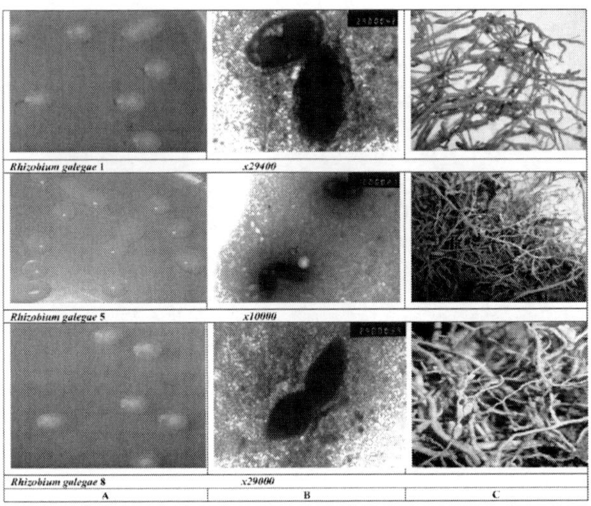

FIGURE 16.1 Morphology of *R. galegae* Strains: A – Colonies, B – Cells, C – Symbiosis With Host Plant (Nodules).

Tests were performed to check the ability of *R. galegae* isolates 1, 5, 8 to infect other legume crops: *Pisum sativum* (peas), *Trifollium repens* (clover), *Phseolus vulgaris* (beans), *Medicago sativa* (alfalfa), *Vicia sativa* (vetch), *Clycine max.* (soybean), *Galega orientalis*, *Galega officinalis*. Formation of nodules was recorded exclusively on *Galega orientalis* roots.

Natural isolates № 1, 5, 8 were identified as *Rhizobium galegae* strains able to produce nodules on the roots of host plant.

Toxicological – sanitary examination of *R. galegae* strains was conducted. Strains *R. galegae* № 1, 5, 8 were deposited at Belarusian collection of nonpathogenic microorganisms, Institute of Microbiology, National Academy of Sciences, Belarus under the respective registration numbers: *Rhizobium galegae* BIM B-436D, *Rhizobium galegae* BIM B-437D, *Rhizobium galegae* BIM B-438D.

Laboratory studies on quantitative and qualitative parameters of symbiotic activity of local *R. galegae* isolates and evaluation of competitive ability of nodulating bacterial strains [2] established that:

- strain *R. galegae1* (BIM B-436D) displays elevated saprophytic and rhizospheric rather than nodulating competitive ability caused by rapid proliferation and colonization of host plant roots (*Galega orientalis* Lam.) promoting adaptation to various soil media and survival in different agrocenoses.
- strain *R. galegae*5 (BIM B-437 D) is characterized by superb saprophytic, rhizospheric, nodulating competitive ability, favoring resistance to extreme environment. This strain exhibits super-viability in ecosystems, intense reproduction and root colonization rate, compatibility with host plant *Galega orientalis*, capacity to yield nitrogen –fixing nodules on root surface.
- strain *R. galegae* 8 (BIM B-438D) prevalently possesses nodulating competitive ability contributing to swift root penetration and virulent activity of the culture. Excellent compatibility with host plant and lack of variety specificity allow setting up efficient symbiosis with diverse biovars of *Galega orientalis*, promoting active reproduction in rhizosphere and survival in various soil-climatic conditions.

16.4 CONCLUSIONS

1. Qualitative evaluation of microbial ability to initiate symbiotic relationship with host plant allowed to screen most efficient and competitive variants among natural isolates of *Rhizobium galegae*. Quantitative analysis proved that each selected variant was distinguished by high saprophytic, rhizospheric, nodulating competitive ability.
2. The isolates of nodulating bacteria tested for symbiotic efficiency were identified as *Rhizobium galegae* strains.

3. Studies on competitive potential of *R. galegae* strains enabled to assess their viability, reproduction rate, root adsorption, which is vital for further soil adaptation and creation of symbiotic relations with *Galega orientalis* Lam.
4. New symbiotically active and competitive *R. galegae* strains will be used as active principle of microbial fertilizers intended for presowing seed treatment of various *Galega orientalis* biovars cultivated in Belarus.

KEYWORDS

- ***Galega orientalis*** **Lam.**
- **nodulating bacteria**
- **symbiosis**

REFERENCES

1. Fyodorov, M. V. (1948). Biological Fixation of Atmospheric Nitrogen, Fyodorov, M. V. Moscow, Selhozgiz Press, 442p (in Russian).
2. England, K. S. (1993). Bacterial Survival in Soil, Effect of Clays and Protozoa, England, K. S., Lee, H. & Trevors, J. T. Soil Biology & Biochemistry, *25*, 525–531.
3. Triplett, E. W. (1993). Crop Rotation Effects on Populations of *Bradyrhizobium Japonicum* and *Rhizobium Meliloti*, Triplett, E. W., Albrecht, K. A. & Oplinger, E. S. Soil Biology & Biochemistry, *25*, 781–784.
4. Indira, B. N. (1996). Herbaspirillum Associated with Forage Grasses, Indira, B. N. & Bagyaraj, D. J. Proceedings of the National Academy of Sciences, India Section B, Biological Sciences, *63(1),* 25–30.
5. Martyniuk, S. (1999). Effect of Agricultural Practices on Populations of *Rhizobium* in Some Field Experiments, Wozniakiwska, A. & Martyniuk, M. (30 Oct 1999). Proc. of the International Conference "Diversity of Microbial Species and Regulation of their Activity", Vilnius, September, Botany Lithuan, Suppl., *3,* 99–102 (in Russian).
6. Radera, G. (2001). Description of Biovars in the *R. Galegae* Species Biovar Orientalis and Biovar Officinalis, Radera, G., Gurgens, G., Niemi, M., Suomiren, L., & Lindstrom, K. Systematic and Applied Microbiology, *24(2),* 192–205.
7. Vankova, A. A. (2001). Efficiency of Inoculating *Galega Orientalis* with Nodulating Bacterial Strains of Genus *Rhizobium*, Vankova, A. A. & Ledovskaya, Yu. A. Agricultural Microbiology in XIX-XXI Centuries, Abstructs of Russian Conference, St Petersburg, 45–46 (in Russian).
8. Dovnar, I. A. (2000). Response of *Galega orientalis* to Agrotechnical Procedures During the First Year of Cultivation, Dovnar, I. A., Reserves of Raising Forage Productivity in Belarus Republic, Proceedings of Scientific Conference, Gorki, 84–87 (in Russian).
9. Zabolotnaya, V. P. (2001). Inoculation of *Galega Orientalis* as a Method Promoting Biomass Yield and Protein Accumulation, Zabolotnaya, V. P. et al. Phyziology and Biochemistry of Cultivars, *33(4),* 313–318 (in Russian).
10. Popov, A. A. (1999). Bacterial Preparations Efficient Agents Increasing Nitrogen Fixing Activity and Productivity of *Galega orientalis,* Popov, A. A. et al. Proceedings of 9 International Symposium on Forage Crops, Syktyvkar, 152–154 (in Russian).

11. Acetylene Method Application Guidelines for Selection of Legume Crops to Intensify Symbiotic Nitrogen Fixation (1982). Issued and Approved at All-Union Research Institute of Agricultural Microbiology, Leningrad, 10pp (in Russian).
12. Onishchuk O. P. Genes controlling nodulating competitive ability of nitrogen –fixing bacteria and selection applications, Onishchuk, O. P. et al. Genetics, *31(3),* 293–303 (in Russian).
13. Fyodorov, S. N. (1987). Production of *Rhizobium Meliloti* Mutants with Altered Symbiotic Properties under the Impact of UV-rays, Fyodorov, S. N., & Simarov, B. V. Agricultural Biology, *9*, 44–49 (in Russian).
14. Araujo, R. S. (1994). Hydrophobic Mutant of *Rhizobium Altered* in Nodulation Competitiveness and Growth in the Rhizosphere, Araujo, R. S., Robleto, E. A., & Handelsman, J. A. Applied and Environmental Microbiology, *60*, 1430–1436.
15. Robleto, E. A. (1998). Effects of Bacterial Antibiotic Production on Rhizosphere Microbial Communities from a Culture-Independent Perspective, Robleto, E. A., Borneman, J., & Triplett, E. W. Applied and Environmental Microbiology, *64,* 5020–5022.
16. Fyodorov, S. N. (1986). Evaluations of Symbiotic Properties of Alfalta Nodulating Bacteria (Rhizobium Meliloti) under Laboratory Conditions, Fydorov, S. N., Fokina, I. G., & Simarov, B. V. Agricultural Biology, *1,* 112–118 (in Russian).
17. Sharypova L.A. Method of comparing competitive ability of efficient *Rhizobium meliloti* strains/ L.A., Sharypova, B.V. Simarov, Collected Papers of All-Union Research Institute of Agricultural Microbiology, *55,* 85–90 (in Russian).
18. Rokitsky, P. F. (1973). Principles of Variation Statistics for Biologists, Rokitsky, P. F. Minsk Belarus State University, 221p (in Russian).

CHAPTER 17

DEGRADATION OF TRIETHYLAMINE BY STRAIN *RHODOCOCCUS QINGSHENGII* B-823D

ROMAN K. NAGORNY and ALISSA S. SAMSONOVA

Institute of Microbiology, National Academy of Sciences, Belarus, d. 2, Kuprevich St., Minsk, 220141, Belarus
E-mail: roman19031988@mail.ru

CONTENTS

ABSTRACT

Strain *Rhodococcus qingshengii* B-823D isolated from natural sources is capable to use triethylamine as a sole source of carbon and nitrogen. In batch culture it degrades the toxicant in concentrations 100, 1000 and 10000 mg/L by 40 min, 240 min and 32 h, respectively. In continuous culture strain B-823D decomposes 1000 mg/L triethylamine levels in the medium at flow rate 100 mL/h.

The cells of strain B-823D are readily immobilized on polyamide fiber (130.2 mg/g carrier) and on polyethyleneterephthalate (88.2 mg/g support). It was shown that adhesive properties of microbial culture were not significantly affected by carbon source in the nutrient media.

Hypothetical pathway of triethylamine catabolism by strain B-823D was proposed. It was demonstrated that triethylamine is oxidized to triethylamine-N-oxide further split up to diethylamine, acetaldehyde and water. Diethylamine is degraded to ethylamine and the latter is cleaved into acetaldehyde, water and ammonium. Acetaldehyde is broken down to carbon dioxide and water. As a result cells of strain B-823D carry out complete triethylamine mineralization.

17.1 INTRODUCTION

Triethylamine (TEA) is widely used in foundry for Cold-box-amine process as the catalyst of polyurethane synthesis reaction. To ensure dosed TEA supply, special gas generators evaporate it, mix with air and feed to the core box [1]. The spent mixture is directed to neutralizer where it is treated with diluted sulfuric acid. The resulting triethylammonium sulfate is discharged into overall production effluents channeled to decontamination unit. As a rule, activated sludge of biological decomposition facilities fails to break down tertiary amines. Discharges of recalcitrant compounds, like TEA, into water reservoirs cause in some cases a critical stress on bioenvironment – plants, animals, humans.

TEA applied as solvent, corrosion inhibitor, feedstock for manufacturing various chemicals, including antibiotics and pesticides [2], is also one of the major pollutants in gaseous discharges of chemical plants [3].

High volatility and chemical reactivity are characteristics referring TEA to potentially hazardous substances that may provoke acute intoxications. TEA vapors irritate mucous membranes of respiratory tract and eyes, induce excitation of central nervous system followed by its depression [4]. Experiments with test animals revealed TEA teratogenic effect [5].

So far TEA disposal in wastewaters was performed predominantly by physical-chemical methods. The technology of TEA photooxidation on TiO_2 plates in presence of O_2, N_2 and H_2O was reported [5]. Physical-chemical methods of TEA removal are expensive and laborious, they are considerably less efficient than biological counterparts.

Biofiltration appears to be one of economically attractive ways to cope with the problem of volatile organic pollutants. Diverse solid materials (peat, compost, wood chips) engaged in the process serve as biofilm support. Contaminated air passing through bioreactor is exposed to biofilm action and the polluting compounds are oxidized by immobilized microorganisms [6–8]. Natural materials carrying microbial cells provide biofilm inhabitants with inorganic substances, ruling out the need of extra feeding. Biofiltration main targets are hydrogen sulfide, ammonia, mercaptans, amines [9–12]. The Iranian researchers used 6 L 3-section biofilter system containing a mixture of compost and wood shavings plus activated sludge to investigate TEA degradation process [13]. They studied efficiency of toxicant elimination from air in production zone of mechanical engineering plant. The results indicated that TEA conversion rate was governed by microbial activity rather than by expense of recycling solution [14].

Methods of local microbial treatment of wastewaters contaminated with TEA from ecological viewpoint are considered as most effective. They engage highly active microbial degraders immobilized on support in bioreactor. A large number of publications evidence clear advantage of immobilized systems where microbial cells are fixed to water-insoluble carrier not subject to wash-out [15–18]. It was shown that attached microorganisms are more resistant to toxic action of the pollutants, are propagating faster than in suspended state and are distinguished by increased metabolic activity [19, 20].

Advanced technology for recovery of TEA-polluted air from production zone of industrial enterprises was designed by the team of Industrial Ecological Systems in cooperation with researchers from Institute of Microbiology, Belarus National Academy of Sciences. A pivotal role is played by performance of absorption biochemical unit (ABU). ABU inclusion into production scheme is economically justified due to low operation costs associated with minimal amount of required expendable materials. Percentage of trapped toxic gases from ventilation air reaches 70–99.9% depending on inlet concentrations and physical-chemical characteristics of the pollutant. ABU maintenance envisages only fan and pump repairs. The units do not cause secondary contamination.

TEA disposal from ventilation discharges using ABU technology is currently practiced at Cheboksary aggregate plant, Mariupol heavy machinery plant, Rovno foundry, Lebedyany mechanical engineering factory, Yaroslavl motor works. ABU performance record evidences their superiority over conventional scrubbers in economic, operational and ecological characteristics. The complex detoxifying ventilation air from casting, cooling and mold stripping units was launched in 2003 at Lebedyany plant. Laboratory analyzes indicated that efficiency of TEA elimination from air of industrial zone equaled 97%.

ABU are unmatched in efficiency, economic and ecological parameters. This statement is supported by the fact that over 50 units developed and manufactured by

Industrial Ecological Systems are operating at various enterprises of former Soviet Union with overall term of continuous service 19 years [1].

One of the main indexes reflecting efficiency of ABU performance is decomposing activity of microorganisms immobilized on support in bioreactor. Search for active microbial cultures TEA-degraders capable of immobilization on fibrous supports seems a high priority task.

German scientists isolated TEA-degrading microbial strains *Pseudomonas citronellolis* RA1 and *Mycobacterium dienhoferi* RA2 carrying out full utilization of the pollutant by 4 days in water solutions with concentration 50 mg/L [21].

Taiwan researchers proposed a mixed bacterial culture for TEA disposal. The culture may grow in the medium containing 650 mg/L TEA, while at 200 mg/L concentration TEA removal efficiency approaches 100% [22].

The team from Nanking agricultural university isolated from activated sludge of decontamination facilities of pharmaceutical company strain *Arthrobacter protophormiae* R4 completely degrading TEA in concentration 100 mg/L by 32 h. Correlation of toxicant decay rate by cells of *Arthrobacter protophormiae* R4 with temperature and pH was investigated. Inhibiting effect of several metal ions (Cu^{2+}, Mn^{2+}, Zn^{2+}, CO_2^{+}, Ni^{2+}, Ag^{+}) on TEA decomposition rate was established. The pathway of pollutant catabolism in *Arthrobacter protophormiae* R4 cells was suggested [23].

Strains responsible for TEA degradation – *Rhodococcus erythropolis* B-301D, *Rhodococcus ruber* B-300D, *Bacillus coagulans* 4NG, *Bacillus subtilis* RT1, *Bacillus subtilis* RT2, *Bacillus thuringiensis* X7 × *Rhodococcus qingshengii* B-823D were isolated at laboratory of xenobiotic degradation and remediation of natural and industrial media, Institute of Microbiology, NAS Belarus. Microbial strains cultured on mineral media with TEA do not require promoters of growth and xenobiotic degradation; they use the pollutant as a sole source of carbon and nitrogen.

Strain *Rh. ruber* B-300D was isolated from activated sludge of biological decontamination unit. Strain *Rh. erythropolis* B-301D was derived by enrichment culture technique from soil contaminated with organic synthesis wastes. Free-flowing cells of strain *Rh. ruber* B-300D in batch culture fully dissimilate TEA concentrations 1000, 3000, 7000 mg/L in 240, 360 and 384 h, respectively. Free-flowing cells of *Rh. erythropolis* B-301D during batch fermentation carry out total break-down of TEA levels 1000, 3000, 7000 mg/L in 240, 312 and 360 h, respectively. Concerted action of strains *Rh. ruber* B-300D and *Rh. erythropolis* B-301D results in more intense TEA decomposition as compared to monoculture. TEA concentrations 1000, 3000 and 7000 mg/L were fully converted by binary association after 96, 240 and 312 h, respectively. Studies on adhesive properties of both microbial strains showed that they were firmly bound to polyamide fiber carrier. Dry weight of *Rh. ruber* B-300D and *Rh. erythropolis* B-301D biomass equaled 300 and 170 mg/g support, respectively. It allowed to examine degrading activity of the strains in laboratory 1 L detoxification model containing 20 g of polyamide fiber as biofilm support.

Continuous mixed culture of immobilized *Rh. ruber* B-300D and *Rh. erythropolis* B-301D cells led to total TEA degradation in concentration 1000 mg/L at flow rate 9 mL/h [24].

Strain *B. coagulans* 4NG was isolated from activated sludge of municipal sewage decontamination station. In batch culture free-flowing cells of strain *B. coagulans* 4NG catabolized TEA concentrations 100, 1000 and 1500 mg/L in 12, 96 and 168 h, respectively; immobilized cells in 8, 48 and 72 h, respectively. TEA levels 2000 and 3000 mg/L were used upon 168 h by free-flowing *B. coagulans* 4NG cells to 70 and 20% degree, respectively, in contrast to immobilized cells efficiency – 100 and 60%, respectively. 48-hour immobilization of *B. coagulans* 4NG cells on polyamide fiber, polyethyleneterephthalate and polypropylene resulted in dry biomass concentrations 65, 80 and 110 mg/g support, respectively.

In continuous culture 100% TEA degradation was achieved by immobilized *B. coagulans* 4NG cells at concentration of the pollutant 1000 mg/L and flow rate 7 mL/h [25].

Strain *B. subtilis* RT1 was isolated from enrichment culture based on soil artificially contaminated with TEA and activated sludge of decontamination facilities at organic synthesis plant. Free-flowing cells of strain *B. subtilis* RT1 degraded TEA concentrations 100, 1000, 1500 mg/L in batch culture by 12, 96 and 168 h, respectively; immobilized cells complete the process in 12, 48 and 72 h, respectively. TEA levels 2000 and 3000 mg/L upon 168 h were partially digested by free-flowing bacterial cells to the degree 71% and 26%, respectively, whereas the percentage for immobilized cells reached 100 and 64%, respectively. During 48 h of *B. subtilis* RT1 cell immobilization polyamide fiber, polyethyleneterephthalate and polypropylene carriers accumulated dry biomass levels 63, 78 and 104 mg/g support, respectively. In continuous culture immobilized cells of strain *B. subtilis* RT1 achieved total decomposition of TEA concentration 1000 mg/L at flow rate 10 mL/h [26].

Strain *B. subtilis* RT2 was recovered by enrichment culture based on soil artificially polluted with TEA and activated sludge of decontamination facilities processing organic synthesis effluents. Free-flowing *B. subtilis* RT2 cells effectively broke down TEA concentrations 500, 1000, 3000 mg/L by 48, 72 and 240 h, respectively, during batch fermentation. *B. subtilis* RT2 cells immobilized on polypropylene and polyamide supports in the course of 48 h process attained fixed dry biomass concentrations 49 and 38 mg/g carrier, respectively.

Strain *B. thuringiensis* X7 was isolated by enrichment culture technique from soil samples taken around Chinese machine-constructing and chemical synthesis plants. Free-flowing *B. thuringiensis* X7 cells in batch culture decomposed TEA concentrations 100, 1000 and 1500 mg/L by 48, 96 and 120 h, respectively. Dry biomass weight of *B. thuringiensis* X7 cells bound to polyamide fiber carrier constituted 150 mg/g support after 48 h of immobilization. In continuous culture immobilized *B. thuringiensis* X7 cells completely degraded TEA levels 1000 mg/L at flow rate 14 mL/h. Acceleration of flow rate to 60 mL/h resulted in 95% pollutant decay.

TEA-degrading strain *Rh. qingshengii* B-823D was isolated by enrichment culture technique from sod-podzol soil sampled in Baranovichy, Brest region. This strain displayed more active growth on mineral medium supplemented with TEA as the only source of carbon and nitrogen than previous isolates described above.

The aim of this study is to investigate the TEA degradation by strain *Rh. qingshengii* B-823D.

17.2 MATERIALS AND METHODOLOGY

Inoculation material of strain B-823D was grown in 1000 mL Erlenmeyer flasks containing 500 mL of liquid medium with TEA as the sole source of carbon and nitrogen. The nutrient medium was composed of (g/L): $NaH_2PO_4 \times 2H_2O$ – 1.6; KH_2PO_4 0.7; $MgSO_4 \times 7H_2O$ – 0.8; ТЭА – 10.0, pH 7. The microbial culture was introduced by slant loop technique and the inoculum was incubated on orbital shaker Infors HT (Ecotron, Switzerland), at agitation rate 160 rpm, temperature 30°C for 72 h. The resulting cells were washed twice and resuspended in phosphate buffer (pH 7) to attain OD_{600} values 1. Phosphate buffer of the following composition was applied (g/L): KH_2PO_4 3.4; $Na_2HPO_4 \times 12H_2O$ – 8.9 [27]. Bacterial growth was monitored by measuring OD_{600} at UV-spectrophotometer 2401 PC (Shimadzu, Japan).

To study cell immobilization of strain B-823D on synthetic support glass column of 1 L volume was used. The carrier (polyamide fiber and polyethylenetherephthalate) were placed into the column in amount 20 g and it was filled with 810 mL of the following medium (g/L): $NaH_2PO_4 \times 2H_2O$ – 1.6; KH_2PO_4 – 0.7; $MgSO_4 \times 7H_2O$ – 0.8 plus 90 mL of strain B-823D cultural liquid with cell titer 1×10^9 CFU/mL. Immobilization was conducted during 48 h at temperature 25°C and intense aeration (20 L air/h). To evaluate weight of immobilized cells the support was removed from the column and dried at temperature 105°C. Dry mass of microbial cells immobilized on polymeric fiber was calculated as the weight difference of support before and after immobilization procedure and expressed in mg/g support.

Degrading potential of free-flowing cells of examined strain was analyzed during batch culture in 500 mL Erlenmeyer flasks, containing 225 mL of liquid medium of the following composition (g/L): $NaH_2PO_4 \times 2H_2O$–1.6; KH_2PO_4–0.7; $MgSO_4 \times 7H_2O$–0.8 supplemented with TEA. pH of the medium was adjusted to 7 with concentrated HCl. 25 mL of B-823D cultural liquid with cell titer 1×10^9 CFU/mL were added to the filled flasks. Cell titer in the fermentation media equaled 1×10^8 CFU/mL, initial TEA concentrations – 100, 1000 and 10 000 mg/L. TEA degradation was studied in orbital shaker culture (Infors HT, Ecotron, Switzerland; 160 rpm, 29°C).

To evaluate decomposing activity of B-823D immobilized cells in continuous culture 810 mL of the following mineral medium was fed into bioreactor with 20 g polymeric support (g/L): $NaH_2PO_4 \times 2H_2O$ – 1.6; KH_2PO_4 0.7; $MgSO_4 \times 7H_2O$ – 0.8

and 90 mL of strain B-823D cultural liquid with cell titer 1×10^9 CFU/mL. Immobilization proceeded during 48 h at temperature 25°C and intense aeration (20 L air/h).

Degrading activity of B-823D cells was assessed via reduction of TEA concentration in fermentation media. Sampling of specimens to check TEA level in cultural broth was performed at regular time intervals (from 5 min to 8 h depending on initial TEA concentration). The samples were sonicated at disintegrating device Sonifier 450 (Branson, USA) using the following parameters: power – 0.05 kW, temperature – 4°C, duration – 600 pulses 0.5 sec each. Cell lysate was centrifuged for 15 min at 15000 g. Residual TEA levels in supernatant were determined at liquid chromatograph Agilent 1200 equipped with mass-detector Agilent 6410 Triple Quad. Separation of components in analyzed samples was conducted in the column Zorbax XDB C18 (4.6 × 50 mm; 1.8 μm) at temperature +25 °C. Injection volume was 2 μL. Mobile phase: A – 0.1% aqueous solution of trifluoroacetic acid, B-acetonitrile.

Isocratic mode of 2% phase B elution was applied. Eluate flow rate – 0.5 mL/min. Ionization interface – electrospray Agilent G1948B API-ES in positive ion mode. Full scanning regime (MS2-Scan) in mass range 30 to 200 kDa was used. Detector operation parameters: temperature of desiccating gas +300 °C, flow rate of desiccating gas 10 l/min, sprayer pressure 30 psi, capillary voltage – 4000 V, fragmentor voltage – 60 V. Analysis of chromatograms and mass-spectra was performed using Agilent Mass Hunter Workstation Software version B.01.03 (Agilent Technologies Inc., США).

17.3 RESULTS AND DISCUSSION

Toxicity of TEA high concentrations for microorganisms is determined by its strong base reactivity. Elevated TEA concentrations in bioreactors of local decontamination units result in decay of microbial degraders or significant loss of activity. Effective solution of the problem might be supply of acids into effluents or absorption solutions to neutralize TEA and produce triethylammonium salts.

ABU application record demonstrates that average TEA concentrations in absorption solutions do not exceed 100 mg/L. At such toxicant levels pH of absorption solution ranges from 8.6 to 9.0 depending on concentration of adjacent substances. To assess pH effect on TEA decomposition rate we compared two fermentation variants:

1) mineral medium was supplemented with TEA as the sole carbon and nitrogen source up to concentration 100 mg/L (pH of the medium 8.8);
2) mineral medium with TEA supply as the only carbon and nitrogen source up to concentration 100 mg/L and subsequent neutralization with HCl. In this case TEA was converted into triethylammonium chloride (pH of the medium 7).

During batch fermentation in medium 1 free-flowing cells of strain B-823D carried out total TEA degradation at concentration 100 mg/L by 80 min, whereas in medium 2 the similar process took 40 min (Fig. 17.1). Degradation efficiency declined 2-fold when pH increased from 7.0 to 8.8.

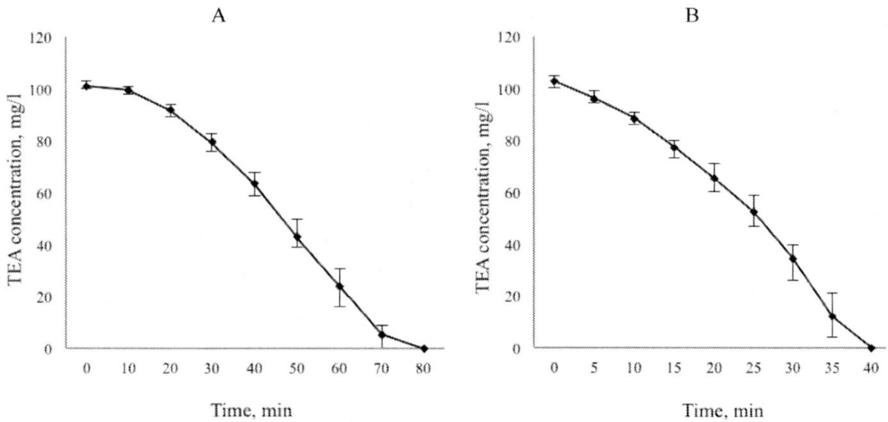

FIGURE 17.1 Degradation of 100 mg/L TEA concentrations by free-flowing cells of strain B-823D at pH 8.8 (A) and 7.0 (B).

The obtained results allow to recommend preconditioning of wastewaters and absorption solutions containing TEA prior to their feeding into bioreactor, namely acid supply to neutralize the toxic pollutant. We found that cells of strain B-823D possess superior TEA-decomposing activity as compared to the cultures isolated earlier by researchers of our laboratory.

Most publications report significant advantages of immobilized systems over free-flowing microorganisms in disposal of contaminating substances. Adhesive potential of bacteria is a vital technological property favoring the choice of specific microbial degrader for biodetoxification processes. Immobilization on various supports promotes retention of microorganisms in bioreactor providing for elevated cell concentrations in small volume and intensifying thereby break-down of the pollutant in the media. Immobilized cells are not exposed to adverse impact of toxic substrates or other factors and they serve longer then free-flowing cells [28].

Microbial adhesion responsible for initial stages of immobilization on the carrier depends considerably on cell properties and characteristics of the support. We tested adhesive capacity of B-823D cells with respect to most common fibrous supports: polyamide fiber and polyethyleneterephtalate (Table 17.1).

TABLE 17.1 Adhesion of B-823D Cells on Diverse Supports

Type of fibrous support	Absolutely dry biomass, mg/g support	
	Cells grown on MPA medium	Cells grown on mineral medium with TEA
Polyamide fiber	129.0 + 1.0	130.2 + 0.8
Polyethyleneterephtalate	88.2 + 2.6	84.8 + 1.8

Note: MPA – meat-peptone agar.

It was found that microbial degraders were binding more actively (by 46.6%) to polyamide fiber. Carbone source in the medium did not influence adhesive efficiency. The obtained findings clearly indicated that further experiments on TEA degradation should involve cells of strain B-823D immobilized on polyamide fiber support.

Water recycling in ABU technological scheme enables to engage both free-flowing and immobilized cells for TEA treatment. Given preliminary neutralization of toxicant in absorption solution and lack of toxic satellite compounds, application of free-flowing cells is expedient since the protective role of the carrier restricting contact area with the media becomes redundant. In this case maximum cell interaction with the media constituents is essential, whereas support interference reduces operation volume of bioreactor.

If preconditioning of absorption solution is not performed, it could result in TEA build-up and pH rise. Literature data evidence higher resistance of immobilized microorganisms to increased pH levels. We studied decomposition of TEA 100 mg/L concentrations by B-823D cells immobilized on polyamide fiber (Fig. 17.2).

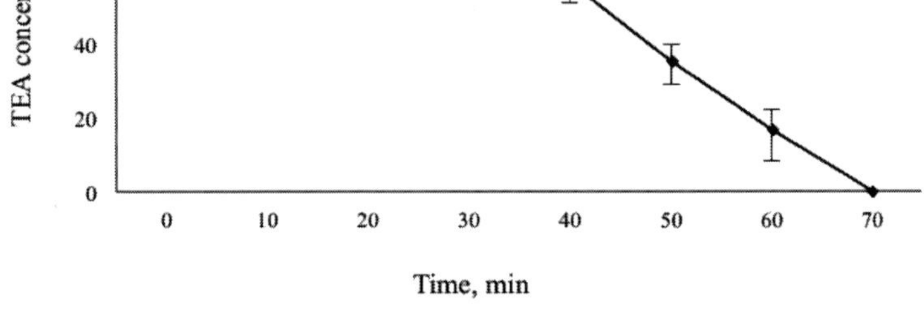

FIGURE 17.2 TEA Degradation at Concentration 100 mg/L by Cells of Strain B-823D Immobilized on Polyamide Fiber (pH 8.8).

Immobilized cells of strain B-823D completely degraded TEA concentrations 100 mg/L by 70 min in batch culture. The resulting efficiency of toxicant decomposition by immobilized microbial cells rose by 14%.

Investigation of TEA cleavage in hyperconcentrations was conducted in neutral pH range. Toxicant levels 1000 and 10000 mg/L were fully used by free-flowing B-823D cells in batch culture during 4 and 32 h, respectively (Figs. 17.3 and 17.4).

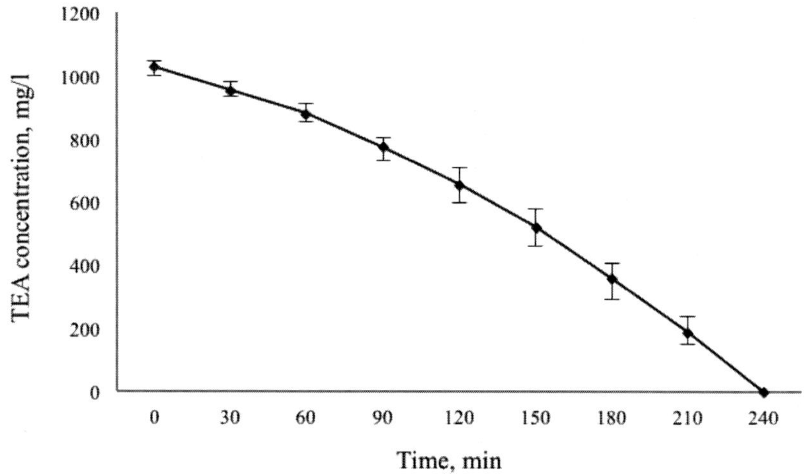

FIGURE 17.3 TEA degradation at 1000 mg/L concentration by free-flowing cells of strain B-823D.

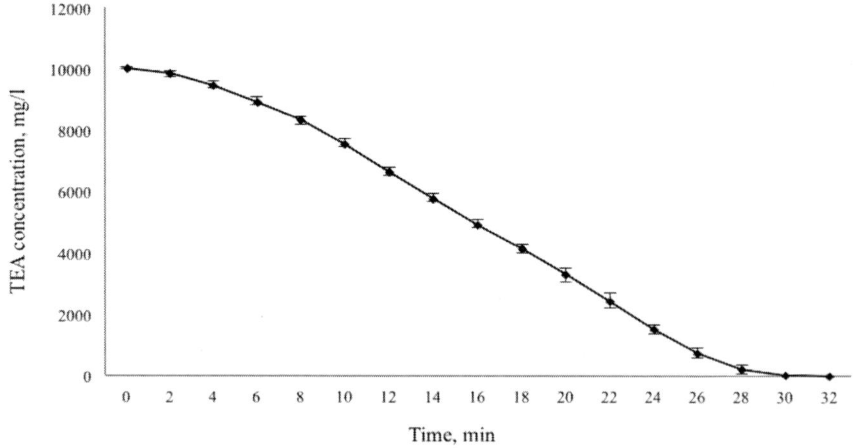

FIGURE 17.4 TEA degradation at 10,000 mg/L concentration by free-flowing cells of strain B-823D.

At TEA concentrations 1000 and 10,000 mg/L in mineral medium pH levels reaches the values severely restraining microbial degrading activity. Emergency TEA discharges into local decontamination unit may lead to inactivation of microorganisms.

It should be noted that the above-cited data were obtained in batch culture using high concentrations of xenobiotic. Accumulation of metabolites inhibiting disruption process prolonged degradation period.

TEA degradation by immobilized cells of strain B-823D in continuous culture was investigated using originally designed laboratory model of decontamination unit incorporating the following elements:

- initial model (absorption) solution vessel;
- bioreactor with biomass of TEA-degrading microorganisms immobilized on 20 g polyamide fiber support;
- container for detoxified model (absorption) solution;
- compressor for pumping air into bioreactor;
- tubing system and valve regulating flow rate of model solution (in the reactor).

The process of TEA degradation monitored at various flow rates of model waste water in bioreactor – from 100 to 1000 mL/h. Significant impact of flow rate on TEA decomposition was established (Fig. 17.5). Full consumption of TEA concentration 1000 mg/L was accomplished at flow rate 100 mL/h. The obtained results may lay the basis for elaboration of semiindustrial technology of TEA removal from wastewaters and for calculation of optimal volume of pilot-plant bioreactor.

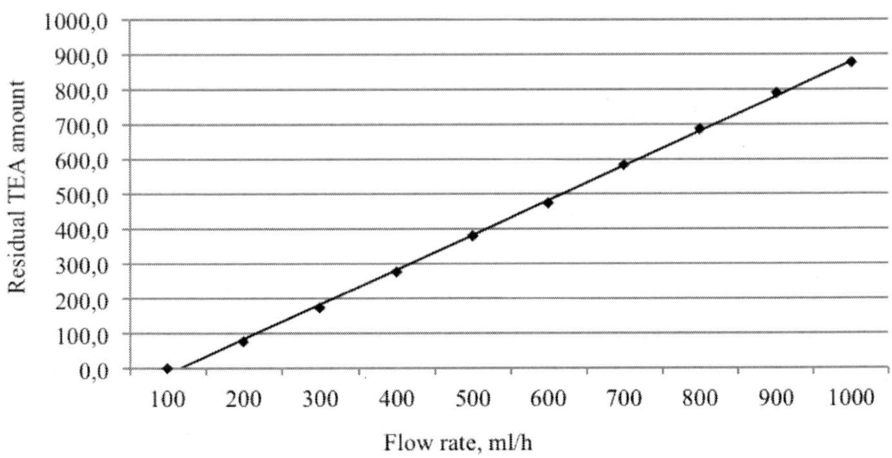

FIGURE 17.5 Correlation of TEA residual amount in wastewater with flow rate (TEA concentration 1000 mg/L).

To optimize TEA decomposition process the conditions beneficial for strain B-823D action were examined. It was found that the highest intensity of degradation process was achieved at aeration rate 20 L air/h, pH 7.0, temperature 30°C.

Production of such hypothetical intermediates of TEA catabolism as triethylamine-N-oxide, diethylamine, diethylamine-N-oxide, ethylamine, and ethylamine-N-oxide is postulated in the course of TEA microbial digestion. The above-mentioned compounds distinguished by high toxicity aggravate the problem of wastewater recovery. Biological TEA treatment may be regarded effective only in case of full mineralization of the pollutant. To validate efficiency of the process we performed chromatographic analysis of cultural liquid after fermentation of strain B-823D on mineral medium with TEA as the sole source of carbon and nitrogen. Examination of chromatograms and mass spectra did not reveal peaks corresponding to substances with molecular weight similar to that of presumed oxidation products. We suggested therefore possibility of intracellular amine oxidation. To check the assumption we conducted chromatographic study of cell disintegration lysates. Analysis of chromatograms and mass spectra enabled to detect peaks corresponding to substances with molecular weight matching those of some mentioned TEA catabolic intermediates.

Chromatography of probable TEA metabolites with molecular weight similar to weights of detected intracellular substances demonstrated in certain cases time coincidence of peak emergence. It appears to be indirect evidence of structural resemblance of examined compounds.

These substances were not detected in TEA-free cultural media. According to our hypothesis, the intracellular compounds are intermediate products of TEA oxidation. The completed studies allowed to propose the following hypothetical pathway of TEA catabolism by strain B-823D (Fig. 17.6).

The cells of B-823D cultured on media with TEA as the sole source of carbon and nitrogen accumulated the following oxidation products: triethylamine-N-oxide, diethylamine, diethylamine-N-oxide, ethylamine, ethylamine-N-oxide, acetaldehyde. Ammonium was released in the medium in the course of TEA utilization. Strain B-823D was not capable to oxidize tertiary amines comprising methyl groups: diethylmethylamine, dimethylethylamine, trimethylamine. It is presumed that TEA is transformed to triethylamine-N-oxide further split up into diethylamine, acetaldehyde and water. Diethylamine is converted to diethylamine-N-oxide subsequently cleaved into ethylamine, acetaldehyde and water. Ethylamine is oxidized to ethylamine-N-oxide yielding upon decomposition acetaldehyde and ammonium. Acetaldehyde is degraded to acetic acid and the latter as acetyl-CoA undergoes full decay in TCA cycle.

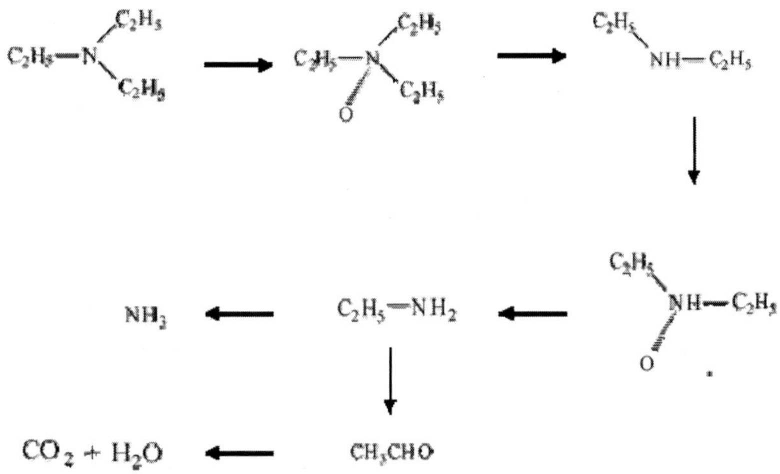

FIGURE 17.6 Pathway of TEA catabolism by strain B-823D.

17.4 CONCLUSIONS

TEA is one of the major pollutants in gaseous discharges of chemical and foundry plants applying Cold-box-amine process. ABU are very effective devices for regeneration of contaminated air in production zone of industrial enterprises. A key parameter defining efficiency of ABU performance is degrading activity of microorganisms immobilized on the support in bioreactor. We isolated from natural sources strain *Rhodococcus qingshengii* B-823D capable to consume TEA as the sole source of carbon and nitrogen.

In batch culture B-823D cells degrades the toxicant in concentrations 100, 1000 and 10,000 mg/L by 40 min, 240 min and 32 h, respectively. In continuous culture strain B-823D decomposes 1000 mg/L triethylamine levels in the medium at flow rate 100 mL/h.

The cells of strain B-823D were readily immobilized on polyamide fiber (130.2 mg/g carrier) and on polyethyleneterephthalate (88.2 mg/g support). It was shown that adhesive properties of microbial culture were not significantly affected by carbon source in the nutrient media.

Hypothetical pathway of triethylamine catabolism by strain B-823D was proposed. It was demonstrated that triethylamine is oxidized to triethylamine-N-oxide further split up to diethylamine, acetaldehyde and water. Diethylamine is degraded to ethylamine and the latter is cleaved into acetaldehyde, water and ammonium. Acetaldehyde is broken down to carbon dioxide and water. As a result cells of strain B-823D carry out complete triethylamine mineralization.

KEYWORDS

- **acetaldehyde**
- **catabolism**
- **ethylamine**
- **triethylammonium**

REFERENCES

1. Shapovalov, Y. P. (2007). Recovery of Ventilation Air a Progressive Choice, Metal-info Journal, *4*, 25–29 (in Russian).
2. Akesson, B., Floren, T., & Skerfving S. (1988). Visual Disturbances after Experimental Human Exposure to Triethylamine, British Journal of Industrial Medicine, *45,* 262–268.
3. Borger, T., Salden, A., & Eigenberger, G. A. (1997). Combined Vacuum and Temperature Swing Adsorption Process for the Recovery of Amine from Foundry Air, Chemical Engineering and Processing, *36*, 231–238.
4. Belin, L., Wass, U., Audunsson, G. & Mathiasson, I. (1983). Amines Possible Causative Agents in the Development of Bronchial Hyper Reactivity in Workers Manufacturing Poly Urethanes from Isocyanates, British Journal of Industrial Medicine, *40*, 251–257.
5. Aimin, H., Cao, L., Jie, C., Spiess, F., Steven, L., Timothy, N., Steve, O., & James, D. (1999). Photo Catalytic Degradation of Tri Ethyl Amine on Titanium Oxide Thin Films, Journal of Catalysis, *188*, 40–47.
6. Adler, S. (2001). Bio Filtration-a Primer, Chemicals Engineering Program, *97*, 33–41.
7. Burgess, J. E., Parsons, S. A., & Stuets, R. M. (2001). Development in Odors Control and Waste Gas Treatment Biotechnology, Biotechnol Advertise, *19*, 35–43.
8. Strikauska, S., Zarina, D., Berzins, A., & Viesturs, U. (1999). Biodegradation of Ammonia by Two Stage Biofiltration System, Environment Eng, Policy, *1*, 175–182.
9. Wani, A., & Lau, A. (1999). Bio Filtration Control of Pulping Odors Hydrogen Sulfide, Performance Macro Kinetics and Co existence Effects of Organo-sulfur Species, J. Chemicals Technol, Biotechnol, *74,* 9–15.
10. Busca, G. & Pistarino, C. (2003). Abatement of Ammonia and Amines from Waste Gases, J. Loss Prevention in the Process Industries, *6*, 157–166.
11. Yoon, I. K. & Park, C. H. (2002). Effects of Gas Flow Rate, Inlet Concentration and Temperature on Bio Filtration of Volatile Organic Compounds in a Peat Packed Bio Filter, J. Bioscience Bioengineering, *93*, 165.
12. Tang, H. & Hwang, S. (1996). Waste Gas Treatment in Bio Filters, J. Air Waste Management Association, *46,* 349–354.
13. Torkian, A., Keshavarzi, H. & Azimi, A. (2005). Effects of Operational Conditions on the Performance of Tri Ethyl Amine Bio Filtration, Iranian Journal of Environmental Health Science and Engineering, *2(2)*, 31–40.
14. Mehrdadi, N., Keshavarzi, H., Azimi, A., & Mirmohammadi, M. (2010). Removal of Tri Ethyl Amine Vapor from Waste Gases by Bio Trickling Filters, Iran J. Environment Health Science Eng., *7(4)*, 337–334.
15. Bucke, C. (1983). Carbohydrate Transformations by Immobilized Cells, Biochemistry Society, Symp, *48*, 25–38.

16. Gvozdyak, P. I. (1985). Decontamination of Industrial Effluents by Immobilized Microorganisms, Chemistry and Technology of Water, *7(1),* 64–68 (in Russian).
17. Pervushin, Y. V. (1990). Efficiency of Bio Detoxification Systems with Immobilized Microorganisms, Plastic Materials, *11,* 81–85 (in Russian).
18. Syomochkina, N. F. (2000). Adhesion of Microorganisms Degrading O-phthalate Esters to Capron Support, Microbiology and Biotechnology on the Eve of XXI Century, Proceedings of International Conference, Minsk, (June 1–2–2000) Institute of Microbiology NAS Belarus, 92–94 (in Russian).
19. Meriem, B. (1994). Degradation of Sodium Anthraquinone Sulphonatc by Free and Immobilized Bacterial Cultures, Applied Microbiology and Biotechnology, *41(1)*, 110–116.
20. Dunny, G. M. (2008). Multi Cellular Behavior in Bacteria Communication, Cooperation, Competition and Cheating, Bioessays, *4(30)*, 296–298.
21. Rappert, S., Ayoub, R., Li, R., Botsch, K., Obi, I., & Muller, R. (2004). Biological Degradation of Odor-Active Individual Materials and Real Fuel Mixtures, Detecting and Minimizing Odors, Messungund Minimierung Von Gerüchen, Hamberburger Berichte, 203–220 (in German).
22. Wang, C., Lu, K., & Chen, X. (2007). Removal of Tri Ethyl Amine from Synthetic Waste Water by Acclimated Mixed Bacteria Cultures, International Bio Deterioration and Biodegradation, *59,* 202–205.
23. Cai, T., Chen, L., Ren, Q., Cai, S., & Zhang J. (2010). The Biodegradation Pathway of Tri Ethyl Amine and its Biodegradation by Immobilized *Arthrobacter Protophormiae* Cells, Journal of Hazardous Materials, *186(1)*, 59–66.
24. Samsonova, A. S., Syomochkina, N. F., Aleschenkova, Z. M. & Petrova, G. M. (2007). Microorganisms-Degraders of Tri Ethyl Amine for Bio Recovery of Waste Waters, Collected Papers, Microbial Bio Technologies Basic and Applied Aspects, Minsk, Logvinov Press, 366–373 (in Russian).
25. Nagorny, R. K. (2011). Strain *Bacillus Subtilis* Tri Ethyl Amine Degrader 15, International Conference of Young Scientists, "Biology Science of XXI Century" Pushchino, April 19–23, *1,* 301 (in Russian).
26. Samsonova, A. S. & Nagorny, R. K. (2012). Strain *Bacillus Special* Degrading Tri Ethyl Amine, 16 International Conference of Young Scientists "Biology Science of XXI Century" Pushchino, April 19–23, *1*, 310 (in Russian).
27. Doson, R., Elliot, D., Elliot, U., & Jones, K. (1986). Data for Biochemical Research 2nd ed., Oxford, Clarendon Press, 554.
28. Wang, J. L., Ye, Y. C. & Wu, W. Z. (2003). Comparison of Di-n-methyl Phthalate Biodegradation by Free and Immobilized Microbial Cells, Biomed, Environment Science, *16(2),* 126–132.

CHAPTER 18

STRAWBERRY POLLEN BREEDING FOR COLD RESISTANCE

IRINA G. PUHACHOVA, GALINA A. EVSTRATOVA, and ROMAN M. PUHACHOV

Education establishment Belarusian state agricultural academy Michurin St. 5, Gorki, Mogilev region, 213407, Belarus; E-mail: puhachova.irina@gmail.com

CONTENTS

ABSTRACT

There are the results of nutrient medium and conditions optimization for strawberry pollen germination under laboratory conditions. The evaluation of 15 strawberry varieties pollen viability was carried out. The mode of low temperature stress (+3° C for 3 h) was developed as the background for selection of male gametophyte with the properties desired. The most resistant to temperature stress pollen grains were used for pollination and hybrid seed production. The regularities of features implementation were identified for strawberry seedlings that had been obtained under the influence of cold treatment at the level of male gametophyte. The maximum effect of pollen selection was observed in the hybrid combination Slonenok × Pegasus (index = 1.09–1.69), as well as in those combinations where father crossing component is Senga Sengana (index = 1.08–8.14). The influence of pollen selection has been significantly proven to increase the features "root length" and "length of hypocotyl" in combination Slonenok × Pegasus. It has also been identified the positive impact of pollen selection on germination percentage and seedling hypocotyl length in hybrid combinations that involve such a variety as Senga Sengana. In three of five studied combinations (Polka × Senga Sengana, Slonenok × Pegasus, Tsaritsa × Senga Sengana) high speed germination was inherent for seeds that were derived from cold-treated pollen pollination.

18.1 INTRODUCTION

Strawberry (*Fragaria xananassa* Duch.) is early ripening crop. Norms of the Institute of Nutrition of the Russian Academy of Medical Sciences established that during the year a person should eat 4–5 kg of strawberries. Strawberry plants are flexible, they can adapt to different soil and climatic conditions. The ease of reproduction provides wide dissemination of this agricultural crop [1, 2].

However, the productivity is reduced due to damage of strawberry by fungal diseases and insect pests, weed suppression, and because of the short period of berries ripening, demand to the length of daylight, winter damage, spring frosts and cold weather [3, 4].

Plants resistance to abiotic and biotic stresses can be improved by plant growth regulators and micronutrients, which are widely used to influence on plant morphogenesis processes *in vitro*, and to regulate the field crop cultivation [4].

For more than 100 years chemical pesticides play an important role in the fight against disease pathogens, pests and weeds [5]. The need for such a struggle is quite obvious when you consider that yield losses caused by them constitute now from 20 to 40% [6].

Application of pesticides and fertilizers with particular acuteness raised the question about the crop quality due to the risk of accumulation in the soil and in the plants themselves of residual amounts of pesticides, nitrates and other harmful to human health substances.

There is another way to lower the human pressure on the environment without reducing crop yields – adaptive breeding, aimed at the creation of resistant varieties [7].

It was suggested to use the pollen after artificial selection for subsequent pollination in order to increase the plants resistance. Gametophyte selection (selection at the level of macro and microspores) is regarded to be one of the most effective ways to improve the sporophyte generation stability to biotic and abiotic stresses [8].

The aim of our research was to develop a strawberry pollen selection method on the basis of cold resistance, as well as estimation of this selection influence on the cold resistance of produced plants. It was necessary to solve the following problems to achieve this goal:

1. Selection of nutrient medium for strawberry pollen germination.
2. Assess the strawberry pollen viability of different varieties.
3. Investigation of cold stress duration influence on pollen germination in conditions of artificial nutrient medium.
4. Estimation of cold resistance pollen selection influence on the strawberry seedlings features.

18.2 MATERIALS AND METHODOLOGY

Investigations were carried out during 2009–2012 years at the department of agricultural biotechnology and ecology of the Belarusian state agricultural academy. The development of the selecting method of cold resistance strawberry pollen consisted of several stages.

Stage 1. The first step in the developing of method was to specify the composition of the nutrient medium (2009–2010 years). The optimal sucrose content (20%) was found after a detailed study of the literature data [9]. We studied the ability of pollen germination at different concentrations of boric acid (0.06% and 0.006%) and at different levels of nutrient medium acidity (from 5 to 9 pH). Pollen germination was carried out at the temperature of 28° C for 3 h. Pollen grains of strawberry variety Senga Sengana were used for germination.

Stage 2. Nutrient medium composition, which was chosen as the result of these experiments, is used to assess the viability of pollen 15 varieties of strawberry: Pegasus, Vima Tarda, Senga Sengana, Vima Rina, Elsanta, Istochnik, Kent, Marmolada, Elkat, Carmen, Vima Zanta, Feyerverk, Slonenok, Tsaritsa, Polka.

Stage 3. The same composition of the nutrient medium was used for the experiment to study the effect of low temperature on the viability of pollen (2010 and 2011 years). Pollen, together with a nutrient medium, was placed at +28° C for 30 min, then for 1, 2, 3, 4, 5 and 6 h at the temperature of +3° C. After influence of low temperature pollen germinate for 3 h at the temperature of +28° C. Pollen viability was determined using a microscope Biolam (increase 7×20). The percentage of pollen grains germination and pollen tube length were determined. Pollen tube length was calculated as the average of the last 20 measurements in each field of vision under the microscope.

Stage 4. The next step was to use cold-treated pollen for pollination of strawberry flowers (variant Cold+). For this purpose pollen, together with the nutrient medium, was left for 30 min at +28°C, then for 3 h +3°C. Pollination was performed by brush. Control flowers were pollinated by common pollen grains (variant Cold–). The following varieties were selected as feminine forms: Slonenok, Tsaritsa, Polka; and Senga Sengana, Pegasus, Elkat as paternal forms. We tried to carry out the topcross scheme. The main criterion for selection of parent pairs was identical time of flowering; in addition, we took into account productivity. Finally we have got a sufficient for further experiments number of seeds in the experimental and control variants for the five hybrid combinations (Polka × Pegasus, Polka × Senga Sengana, Slonenok × Pegasus, Slonenok × Elkat, Tsaritsa × Senga Sengana).

Seeds obtained by this method in June 2011, were used for germination in February 2012. For each combination were prepared 600 seeds. All seeds were treated with concentrated sulfuric acid for 6 min to accelerate germination. After acid treatment, the seeds were washed with water (first piped from tap, then – with distilled) and placed in Petri dishes on moistened by distilled water filter paper (100 seeds in each Petri dish). Three Petri dishes for each combination were placed at the temperature of 22–24° C (optimal) and other three at the temperature of 10–12° C (low). After seed germination the average percentage of germination, seedling root length and seedling hypocotyl length were calculated. Data were processed by variance analysis program NCSS&PASS 2000 (Duncan's Test).

After measurements performed the seedlings were planted in containers and placed into the cultivating room (temperature 22–24°C; humidity 70%; photoperiod 16:8). After the appearance of true leaves plants were transplanted into the cassette. Then July 5th, 2012 they were moved to the open ground.

18.3 RESULTS AND DISCUSSION

STAGE 1. SELECTION OF NUTRIENT MEDIUM FOR STRAWBERRY POLLEN GERMINATION

Pollen of many plant species is able to germinate not only on stigma, but also on artificial media. Moreover, G. Mole germinates pollen even in a drop of distilled water in 1834 (cited by Golubinsky [9]). This ability of pollen grains to germinate in artificial conditions greatly simplifies the research and provides the ability to determine their viability quickly, examine the dynamics of the germination and growth of pollen tubes, depending on the genotype and cultivation conditions. Stepanov [10] notes, that the pollen selection of resistant plants should be realized by making assessment of pollen viability and pollen tube length, and it is much more productively than the score on only one of these indicators.

According to published data the composition of nutrient medium has been chosen (20% sucrose and 0.06% boric acid). It is suitable for germination of different plant pollen. However, the first attempts of pollen germination of our samples were unsuccessful. Then it was decided to make experiments with the culture conditions, and particularly with the medium acidity. Results are presented in Table 18.1.

TABLE 18.1 Evaluation of Strawberry Pollen Viability Depends on Nutrient Medium pH Level

pH	Germination of pollen, %			Pollen tube length, in the pollen grain diameter		
	2009	2010	average	2009	2010	average
9	9.0	5.0	7.0	1.0	2.0	1.5
8	25.0	16.0	20.5	2.0	3.0	2.5
7	37.0	23.0	30.0	5.0	7.0	6.0
6	85.0	80.0	82.5	4.0	3.0	3.5
5	68.0	53.0	60.5	2.0	3.0	2.5
LSD_{05}	11.32	14.91		1.14	2.38	

In 2009, the greatest number of Senga Sengana strawberry variety pollen grains (68–85%) germinated at pH level of 5 and 6. Neutral pH, and fortiori alkalescent, significantly reduced the percentage of germination. In 2010 the results of research confirmed installed pattern. It should be noted, that the level of pollen germination in 2010 was lower compared to 2009 (by 4–15%). The reasons, probably, are the different environmental conditions (more hot and dry weather in 2010).

One way to determine the length of pollen tubes is to express it in the pollen grains diameter. In 2009, maximum pollen tubes length was obtained at pH=6 and pH=7. It was on the level of 4.0 and 5.0 diameters of pollen grain, respectively. In 2010 the length of pollen tubes, formed during the pollen germination, was exceeded the previous year level and has reached a maximum (7.0 diameters of pollen grain) at pH=7.

As a result of two-year study, according to the obtained percentage of pollen germination and pollen tubes length, it was found that the most favorable pH level is 6. It allows to grow about 80% of pollen grains, their pollen tubes length is about 3.5 diameters of pollen grain.

At the same time the investigation of strawberry pollen germination at different concentrations of boric acid was conducted (Table 18.2).

TABLE 18.2 Evaluation of Strawberry Pollen Viability Depends on the Concentration of Boric Acid

Boric acid, %	Germination of pollen, %		Pollen tube length, in the pollen grain diameter	
	2009	2010	2009	2010
0.006	79	83	3	3
0.06	70	70	2	2

In the experiments that were held in 2009 and 2010 years showed no significant differences in the percentage of germination and pollen tube length on nutrient media containing different concentrations of boric acid. It was decided to use 0.006% boric acid concentration for further research, according to the reasons of economy.

Thus, as the optimal composition of nutrient medium for strawberry pollen germination may be used 20% sucrose and 0.006% of boric acid at pH=6.

STAGE 2. ASSESS THE STRAWBERRY POLLEN VIABILITY OF DIFFERENT VARIETIES

The next stage of our research was to apply the chosen composition of nutrient medium for evaluating of strawberry varietal differences on pollen viability (Table 18.3).

TABLE 3 Pollen viability of strawberry different varieties

Variety	Pollen germination, %		The average percentage of pollen germination
	2010	2011	
Pegasus	89.6	7.8	48.7
Vima Tarda	77.0	54.7	65.9
Senga Sengana	74.7	75.0	74.9
Vima Rina	95.5	77.3	86.4
Elsanta	20.0	21.2	20.6
Istochnik	19.9	90.3	55.1
Kent	7.3	77.7	42.5
Marmolada	95.3	5.2	50.3
Elkat	59.0	43.5	51.3
Carmen	94.1	89.7	91.9
Vima Zanta	23.5	51.3	37.4

TABLE 3 *(Continued)*

Variety	Pollen germination, %		The average percentage of pollen germination
	2010	2011	
Feyerverk	94.6	86.7	90.7
Slonenok	78.4	57.4	67.9
Tsaritsa	76.1	51.8	64.0
Polka	82.3	76.2	79.3
LSD_{05}	14.94	12.36	

In 2010, the highest percentage of germination (89.6–95.5%) was observed in Pegasus, Vima Rina, Marmolada, Carmen and Feyerverk varieties. The same feature of other studied varieties (Vima Tarda, Senga Sengana and Elkat) was significantly worse than the level of previous ones (by 31–37%). There is also a group of varieties (Kent, Istochnik, Elsanta and Vima Zanta), pollen of which has poor germination in the conditions we have created: percentage of germination was about 7–24%.

The studying of pollen viability in 2011 allows dividing the varieties into two groups: with high and low value of the feature. The first group includes varieties Istochnik, Carmen, Feyerverk (86.7–90.3%). The second group includes Pegasus, Vima Tarda, Elsanta, Marmolada, Elkat, Vima Zanta, which significantly inferior to varieties of the first group on the pollen germination.

According to the results of two years studying, Carmen and Feyerverk varieties formed stably high pollen viability (90.7–91.9%). Low percentage of pollen germination was achieved in Elsanta, Elkat, Vima Zanta varieties (20.6–51.3%). Pollen viability of Pegasus, Kent, Istochnik, Marmolada varieties over years of research strongly dependent on external conditions differed from very low (5.2%) to high (94.6%).

STAGE3. INVESTIGATION OF COLD STRESS DURATION INFLUENCE ON POLLEN GERMINATION IN CONDITIONS OF ARTIFICIAL NUTRIENT MEDIUM

Dry mature pollen of most plants is resistant to the effect of low positive temperatures and can survive for a long time, even during freezing [9]. Given this fact, it was necessary to find an effective way of cold treatment at the beginning of pollen grains germination. Mascarencas et al. [11] indicate the higher efficiency of selection on this stage of pollen development. On the ground of this fact, we decided to use for cold treatment the pollen grains, which were staying in the nutrient medium at the optimum temperature during 30 min, swollen and start to germinate. Chesnokov [12] used a similar approach while working with the pollen of maize and Kilchevsky, and Pugacheva [13] while working with tomato pollen.

At this stage of method development to create the cold-resistant male gametes special experiments were held. Their aim was to study the basic characteristics of the pollen grains viability on an artificial medium after cold treatment. The degree of cold stress hardness was determined on the base of reducing of the percentage of germination and pollen tube length concerning the control (average feature value in 2010 and 2011 years). Table 18.4 presents the results of experiments carried out in 2010 and 2011 with Senga Sengana variety pollen grains.

TABLE 18.4 Evaluation of Cold Stress Duration Influence on Pollen Viability on an Artificial Nutrient Medium

Processing time at temperature (+3°C), hour	The percentage of pollen germination		Average reduction of germination relative to the control, %	Pollen tube length, in the pollen grain, diameter		Reduction of pollen tube length relative to the control, %
	2010	2011		2010	2011	
0 (control)	74.7	75.0	-	7.2	6.5	-
1	68.5	62.1	87.2	6.3	6.1	90.5
2	54.6	49.7	69.7	5.8	5.5	82.5
3	28.1	25.5	35.8	4.0	3.7	56.2
4	22.4	18.7	27.5	2.5	2.3	35.0
5	13.6	10.0	15.8	1.5	1.1	19.0
6	5.3	3.2	5.7	1.2	0.8	14.6
LSD_{05}	21.38	18.44		0.13	0.21	

Percentage of germination and pollen tube length in optimal conditions (control) are rather high. This fact indicates the high quality and maturity of pollen grains. Treatment at the temperature +3°C reduced the viability of pollen from 62. 1–68.5% at one-hour treatment to 3.2–5.3% at six-hour stress. Length of pollen tubes was reduced to a value between 0.4–0.9 to 5.7–6.0 pollen grain diameter, depending on the duration of treatment.

Cold stress during 3–4 h significantly reduced the percentage of germination to the level of 27.5–35.8%. Length of pollen tubes varied similarly and reached 2.4–3.9 pollen grain diameter. Therefore, we have chosen the cold stress treatment at +3°C for 3 h as sufficiently effective regime for the selection of the most cold-resistant pollen grains. Essential element of our method is the suggestion to treat by low temperature pollen grains at the most vulnerable stage of development (start of pollen tubes germination) in suspension with the nutrient medium.

STAGE 4. ESTIMATION OF COLD RESISTANCE POLLEN SELECTION INFLUENCE ON THE STRAWBERRY SEEDLINGS FEATURES

Since the moment when pollen falls on the stigma, the complex of physiological and biochemical interaction of male gametophyte and diploid sporophyte tissues with very high biological activity begins on the mother plant.

Macromolecular compounds, which are necessary for the pollen tube growth can be synthesized in the pollen grains during the process of development and maturation. Pollen grains integument substances have sporophyte and gametophyte origin. They are used not only for pollen germination and pollen tube growth, but also to control the breeding system. Sometime after germination the pollen tube begins to metabolize the stigma and pistil tissue exudate substances very actively. Due to these substances pollen tube grows to the embryo sac.

The most accurate way to evaluate the plants male gametophyte viability is to determinate the percentage of having started to growth pollen grains during germination on artificial nutrient medium. Availability of controlled conditions (fixed temperature, humidity, light) helps to protect the pollen grains, which have stood the cold stress pressure, from the adverse effects during germination. However, in vivo pestle tissues and environment have a direct impact on the pollen and pollen tube growth. Therefore, testing of the laboratory experiments results in actual field conditions is a critical step of the experiments.

Hybridization was carried out between strawberry's varieties previously. We have received enough seeds in five hybrid combinations. Parent varieties are characterized by economically valuable traits, potential yield and suitable time of flowering. Seeds obtained from normal pollination with dry pollen by brush and after cold stress treatment were grown at optimal (22–24° C) and stress (10–12° C) temperature. The results of the final germination percentage, root length and seedling hypocotyl evaluation are shown in Table 18.5.

TABLE 18.5 Signs of Strawberry Seedlings, Depending on Temperature and Method of Pollen Processing

Character	Crossing combination	Temperature of seed germination			
		Optimal (22–24°C)		Low (10–12°C)	
		The method of pollen treatment		The method of pollen treatment	
		Cold+	Cold–	Cold+	Cold–
Germination of seeds, %	Polka × Pegasus	32.0	77.0	14.0	65.0
	Polka × Senga Sengana	57.0	7.0	41.0	9.0
	Slonenok × Elkat	58.0	87.0	71.0	73.0
	Slonenok × Pegasus	88.0	63.0	57.0	59.0
	Tsaritsa × Senga Sengana	62.0	47.0	50.0	23.0
	Average	56.0	54.5	46.6	45.8
Root length, mm	Polka × Pegasus	5.8	6.5	5.9	9.8
	Polka × Senga Sengana	6.9	2.7	12.8	7.7
	Slonenok × Elkat	7.2	6.5	9.2	8.9
	Slonenok × Pegasus	8.6	5.1	8.5	6.7
	Tsaritsa × Senga Sengana	7.6	6.7	11.6	9.0
	Average	7.2	5.5	9.6	8.4
Hypocotyl length, mm	Polka × Pegasus	7.2	9.4	8.3	10.6
	Polka × Senga Sengana	6.8	6.3	10.8	14.6
	Slonenok × Elkat	10.3	12.8	11.5	12.6
	Slonenok × Pegasus	14.7	11.1	13.9	12.7
	Tsaritsa × Senga Sengana	12.4	11.2	13.0	11.7
	Average	10.3	10.2	11.5	12.4

Note: "Cold+" – seeds obtained by pollination with cold treated pollen grains. "Cold–" – seeds obtained after pollination with normal pollen grains.

Seeds final germination percentage was determinated at a greater extent by the temperature of germination than by seeds production method. Under the conditions of 10–12°C the average germination percentage was about 8.7–9.4% lower than at 22–24°C.

Average root length of seedlings, which were obtained by using selection for cold resistance at the male gametophyte level, was on 1.2–1.7 mm more than the same feature at ordinary pollen pollination.

Under optimal temperature conditions the hypocotyls average length of seedlings, which were produced by different methods, varied between 10.2 to 10.3 mm. It was slightly higher at the temperature of 10–12°C than at 22–24°C.

The meanings of hybrid seedlings characteristics were more likely to be higher if the seeds were obtained from cold-treated pollen pollination.

We calculated indexes that reflect the efficiency of pollen cold treatment (Table 18.6). To evaluate the impact of pollen selection, we have found the ratio of the features value of seedlings, obtained by using of pollen cold treatment and without it.

TABLE 18.6 Indexes, That Reflect the Effectiveness of Pollen Cold Treatment

Crossing combination	The method of pollen treatment	the percentage of seeds germination, %		Root length, mm		Hypocotyl length, mm	
		22–24°C	10–12°C	22–24°C	10–12°C	22–24°C	10–12°C
Polka × Pegasus	Cold+	32	14	5.8	5.9	7.2	8.3
	Cold–	77	65	6.5	9.8	9.4	10.6
	index	*0.42*	*0.22*	*0.89*	*0.60*	*0.77*	*0.78*
Polka × Senga Sengana	Cold+	57	41	6.9	12.8	6.8	10.8
	Cold–	7	9	2.7	7.7	6.3	14.6
	index	*8.14*	*4.56*	*2.56*	*1.66*	*1.08*	*0.74*
Slonenok × Elkat	Cold+	58	71	7.2	9.2	10.3	11.5
	Cold–	87	73	6.5	9.2	12.8	12.6
	index	*0.67*	*0.97*	*1.11*	*1.0*	*0.81*	*0.91*
Slonenok × Pegasus	Cold+	88	57	8.6	8.5	14.7	13.9
	Cold–	63	59	5.1	6.7	11.1	12.7
	index	*1.39*	*0.97*	*1.69*	*1.27*	*1.32*	*1.09*
Tsaritsa × Senga Sengana	Cold+	62	50	7.6	11.6	12.4	13.0
	Cold–	47	23	6.7	9.0	11.2	11.7
	index	*1.32*	*2.17*	*1.13*	*1.29*	*1.11*	*1.11*

Note: "Cold+" – seeds obtained by pollination with cold treated pollen grains. "Cold–" – seeds obtained after pollination with normal pollen grains.

If the index is greater than 1.0, consequently the features value of seedlings, obtained by pollination with cold treated pollen, exceeds the features value of seedlings from normal pollen pollination. Thus, the positive effect of male gametophyte selection is shown.

In our research the greatest advantage of pollen selection was manifested in crossing combination Slonenok × Pegasus (index 1.09–1.69), as well as in those combinations where its father crossing component was Senga Sengana (index 1.08–8.14).

Statistical data processing was carried out on the basis of the ANOVA (Duncan's Test) using software NCSS&PASS 2000 (Table 18.7).

TABLE 18.7 Results of Statistical Processing of Seedlings Features, Obtained by Using of Pollen Cold Treatment and without it

Character	Variant	Polka × Pegasus	Polka × Senga Sengana	Slonenok × Elkat	Slonenok × Pegasus	Tsaritsa × Senga Sengana
the percentage of seeds germination, %	1 (Cold+)	14 [b]	41 [a]	71 [ab]	57	50 [a]
	2 (Cold)	32 [b]	57 [a]	58 [b]	88	62 [a]
	3 (Cold+)	65 [a]	9 [b]	73 [ab]	59	23 [b]
	4 (Cold–)	77 [a]	7 [b]	87 [a]	63	47 [ab]
Root length, mm	1 (Cold+)	5.9 [b]	12.8 [a]	9.2 [a]	8.5 [a]	11.6 [a]
	2 (Cold–)	5.8 [b]	6.9 [b]	7.2 [b]	8.6 [a]	7.6 [bc]
	3 (Cold+)	9.8 [a]	7.7 [b]	8.9 [a]	6.7 [b]	9.0 [b]
	4 (Cold–)	6.5 [b]	2.7 [c]	6.5 [b]	5.1 [c]	6.7 [c]
without rarefactions length, mm	1 (Cold+)	8.3	10.8	11.5	13.9 [a]	13.0
	2 (Cold–)	7.2	6.8	10.3	14.7 [a]	12.4
	3 (Cold+)	10.6	14.6	12.6	12.7 [ab]	11.7
	4 (Cold–)	9.4	6.3	12.8	11.1 [b]	11.2

Note: variants 1 and 3 – seedlings germinated at 10–12 °C, variants 2 and 4 – seedlings germinated at 22–24°C; means with the same letter in the terms of each hybrid and feature are not significantly different ($P<0.05$) using Duncan multiple comparison test.

During the results analysis, it was noted that on the grounds of feature "the average percentage of germination" 2 out of 5 crossing combinations (Polka × Senga Sengana and Tsaritsa × Senga Sengana) had a significant advantage for seeds, obtained after pollination with cold treated pollen, when cultured under optimal and stressful conditions. In the crossing combination Polka × Pegasus was installed an inverse trend. In the crossing combination Slonenok × Elkat significantly greater percentage of seeds germination was received when seedlings obtained by pollination with normal pollen and germinated at 22–24° C. Significant differences were not detected between the analyzed variants in combination Slonenok × Pegasus.

On the basis of "root length" in hybrid combinations Polka × Pegasus and Slonenok × Elkat seedlings had the highest values, when they were obtained after pollination with normal pollen and were germinated in a stressful environment. There was also high value of feature in the hybrid combination Slonenok × Elkat, when the seedlings were obtained by pollination with cold treated pollen and were cultivated under stressful conditions. Such regularity can be seen in hybrid combinations Polka × Senga Sengana and Tsaritsa × Senga Sengana. In combination Slonenok × Pegasus the highest characteristic value was observed in seedlings obtained by pollination with cold treated pollen, regardless of the cultivation conditions.

Significant differences on the basis of "length of hypocotyl" were detected only in hybrid combination Slonenok × Pegasus. This fact shows the superiority of seedlings, which were obtained by pollination with cold treated pollen grains.

Figures 18.1–18.5 reflects the dynamics of strawberry seed germination, the seeds were produced by different methods (see footnote and variants in Table 18.7).

The seeds of Polka × Pegasus hybrid combination (Fig. 18.1) germinated more intensively if were obtained with normal pollen pollination. Such regularity was observed in optimal and stress conditions. At 22–24° C the seed germination percentage was about 45% higher, if the seeds were obtained by normal pollen pollination, than for seeds which were produced by cold treated pollen pollination, and at 10–12° C – about 51% higher.

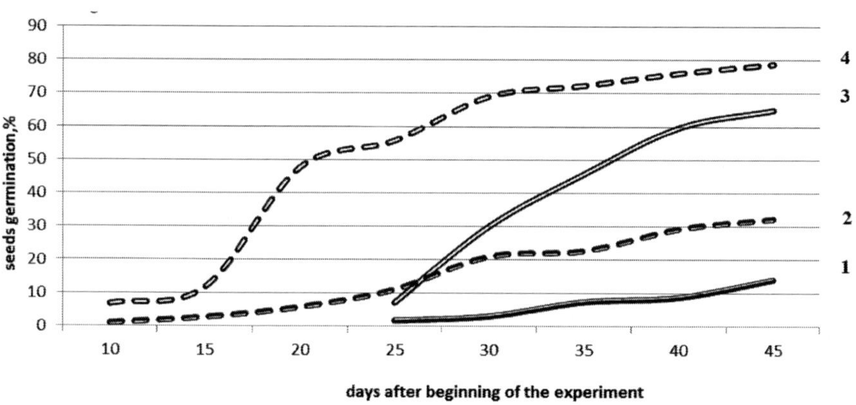

FIGURE 18.1 Dynamics of seed germination hybrid combination Polka × Pegasus.

The same regularity can be seen in hybrid combination Slonenok × Elkat (Fig. 18.2), normal pollen advantage in optimal conditions was 29%, and in stressful – 2%.

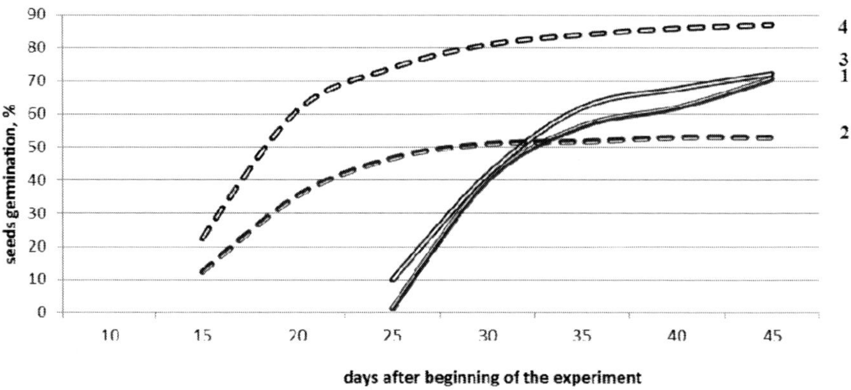

FIGURE 18.2 Dynamics of seed germination hybrid combination Slonenok × Elkat.

In other crossing combinations the opposite regularity was observed. The highest germination speed was typical to the seeds, which were obtained from cold-treated pollen pollination.

In the hybrid combination Slonenok × Pegasus (Fig. 18.3) a higher percentage of germination (8% more) at optimum temperature was observed in seedlings, which were obtained after cold treatment of pollen. Cold treatment on the pollen level virtually no impact on the analyzed feature of seedlings under stressful conditions.

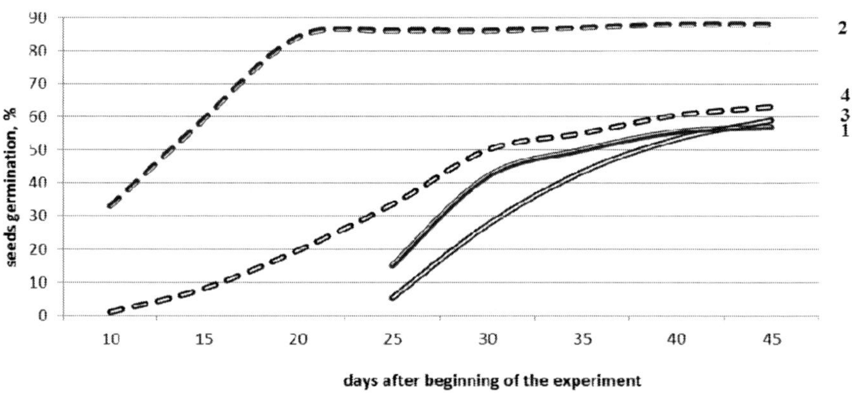

FIGURE 18.3 Dynamics of seed germination hybrid combination Slonenok × Pegasus.

Percentage of seed germination in hybrid combination Polka × Senga Sengana (Fig. 18.4) was on 50 and 32% higher if the seeds were obtained from cold-treated pollen pollination, than the seeds were obtained without pollen selection.

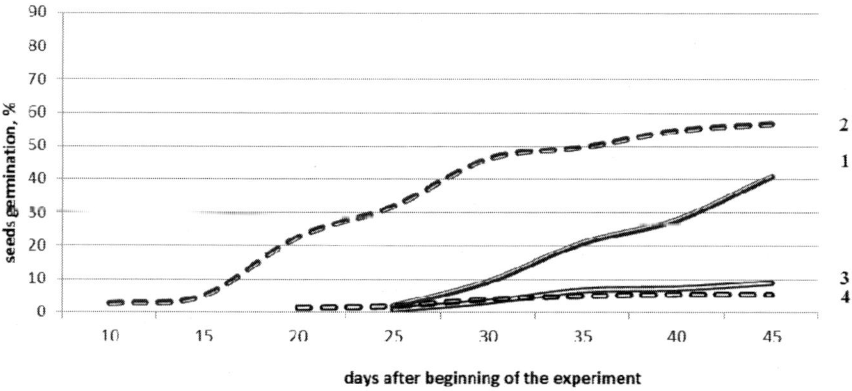

FIGURE 18.4 Dynamics of seed germination hybrid combination Polka × Senga Sengana.

In the hybrid combination Tsaritsa × Senga Sengana (Fig. 18.5), the germination percentage of seeds derived from cold-treated pollen pollination was 15–27% higher in stress and optimal conditions than derived from normal pollen pollination.

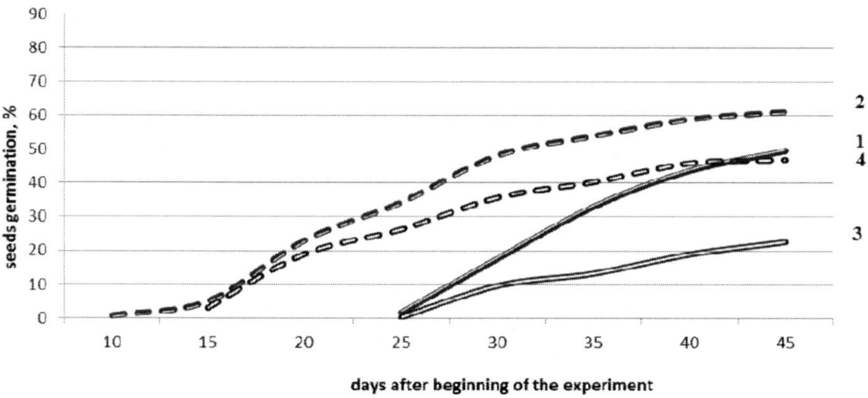

FIGURE 18.5 Dynamics of seed germination hybrid combination Tsaritsa × Senga Sengana.

As a result of studying the dynamics of germination we found out that on condition of normal pollen pollination the seeds in two of five investigated combinations (Polka × Pegasus and Slonenok × Elkat) increased the intensity of germination. However, in other combinations (Polka × Senga Sengana, Slonenok × Pegasus, Tsaritsa × Senga Sengana) high speed germination was typical for seeds derived from cold-treated pollen pollination. Thus, the results of the research indicate the presence of defined positive impact of the cold-resistant pollen selection on increasing of seedlings cold resistance.

It was not possible to evaluate economically valuable signs of hybrid combination plants, which were produced by different methods, due to lack of flowering occurrence in the first year of development.

18.4 CONCLUSIONS

1. Nutrient medium with 20% of sucrose and 0.006% of boric acid at pH=6.0 should be used as optimal composition for strawberry pollen germination.
2. During the pollen viability investigation it was found that Carmen and Feyerverk varieties formed stably high pollen viability (90.7–91.9%). Low percentage of pollen germination was achieved in Elsanta, Elkat, Vima Zanta varieties (20.6–51.3%). Pollen viability of Pegasus, Kent, Istochnik, Marmolada varieties over years of research strongly dependent on external conditions differed from very low (5.2%) to high (94.6%).
3. As a sufficiently effective way to select the most cold-resistant pollen grains is proposed to use mode +3 °C for 3 h.
4. In our investigations the maximum advantage of pollen selection was manifested in the hybrid combination Slonenok × Pegasus (index = 1.09–1.69), as well as in those combinations where fathers crossing component is Senga Sengana (index = 1.08–8.14).
5. Effect of pollen selection have been significantly proven to increase the features "root length" and "length of hypocotyl" in combination Slonenok × Pegasus. We also identified the positive impact of pollen selection on germination percentage and seedling hypocotyl length in hybrid combinations involving variety Senga Sengana.
6. As a result of studying the dynamics of germination we found out that two of the five investigated combinations (Polka × Pegasus and Slonenok × Elkat) increased the intensity of seed germination if the last were obtained from normal pollen pollination. However, in other combinations (Polka × Senga Sengana, Slonenok × Pegasus, Tsaritsa × Senga Sengana) high speed germination was typical for seeds derived from cold-treated pollen pollination. Thus, the results of the research indicate the presence of defined positive impact of the cold-resistant pollen selection on increasing of seedlings cold resistance.

KEYWORDS

- **abiotic stress**
- **hybrid *Fragaria xananassa***
- **hypocotyl**
- **male gametophyte**
- **pollination**
- **selection**

REFERENCES

1. Govorova, G. F. (2004). Strawberry Past, Present, Future, Govorova, G. F., & Govorov, D. N. Moscow Ros inform agroteh, 348p (in Russian).
2. Klakotskaya, N. V. (2009). Economic and Biological Characteristics of the New Strawberry Collection in Belarus, Thesis of the Dissertation for the Degree of PhD on Agricultural Sciences Specialty (06.01.05) Plant Breeding and Seed Production, Klakotskaya, N. V. Samokhvalovichy, Scientific Research Institute of Horticulture, 20p (in Russian).
3. Popova, I. V. (1980). Breeding of Strawberries for Wilt Resistance, Popova, I. V. & Konstantinova, A. F. Program and Method of Fruit, Berry and Nut Crops Breeding, Michurinsk, Publisher of Russian Scientific Research Institute of Fruit Plants Breeding, 204–206 (in Russian).
4. Zubov, A. A. (2004). Theoretical Basis of Strawberry Breeding, Zubov, A. A. Michurinsk Publisher of Russian Scientific Research Institute of Fruit Plants Genetics and Breeding Behalf Michurin, 196p (in Russian).
5. Poliksenova, V. D. (2009). Individual Plant Resistance to Pathogens and Abiotic Stresses (on Tomato), Bulletin of the Belarusian State University, *1*, 48–60 (in Russian).
6. Global Pact against Plant Pests Marks 60years in Action, Food and Agriculture Organization of the United Nations, URL: http: //www.fao.org/news/story/en/item/131114/icode/ (Accessed 12 February 2014).
7. Methodological Guidelines on Agricultural Plants Gamete Breeding [Ed. by Pivovarov, V. F. (2001)] Moscow Russian Scientific Research Institute of Breeding and Seed Growing of Vegetable Crops, 195p (in Russian).
8. Mulcahy, D. L. & Mulcahy, G. B. (1983). Pollen Selection an Overview Pollen Biology and Implications for Plant Breeding, [Mulcahy, D. L. Ed] Amsterdam Elsevier, 15–17.
9. Golubinsky, I. N. (1974). Biology of Pollen Germination, Kiyev, Naukova Dumka, 368p (in Russian).
10. Stepanov, V. A. (1998). Daikon Radish Starting Material for Plant Breeding and Seed Production in Central Nechernozemie Region, Thesis of the Dissertation for the Degree of Ph.D on Agricultural Sciences, Specialty (06.01.05) Plant Breeding and Seed Production, Moscow, 28p (in Russian).
11. Mascarenhas, J. P., & Altschuler, M. (1983). The Response of Pollen to High Temperatures and its Potential Applications, American Science, *75*, 3–8.
12. Chesnokov, U. V. (1999). Genetic Changes, Induced by the Exogenous DNA Transfer in Higher Plants by Germinating Pollen, Thesis of the Dissertation for the Degree of Doctor on Biological Sciences, Specialty (03.00.15) Genetics Kishinev, 48p (in Russian).

13. Kilchevsky, A. V. (2002). Tomato Gametophyte Breeding for Cold Resistance, Kilchevsky, A. V. & Pugacheva, I. G. Proceedings of the National Academy of Sciences of Belarus Series of Agrarian Sciences, 4, 35–39 (in Russian).

PART VI

INNOVATIVE TECHNOLOGIES OF CULTIVATION OF AGRICULTURAL CROPS

CHAPTER 19

APPLICATION OF NATURAL SPRINGS OF MOUNTAIN ZONE OF NORTH KAUKAZUS FOR INCREASE OF SELENIUM CONTENT IN GARLIC

SARRA A. BEKUZAROVA and ALAN T. KESAYEV

Gorsky State Agrarian University, d. 37, Kirov St., Vladikavkaz, Republic of North Ossetia Alania, 362040, Russia; E-mail: bekos37@mail.ru

CONTENTS

ABSTRACT

Our research shows an alternative, ecologically secure method of garlic plants enrichment of garlic plants by selenium, based on the application of natural growth stimulators in their cultivation. In the present study we used hydrogen sulfide mineral water from mountain springs and zeolite-containing clay dialbeculit. Other presently applied methods of garlic cultivation present a certain danger because the microelements inputted pollute the environment. In these conditions plants accumulate not more than 10% of applied dose. In the paper the peculiarities of garlic cultivation in the mountain and foothill conditions of North Ossetia and its useful properties are described as well.

19.1 INTRODUCTION

In the production of ecologically secure foodstuff microelements play an important role, as they determine the resistance of living organisms to stress factors [1–3]. Selenium, presented in some vegetables, including garlic (*Allium sativum* L.), is one of such elements [4]. Taking it into account, this selenium-containing culture is cultivated in many countries. Russia takes the fourth place in the world in garlic production after China, India and South Korea [5]. General need in garlic in our country is about 300,000 tons per year, but the real production is 254,000 tons. Low capacity to strike root under new growth conditions is a weak point in garlic cultivation [5]. Therefore in mountainous conditions local populations adapted to a given region are cultivated during many centuries. Garlic contain vitamins B_1, B_2, B_9, ß-carotene, vitamin PP; macroelements (mg/kg): potassium (260), sodium (80), calcium (60), phosphorus (100), magnesium (30), chlorine (30); microelements: iron (1500), copper (130), cobalt (9), iodine (9), manganese (810), zinc (1025); phytoncides, that is, natural antibiotics. Besides, garlic is the main supplier to human organism of so important element as selenium [6–8]. The methods of enrichment of garlic and vegetables by selenium are developed [2–4]. Selenium, beside of zinc, calcium and potassium, is included in more than 200 hormones and ferments regulating the functioning of human organs and systems. Plants are able to absorb and transform different forms of selenium: inorganic into organic ones and vice versa [7].

The ability to secrete phytoncides during a long time, about 70 h after pounding, is an important peculiarity of garlic [5, 6]. In this connection garlic is widely used in medicine [8, 9].

Selenium takes part in the formation of 80% of energy in human body, slowing of senescence, activation of stem cells, start of antioxidant protection and of many others vitally important processes [9]. Among different ways of plants enrichment by selenium the most common are the application of sodium selenate or selenite in the soil [10] and sparing of the crop by solutions of selenium salts [11, 12]. Sodium selenite (Na_2SeO_3) in the concentration 0.1% of water solution brought in the clay with high absorbing capacity during 6–10 days enriches it by selenium, which takes part in chlorophyll formation and is antagonist of heavy metals. It contributes to the accumulation of vitamins (C, E, B group), increases garlic plant resistance to water stress, salinization and drought [4, 11].

19.2 MATERIAL AND METHODOLOGY

In order to increase the productivity of winter garlic and the selenium content in it, chives garlic were soaked in the mixture of Zakinskaya mineral water and dialbeculit clay (originated from the floodplain of Urdson river, the tributary of Terek river, North Ossetia-Alania Republic) in the ratio 2:1, with further mulching after planting by the same mixture. Between the rows (30 cm) annual clover (*Trifolium apertum* L.) was sown.

The experiments were performed in the mountain zone at the altitude 2000 m a.s.l. (Zakka village, North Ossetia-Alania). The mixture was prepared in the proportion 50 kg of dialbeculit clay and 25 L of Zakinskaya mineral water per 1 hectare. The components were mixed and kept 2–3 h for full absorption. After 2–3 h seed material (chives or bulbs of garlic) was immersed into the mixture. The concentration of selenium in garlic was determined in the Institute of Agroecology of Gorsky State Agrarian University. After the planting of chives or bulbs the mulching was performed by the same mixture by the layer of 3–5 cm.

The next method is the cultivation of garlic on the slopes. Perennial legumes and cereals were sown in the proportion 5 kg of meadow timothy-grass and 15 kg of meadow clover. This mixture is sown across the slope continuously. At the end of vegetation period they are mown off for hay. By the end of the first year of life herbs consolidate their grip on the slopes due to the strong root system. By the spring of the first year of life by first hay harvest (2nd ten days of May) herbs reach the height of 50–70 cm, what considerably decreases soil erosion processes. Green mass is mown for green forage and after hay harvest wide-row crop (with row-spacing of 45 cm) is formed from continuous one by cultivator. The shrub-shaped haricot with short vegetation period is sown in row-spacing. By the autumn the haricot ripens and frees the space for winter planting of garlic.

Before planting of winter culture row-spacing is loosened with the formation of sulci, were zeolite-containing clay dialbeculit is inputted by the layer of 5–6 cm. Previously, 6–10 before planting the clay is crumbled up and soaked in the water solution of sodium selenite in the concentration 0.1%.

19.3 RESULTS AND DISCUSSION

Dialbeculit clay relatively to other known zeolite-containing clays (irlit, lexenit, alanit), is characterized by lower density (1.4–1.45 g cm^{-3}), which conditions the high content of hydromicas in it. Dialbeculit contains considerable amount of water-soluble salts, approaching according to this character to low mineralized silt sulfide moods with high absorbing capacity. Dialbeculit contains (in %): silicon – 46.5; iron – 7.1; calcium – 37; zinc – 1.1; potassium – 1.1; nickel – 1.7; phosphorus – 1.7; cobalt – 0.1. The clay reaction is alkaline (pH 9.1) because of high content of calcium.

Dialbeculit mixed with Zakinskaya water has high sorption capacity and is enriched by substances, which are contained in mineral water (in mg/L): calcium 320; magnesium 96; sulfates – 118; chlorides – 180; nitrates – 89.4; hydrogen sulfide – 120; potassium – 2.1, pH – 6.14 [13].

When combining two components with different reaction (pH 9.1 and pH 6.1) pH of the mixture becomes 7.6.

Sulfur contained in water (as a part of hydrogen sulfide) first of all blocks diseases of garlic. It is known [6, 7] that selenium replaces sulfur in amino-acids methionine and cysteine. First selenium methionine is synthesized, than it is transformed into selenium cysteine, selenium cystathion or, under demethylation, into selenium-methyl-selenium-cysteine. The last one, when interacting with glutamin, forms glutamil-selenium- methionine-selenium- cysteine. Active synthesis of these nonprotein amino-acids is the most typical for plants cumulating selenium (garlic in the present case). Beside of above-mentioned compounds, in the plants cumulating ultrahigh selenium concentrations were also detected [6, 7]: selenium-methyl-cysteine, selenium-homo-cysteine, glutamine-selenium-methyl-selenium-cysteine, di-methyl-di-selenium etc. Selenium is presented in a set of redox ferments together with iron, which is contained in dialbeculit in a quantity 7.1%.

Selenium participates in the reactions of chlorophyll formation, synthesis of tricarboxylic acids, as well as in the metabolism of long-chain fatty acids. All of this reveals active participation of sulfur (from Zakinskaya mineral water) in the photosynthesis process. The high contribution of sulfur and selenium to the formation of tocopherol is observed in garlic plants [7].

The function of sulfur in plant organism consists in the maintenance of the level of cell oxidation-reduction potential by means of reversibility of reactions cysteine-cystine-SH-glutathione-SS-glutathione. Sulfur is also a component of coenzyme A and vitamins (lipoic acid, biotine, thiamine), which play an important role in the respiration and lipid exchange [6, 7].

The basis of selected parameters (soaking in Zaskinskaya water mixed with dialbeculit clay during 2–3 h) is conditioned by the high sorption capacity of clay under interaction with mineral water and its subsequent prolonged effect in the soil, which provides the synergism of all elements necessary for plant nutrition.

Consequently, encapsulation by the mixture of dialbeculit clay and mineral water of local origin not only decreases expenses for fertilizers, used in other methods, but also contributes to normal physiological processes in plants leading to garlic enrichment by selenium.

The complex of macro and microelements contributes to better plants taking, allows high synergism of the effect. Alkaline reaction of the medium (mean pH=7.6) provides better plants taking, especially in acid soils of mountain zone.

As dialbeculit clay has high heat capacity, it allows to protect garlic chives from winter frost during the period of germination and to keep moisture under spring drought.

Selection of parameters of mulching layer after planting (3–5 cm) is determined by the need to protect the planted chives from the effect of low temperatures, which occur in mountain conditions, as well as to supply the soil with necessary nutrients and to prevent the evaporation of soil water, to preserve microelements, to prevent *Fusarium* disease and neck mold.

The mixture of mineral water and dialbeculit clay play the role of prolongation, providing the plant material by necessary elements. The mixture on seed capsule and mulch at the soil surface due to their high heat capacity provide favorable conditions for garlic development and selenium accumulation [13, 14].

Due to the prepared mixture and mulch the temperature in the zone of seed bed was higher than at the soil surface by 1.5–2°C. Hence, in winter time such crops are less exposed to the effect of low temperatures.

The annual clover species sowed in row-spacing of garlic provides the plant supply by biological nitrogen by means of tuber bacteria situated in the root system.

The Table 19.1 shows that the complex of macro and microelements, mulching and clover sowing in row-spacing provide high synergism in increase of selenium content in garlic. According to the results of our experiments, the proposed methods provide the increase of garlic productivity and quality (Table 19.1).

TABLE 19.1 Effect of Zakinskaya Mineral Water and Dialbeculit Clay on the Productivity and Quality of Winter Garlic

Variant of experiment	Bulb mass, g	Morbidity, %		Selenium content, mg/kg	Winter hardiness, %
		Fusarium disease	Neck mold		
Garlic planting by chives – control	29.2	4.6	5.2	7.2	82.4
Soaking of chives in mineral water 2–3 h	26.7	4.1	4.8	5.6	80.5
Incapsulation of chives by dialbeculit clay	23.8	3.8	4.2	4.9	84.8
Sowing of clover in garlic row-spacing	19.8	5.2	5.8	3.6	76.6
Mineral water + dialbeculit clay without exposition	27.5	4.8	5.6	7.1	82.8

TABLE 19.1 *(Continued)*

Variant of experiment	Bulb mass, g	Morbidity, %		Selenium content, mg/kg	Winter hardiness, %
		Fusarium disease	Neck mold		
Mineral water + dialbeculit clay + exposition 2–3 h	28.6	4.0	3.8	8.2	8.4
Mineral water + dialbeculit clay + exposition during 2–3 h + mulching	32.6	3.0	2.4	12.6	89.5
Mineral water + dialbeculit clay + mulching + sowing of clover in garlic row-spacing	39.2	2.4	1.8	16.8	92.8
Soaking in mineral water and clay for 1–1.5 h + mulching	30.5	3.2	2.8	11.6	87.3
Soaking in mineral water and dialbeculit clay for 3.5–4 h + mulching	31.5	2.2	1.2	13.8	90.9

It follows from the Table 19.1 that under exposition in Zakinskaya mineral water and dialbeculit clay during 2–3 and 3.5–4 h garlic morbidity decreases relatively to control; selenium content increases from 7.2 in control to 16.8 mg/kg in the optimal variant. Winter hardiness and bulb mass also increase in the optimal variant from 84.8% to 92.8%, and from 29.2 to 39.2 g, respectively. The results of our experiments show that the methods, which we propose, provide the increase of garlic productivity and quality.

In one variant of experiment the mulching clay layer of 5–6 cm, saturated by natrium selenite, provides the planted garlic chives by nutrients. Haricot, sowed before garlic planting, extracts colins, preventing chives affection by *Fusarium* disease and hook-worm.

Remains of haricot after harvesting and mowed biomass of herbs, used as mulch, surely protect planted chives from low winter temperatures and also protect crop from erosion.

19.4 CONCLUSIONS

1. The planting of garlic in mountain zone with application of natural sources of stuff, namely mineral water and zeolite-containing clay leads to the decrease of morbidity and increase of the productivity of garlic.

2. Selenium content in bulbs increased from 7.2 to 16.8 mg/kg in the optimal variant. The morbidity by *Fusarium* disease decreased from 4.6 to 2.2%
3. On the mountain slopes the applied method consisted in sowing of perennial legume herbs, its further thinning for establishment of haricot culture with subsequent autumn planting of winter garlic using haricot remains after harvesting for mulching provides not only increase of crop, but also its quality, protecting at the same time soil fertility by means of decrease of erosion processes.

ACKNOWLEDGEMENTS

Authors would like to thank Fyodor A. Tatarinov for the translation of this chapter from Russian.

KEYWORDS

- **dialbeculit**
- **mulching**
- **Zakinskaya mineral water**

REFERENCES

1. Seregina, I. N. Productivity and Adaptive Capacity of Agricultural Crops under Application of Microelements and Growth Regulators, PhD thes is in Biological Sciences, Author's Abstract http://www.dissercat.com/content/produktivnost-i-adaptivnaya-sposobnost-selskokhozyaistvennykh-kultur-pri-primenenii-mikroele#ixzz2s4JepSk6 (in Russian).
2. Seregina, I. I., & Nilovskaya, T. N. (2002). Biological Role of Selenium in Plants, Agricultural Chemistry, *10*, 76–85 (in Russian).
3. Golubkina, N. A., Konofeyeva, N. I., Pavlov, L. V., et al. (20.12.2003). Method of Vegetables Enrichment by Selenium, Patent of RF №2218764 (in Russian).
4. Golubkina, N. A., Sokolov, Ya A., Sokolova, Ya A. et al. (20.09.2002). Method of Garlic and Edible Roots by Selenium, Patent of Russian Federation, № (2189155), (in Russian).
5. Suzan, V. G., Grinberg, E. G., & Shtaynert, T. V. (2013). Production of Garlic in Siberia and Ural Problems and Perspectives, Potatoes and Vegetables, *9*, 9–11 (in Russian).
6. Sheludzhen, Kh A., Lebedovsky, I. A. & Bondareva, T. N. (2013). Biogeochemistry and Agro Chemistry of Selenium, Scientific Journal of the Kuban State University, *92(8)*, 1–11 (in Russian).
7. Sheludzhen, Kh A. (2003). Biogeochemistry, Maikop, Publishing House "Adygea", 1028p (in Russian).
8. Anisimova, O. (2007). Firest-Aid Set from Nature Garlic, Russian Pharmacies Magazine, Moscow, Group of Companies "Remedium", *22* (in Russian).
9. Gmoshinsky, I. V. (2001). Selenium in Human Feeding, Proceedings of XVI Session of Ugolev, A. M. Academic School-Seminar "Modern Problems of Physiology and Pathology

of Digestion" Appendix, *14* to Russian Journal of Gastroenterology, Hepatology, Coloproctology, XI, *4*, 121–127 (in Russian).

10. Golubkina, N. A., Nikulshin, V. P., & Khrykina, Yu A. (2007). Pecularities Extra Root Method of Enrichment of Garlic Plants by Selenium, Agricultural Biology, Plant Biology Series, *1*, 82–85 (in Russian).
11. Yli, Halla, M. (2005). Influence of Selenium Fertilization on Soil Selenium Status, Proceedings Twenty Years of Selenium Fertilization (September 8–9, Helsinki, Finland), Agrifood Research Reports, *69*, MTT Agrifood Research Finland, Jokioinen, 25–32.
12. Workshop Twenty Years of Selenium Fertilization, Proceedings Twenty Years of Selenium Fertilization (September 8–9, Helsinki (2005) Finland), Agrifood Research Reports, *69*, MTT Agrifood Research Finland, Jokioinen.
13. Bekuzarova, S. A., Alborov, I. D., Kesayeva, Z. A. et al. (10.10.2013). Method of Increase of Selenium in Garlic of Mountain Zone, Patent of RF, № 2494593 (in Russian).
14. Kachmazov, D. G. (2010). Pecularities of Winter Garlic Cultivation in Mountain and Foot Hill Conditions of North Ossetia-Alania Republic PhD Thesis Author's Abstract, Vladikavkaz, 22p (in Russian).
15. Tsabolov, Kh P. (2012). The Influence of Phyto Regulators on the Productivity and Quality of Garlic Bulbs, Proceedings Gorsky State Agrarian University, *49,* Part 3, 37–40 (in Russian).

CHAPTER 20

STIMULATORS OF INCREASE OF LUCERNE SEEDS GERMINATION

SARRA A. BEKUZAROVA[1], IRINA M. HANIYEVA[2], and NURBEK L. ADAYEV[3]

[1]Gorsky State Agrarian University, d. 37, Kirov St., Vladikavkaz, Republic of North Ossetia Alania, 362040, Russia; E-mail: bekos37@mail.ru;

[1]The North Caucasus research institute of mountain and foothill agriculture, village Mikhailovskoe, d. 1Williams St., Suburban district, Republic of North Ossetia Alania, 363110, Russia

[2]V.M. Kokov's Kabardino-Balkar State Agrarian University, d. 1v, Lenin Avenue, Nalchik, Republic Kabardino-Balkaria, 360030, Russia; E-mail: imhanieva@mail.ru

[3]Chechen Research Institute of Agriculture, d. 1, Lenin st., Grozny district, Gikalo settlement, Chechen Republic, 366021, Russia; E-mail: chechniish@mail.ru

CONTENTS

ABSTRACT

To increase the capacity of the seeds to germinate, we used mineral water Sernovodskaya, molasses, argillaceous deposits and local baterialnye fertilizers in different expositions and combinations. We have determined the optimal variants of their application.

20.1 INTRODUCTION

Presowing treatment of lucerne (alfalfa) seeds evokes the activation of metabolic processes in plant organism, increases plant capacity to resist environmental stress factors and pathogens, which are very important conditions of increase of seed productivity [1]. The microfertilizers, which are usually applied are very toxic to soil microorganisms [1–4]. A number of studies is devoted to presowing treatments of seeds of cereals, fodder and technical crops, when chemical substances are applied, namely boron, molybdenum, vanadium and others [4]. When exceeding acceptable bounds of these elements the plant metabolism is affected.

Our studies are aimed to find ecologically pure ways of cultivation of crops. The works on the seeds inoculation by bacterial fertilizers have a high importance for the increase of legumes growth, the improvement of their productivity and improvement of soil fertility by means of bacterial fertilizers [5, 6].

Lucerne seeds, as well as seeds of many other legumes, in ordinary conditions have hard, water and air-proof capsule. Mechanical damage of seed capsules (scarification) is one of the ways to increase germination of seeds. The method of seeds scarification increases the vitality and productivity of plants, leads to the development of strong root system and to increasing the number of seeds per plant [7].

After the damage of hard capsule (scarification) water and air get access to the seeds. Seed material with undamaged capsule passes the dormancy period and does not consume nutrients for respiration [5, 8].

The amount of hard seeds of lucerne strongly varies in dependence on the method of scarification and on mechanisms applied for it: namely if it was performed by means of threshing or wipe. Under scarification seeds lose their impermeability and get the capacity of normal swelling and germination [4, 9].

Without scarification the germination of hard seeds (of lucerne, clover, sainfoin) may delay from one to three or more days after sowing [1]. Scarification of seeds is particularly important for the renewal of perennial herbs under their undersow on degraded pastures. This method allows permanent improvement of herbage because of the regular appearance of young plants from germinating scarified seeds. The field germination capacity of mechanically processed seeds increases from 60 to 90% [5, 8].

20.2 MATERIALS AND METHODOLOGY

The study aimed to decrease the hardness of seeds and to improve their ability to germinate was carried out at the experimental bases of North Caucasus research institute of mountain and foothill agriculture, Kabardino-Balkarsky State Agrarian University and Chechen Research Institute of Agriculture.

The mineral water Sernovodskaya was applied as stimulator for ordinary (not scarified) seeds, which were soaked in this mineral water 5 7 or 8–10 h according to the following scheme (experiment 1):

- soaking of seeds in water (control);
- soaking of seeds in the mineral water Sernovodskaya for 5–7 h;
- soaking of seeds in the mineral water Sernovodskaya for 8–10 h;
- soaking of seeds in the mineral water for 8–10 h + molasses (waste product of sugar production);
- sowing of seeds with molasses 8–10 kg/ha – molasses + argillaceous deposits 20–25 kg/ha

Experiments were performed with triple replications; the sample plot area during the years of the research was from 20 to 25 m^2.

The laboratory and field germination capacity and germination energy of nonscarified seeds were determined.

In the same variants germination capacity and germination energy were also determined on the background of scarification. Seeds were scarified by flint-paper just before sowing. Before sowing the development intensity of plants and their reproductive peculiarities under ripening were determined in the laboratory.

The mineral water Sernovodskaya includes (mg/dm^3) natrium 1351, magnesium 30, bromine 0.18, sulfate 1292, calcium 40, chloride 467, iodine 0.06, hydrogen carbonates 1280, silicic acid 20; mineralization of water is 4–5 g/L.

Beet bin molasses (all-Russian State Standard 52304–2005) includes 75% of dry substances, 44% of sugar with direct polarization, 1% of reducting substances, 1.5% of calcium salt, 46% of fermenting sugars, pH was 6.5–8.

Argillaceous deposits (open pit near Terek river) include (%): silicon (SiO_2) – 57, 6; aluminum (Al_2O_3) – 20.7; calcium (CaO) – 0.60; magnesium (MgO) – 2.25; iron (Fe_2O_3) – 4.03, (FeO) – 2.41; manganese (MnO) – 0.1; potassium (K_2O) – 2.80; natrium (Na_2O) – 0.91; organic matter – 2.8.

Argillaceous deposits were mixed with scarified seeds in the proportion 30–40 kg/ha.

In the experiment №2 before sowing seeds were treated by sodium selenate in the concentration 0.01% of water solution and waste product of starch-treacle production (corn extract) in the proportion 5–7 kg per hectare rate of seeds.

Corn extract includes 20% of dry protein, 2.76% of fat, 30% of biologically effective substance (BEV) 0.69% of fresh cellulose, 0.6% of phosphorus, 6.4% of nitrogen. According to all-Russian State Standard (GOST) extract includes soluble carbohydrates (12–17%), starch (0.5%) and other substances. The high biological activity of extract makes it indispensable raw material for feeding of soil microflora, being at the same time an excellent substrate for seeds.

The determination of sowing qualities of seeds was performed according to the following requirements: purity – GOST 12037–82; germination capacity and energy and percent of hard seeds – GOST 12038–84, mass of 1000 seeds – GOST 12042–84.

For seeds inoculation we applied local bacterial fertilizers, prepared from the roots of old lucerne plants, which had tuber bacteria in the layer 0–20 cm on the developed roots of 2–4th year of life [5, 6]. Roots were reduced to fine particles in the period of spring growth in the proportion 7–8 plants per hectare rate of seeds (8–10 kg for wide-row sowing).

Two varieties of lucerne were applied in the experiment: Kevsala and Vega 87.

20.3 RESULTS AND DISCUSSION

The experiments with the preparation of seeds to the sowing showed that beside of scarification, the soaking of seeds in the mineral water Sernovodskaya also has positive effect. Micro- and macroelements contained in the mineral water contribute to the improvement of metabolism in the seeds and hence intensify the process of embryo germination.

Presowing processing of seeds positively affected their germination energy and germination capacity (Table 20.1).

TABLE 20.1 Germination Capacity and Germination Energy of Lucerne Seeds in Dependence on the Presowing Processing of Nonscarified Seeds (Average Data for Two Varieties)

Method of seeds treatment	Germination energy, %	Seed hardiness, %	Germination capacity, %	
			Laboratory	Field
Soaking in water – control	57	13.5	62	58
Soaking in the mineral water Sernovodskaya during 5–7 h	65	12.2	68	63
Soaking in the mineral water Sernovodskaya during 8–10 h	72	9.2	76	70
Soaking in the mineral water Sernovodskaya during 8–10 h + molasses	78	7.0	84	78
Sowing of seeds with molasses	76	11.5	82	76
Molasses + argillaceous deposits	78	12.0	84	80

It follows from the data of Table 20.1 that under soaking of seeds in Sernovodskaya mineral water the hardiness of seeds decrease by 4–6% in dependence on the duration of exposition. The laboratory germination capacity increases from 62% in the control to 84% in the optimal variant, what can be explained by the unique composition of Sernovodskaya, which includes sodium, magnesium, bromine, calcium, chloride, sulfate, iodine, hydrogen carbonate and other microelements, stimulating the germination of seeds and respiration rate. The best effect was observed when soaking seeds in the mineral water Sernovodskaya during 8–10 h with addition of molasses. High results were also obtained when applying of molasses mixed with argillaceous deposits.

Further it was found that the amount of hard seeds in the seed material of lucerne highly varies in dependence on the way of thresh and wipe of seeds from pods, as well as on the conditions of storing.

Under combine harvesting of seeds for sowing a high portion of hard seeds in the yield often occurs, because threshing apparatus of most of combine harvesters insufficiently damage capsules of seeds. Taking into account considerable portion of hard legume seeds under combine harvesting, some authors recommend to perform mandatorily seed scarification before sowing [8, 9].

It is known that scarification and damage of seed capsules may evoke fungi diseases. Such seeds can be also more damaged by soil pests. That is why we applied Sernovodskaya water, argillaceous deposits and molasses, which increase plant immunity and at the same time improve germination capacity of seeds (Table 20.2).

TABLE 20.2 The Effect of Scarification and of the Treatment by Argillaceous Deposits With Bacterial Fertilizers on the Germination Capacity of Lucerne Seeds (Average Data For Two Varieties)

Method of seeds treatment	Laboratory germination capacity, %	Germination energy, %	Field germination capacity, %
Control (without treatment)	82	70	76
Scarification + molasses	84	80	82
Scarification + argillaceous deposits	88	84	90
Scarification	86	76	82
Scarification + argillaceous deposits + local bacterial fertilizers	90	84	88

When analyzing the obtained results (Table 20.2), it is possible to conclude that maximum germination capacity was observed in the variant with scarification with further encapsulation by argillaceous deposits and bacterial fertilizers, obtained from the root system of old plants [10].

Scarification of seeds, their treatment by argillaceous deposits and bacterial fertilizers revealed differences in the reproductive peculiarities of lucerne (Table 20.3).

TABLE 20.3 The Effect of Seed Scarification, Argillaceous Deposits and Bacterial Fertilizers on the Reproductive Peculiarities of Lucerne (Average Data For Two Varieties)

Method of seeds treatment	Number of beans in one plant, items	Number of seeds in pods, %	Biological yield of seeds	
			g/ m²	% of control
Control (without treatment)	578	42.0	17.5	-
Scarification + molasses	682	48.8	21.8	124.5
Scarification + argillaceous deposits	705	52.4	24.5	140.0
Scarification	625	45.6	19.8	113.1
Scarification + argillaceous deposits + local bacterial fertilizers	748	54.8	28.6	163.4
Scarification + local bacterial fertilizers	712	54.5	26.6	134.8
LSD_{05}	17.0	2.6	2.3	

Data of Table 20.3 show that maximum yield of seeds was obtained in the variant with seed material treatment using scarification of seeds with adding of argillaceous deposits mixed with local bacterial fertilizers. In the best variant biological yield reached 28.6 g/m^2, which was by 63.4% higher than control.

The analysis of the obtained data shows that in all variants of experiment the biological yield was reliably above the control by 24.5–63.4%, that confirms the stimulating effect of argillaceous deposits and local bacterial fertilizers on the background of scarification.

The complex of macro and microelements presented in the argillaceous deposits provided the increase of germination capacity and productivity of green mass, as well as nitrogen fixation capacity of lucerne (Table 20.4).

TABLE 20.4 The Effect of Seed Scarification, Argillaceous Deposits and Bacterial Fertilizers on the Productivity of Lucerne Variety Vega 87

Method of seeds treatment	Field germination capacity, %	Deviation from control, %	Amount of tubercles per plant in the phase of flowering		Yield of green mass	
			Pieces	% of control	t/ha	% of control
Control (without treatment)	68	-	75	-	17.2	-
Scarification	82	+14	98	130.6	18.6	108.1
Scarification + argillaceous deposits	88	+20	126	168.0	20.2	117.4
Argillaceous deposits + local bacterial fertilizers	86	+18	132	176.0	21.5	125.0

The results of experiments reported in Table 4 for the variety Vega 87 show the positive effect of argillaceous deposits on the background of scarification and local bacterial fertilizers. The amount of tubercles per plant increases by 68–76%, the yield of green mass increases by 17.2–25%.

The silicon presented in clay is accumulated in the epidermal tissues of plants and creates a strong protective barrier against insects-pests and diseases. The clay contains above 60% of silicon compounds. The application of silicon fertilizer increases not only the germination capacity of treated plants, but also their qualitative characteristics. Seeds treated by argillaceous deposits gave the increase of yield of fodder mass and improved the quality of varieties under study, namely the concentrations of proteins, ashes, dry substance, fat, what can be explained by the intensification of cell metabolism (Fig. 20.1).

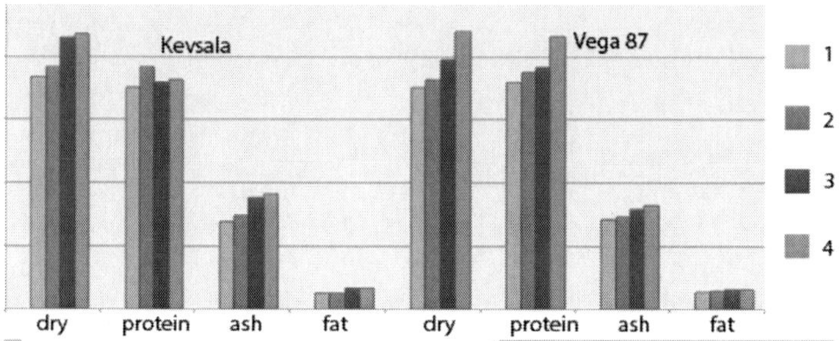

FIGURE 20.1 Concentration of nutrients in dry mass (%) of lucerne varieties Kevsala and Vega 87 in the phase of flowering after scarification of seeds, their treatment by bacterial fertilizers and clay (1 – control, 2 – scarification, 3 – scarification + argillaceous deposits, 4 – argillaceous deposits + local bacterial fertilizers).

Besides of stimulating effect, silicon compounds added with argillaceous deposits have a high absorbing capacity [11] and hence they are able to retain bacterial fertilizers, obtained from lucerne plants growing at the territory, in the root-inhabited layer.

One of the important functions of silicon presented in the argillaceous deposits is the increase of plant resistance to unfavorable environmental factors, in particular the increase of winter hardiness.

Selenium, apart from its direct catalytic effect, is able to affect some regulation processes, for example, by indirect regulation of biological synthesis of sulfur-containing amino acids and some enzymes [11]. Selenium evidently plays a key role in the intensification of metabolism at the beginning of plant development. We showed at the example of lucerne [12] that the treatment of seeds by water solution of sodium selenate (Na_2SeO_4) in the concentration 0.01% with further encapsulation of seeds by corn extract in the concentration 3–4 kg/ha contributes to the increase of seed germination energy, laboratory and field germination capacity by 5–23% (Table 20.5). This is because of selenium affects also the intensity of gaze exchange [11].

TABLE 20.5 The Effect of Sodium Selenate on the Germination Capacity of Seeds, %

Method of seeds treatment	Germination energy	Laboratory germination capacity	Field germination capacity
Soaking in water (control)	75	82	78
Soaking in sodium selenate, 0.001%	80	86	83
Treatment by corn extract	82	88	85
Soaking in sodium selenate mixed with corn extract	90	98	96

We suppose that the hardiness of varieties Kevsala and Vega 87 under study increases after the scarification of seeds, their treatment by bacterial fertilizers and clay, containing silicon. This will be the topic of our further research.

20.4 CONCLUSIONS

1. The value of the given study consists in the successful application of natural raw materials: clay, molasses, corn extract, Sernovodskaya mineral water and local bacterial fertilizers, obtained from the roots of old lucerne plants, which improved the biological potential of lucerne.
2. The corn extract, because of its high viscosity and high concentration of solid, acts as glue for seeds and preserves a part of water solution of sodium selenate.

3. The addition of the complex of macro and microelements, which are contained in the mineral water, clay and corn extract, provides higher characteristics of germination energy and germination capacity of seeds.
4. Presowing treatment of seeds by different stimulators at the background of seed scarification and their soaking in water solutions with microelements increase germination energy of seeds, growth and development of Lucerne plants.
5. Sodium selenate in the concentration 0.01% of water solution contributes to the germination energy, laboratory and field germination capacity of seeds by 5–23%.

ACKNOWLEDGEMENTS

Authors would like to thank Fyodor A. Tatarinov for the translation of this paper from Russian and A. Bukhonin for helping in preparing the illustration.

KEYWORDS

- **alfalfa**
- **argillaceous deposits**
- **fertilizers**
- **microelements**
- **mineral water Sernovodskaya**
- **mossales**
- **scarification**
- **selenium**

REFERENCES

1. Lucerne Tarkovsky, M. I. (1974). CompilerMoscow, Kolos, 240p (in Russian).
2. Medvedev, G. A. (2003). Biological Basis for Improving Seed Production Lucerne, Volgograd, Publisher Volgograd Agricultural Academy168p (in Russian).
3. Goncharov, P. L. & Lubenets, P. A. (1985). Biological Aspects of Growing Lucerne, Novosibirsk, Science Siberian Division, 253p (in Russian).
4. Statsenko, A. A. (2001). Stimulation of Germination crops with Large Seeds Advances in Science and Technology of Agriculture, *12*, 33–34 (in Russian).
5. Bekuzarova, S. A., Gazdanov, A. U., Bassiev, S. S., et al. (27.06.2000). The Method to Stimulate Germination Hard Seeds of Legumes, Patent №2151479, Bull, *18* (in Russian).
6. Farniev, A. T., Kozyrev, Kh A., Gerasimenko, M. V., & Kalitseva, D. T. (27.05.2001). The Method of Inoculation of Seeds of Lucerne, Patent №2167509, Bull, *9,* (in Russian).

7. Bzikov, M. A., Bekuzarova, S. A., Abiyev, V. B. et al. (27.02.2008). The Pre-Processing Method of Seeds, Patent №2317669, Bull, *6,* (in Russian).
8. Bekuzarova, S. A., Gridnev, N. I., & Kshnikatkina, A. N. (10.04.2009). The Pre-Processing Method of Nectar Producing seeds of Herbs, Patent№ 2151479, Bull, *10,* (in Russian).
9. Zherukov, B. H., Hanieva, I. M., Haniev, M. H. et al. (27.04.2013). The Pre-sowing Method Treatment of Lucerne Seeds, Patent № 2479974, Bull, *12* (in Russian).
10. Bekuzarova, S. A., Khubaeva, G. P., & Lushchenko, G. V. (20.09.2009). Substrate for Legumes from Industrial Waste, Patent № 2461184, Bull, *26* (in Russian).
11. Blinohvatov, A. F. (2001). Seleniumin the Biosphere, Penza, Publisher Penza Agricultural Academy, 365p (in Russian).
12. Bekuzarova, S. A., Abiyeva, T. F., & Tedeyeva, A. A. (27.02.2006). Method of Pre-sowing Treatment of Seeds, Patent № 2270548, Bull, *6* (in Russian).

CHAPTER 21

OIL FLAX IN THE FOOTHILLS OF NORTHERN CAUCASUS

SARRA A. BEKUZAROVA[1], IRINA M. HANIYEVA[2], and SVETLANA S. KAZIEVA[1]

[1]Gorsky State Agrarian University, Vladikavkaz, Kirov St., 37, Republic of North Ossetia Alania, 362040, Russia; E-mail: bekos37@mail.ru

[2]V.M. Kokov's Kabardino-Balkarsky State Agrarian University, d. 1v, Lenin Avenue, Nalchik, Kabardino-Balkaria, 360000, Russia; E-mail: imhanieva@mail.ru

CONTENTS

ABSTRACT

In order to increase the yield of seeds of oil flax we studied the terms and norms of sowing and the presowing treatment of seeds by natural substances: therapeutic mud used in medical treatment mixed with molasses, Centurion herbicide and biological preparation Ambiol. Additional fertilizing was performed during the vegetation period when the plant takes a form of a small fir-tree. Maximum productivity of flax was reached under the sowing in the second ten days of April with the rate of 7 millions of seeds per hectare.

21.1 INTRODUCTION

Oil flax is a valuable technical culture. Russia kept leading position among countries producing seeds of this important culture. However, the situation, which occurred in the agroindustrial complex at the end of XX and the beginning of XXI centuries, affected the cultivation of oil flax, which crops decreased from 43,000 ha in 1990 down to the critical level of 4000 ha in 1997. By 2010, flax crop area increased to 120,000 ha, mostly in Southern, Volga and Siberian regions [1].

The practice shows [2–4], that the yield capacity of oil flax is determined by the applied technology of its cultivation. Such technological features as terms and rates of sowing, supply of plants by nutrition elements and variety characters considerably influence the productivity and quality of seeds.

In the present study we evaluated the effect of different technological methods of the cultivation of oil flax of varieties Rucheyok, VNIIMK 620 and VNIIMK 630, included in the State register, and permitted for use in the Northern Caucasus. We determined the crop productivity and seed quality in the soil and climatic conditions of experimental stations of North Caucasus research institute of mountain and foothill agriculture in the foothills zone and of Kabardino-Balkarsky State Agrarian University.

21.2 MATERIALS AND METHODOLOGY

The scheme of the experiment was the following.

1. We determined productivity of varieties in dependence on soil temperature under different sowing terms:
   a) under soil heating up to 6–8°C (first ten days of April);
   b) under soil heating up to 8–10°C (second ten days of April);
   c) under soil heating up to 10–12°C (third ten days of April).
2. We determined productivity and technological properties of varieties in dependence on sowing rate:
   a) 3 millions of germinating seeds per ha (20 kg/ha)
   b) 5 millions of germinating seeds per ha (40 kg /ha)

c) 7 millions of germinating seeds per ha (60 kg /ha)
d) 9 millions of germinating seeds per ha (80 kg /ha).

3. Before sowing seeds were covered by used Tambukan mud in the proportion 2:1, during the vegetation period when the plant takes a form of a small fir-tree they were additionally fertilized by the mixture of used Tambukan therapeutic muds and molasses in the proportion 1:1 with Centurion and biological preparation Ambiol (code 244690–1200 of the All-Russian classifier of products) using the methodology, which we elaborated earlier [5]. Tambukan therapeutic muds are got from the Tambukan Lake, which is situated in a hollow near the road Piatigorsk – Nalchik. Initial Tambukan mud includes 33% of fatty acids, 11.3% of phospholipids, 3.48% of sterols, 3.14% of ethers of sterols, 12.2% of glycerides, 16.15% of pigments, 5.54% of sulfur containing substances [6].

In particular, the mud includes cyanobacteria containing up to 60% of amino acids and proteins, carotenes (they contains 25 times more beta-carotene than carrot), vitamin E_1 (more than fresh beef liver) and humic acid (up to 5%). Cyanobacteria of Tambukan mud contain 50 times more iron than wheat germ [6]. Cyanobacteria of Tambukan mud are valuable source of natural antioxidants, including vitamins B_1, B_5 and B_6, minerals Ca, Mg, Cu, Fe, Zn, Se and other microelements, they have also high concentration of chlorophyll. All substances contained in the cyanobacteria of Tambukan mud are easily assimilated by living organisms and, when getting to the soil, they sustain and keep soil microflora and at the same time they clean it from radioactive nuclides and heavy metals [6, 7].

Tambukan mud is used for the cure of wide spectrum of human diseases [7]. After the use in the cure the mud is stored as used material or thrown down in the Tambukan lake. In the mud used for cure the concentration of iron sulfides decreases two times, the concentration of organic substance decreases by 50%, mineralization decreases from 72 to 43 g/l, the amount of anaerobic organisms also decreases. But nitrobacteria, both aerobic and anaerobic, decomposing cellulose, and other microorganisms taking part in the decomposition of root remains, are preserved in it [6, 7].

Molasses are waste product of sugar production. According to all-Russian State Standard 52304–2005 molasses contain 9% of nitrogen compounds, mainly amids, 58–60% carbohydrates, mainly sugar, and 7–10% of ashes, calcium salts (Ca_2O – 1.5%), pH 6.5–8. As predecessor we used alfalfa crop. Before plowing mineral fertilizers ($P_{30}K_{30}$) were applied in the amount of one half of standard. Nitrogen was not applied because alfalfa cumulates it in sufficient amount by means of tuber bacteria situated in the rhizosphere. Presowing soil processing and plot leveling were also performed.

21.3 RESULTS AND DISCUSSION

Productivity of varieties, namely yield capacity, the mass of 1000 seeds and oil production, depended on the terms of sowing and consequently on the soil temperature (Table 21.1).

TABLE 21.1 Productivity of Oil Flax Varieties in Dependence on the Terms of Sowing

	Rucheyok				VNIIMK 620				VNIIMK 630			
Dates of sowing	Yield, centners/ha*	Mass of 1000 seeds, g	Oil content, %	Oil yield, centners/ha	Yield, centners/ha*	Mass of 1000 seeds, g	Oil content, %	Oil yield, centners/ha	Yield, centners/ha*	Mass of 1000 seeds, g	Oil content, %	Oil yield, centners/ha
05–10	11.4	7.0	54.3	6.2	12.2	8.6	42.6	5.2	12.8	7.8	47.6	6.1
15–20	12.6	7.0	54.7	6.9	13.0	8.5	44.6	5.8	14.0	7.4	53.1	6.8
25–30	8.7	6.7	52.8	4.6	10.3	7.8	38.8	4.0	10.1	6.9	47.5	4.8
HCP_{05-} 0.38 centners/ha												
Error of experiment S_x – 2.5%												

Sowing was performed in April.
*Metric centners (100 kg) are used.

Table 21.1 shows that the highest yield and oil crop values were obtained for all three varieties when sowing in the second ten days of April.

Under too early sowing sprouts are disjointed what leads to the decrease of yield and oil production per 1 ha. Ander late sowing yield capacity dramatically decreases and even the application of fertilizers could not compensate losses.

In other experiment we studied the productivity of oil flax in dependence on sowing rate (Table 21.2).

TABLE 21.2 Productivity of Oil Flax Varieties in Dependence on the Amount of Seeds Sowed Per 1 Hectare

Sowing rate per 1 ha	Rucheyok				VNIIMK 620				VNIIMK 630			
	Yield, centners/ha	Mass of 1000 seeds, g	Oil content. %	Oil yield, centners/ha	Yield, centners/ha	Mass of 1000 seeds, g	Oil content, %	Oil yield, centners/ha	Yield, centners/ha	Mass of 1000 seeds, g	Oil content, %	Oil yield, centners/ha
3 millions	12.2	6.8	45.9	5.6	10.9	8.2	49.5	5.4	12.3	6.5	48.7	6.0
5 millions	9.8	6.8	62.2	6.1	12.6	8,2	48.4	6.1	13.8	6.5	47.8	6.6
7 millions	10.6	7.0	55.5	7.0	14.0	8.2	47.8	6.7	15.2	7.0	50.0	7.6
9 millions	10.2	6.5	56.8	5.8	11.6	8.0	43.9	5.1	11.8	6.2	47.4	5.6

HCP_{05-}0.78 centners/ha

Error of experiment S_x – 1.65%

The highest yield was reached when sowing 3 millions of seeds per hectare for Rucheyok variety – 12.2 centners/ha; when sowing 7 millions, the highest yield was obtained for varieties VNIIMK 620–14.0 centners/ha and VNIIMK 630–15.2 centners/ha. Oil yield was higher when sowing 7 millions of seeds per hectare for varieties VNIIMK 620 and VNIIMK 630. The oil percentage varied under different sowing rates.

Further increase of sowing rate led to the decrease of yield, oil yields and oil percentage of seeds (Table 21.2). This decrease happened because under too high stocking density plants are suppressed because of competition for light, water and nutrients.

In the third experiment (Table 21.3) we determined the productivity of VNIIMK 620 variety after the mixing of seeds with moist mud in the proportion: 100 kg/ha of used Tambukan mud and 50 kg/ha of seeds.

Higher amount of the mud leads to the appearance of coarse lumps when drying what complicates homogenous sowing by sowing-machine. The increase of amount of Tambukan mud complicates the functioning of hose syringe.

TABLE 21.3 Productivity Oil Flax Variety VNIIMK 620 in Dependence on Presowing Seed Treatment

Variant of experiment	Yield of seeds, centners/ha	Mass of 1000 seeds, g	Oil content, %	Yield of oil, centners/ha
Control (without treatment)	13.8	7.0	51.5	7.1
Presowing seed treatment by Tambukan mud	14.5	7.2	51.7	7.5
Additional fertilizing of plants by Tambukan mud without presowing seed treatment	16.0	7.3	52.0	8.3
Additional fertilizing of plants by Tambukan mud Adtionaltilizing by Tambukan mud after presowing treatment	15.6	7.1	51.8	8.1
Additional fertilizing by molasses after presowing treatment by Tambukan mud	16.2	7.2	52.0	8.4
Additional fertilizing by Ambiol after presowing treatment by Tambukan mud	17.6	7.2	52.0	9.1
Treatment of seeds by Centurion herbicide	16.5	7.1	51.6	8.5
Seed treatment: mixture Tambukan mud and molasses + additional fertilizing: Centurion and Ambiol	18.8	8.2	53.5	10.1

In other variants of the experiment seeds were treated by the mixture of 200 g/ha of Centurion herbicide and 10 mg/ha of growth regulator Ambiol used in Russia. The mixture was dissolved in 300 l of water. After drying the seeds are easily sown by narrow-row flax sowing-machine (type SZL-3.6) with row-spacing of 7.5 cm. In the proposed mixture molasses are used in the proportion 50 kg/ha.

When the height of flax plants reached 5–15 cm, the crop was treated by the mixture of Tambukan mud (50 kg/ha), Centurion herbicide (0.2 L/ha) and growth regulator Ambiol (10 mg/ha).

The presowing encapsulation of seeds by used Tambukan mud supplies them with moisture and nutrients, which allows decreasing the dosage of fertilizers applied [5, 6].

Under sowing rate of 50–60 kg of seeds per 1 hectare 100–120 kg/ha of used therapeutic muds were applied. In such mixture molasses is used as adhesive ingredient and nutrient medium for plants and soil microflora.

Centurion herbicide and Ambiol biological preparation are applied in doses recommended for use. This mixture allows not only eliminate weeds, but also stimulate plant growth and development.

The data of Table 3 show that all variants of presowing treatment and additional fertilizing lead to the increase of variables in comparison to control. The best result was obtained under seed encapsulation by Tambukan mud mixed with molasses in combination with additional fertilizing by mixture of herbicide Centurion and biological preparation Ambiol. In this variant the yield of seeds increased by 5 centners/ha, the mass of 1000 seeds increased by 1.2 g, oil concentration of seeds increased by 2%, oil yield increased by 3 centners/ha.

21.4 CONCLUSIONS

The obtained results give the full certitude in the possibility to cultivate oil flax very successfully in the conditions of foothills zone of Northern Caucasus beside of the main oil culture – sunflower.

Presowing treatment of seeds by used therapeutic muds and additional fertilizing by the mixture of used products (molasses and mud) with herbicide and biological preparation provide the increase of yield by 5 centners/ha relatively to control, which consists 30%; at the same time oil yield increases by 42.5% without additional expenses for fertilizers and stimulating substances.

Our experiments allow giving the following recommendations for oil flax cultivation in the conditions of foothills zone of Northern Caucasus.

1. The optimal term of sowing is the second ten days of April, which permits to get 12.6–13.0 centners/ha of seeds in dependence on the variety and the oil yield within the limits 5.8–6.9 centners/ha. In contrast, the earlier sowing does not increase, but decreases yield capacity, at the same time the vegetation period is delayed. Under later sowing (at the end of April) yield capacity dramatically falls and even the application of fertilizers does not compensate yield losses.
2. In order to obtain high yield of oil flax it is necessary to sow it with the rate of 7 millions of germinating seeds per 1 ha, which allows to get 12.6–15.2 centners/ha in dependence of variety, oil yield can reach 6.7–7.6 centners/ha. Densification or thinning of crops leads to the decrease of yield of seeds and oil.

3. In the conditions of foothills zone of Northern Caucasus the varieties VNIIMK 620 and VNIIMK 630 are perspective from the point of view of seed yield and oil production.
4. The results of our research show that the oil concentration depends on the particularities of varieties and varies within the limits 38.8–62.2% in dependence on the applied methods.
5. Presowing seed treatment by used Tambukan mud mixed with molasses in the proportion 1:1 with addition of the herbicide Centurion and biological preparation Ambiol provides the increase of yield by 30% and oil production by 42.5% without additional expenses for fertilizers and other stimulating substances.

ACKNOWLEDGEMENTS

Authors would like to thank Fyodor A. Tatarinov for the translation of this paper from Russian and the help in its editing.

KEYWORDS

- **Ambiol**
- **Centurion**
- **molasses**
- **soil**
- **Tambukan therapeutic mud**

REFERENCES

1. Lukomets, V. M., Bocharev, N. I. & Gorlov, S. L. Eds (2010). Perspective Resource-Saving Technology of Oil Flax Production, Methodic Recommendations, Rosinform teh, 50p (in Russian).
2. Kolomeychenko, V. V. (2007). Plants Breeding Manual for Students of Higher Education Studying the Specialty of 10200, "Agronomy" Moscow, "Agrobisneszentr" 596p (in Russian).
3. Khorkhorin, B. G. (10.03.1996). Method of Cultivation of Long-Fibred Flax, Patent RF № 2055462
4. Rusakova, G. G., Beletskaya, M. N., Korolyova, T. A., & Maksyuta, V. N. (10.08.1999). Means for Extra-Root Additional Fertilization of Oil Cultures, Patent RF № 2134252 (in Russian).
5. Zherukov, Kh, B., Hanieva, I. M., Haniev, M. X., Bekuzarova, S. A., & Boziyev, A. L. (10.07.2013). Method of Cultivation of Oil Flax, Patent RF № 2486734, Bulletin, *19* (in Russian).
6. Tambukan Lake (1999) Ancient Source of Healing Power, "Bivitex" Company Publishers, 12p (in Russian).

7. Karaulov, Kh.G. Development of Low-Wasted Technology of Medical Products of Peloids of Tambukan Lake, their Study and Standardization, Abstract of Thesis for the Degree of Candidate of pharmacological Science, Pyatigorsk (in Russian) http://medical-diss.com/farmakologiya/razrabotka-maloothodnoy-tehnologii-lekarstvennyh-preparatov-peloidov-tambukanskogo-ozera-ih-issledov.

CHAPTER 22

IMPROVEMENT OF CULTIVATION AND HARVESTING OF FESTULOLIUM SEED IN THE FOREST-STEPPE OF CENTRAL CHERNNOZEM ZONE OF RUSSIA

VLADIMIR N. OBRAZTSOV, DIANA I. SHCHEDRINA, VASILIY AN. FEDOTOV, SABIR V. KADYROV, and VLADIMIR V. KONDRATOV

Voronezh State Agricultural University, d. 1, Mitchurina St., Voronezh, 394087, Russia; E-mail: ovennn@mail.ru

CONTENTS

ABSTRACT

Festulolium (*x Festulolium* F. Aschers. et Graebn.) is a new perennial grasses from Poaceae family. Among its advantages is capacity to good regrowth, improved sugars content and higher winter hardiness. It is a good component of herb mixtures when creating cultivated hayfields and pastures.

Wide application of this new culture for forage production is restrained by the deficit of seeds, caused by the imperfection of technologies of their production. There are still very few of studies on the biology and technology of festulolium cultivation in the forest-steppe of Central Chernozem Region of Russia (CCR). Our study is the first where the biological peculiarities of festulolium were researched in the conditions of the forest-steppe of CCR, the main technological methods of festulolium cultivation and seeds harvesting are elaborated, including the optimal dosage of mineral fertilizers, herbicides, application of film forming tools for preharvesting crop processing. Their reasonable application allows to decrease energy costs and to get in average 590–620 kg/ha of conditioned seeds.

22.1 INTRODUCTION

It is known that successful development of animal breeding is closely related with the solution of the problem how to obtain sufficient amount of high-quality feeds. The problem of the increase of feed crops productivity and feed quality improvement was earlier and remains now very relevant [1, 2].

Perennial legumes and grain crops are effective feed cultures. They allow to solve the problem of protein-balanced feeds, provide the protection of soil fertility, increasing of ecological security and stability of forage production [3–5]. However, traditional grass species have insufficient amount of soluble carbohydrates, extensive regrowth after harvesting circles, summer growth depression.

The artificially created intergeneric hybrid festulolium is the better feed crop with high energy and protein sustenance, which can be used as green forage when creating cultivated hayfields and pastures. The advantages of this new culture are good regrowth capacity, elevated sugar content and higher winter hardiness [6–10].

The portion of festulolium (together with perennial ryegrass) among the stock of seeds perennial grain crops increased during recent years from 2 to 10% and continues to increase. This conditions the increase of crops of these valuable feed cultures, in particular for creation of cultivated pastures and hayfields.

The creation of specialized festulolium varieties (Aelita, VIK-90, Viknel, Izumrudny, Debut, Sinta) allowed to enlarge its cultivation in the northern regions of Russia (Arkhangelsk, Syktyvkar, Perm, Yekaterinburg). Favorable conditions for festulolium cultivation for seeds and forage purposes develop in Central, Central-Chernozem and North-western regions of Russia [11].

However, festulolium is not yet widely spread in forage production in Central Chernozem region because of insufficient seed production.

In dependence on morphological type of parent forms (fescue and ryegrass), festulolium inherits a certain combination of valuable characters, which allow its use both in creation of cultivated pastures, as well as hayfields. However, technology of festulolium cultivation is not studied enough. The necessity arose of scientific elaboration of technology of its cultivation, aimed to the complete realization of its biological potential and obtaining of stable crops of this new culture. Such was the aim of our research.

22.2 MATERIALS AND METHODOLOGY

Experimental part of the study was performed of the department of plant breeding, forage production and agrotechnologies of Voronezh State Agrarian University on the fields of the training, research and technological center "Agrotechnology" in 2009–2011.

The soil of experimental site is leached loamy chernozem. Humus content in the topsoil is from 4.56 to 5.50%, pH_{solt} is 4.9, soil saturation by bases is 74–86%. The amount of mobile phosphorus (P_2O_5) and exchange potassium (K_2O) was 78–129 and 109–118 mg/kg of soil, the total of absorbed bases was from 21.3 to 22.2 mg/eV per 100 g of soil.

The variation of meteorological conditions during the years of experiments allowed to evaluate objectively their results. The year 2009 was more favorable for festulolium although the precipitation total was below long-term average. The year 2010 was abnormally dry and warm, the air temperature on the soil surface in hot summer days reached 60°C, mean monthly temperatures of July and August were above 25°C. The conditions of vegetation season of 2011 were relatively favorable.

The festulolium variety VIK-90 was applied in the experiments. Festulolium crop was preceded by the mixture of vetch-oat mixture, harvested for green forage. The soil preparation to sowing was common for seed crops of perennial grasses, worked out for Central-Chernozem region. After precursor harvesting the eddish hulling until the depth of 8–10 cm was performed with further (two weeks later) plowing with colters to the depth of 27–30 cm. In spring, as soil ripened, early spring harrowing in two traces and then presowing cultivation and soil packing before and after sowing were performed. The sowing was performed nakedly to the depth of 0.5–1.0 cm by wide-row (45 cm) method with sowing rate 6.0 kg of seeds per ha. The area of plot was 20 m^2. Experiment was performed with 4-times repetition; the distribution of sample plots was randomized. Attendant measurements and observations were performed according to the methods [12] common in the seed-farming of perennial grasses.

22.3 RESULTS AND DISCUSSION

During the first year of life of festulolium plants the meteorological conditions of vegetation season, especially during the germination, have a strong effect on the plant growth and development. Conditions during the period of the research showed to be favorable for festulolium seeds germination, their mean field germinating capacity reached 75.8%.

In 2009 and 2010 conditions for the germination of this grass were more favorable: field germinating capacity averaged for all variants reached the level of 76.1–81.8%. In 2011 dry weather occurred during sowing and germination period and field germinating capacity decreased to 69.5%.

In 2010, the second part of the summer hot and dry weather with air temperature above 25°C continued during 24 days. This negatively affected the safety of plants during the first year of life. The mean plant mortality during vegetation was the highest – 15.2%, whereas in other years it did not exceed 8.3–9.2%.

In the year of sowing festulolium grows slowly and does not form generative shoots. They are formed during the second and subsequent years, both from wintered shoots as well as from shoots newly appeared in spring.

After 4–6 days the appearance of festulolium young growth the first leaf develops, after 5–6 days the second one opens. In the phase of 3–4 leaves the first side shoot of bushing out appears. Festulolium of the first year of life vigorously clusters during the whole summer period. The amount of well developed shoots in the first year of life directly affects wintering and the crop amount.

Festulolium is a cross-pollination plant. The flowering of one plant continues from 2 to 4 days, within the population the flowering is extended up to 8–11 days. During the second year of life the period from spring growing again until full seeds maturity continued 99–107 days.

22.3.1 FERTILIZATION OF SEED GRASS

An important role in the system of agricultural methods providing the high seed productivity of festulolium belongs to fertilizers. The studies of many researchers [13–18] on the effect of mineral fertilizers on the seed productivity of cereal grasses prove the key importance of nitrogen fertilizers.

Unlike the cultivation of herbs for forage, when it is necessary to obtain the highest yield of vegetative mass, in the seed crops fertilizers should contribute to the formation of nonlodging or slightly lodging crops, provide the maximum formation of generative organs, as well as synchronic ripening of seeds.

Nitrogen fertilizers increase the vegetation period of plants, especially the time of bushing out and booting. In the second year of life the vegetation period of plants increased relatively to control without fertilization by 3–7 days and 3–8 days under additional fertilizing by ammonium nitrate (N) and nitroammophoska (ammonium nitrate phosphate fertilizer, NPK), respectively (Table 22.1).

TABLE 22.1 The Effect of Fertilizers on the Duration of Periods From the Vegetation Renewal of Festulolium of the Second Year of Life to Booting, Earing, Flowering and Seeds Ripening (Average for 2009–2011)

Fertilizer	Dosage of fertilizer, kg/ha of active ingredient (AI)	Duration of period from spring vegetation renewal to the beginning of the phase			
		Booting	Earing	Flowering	Full seeds maturity
Control (no fertilizers)		32	46	63	99
Ammonium nitrate	N_{45} – in autumn	34	49	66	100
	N_{60} – in autumn	34	51	68	103
	N_{75} – in autumn	35	51	69	104
	N_{90} – in autumn	36	52	71	105
	N_{30} – in autumn +N_{30} – in spring	36	53	71	106
	N_{45} – in autumn +N_{45} – in spring	37	54	72	107
Nitroammophoska 16:16:16	$(NPK)_{45}$ – in autumn	35	50	69	101
	$(NPK)_{60}$ – in autumn	35	51	70	104
	$(NPK)_{75}$ – in autumn	36	52	71	105
	$(NPK)_{90}$ – in autumn	37	53	73	106
	$(NPK)_{30}$ – in autumn + $(NPK)_{30}$ – in spring	37	52	72	107
	$(NPK)_{30}$ – in autumn + $(NPK)_{45}$ –in spring	38	53	74	107

Additional nitrogen fertilization increases the height of plants. In the second year of life in the control variant mean height of generative shoots at seed plots reached 41.8 cm, and when applying additional fertilization by N_{45} and N_{90} it increased by 7.3 and 18 cm, respectively.

The increase of fertilization dosage from 75 to 90 kg/ha of active ingredient (AI) contributed to the intensification of plant growth at early stages of vegetation leading to high lodging of plants, which negatively affected cross-pollination of flowers, seed formation and ripening and also worsened the conditions of their harvesting. E.g., in wet year 2009 under the application of N_{45} and N_{60} lodging reached 16.8 and 35.2%, and when applying N_{75} and N_{90} it reached 39 and 46.8%, respectively. In the year 2010, characterized by low amount of precipitation in the second half of vegetation season, only low grass lodging was observed. It was found that in the lodged crop the phase of seeds maturation was delayed by 2–4 days relatively to control and the amount of nonfilled (frl) weevils in inflorescences increased.

Under application of nitroammophoska the lowest measure of lodging of seed crops (14.9%) was observed when applying $(NPK)_{45}$. Under application of N_{90} and $(NPK)_{90}$ grass lodging reached in average 33.6 and 34.8%.

It was found that festulolium has high winter hardiness. The wintering rate during all years of study under the application of mineral fertilizers reached 84.9–88.9%. The shoots mortality during winter period was low - 11.1–15.1%.

Favorable conditions for festulolium wintering occurred in the winter 2010/2011 with high snowing. In all variants plant mortality in this period was low (4–8%). Better plant safety (95.9–97.8%) was observed in the variants with elevated doses of fertilizers, less plants (92%) survived without fertilizers.

Conditions of winter 2009/2010 were the most adverse. Strong frost and long-term absence of snow cover led to the considerable plant mortality. In the control variant before start of winter density of shoots was 1030 items/m^2, after wintering it was only 734 items/m^2, the mortality reached 28.7%. The application of ammonium nitrate and nitroammophoska increased the amount of wintered festulolium plants to 77–81.8%, which is by 5.7–10.5% higher than under control.

Festulolium crops have high nutrition quality and high sustenance, quite satisfying the needs of cows. The protein and fat content reached 11.3–22.0% and 3.4–5.7%, respectively. The cellulose content was low, namely 18.2–25.8%. The concentration of soluble sugars under application of fertilizers reached 16.1–20.5%, which is by 1.3–5.7% higher than in control.

The forage was more nutritious in the variants with N_{90} and $(NPK)_{90}$, when the content of fodder units per 1 kg of dry substance reached 0.96, whereas in other variants it was by 0.02–0.13 lower.

The exchange energy content of fodder was from 9.1 to 11.7 MJ/kg. Green mass contained sufficient amount of phosphorus (2.4–2.9 g) and calcium (4.3–5.8 g).

The density of festulolium generative shoots in the second year of life was higher in the variants with fertilization - under the application of ammonium nitrate and nitroammophoska it reached 762–853 and 775–868 items/m^2, respectively; in the control it was by 16–26% lower. However, the increase of doses of both, nitrogen, as well as complex fertilizer above 60 kg/ha of AI, did not increase any more the amount of generative shoots (Table 22.2).

TABLE 22.2 Elements of Crop Structure of Seed Festulolium of Second Year of Life in Dependence on Fertilizers (Mean for 2009–2011)

Fertilizer	Dosage of fertilizer, kg/ha of AI	Ear length, cm	Number of generative shoots, 1/m^2	Number of spikelets in ear	Number of seeds в ear
Control (no fertilizers)		16.0	639.3	14.4	48.7
Ammonium nitrate	N_{45} – in autumn	17.1	762.2	15.8	57.8
	N_{60} – in autumn	18.3	853.6	16.7	65.2
	N_{75} – in autumn	18.4	813.1	17.9	66.7
	N_{90} – in autumn	19.7	793.6	18.3	67.1
	N_{30} – in autumn+N_{30} – in spring	17.2	834.0	15.9	63.7
	N_{45} – in autumn+N_{45} – in spring	18.0	815.1	16.6	64.9
Nitroammophoska 16:16:16	$(NPK)_{45}$ – in autumn	18.0	775.1	16.0	61.3
	$(NPK)_{60}$ – in autumn	18.9	868.0	17.2	67.1
	$(NPK)_{75}$ – in autumn	20.0	827.2	18.6	67.1
	$(NPK)_{90}$ – in autumn	21.0	805.8	19.2	66.9
	$(NPK)_{30}$ – in autumn + $(NPK)_{30}$ – in spring	17.8	844.7	17.1	62.8
	$(NPK)_{45}$ – in autumn + $(NPK)_{45}$ – in spring	18.5	822.2	17.6	64.0
LSD_{05} for fertilizer type		0.8	37.2	0.3	1.8
LSD_{05} for fertilizer dose		1.5	24.3	0.5	2.2

Mineral fertilizers increased the size of ears. Their length in control variant was 16 cm, whereas under fertilization by ammonium nitrate and nitroammophoska it reached 17.1–19.7 cm and 18–21 cm, respectively, that is, ears were by 11.4–13.2% longer. This trend was observed for all years of life of seed crop.

The application of fertilizers increased the amount of spikelets in ear by 1.4–4.8 because of more intensive nutrition of generative organs during the period of their growth, flowering and seed formation.

The application of nitrogen fertilizers in the rate of 60 kg/ha of AI in autumn of the year of sowing or after the harvest of vegetative mass, that is, in the mid September, showed to be optimal for the formation of highly productive festulolium seed crop with low lodging.

The crop of festulolium seeds directly depended on the amount of generative shoots per unit crop area and on the seed content of inflorescences (Table 22.3).

TABLE 22.3 Crop of Festulolium Seeds in the Crops of Different Year of Life in Dependence on Fertilizers, kg/ha (Mean for 2009–2011)

Fertilizer type	Fertilizer dose, kg/ha of AI	Year of life		
		2nd	3rd	4th
Control (without fertilizers)		410.7	192.0	154.6
Ammonium nitrate	N_{45} – in autumn	489.4	206.4	180.5
	N_{60} – in autumn	591.4	253.3	223.4
	N_{75} – in autumn	540.3	224.5	207.9
	N_{90} – in autumn	514.3	212.6	200.8
	N_{30} – in autumn+N_{30} – in spring	559.7	241.5	208.8
	N_{45} – in autumn+N_{45} – in spring	568.3	246.9	213.8
Nitroammophoska 16:16:16	$(NPK)_{45}$ – in autumn	513.4	247.0	197.9
	$(NPK)_{60}$ – in autumn	620.5	272.5	240.9
	$(NPK)_{75}$ – in autumn	565.8	244.0	227.5
	$(NPK)_{90}$ – in autumn	538.5	230.1	222.6
	$(NPK)_{30}$ – in autumn + $(NPK)_{30}$ – in spring	586.1	260.0	228.0
	$(NPK)_{45}$ – in autumn + $(NPK)_{45}$ – in spring	591.7	271.9	230.8
LSD_{05} per fertilizer type		27.9	14.3	15.7
LSD_{05} per fertilizer dose		24.1	12.1	12.9

The crop of seeds in the variant without fertilizers was 410.7 kg/ha. Under autumn application of nitrogen fertilizers (N_{30}–N_{90}) crop of seeds of festulolium of second year of life increased relatively to control by 16.1–30.6% and formed from 489.4 to 591.4 kg/ha. It was even bigger (by 19.9–35.5% relatively to control) in the variants with the application of nitroammophoska and formed from 513,4 to 620.5 kg/ha.

The applied fertilizers practically did not affect sowing quality of festulolium seeds. Laboratory germinating capacity varied from 93 to 95% and the mass of 1000 seeds formed 2.91–2.99 g.

The prime cost of 100 kg of festulolium seeds was the lowest (43000 rubles) and the profitability level of their production was the highest (179%) in the variant N_{60}. It was also rather high (145%) under application of $(NPK)_{60}$.

22.3.2 HERBICIDES IN FESTULOLIUM CROPS

The young growth of festulolium is considerably suppressed by vigorous weeds [19]. However, herbicides, allowed for the application in festulolium crops are still missing.

In this occasion the need arose to find the dosage of perspective herbicides against the weeds in festulolium crops. The scheme of experiment included the control variant (no application of herbicides), and the application of three herbicides in different concentrations in the first year of life of seed crop: Lontrel Grand (0.12; 0.125; 0.13 g/ha), Dikamba (0.1; 0.15; 0.2 L/ha), Avrorex (0.5; 0.55; 0.6 L/ha). The application of herbicides was performed once in the phase of bushing out.

The choking up of festulolium seed crops by weeds was determined in the first and second years of life. In the year of creation of festulolium seed crop more than 16 weed species occurred in the crop, belonging to different families: Chenopodiaceae, Poaceae, Asteraceae, Brassicaceae, Convolvulaceae, Geraniaceae, Euphorbiaceae. In different years the following weeds dominated: perennial – *Convolvulus arvensis* L., *Cirsium arvense* L., *Sonchus arvensis* L., *Agropyron repens* L., *Taraxacum officinale* Wigg., *Barbarea vulgaris* R. Br., *Euphorbia virgata* Waldst. & Kit., *Artemisia absinthium* L.; spring crops – *Chenopodium album* L., *Capsella bursa-pastoris* L., *Panicum crus galli* L., *Thlaspi arvense* L., *Galium aparine* L., *Setaria glauca* L.; annual and biennial wintering – *Matricaria inodora* L., *Erodium cicutarium* L.

It was found that in 2009 annual and biennial weeds represented 54.1% of weeds with the domination (25.9%) of wintering species (Table 22.4).

TABLE 22.4 Groups of Weeds in the Festulolium Herbage of First Year of Life, % (2009–2011)

Groups of weed species	2009	2010	2011
Annual and biennial, totally including:	54.1	60.3	54.4
spring early	7.7	8.6	5.1
spring early	20.5	29.4	25.1
Wintering	25.9	22.3	24.2
Perennial, totally including:	45.9	39.7	45.6
sprouting	36.2	30.6	34.5
rhizome	3.3	1.2	2.2
taproot	6.4	7.9	8.9

In 2010, annual and biennial weed species, mainly *Matricaria inodora* L., *Panicum crus galli* L., *Thlaspi arvense* L., *Capsella bursa-pastoris* L., dominated (60.3%). Among perennial species *Sonchus arvensis* L. and *Convolvulus arvensis* L. dominated. Their density varied during the vegetation season. The amount of annual and biennial weeds increased by 14.1%, whereas the amount of perennial weeds increased by only 4.1%. Low increase of weeds amount in 2010 is related with drought, especially in the 2nd and 3rd 10-days periods of May.

In the conditions of 2011–45.6% of weeds were perennial and 54.4% were annual and biennial species. Annual and biennial weeds were dominated by: *Thlaspi arvense* L., *Setaria glauca* L., *Matricaria inodora* L., and perennial weeds were dominated by *Convolvulus arvensis* L., *Cirsium arvense* L. and *Taraxacum officinale* Wigg. Our research showed that the application of Avrorex herbicide in the dose 0.5 L/ha already 30 days later provided the decrease of total amount of weeds by 56.2% (the amount of perennial species decreased by 57.6%). Effectiveness of this herbicide against annual and biennial weeds during the first period after treatment was not very high (61.2%) because of the presence of Gramineae weeds resistant to this herbicide in *festulolium* crops, namely *Panicum crus galli* L., *Setaria glauca* L. and *Agropyron repens* L. After 45 the density of annual and biennial weeds decreased from 75.4 to 30.8 plants/m^2, that is, by 59.2%, and at the end of vegetation the density of weeds in this variant reached 49 plants/m^2, or by 63.4% less than in control variant.

Avrorex in the dose of 0.55 L/ha decreased total weeds density down to 60.5 plants/m^2 after 30 days, and down to 36.1 plants/m^2 before wintering. Total density of weeds decreased after 30 days by 64.3%, and at the end of vegetation season by 73% relatively to control. The highest mortality (64.7%) was observed in perennial weeds after 30 days and before wintering it reached 70%. The mortality of annual and biennial weeds was slightly lower relatively to perennial, but however it was rather high, namely 64.1% after 30 days and 71.1% before wintering. Total weeds mortality reached 64.3–70.6%. The application of herbicide Avrorex in the dose of 0.6 L/ha provided the highest herbicide effect. Total decrease of amount of annual, biennial and perennial weeds in the crops reached 69.0–77.5%. Maximum herbicide effectivity against perennial species of weeds formed 75.3%.

Herbicide Dikamba in the dose of 0.1 L/ha 30 days after the application decreased the amount of all weed species by 50%. During the vegetation weeds mortality increased and reached 57.5% before wintering. Higher decrease was observed for perennial weeds – by 46.9% and 52.4% until the first inventory period and until the end of vegetation season, respectively. 30 days after the treatment of crops by Dikamba herbicide in the dose of 0.15 L/ha the density of annual and perennial weeds decreased by 66.8 and 61.1%, which was considerably more effective than the dose of 0.1 L/ha.

When increasing the dose of herbicide Dikamba to 0.2 L/ha total weeds mortality increased once more and reached 30 days after the application and at the end of vegetation season 68.9% and 74%, respectively. It is much more effective compared to the dose of 0.15 L/ha. The highest decrease of density was observed in annual and biennial weeds, namely by 72.2% and 77.1% 30 days after treatment and before wintering, respectively.

The increase of dose of Dikamba up to 0.2 L/ha led to further decrease of density of *Sonchus arvensis* L., *Taraxacum officinale* Wigg., *Cirsium arvense* L. At the end of vegetation season full mortality of *Convolvulus arvensis* L. was observed.

The mass of weeds decreases in parallel with their density. Herbicide Dikamba in the dose of 0.10 L/ha decreased the mass of remained weeds by 50.7% and 57.3% 30 and 45 days after treatment, respectively. The decrease of mass of annual and biennial weeds was the highest, namely by 58.7% after 30 days and by 60.3% after 45 days. The Dikamba dose of 0.15 L/ha was more effective than its dose of 0.10 L/ha, the decrease of mass of annual + biennial and perennial weeds formed before the wintering 65 and 67.8%, respectively. Dikamba under dose of 0.2 L/ha once more decreased the mass of weeds, namely by 71.3% until the first inventory after treatment.

Herbicide Lontrel Grandin the dose of 0.12 **kg**/ha decreased the amount of annual and biennial weeds in festulolium crop 30 days after treatment by 26%, the amount of perennial weeds decreased at the same time by 39.9%. This herbicide is highly effective against *Sonchus arvensis* L., *Taraxacum officinale* Wigg., *Convolvulus arvensis* L., *Cirsium arvense* L. In the dose of 0.125 **kg**/ha its effectiveness increased significantly. The total mortality of weeds 30 days after the treatment and at the end of vegetation season reached 38.1% and 47.1%, respectively. The increase of its dose up to 0.13 **kg**/ha once more increased its effectiveness.

The herbicide Avrorex showed to be the most effective in festulolium crops. The productivity structure of seed crop in the variants with application of this herbicide in the year of seeds harvesting considerably improved (Table 22.5).

TABLE 22.5 The Effect of Herbicides on the Structure of Festulolium Seed Crop and Yield of Seeds in the Second Year of Life (Mean for 2010–2011)

Herbicide	Dose, L/ha, kg/ha	Density, 1/m²		Ear length, cm	Number of seeds per ear	Seeds crop, kg/ha
		Weeds	Generative shoots			
Control	0.00	66	739	15.7	60.0	432.7
Avrorex (AI 21 g/L of carfentrazonetyl + 500 g/L of ether 2.4 D)	0.50	45	798	16.0	59.5	480.5
	0.55	36	810	16.1	57.0	496.4
	0.60	32	785	15.8	59.0	472.0
Dikamba (AI 480 g/L dikamba acid)	0.10	52	766	16.1	57.5	459.5
	0.15	44	774	15.8	60.0	464.8
	0.20	38	756	15.7	56.5	449.6
Lontrel Grand (AI klopyralid 750 g/kg)	0.120	29	741	15.9	59.5	433.3
	0.125	24	729	15.8	55.5	428.8
	0.130	21	716	15.4	56.5	419.5
LSD_{05}		1.4	27.1	0.14	3.3	19.5

Thus, the amount of generative shoots in the variant with Avrorex application formed in average for three years 785–810 per m^2, whereas in control it was 739 per m^2. This herbicide was characterized by broad spectrum of effect and successfully eliminated weeds presented in the crop when applying in the dose of 0.55 L/ha.

The amount of vegetating weeds decreased by 45%, which leaded to the highest yield of festulolium seeds, namely 496.4 kg/ha. Herbicide Dikamba is effective against weeds in the dose of 0.15 L/ha. It provided 67.2% mortality of weeds and reliably increased festulolium productivity, which reached 464.8 kg/ha.

The crop treatment by Lontrel Grand leaded to the noticeable depression of festulolium plants, especially in the first days and the measure of their suppression increased with increasing dose of the herbicide. However, in the subsequent period plants developed normally. Nevertheless, under the application of this herbicide in the doses of 0.125 and 0.13 kg/ha a certain trend of the festulolium productivity decrease relatively to control was observed, although it was not considerable.

Sowing qualities of festulolium seeds harvested from the crops treated by herbicides in the year of sowing did not differ considerably from control. The mass of 1000 seeds and the laboratory germination capacity varied in dependence on the variant of experiment from 2.98 to 3.05 g and from 90.5 to 94.5%, respectively.

Thus, in order to form highly productive festulolium seed crop it is necessary to apply herbicides in the year of sowing in order to eliminate dicotyledon weeds. The application of Avrorex in the dose of 0.55 L/ha or Diktamba in the dose of 0.15 L/ha showed to be the most effective. It provided the mortality of 69.5–73% of weeds and the yield of 464.8–496.4 kg/ha of high quality festulolium seeds.

22.3.3 DECREASE OF SEEDS SHEDDING

The harvesting of seed crops of almost all species of Gramineae grasses remains the most complicated technological operation within the cultivation process, because it always includes the element of risk related first of all with weather conditions during harvesting, uneven ripening of seeds and their small size [20].

The terms of harvesting of Gramineae grasses seed crops are set according to the biological peculiarities of the species and its inclination to shedding. Now when cultivating easily shedding perennial Gramineae grasses (pasturable and multimowing ryegrass, meadow fescue, etc.) it is recommended to perform the harvesting under seeds humidity of 40–45% (before the beginning of mass shedding of seeds from inflorescence). In this period the accumulation of dry substance in the seeds stabilizes and reach maximum [21–23].

The festulolium variety VIK-90 is characterized by high crop of green mass (74.3–80 t/ha), dry substance (14.0–15.5 t/ha), longevity, higher winter hardiness evenness of coming in of feed mass during vegetation period, rapid regeneration in spring and after pasture. However, it is characterized also by some disadvantage, presented in its parent forms. The most important of them are uneven ripening of seeds and their easy shedding in the preharvesting period [24, 25]. Any tardiness of harvesting always leads to the considerable loss of seeds (up to 40–55% of yield). This causes low real productivity of festulolium under cultivation for seeds and suppresses its broad spreading in the production conditions of Central Chernozem region [26].

In order to find the ways to decrease losses of festulolium seeds of variety VIK-90, we performed during the ripening period and crop harvesting in 2009–2011 the experiment, which consisted in the application of adhesive preparations, decreasing seed shedding: Elastic (0.8–1.2 L/ha) and Bifactor (0.8–1.2 L/ha).

The adhesive preparations were applied when festulolium weevils reached the stage of milky ripeness (seed humidity not less than 60–65%). Consumption of the solution reached 200 L/ha. After crop treatment the natural afflux of plastic substances from plant vegetative organs into seeds and gradual decrease of seed humidity continued.

After crop treatment a thin cellular film with diffusion effect was formed at festulolium ears, which covers the whole plant surface, but does not close stomata and does not affect gaze exchange, prevents seeds from shedding and allows to harvest them by direct combining under their moisture of 20–25% without high losses, at the same time the expenses for drying of seeds considerably decrease [27]. Most of the leaves at that time remained green.

As alternative to the application of Elastic and Bifactor preparations, we tested the Methylan Universal Premium glue in the dosage from 1.4 to 3.8 **kg**/ha. This preparation consists of modified ethers of starch with antifungal substances.

Seed losses from natural shedding were determined by means of seed harvesting in the special vessels with total surface of 1 m^2, placed in row-spacing of the grass.

Direct combining under seed moisture of 22–25% performed seed harvesting. The height of mowing was increased in order to exclude cutting of green leaves of festulolium plants.

Pre-harvesting treatment of festulolium crops by adhesive preparations allows to perform the harvesting under lower seed humidity and with minimum losses (Table 22.6).

TABLE 22.6 Productivity and the Measure of Shedding of Festulolium Seeds in Dependence on Concentration of Adhesive Preparations (Mean for 2009–2011)

Preparation	**Dose, kg/ha (l/ha)**	**Productivity, kg/ha**		**Rate of shedding**	
		Biological	**Real**	**kg/ha**	**%**
Water (control)		575.6	214.8	360.7	62.0
	1.4	584.3	284.1	300.1	51.0
	1.8	589.5	342.4	247.1	42.9
	2.2	581.3	382.3	199.0	35.4
Methylan Universal Premium	2.6	585.5	448.9	136.6	24.1

TABLE 22.6 *(Continued)*

Preparation	Dose, kg/ha (l/ha)	Productivity, kg/ha		Rate of shedding	
		Biological	Real	kg/ha	%
	3.0	578.7	495.2	83.5	14.8
	3.4	590.0	490.1	99.8	17.6
	3.8	593.4	453.9	139.5	24.1
	0.8	620.3	522.1	98.3	16.8
Elastic	1.0	627.2	543.9	83.4	14.0
	1.2	630.2	563.9	66.4	11.2
	0.8	615.6	524.8	90.8	15.4
Bifactor	1.0	617.4	547.4	70.0	11.9
	1.2	619.7	563.5	56.2	9.7
LSD_{05}		9.8	5.1	3.1	1.8

High effect of preparations under study on festulolium was observed. In the control variant real yield of seeds formed 214.8 kg/ha, 360.7 kg/ha was lost because of natural shedding during ripening process.

Elastic and Bifactor preparations decreased seed shedding and therefore increased real yield of seeds festulolium, which reached 522.1–563.5 kg/ha. Film-forming substances decreased harvesting losses from 62% (in control) down to 9.7–16.8%.

The application of Methylan Universal Premium glue in the dosage under study also provided significant increase of seed yield. The highest yield of seeds (490.1–495.2 kg/ha) was obtained under application of the glue in the dose of 3.0–3.4 kg/ha, which decreased seed shedding down to 14.8–17.6%.

Sowing qualities of festulolium seeds did not worsen because of the application of adhesive preparations. In the control variant laboratory germination capacity amounted 92.2%, whereas in the experimental variants it was 92.2–95.7%. The average mass of 1000 seeds amounted 2.88–2.92 g.

Financial costs of festulolium cultivation (control variant) amounted 29.3 thousand rubles per ha. In the application of adhesive preparations this value amounted 29.9–31.8–3 thousand rubles per ha. The prime cost of 100 kg of seeds in the variants with Elastic and Bifactor preparations was lower, namely 5.4–5.9 thousand rubles per ha and the profitability level was higher, namely 104–123%, in the control variant rate of return was only 12%.

Our study resulted in the patent of Russian Federation "Method of preharvesting processing of festulolium seed crops" (RU 2420050 C1) [28].

22.4 CONCLUSIONS

The following conclusions can be derived on the base of our study of winter hardiness, growth and seed productivity of festulolium in dependence on fertilizers, herbicides and adhesive preparations during seed harvesting in the conditions of forest-steppe of Central Chernozem Region, conducted in 2009–2011:

1. Agroclimatic conditions of forest-steppe of Central Chernozem Region are favorable for the production of festulolium seeds of the variety VIK-90. Plant wintering rate amounted 84.9–88.9%. Fertilizers, including nitrogen ones, improved winter hardiness of festulolium plants by 3.2–7.2%.
2. Festulolium is characterized by slow growth and development in the first year of life, no forming of generative shoots occurred. Nitrogen and complex fertilizers increase the duration of vegetation period in the second year of life by 3–8 days, especially in the period of bushing out and booting. Generally the period from spring regeneration to full ripening in the second year of life amounted 99–107 days, in the third and fourth years of life it continued 95–103 days.
3. Festulolium green mass is highly nutritious fodder of high quality, fully satisfying the needs of milky cattle. It contains in 1 kg of dry mass about 0.96 fodder units, 11.3–22% of fresh protein, 3.4–5.7% of fresh fat, 18.2–25.8% of fresh cellulose, 16.1–20.5% of soluble sugars, from 9.1 to 11.7 MJ of exchange energy, 2.4–2.9 g of phosphorus and 4.3–5.8 g of calcium.
4. Autumn application of ammonium nitrate (N_{60}) or nitroammophoska (NPK_{60}) provides the formation of higher yield of festulolium seeds in the second year of life, from 591.4 to 620.5 kg/ha. In the subsequent years seed productivity of festulolium decreases 2.3–2.7 times. The application of mineral nitrogen in the dose of 30 kg of AI before autumn and spring plant bushing out allowed increase the yield of seeds by 31.4–37.2%.
5. The application of herbicide Avrorex in the dose of 0.55 L/ha in festulolium crops provided elimination of 73% of weeds. Weed mass decreases by 67.3%. The yield of festulolium seeds exceeded control variant by 63.7 kg/ha, which provided the highest economic effectiveness (conditionally net profit amounted 29.2 thousand rubles per ha, prime cost of 100 kg of seeds did not exceed 6.1 thousand rubles, profitability reached 97%).
6. The application of adhesive preparations during preharvesting period allows to decrease losses from natural shedding of festulolium seeds by 49–90%. Higher real productivity (yield) of seeds was reached under treatment of crop by Bifactor glue; it amounted in average for three years from 524.8 to 563.5 kg/ha.

   Methylan Universal Premium glue also provided reliable increase of yield because of the decrease of losses from shedding. The yield of seeds of 490.1–495.2 kg/ha was reached in the variants with glue dosage of 3.0–3.4 kg/ha, when the rate of seeds shedding did not exceed 14.8–17.6%.

7. From the economic point of view the cultivation of festulolium with single autumn application of ammonium nitrate in the dose of 60 kg/ha of AI showed to be more expedient. In this variant the highest conditional net profit of 45.5 thousand rubles per ha under profitability level of 179% was reached. The preharvesting crop treatment by Bifactor preparation showed to be the most economically effective (prime cost of 100 kg of seeds was 5.5 thousand rubles, profitability of production amounted 118%). When applying Methylan Universal Premium glue the prime cost of seeds amounted 6.2–8.8 thousand rubles with profitability level of 14–93%.

KEYWORDS

- **fertilizers**
- **glue**
- **herbicides**
- **weeds**

REFERENCES

1. Kosolapov, V. M. (2012). Complex Comparative Evaluation of Chemical Composition and Productive Effect of Festulolium VIK-90, Adaptive Fodder Production, *3*, 26–28 (in Russian).
2. Kosolapov, V. M. (2010). Problems of Fodder Production and Ways of their Solution at the Actual Stage, Progress of Science and Technique in AIC (Agro-Industrial Complex), *11P*, 23–25 (in Russian).
3. Lazarev, N. N., Avdeyev, S. M., & Dyomina, Yu L. (2007). Accumulation of Gross Energy by Legumes-Gramineae Agrophytocenoses, Proceedings of TAA (Timiryazev Agricultural Academy), *279*, Part 1, 374–377 (in Russian).
4. Perepravo, N. I. (2007). Agro Ecological and Technological Aspects of Seed-Farming of Perennial Grasses, Proceedings of TAA (Timiryazev Agricultural Academy), *279*, Part 1, 331–334 (in Russian).
5. Shamsutdinov, Z. M. (2010). Progress and Development Strategy of the Selection of Fodder Cultures, Fodder Production, *8*, 25–27 (in Russian).
6. Perepravo, N. I., Ryabova, V. E., & Kulikov, Z. A. (2011). Agro-Biological Peculiarities of Seed-Farming of Intergeneric Hybrides of Festulolium (*Festulolium*), Perspectives of Development of Adaptive Fodder Production, Proceedings of Scientific and Practical Conference, Williams, V. R. All-Russian Research Institute of Fodders of Russian Agricultural Academy Moscow-Astana, 96–100 (in Russian).
7. Kutuzova, A. A., Zotov, A. A., & Kuleshov, G. G. (2000). Perspective Directions of Creation of Cultivated Pastures in Russia, Fodder Production, *8*, 12–15 (in Russian).
8. Mikhaylichenko, B. P., Perepravo, N. I., & Zolotaryov, V. N. (1999). Scientific Foundations of Zonal Seed-Farming of Perennial Grasses, Selection and Seed-Farming, *4*, 38–42 (in Russian).
9. Privalova, K. N., & Karimov, R. R. (2010). Effectiveness of Perspective Ryegrass and Festulolium Crops, Fodder Production, *8*, 21–24 (in Russian).

10. Obraztsov, V. N., Schedrina, D. I., & Kondratov, V. V. (2013). Seed Productivity of Festulolium in Dependence on the Cultivation Methods in the Forest-Steppe of Central Chernozem Region, Fodder Production, *7,* 28–30 (in Russian).
11. Perepravo, N. I., & Melnikova, T. E. (2009). Development of Fodder Cultures and Peculiarities of their Commercial Seed-Farming, Proceedings of TAA (Timiryazev Agricultural Academy), *281*, 73–76 (in Russian).
12. Methodical Directions on the Conducting of Research in Seed-Farming of Perennial Grasses, Moscow, VIK, Williams, V. R. (1986) All-Russian Research Institute of Fodders, 135p (in Russian).
13. Lazarev, N. N. (2007). Productive Longevity of Grasses on Hayfields and Pastures, Proceedings of TAA (Timiryazev Agricultural Academy), *279,* 353–356 (in Russian).
14. Zolotaryov, V. N. (1999). Nitrogen Fertilizers Increase Seed Productivity of Ryegrass, Agriculture, *1*, 25 (in Russian).
15. Perepravo, N. I. & Lebedeva, N. N. (2006). Effect of Nitrogen Fertilizers on Seed Productivity of Meadow Fescue New Varieties of Intensive Method of Use, Fodder Production, *8,* 18–20 (in Russian).
16. Bekuzarova, S. A. & Gasiyev, V. I. (2010). Effect of Fertilizers on Festulolium Productivity, Landscape Ecological Foundations of Development of Agriculture Systems in the Agro-Industrial Complex of Mountainous and Foot Hills Regions of Northern Caucasus Collection of Proceedings, Vladikavkaz, North Caucasus Research Institute of Mountain and Foot Hill Agriculture, 127–129 (in Russian).
17. Trukhan, O. V. & Perepravo, N. I. (2010). Effect of Nitrogen Fertilizers on Seed Productivity of Red Fescue of New Variety Sigma, Fodder Production, *7,* 31–35 (in Russian).
18. Obraztsov, V. N., Schedrina, D. I. & Kondratov, V. V. (2012). Effect of Mineral Fertilizers on Seed Productivity of Festulolium in Forest-Steppe of Central-Chernozem Region, Bulletin of Voronezh State Agrarian University, *35(4)*, 44–49 (in Russian).
19. Zolotaryov, V. N. (1991). Effectiveness of Chemical Weeding, Agriculture, *10*, 80 (in Russian).
20. Kulikov, Z. A. (2010). Determination of Optimal Terms of Harvesting of Seed Crops of New Fodder Culture Festulolium (*Festulolium* F. Aschers et Graebn), Proceedings of IX International Scientific Methodological Conference "Introduction of Non-Traditional and Rare Plants" Michurinsk–Naukograd, Michurin State Agrarian University, *1*, 109–112 (in Russian).
21. Mikhaylichenko, B. P., Ryabova, V. E. & Pshonkin, Yu M. (1994). Peculiarities of Pasture Ryegrass Cultivation for Seeds, Selection and Seed-Farming, *3,* 47–49.
22. Perepravo, N. I., Zolotaryov, V. N., Ryabova, V. E. et al. (2008). Cultivation of Perennial Grasses for Seeds in Central Chernozem Regions (Рекомендац××), Moscow Williams, V. R., All-Russian Research Institute of Fodders, 44p (in Russian).
23. Schedrina, D. I., Fedotov, V. A., Obraztsov, V. N. et al. (2009). Cultivation of Perennial Grasses for Seeds in Conditions of Voronezh Regions (Practical Recomendations), Voronezh, E. A., Bolkhovitinov Publishing, 5 p (in Russian).
24. Vasko, P. P., Kozlovskaya, Z. G., Stolepchenko, V. A., Olshevskaya, N. B., & Korolyok, V. V. (2010). Productivity of Different Morpho-Types of Festulolium under Pasturable use of Grass Crops, Agriculture and Defence of Plants, *4,* 18–20 (in Russian).
25. Perepravo, N. I., & Ryabova, V. E. (2003). Biological and Technical Peculiarities of Seed-Farming of Fodder Culture *Festulolium,* Proceedings of V International Symposium "New and Non-Traditional Plants and Perspectives of Their Use," Moscow, Russian University of International Friendship, *2,* 120–122 (in Russian).
26. Obraztsov, V. N., Schedrina, D. I., & Kondratov, V. V. (2013). Application of Film-Forming Reagents on Seed crops of Festulolium in the Forest-Steppe of Central Chernozem Region, Fodder Production, *4P*, 21–23 (in Russian).

27. Fedotov, V. A., Goncharov, S. V., & Savenkov, V. P. (2008). Raps of Russia, Moscow Agro-League of Russia, 336p (in Russian).
28. Obraztsov, V. N., Schedrina, D. I., Kadyrov, S. V. et al. (2011). Methods of Pre-harvesting Treatment of Festulolium Seed Crops, Patent № 2420050 of Russian Federation, Bullet in, *16*, 4p (in Russian).

CHAPTER 23

ELEMENTS CULTIVATION TECHNOLOGY OPTIMIZATION OF *MELILOTUS ALBUS* MEDIK. IN THE MIDDLE VOLGA REGION

VLADIMIR F. KAZARIN* and ALEXANDRA V. KAZARINA

The P.N. Konstantinov Volga Selective Breeding and Seed Production Research Institute, 76, Shosseynaya St., Samara region, Kinel, village Ust-Kinel, 446442, Russia; *E-mail: KazarinVF@mail.ru

CONTENTS

ABSTRACT

Productivity level of *Melilotus albus* Medik. in the Middle Volga Region nonirrigated conditions largely depends on the conditions of the natural moisture. Amount of water precipitation during the critical plants reproductive organs formation period is prerequisite for the seed yield. The results of annual white melilot agroassociation of drought resistance evaluation are adduced. Thus, annual white melilot common biological features as a crop with moderate requirements to moisture, its varieties can differ substantially in adaptability to drought. This culture has valuable next genetically determined adaptive morphological and physiological characteristics associated both with the ability to make fuller soil moisture reserves using by means of more vigorous root system and its ability to reduce evaporation through the leaves by their shape and direction of leaves arrangement, pubescent leaf blade and other signs differently expressed in some varieties. Thus, the accumulated scientific potential and best practices of *Melilotus albus* Medik. cultivation in nonirrigated chernozem steppe zone of the Middle Volga Region allows us to conclude that *Melilotus albus* Medik. is fairly drought resistant and reliable crop for its cultivation in the cereal field rotations in the region.

This paper proposes the set of cultural practices enable to get guaranteed seed yield at least 0.5t/ga under dry conditions. These include – accommodation clover crops after winter ones using after their harvesting the main improved plowed fields processing type; minimizing spring tillage both their number and the depth of treated layer, use for sowing crop varieties that differ in length of vegetation; maneuvering of clover sowing terms in the fields from early to late optimally within reasonable period; creation of fine lumpy soil surface by row treatments during crops maintenance; purity crops from weeds maintaining by means of efficient agricultural methods and herbicides.

Such agricomplex best accumulation focus, saving and most economical use of natural water resources – the basis of consistently high yields both green mass and *Melilotus albus* Medik. seed in nonirrigated conditions of the Middle Volga Region.

23.1 INTRODUCTION

In the Volga Region recurrent droughts in summer is an objective factor, which has a great influence for the stability of feed production. In this case, the stability of agriculture determined by the correct selection of drought-resistant crops and varieties capable for stable high yields forming under any weather conditions.

The annual white melilot (*Melilotus albus* Medik.) is one of the valuable forage, honey and vegetable phytomeliorative objects. It is characterized by high ecological plasticity, resistance to stressful environmental factors, the ability to economically and efficiently agroclimatic resources use. Nutritionally forage close to clover and alfalfa. In practice, it proved promising use for hay, silage, green forage as pasture plants. Complex of economically important features of *Melilotus albus* Medik. suggests the possibility of its widespread use in agricultural production as for feed purposes, and as green manure crops. Therefore, improving the methods of its cultivation for feed purposes and seeds is an important condition for its introduction into the Middle Volga steppe forage. This determined the feasibility and relevance of our research theme choice [1].

23.2 MATERIALS AND METHODOLOGY

Experiments were laid in seed selection breeding and seed production rotation of forage crops laboratory the P.N. Konstantinov Volga Selective Breeding and Seed Production Research Institute in 2005–2013. Fourfold repetition, plots area – 100 m^2. Soil experimental plot shows a typical solum chernozem soil, medium heavy loam texture. Hydrolyzable nitrogen content in the topsoil 11.6–13.2 mg; mobile phosphorus – 15.8–19.5 mg and potassium – 14.5–20.1 mg per 100 g soil. The object of study were annual white melilot varieties Kinelsky, Volzhskiy, Srednevolzhskiy.

Field experiments were accompanied by the necessary supervision, accounting and analysis, which were performed in accordance with generally accepted methodological guidelines [2–4]. Weather conditions during the studies differed sharply, allowing for more complete assessment of the biological properties of selective breeding material both in favorable humidification and in dry and high dry conditions (Table 23.1).

TABLE 23.1 Hydrothermal Coefficient by Months in the *Melilotus albus* Medik. Growing Season (2005–2013 Years)

Years	May	June	July	August	September	Average during the growing season
2005	0.4	0.6	0.4	0.1	1.7	0.6
2006	0.8	0.8	1.2	1.9	1.1	1.2
2007	0.3	1.4	1.4	0.1	1.1	0.9
2008	0.6	1.4	1.0	0.1	1.3	0.9
2009	0.4	0.3	0.7	1.0	0.0	0.5
2010	0.6	0.1	0.1	0.0	0.2	0.2
2011	0.9	1.9	0.1	0.9	2.1	1.2
2012	0.1	0.9	0.3	0.7	0.9	0.4
2013	0.4	0.2	0.5	1.7	2.0	1.0

23.3 RESULTS AND DISCUSSION

Productivity level of *Melilotus albus* Medik. in the Middle Volga Region nonirrigated conditions largely depends on the conditions of the natural moisture. Amount of water precipitation during the critical plants reproductive organs formation period (July) is prerequisite for the seed yield.

Such dependence the *Melilotus albus* Medik. yield on summer precipitation confirmed by many years of this crop cultivation experience in Samara Region. If in adequate moisture years (2006–2008, 2011, 2013) the green mass years of two mowings was 31.6–38.4 t/ga, in dry years (2005, 2009, 2010, 2013), when summer months rainfall fell significantly less long-term average value was only 8.9–17.7 t/ga.

Particularly unfavorable yield formation conditions were formed in high dry 2010, when throughout the growing period had practically no rain, and there was dry hot weather, hydrothermal coefficient (HC) was 0.2. In such extreme conditions of high temperature (35–40°C) and low relative humidity (30–45°C) with soil moisture reserves depletion, sharp productivity decline *Melilotus albus* Medik. agrocenose both green mass (7.0–10.7 t/ga) and seeds (89.1–123.3 kg/ga).

Correlation analysis confirms the close relationship with the magnitude of *Melilotus albus* Medik. productivity hydrothermal coefficient, so seed yield depending on the type r = 0.775–0.897, dry matter yield correlation coefficient was 0.956–0.990.

The main condition for drought-resistant *Melilotus albus* Medik. agrocenose formation is focusing of crop cultivation entire complex for rational use of natural moisture resources. Soil moisture reserves accumulation adjustment for crop rotation is multi-important. By means of root consumption annual white melilot spends moisture in Middle Volga Region loam chernozem steppe from 1.50 m depth. Because in this soil layer moisture reserves, usually consumed during the preceding crop vegetation period, are replenished by autumn-winter – spring precipitation *Melilotus albus* Medik. can be placed in the rotation after deep root crops. But intense moisture accumulation was after winter and early spring cereals, which are considered the best precursors for *Melilotus albus* Medik.

Water consumption in this crop depending on the biological characteristics of cultivars, agrophysical condition of rooting subsoil, prevailing weather conditions ranges from 2890 to 3151 m³. Three years average research data the least amount of water consumption per unit of crop was obtained from ordinary row annual white melilot with seeding rate 5.0–4.0 million units/ha (Table 23.2). Maximum amount of moisture were needed for wide row (0.45 m) crops with seeding rate 1.0 million units/ha.

Regarding the impact of *Melilotus albus* Medik. sowing architectonics for its moisture-saving properties, the definite relationship between the physical and biological water consumption is traced. In wide row crop sowing more moisture is lost for evaporation from the soil surface between the closing rows; in unbroken crops sowing the soil surface faster shaded by crop canopy and crop transpiration extra-predominates over physical loss of moisture. But because of more condensed the soil moisture reserves are rapidly depleted and plants often have a lack of it during the seeds formation.

TABLE 23.2 Effect of Povolzhskiy *Melilotus albus* Medik. Sowing Methods and Seeding Rates for the Total Water Consumption in 0–1.5 m Layer (the Average for 2007–2009 years)

Between row space, m	Seeding mln pieces/ ha	Moisture resources in soil, mm		Moisture using from soil, mm	Water falls*, mm	Total water consumption, mm	Green mass yield, t/ ha	Water consumption for 1t	
		Sowing period	Harvest period					Green mass, M^3	Dry matter, M^3
0.15	2.0	228.1	121.5	106.6	185.2	291.8	25.2	115.8	463.2
	3.0	228.1	101.5	126.6	185.2	311.8	30.5	102.2	408.7
	4.0	228.1	98.2	129.9	185.2	315.1	30.8	102.3	409.2
0.45	1.0	228.1	124.3	103.8	185.2	289.0	9.4	307.5	1024.8
	1.5	228.1	109.0	119.1	185.2	304.3	15.2	200.2	770.4
	2.0	228.1	100.2	120.3	185.2	305.5	21.0	145.5	559.5

*Falls using precipitation – 0.7.

Improve drought *Melilotus albus* Medik. agrocenose can also by sowing dates controlling. Meaning the highly uneven and unpredictable rainfalls for reducing the risk of drought negative impact for *Melilotus albus* Medik. seeds formation it is advisable to sow the early maturing varieties in the early period when the optimal seed layer warming up to 12–14 °C (mid-April), and middle-maturing varieties to mid-May, when soil optimum level warming – 18–20 °C. But the most reliable for *Melilotus albus* Medik. seeds production stabilization to use these two terms for each group of precocity. In this case the preconditions for reduce of stress-arid conditions coincidence probability with water consumption critical period for crop seeds formation. However, at later sowing stages there is danger of non amicable shoots due to of the upper soil layer drying, which leads to sow after rain when the soil has physical maturity. But should not neglect the late dates of sowing as *Melilotus albus* Medik. has biological feature to accelerate development in delay sowing without any noticeable reduction in yield. In our experiments, middle- maturing variety Povolzhskyi with late sowing in the third decade of May gave the same yield as in the early – in April end and is guaranteed matured in mid-September. This confirms the fact that *Melilotus albus* Medik. quite adaptive crop to the sowing time and optimum period can last around 30 days.

The variety role for weather conditions resistance adverse is well known. With total *Melilotus albus* Medik. biological features as a crop with moderate requirements to moisture, its varieties can differ substantially in adaptability to drought.

This crop has number genetically determined adaptive morphological and physiological features associated both with the ability to make fuller use of soil moisture reserves by means of more vigorous root system formation and its ability to reduce evaporation through the leaves by their shape and direction of leaves arrangement, pubescent leaf blade and differently expressed in some varieties [5].

In dry 2010 new early maturing variety Srednevolzhskiy showed high drought resistance on 3.7–4.3 t/ha, which exceeded middle- and late-maturing varieties by the yield of green mass (Table 23.3). On average, during 6 years studying, new variety exceeded the standard both by harvesting from dry matter area unit, digestible protein, metabolizable energy and seed yield.

TABLE 23.3 Efficiency of *Melilotus albus* Medik. Different Ripening Varieties (2005–2010 years)

Variety (rape group), days	Data name	Years						Overage 2005–2010 years
		2005	2006	2007	2008	2009	2010	
Kinelskiy (late) 120–130	green mass, t/ha	14.4	26.4	30.8	28.3	10.9	6.4	18.0
	dry matterm, t/ha	3.9	7.1	8.0	7.5	3.0	1.8	5.3
	digestible protein, t/ha	0.4	0.7	0.8	0.8	0.3	0.2	0.5
	energy, Gj/ha	46.8	85.2	96.0	90.0	36.0	21.6	62.6
	seed yield, t/h	0.29	0.44	0.51	0.50	0.27	0.08	0.35
Povolzhskiy (mid) 110–119	green mass, t/ha	16.4	32.3	34.6	28.1	10.8	7.0	21.5
	dry matter, t/ha	4.2	8.7	9.0	7.3	2.9	2.0	5.7
	digestible protein, t/ha	0.4	0.9	0.9	0.8	0.3	0.2	0.6
	energy, Gj/ha	50.4	87.0	108.0	90.0	34.8	24.0	65.7
	seed yield, t/h	0.35	0.45	0.56	0.65	0.26	0.09	0.39
Srednevolzhskiy (early) 110–119	green mass, t/ha	19.0	44.8	42.2	37.3	12.9	10.7	27.8
	dry matter, t/ha	4.9	11.7	11.0	9.7	3.5	2.9	7.3
	digestible protein, t/ha	0.5	1.2	1.1	1.0	0.3	0.3	0.7
	energy, Gj/ha	58.8	117.0	110.0	97.0	42.0	34.8	76.6
	seed yield, t/h	0.37	0.46	0.67	0.68	0.29	0.12	0.43

Because the weather conditions in the Middle Volga Region differ sharply years by years and was not significantly predictable for guaranteed stable *Melilotus albus* Medik. yields, especially seeds, it is advisable in every farm cultivate different varieties by vegetation. In years with sufficient precipitation in early summer early maturing varieties have advantages, achievers use the resources of soil moisture and precipitation in the first half of the summer for reproductive organs formation. In years when more rain falls are during the second half of the summer (July, August) middle maturing varieties are more productive, because critical period of water consumption (blooming – bud forming and seeds pouring) in such cases coincides with heavy rains.

It is known that presowing seed treatment by physiologically active substances in the plant organism causes the activation of metabolic processes that can protect it from the stressful effects of environmental factors and pathogens, and it is important for optimizing the crop formation conditions.

In 2011–2013, research of physiologically active agents influence for the *Melilotus albus* Medik. productivity was conducted.

According to our research all studied drugs have varying degrees positive impact for productivity and biomass of seeds (Table 23.4). The highest yield of green mass and dry matter yield of 3-year study was observed in variants with treatment by Aminokat 30. High rates were also obtained by Appin.

TABLE 23.4 Effect of Physiologically Active Preparations for *Melilotus albus* Medik. Productivity (t/ha) (2011–2013 years)

Variant	Dining		Flowering		Seeds	Aftermath from control
	Basic mowing + otava	Dry matter	Basic mowing + otava	Dry matter		
Control	18.7	4.9	19.6	5.9	0.61	-
Aminokat 30	22.3	5.8	22.9	6.9	0.84	0.23
LFSC-2	20.8	5.4	21.0	6.2	0.80	0.19
Appin	20.8	5.4	24.0	7.2	0.41	0.10
Albite	20.1	5.7	20.4	6.1	0.81	0.20
Zircon	19.9	5.6	22.9	6.8	0.77	0.16
LSD05	0.43	0.17	0.59	0.25	0.05	

Note: LFSC – liquid fertilizing-stimulating composition, LSD – the least significant difference.

Aminokat 30 and the Albite influenced mostly for seed production, seed yield increase, relative to the control, was 32.8–37.7%.

For better moisture accumulation in the soil and its losses reduce by physical evaporation it is important to optimize the structure of the arable layer by processing techniques. If for soil permeability increasing its structural looseness is paramountly important then to reduce evaporation loss is tighter addition [6]. This implies the important role of deep plowing or subsoiling and autumn, winter and melt water absorption and minimizing the spring and summer treatments to reduce the moisture loss from the upper layer of the soil. In the steppe regions of the Middle Volga for *Melilotus albus* Medik. after grain predecessors the best is improved plowing, which achieves by surface stubbing create top layer of mulch that protects the soil from moisture loss and promotes better absorption of precipitation in August and September. Followed plowing conditions for the cold period accumulation of sediments are made, when the evaporation process is slowed down or not happening, as the moisture is absorbed well in the loosened layer. To retain moisture in the soil in autumn positive role played by surface mulching field by chopped straw, salting by combine during harvesting by device PUN-5 (the universal device attached). This allows to accumulate moisture more than 300–350 m³/ha, than by stubble background.

Spring minimizing soil cultivating is preferred both as from moisture resources saving and more covered weed reducing. Early spring "moisture covering" by harrowing device, adaptable for poor structure soils, where in spring intensive moisture loosing by means of its drops current up to soil surface is not always adapted for Middle Volga structure chernozem where natural combine plowing field create moisture protecting mulcha level, saving water loosing from deeper layers.

In spring on fall aligned plowing on the field and if wintering weeds is absent *Melilotus albus* Medik. needs only one precultivating on seed depth of 3–5 cm. To reduce the loss of moisture from the soil and retain moisture in the seed layer if needed 2–3 spring cultivating (for early weeds sprouting, herbicides termination) should not cultivate more deeply than parameters to retain moisture to swell and produce amicable seed germination. These requirements apply also to between rows: they must also be cultivated on shallow depth (up to 5–6 cm) in order not to lose moisture from the soil by evaporation of the loosened layer.

An important factor of *Melilotus albus* Medik. agrocenosise stability to drought is keeping them weed-free. Weeds, especially perennial and tall annual (*Sonchus arvensis* L., *Abutilon theophrasti* L., *Echinochloa crus-galli* L., *Xantrium strumarium* L.), developing strong root system deplete soil moisture reserves and they are malicious competitors of crop for this life factor. It was noted that special harmfulness of weeds is evident in dry years because of competition for scarce moisture. With high weed degree, according to our observations, in dry years the *Melilotus albus* Medik. yield reduced in 2 times. Therefore, the full suppression of weeds by mechanical effective methods and chemical means is the main condition for agrocenose drought tolerance and *Melilotus albus* Medik. yield improving. Thus, the accumulated scientific potential and best practices of *Melilotus albus* Medik. cultivation in nonirrigated chernozem steppe zone of Middle Volga Region allows us to conclude that *Melilotus albus* Medik. is fairly drought resistant and reliable crop for its cultivation in the field of cereal rotations if this region.

23.4 CONCLUSIONS

To improve *Melilotus albus* Medik. agrocenose drought tolerance to guarantee seed yield receiving at least 0.5 t/ha even in dry years can by whole simple of agronomic technologies:

1. *Melilotus albus* Medik. should be placed after winter crops using main fields processing plowed by improved type after grain crops harvesting;
2. Minimization the spring tillage both by their number and depth of cultivated layer;
3. Use for sowing differ in length of vegetation varieties;
4. Monitoring and changing the *Melilotus albus* Medik. sowing terms in the fields from early to late within reasonable optimal period (from mid-April to late May);
5. The creation of fine lumpy soil surface by row treatments during crops growing;
6. Maintaining purity crops from weeds by effective agromethods and herbicides.

Such agricomplex best accumulation focus, saving and most economical use of natural water resources – the basis of consistently high yields both green mass and *Melilotus albus* Medik. seed in nonirrigated conditions of the Middle Volga Region.

KEYWORDS

- **chernozem**
- **drought resistance**
- **melilot**
- **water consumption**

REFERENCES

1. Glukhovtsev, V. V. (2005). Introduction of Non-Traditional Plants in the Forest of the Middle Volga, Gluhovtsev, V. V. & Kazarin,V. F., Agricultural Science, *4,* 13–14.
2. Glukhovtsev, V. V. (2006). Practicum on the Basics of Research in Agronomy, Glukhovtsev, V. V., Kirichenko, V. G., & Zudilin, S. N., Moscow Kolos, 248p.
3. Dospehov, B. A. (1985) Methodology of Field Skills, Moscow, Agropromizdat, 351p.
4. Crops State Strain Testing Methodology (1971). M, *1*, 225p
5. Kazarin, V. F. (2003). Evaluation of *Melilotus Albus* Medik Accessions of Different Origins in the South of the Middle Volga Steppe, Kazarin, V. F., Agliullina, L. K., & Kazarina, A. V. Collected Materials International Scientific Practical Conference, "Improving the Sustainability of Bio-Resources on the Basis of Adaptive-landscape" Part 2 Orenburg, 68–71.

6. Burov, D. I. (1968). Tillaging as a Factor of Improving the Structural Characteristics and the Structures of the Arable Layers of Zavolga Chernozem, Theoretical Tillage Problems, Leningrad, 19–24.

PART VII

ADAPTIVE REACTIONS OF CULTURAL PLANTS TO STRESS FACTORS

CHAPTER 24

ECOLOGICAL AND GENETIC POTENTIAL OF SOFT SPRING WHEAT IN CONDITIONS OF THE NORTHERN ZAURALYE

NINA A. BOME[1], ALEXANDR Y. BOME[2], and ELENA I. RIPBERGER[1]

[1]Tyumen State University Ministry of Education of Russia, d.10, Semakova St., Tyumen, 625003, Russia, E-mail: bomena@mail.ru

[2]Tyumen basing point of the N.I. Vavilov All-Russia Research Institute of Plant Growing, d. 42-44, Bolshaya Morskaya St., Petersburg, 190000, Russia.

CONTENTS

ABSTRACT

Presented the results several years of research on the selection and generation of source material, to maintain genetic diversity of soft spring wheat to soil and climatic conditions of Northern Zauralye. Experiments carried out under natural field conditions and climatic chamber with the given parameters, allow selection of forms of plants with improved traits. The possibility of applying the methods of hybridization and experimental mutagenesis to expand genetic diversity, increase the scope of variability of quantitative traits and the detection rate forms with increased productivity. Reaction of spring wheat to stress factors (salinity, low temperatures, exposure to pathogenic fungi, etc.), are highlighted in the samples with high ecological plasticity based on laboratory rapid diagnosis in early ontogeny and methods of field assessment.

24.1 INTRODUCTION

Plant genetic resources are the basis for the ecological and qualitative improvement of the environment; ensure food security and bio resource. In recent years, increasingly appears disturbing information about increasing genetic erosion, which can reach 20–35% of the known plant diversity [1].

According to projections, by the middle of the twenty-first century could be lost to 60% of plant species composition [2]. It is known that the loss of components of plant resources is seen as one of the factors of the ecological crisis of the biosphere (the UN Convention on Biological Diversity, 1992, ratified by the Russian Federation in 1995) [3].

The share of cultivated plants in the world does not exceed 3% of the total number of higher plant species of the world flora, but they are characterized by a huge genetic diversity [4]. Plant genetic diversity increases the possibility of choice and provides protection against future adverse conditions such as extreme and volatile environment.

In accordance with the UN Convention on Biological Diversity [3], or being developed by various countries assessment methodology components of genetic resources, based on the use of complex databases. Along with quite extensive information resources (soil, water, vegetation, climatic and agro-climatic zones) more research is needed. For example in relation to ecosystem functions of agricultural biodiversity for understanding of the causes and consequences of their loss is limited.

Climate change is manifested in the increase of the average surface air temperature at 0.7–°C on all continents in the last quarter of the XX century [5]; indicate the need for adjustments in strategy selection of species and varieties of cultivated plants. It should be noted that in Russia during the same time warmer to 1.5°C. Most intensively, this process is almost the entire European part of Russia, in the south of Western Siberia, Baikal, Trans-Baikal region and the north-eastern Yakutia Republic 7 [6].

Our analyzes of statistical data hydrometeorology center of Tyumen on Environmental Monitoring for the period 1993–2012 showed that moisture content of 20 analyzed vegetation periods only 4 (20%) were close to historical averages. Moisture deficit was observed for 8 (40%) field seasons, with the amount of precipitation ranged from 163 mm to 276 mm at a rate of 281 mm (minimum rainfall in 1997 and 2012). Win wet years and was 40% by varying the amount of precipitation data within 322–423 mm.

In terms of average daily air temperature of 9 years (45%) above the norm, and only 2 were cool. Growing seasons can be divided into 3 groups: close to a rate of heat and moisture (1993, 1996, 2001); cool and wet (1999 and 2002), dry and hot (1998, 2003, 2005, 2010, 2011, and 2012).

In this regard, development of scientific bases conservation, enhancement and use of biological diversity of wheat, forming collections for specific soil and climatic conditions is of paramount importance.

Our research focuses on the formation of theoretical knowledge and practical approaches for assessing the impact of a modern agricultural land use on the structural and functional state of the genetic resources of soft spring wheat, taking into account climatic and soil factors south of the Tyumen region.

Tyumen region on the hydrological and climatic conditions can be divided into three main zones: very excessive moisture and lack of heat supply (southern boundary of the zone extends between the Khanty-Mansi and Tobolsk), excessive moisture and lack of heat supply (southern boundary extends beyond Tobolsk), optimum moisture and heat fluxes (southern boundary runs along the line – Ishim, Yalutorovsk, Omsk) [7].

All land funds for use is divided into two parts: the northern and southern. In the northern part of the district includes two national as Khanty-Mansi and Yamal-Nenets, where spring wheat is not currently cultivated. On the southern part of soil and climatic conditions are divided into four zones: taiga and forest subtaiga on the north and south. The main areas of focus for spring wheat in the forest steppe zone and occupy about 30% of the southern region, where the most favorable conditions for the cultivation of food grains. The soils in the northern and southern are forest-steppe humus, gray forest, meadow and meadow-humus.

In general bioclimatic potential of the southern Tyumen region is sufficient for growing a number of crops, including major cereals – wheat, oats and barley. At the same time the stress conditions (drought, excess moisture, harsh winters, and reduced high temperature, etc.) can occur both within one or a few years. All this shows the importance of selecting varieties can favorably interact with the environment and ensure high productivity.

24.2 MATERIALS AND METHODS

The objects of our study were samples of soft spring wheat of different ecological and geographical origin from the world collection the All-Russian Research Institute of Plant Industry by N.I. Vavilov, other research institutions, as well as hybrid and we obtained mutant forms.

Field studies were conducted in two agro-climatic zones of the Tyumen region (northern forest and subtaiga), were the climate is continental with sharp fluctuations in air temperature, moderately moist, hydrothermal coefficient = 1.2–1.3. The sum of daily air temperatures, exceeding 10°C for the studied period was 1700–900°C, lengths of period was 141–124 days. Shortfall during the year rainfall (310–470 mm) is unevenly distributed (70–80% of them are in a warm period), the rainiest month on the average long-term data is July. At the beginning of the growing season plants suffer from lack of moisture and excess of it in the second half of the period leads to the lodging of the wheat plants. There have been quite frequent droughts of mild to moderate intensity [8].

Experiments were carried out according to the Methodological guidance on the study of the world collection of wheat [9], the morphological features and biological properties described on the International Classifier genus *Triticum L.* [10]. Additional field and laboratory studies were carried out on the relevant procedures and guests.

For the identification of pathogenic fungi – pathogens used methods of microscopy and wet chambers (Axiostar Plus "Karl Zeiss," Germany), laminar flow hood, NU-425–400E ("Binder," Germany).

In a laboratory's experiment, the determination of resistance to pathogens *Fusarium* blight methods used: seeds phytopathological analysis with the calculation of the index of the disease [11–13] and of evaluation of plant resistance to root rot [14, 15].

Salt tolerance of spring wheat evaluated on seed germination and seedling characteristics [16]. The substrate was taken calcined sand, moistened to 60% field capacity, which was filled with an inert material of the cell. For 50 seeds collected each sample, replicates were fourfold. NaCl solutions to the level of the osmotic pressure of 7 atm were added.

To assess the stability of the samples to a low temperature experienced embodiment seeds are germinated in a refrigeration chamber at a temperature of 0°C, the control embodiment was under standard conditions in an incubator at a temperature of 22°C, four replications.

Production of hybrid seeds was performed by methods Lukyanenko [17] and Dorofeyev et al. [18]. To clearly identify the genetic parameters were selected varieties, differing in economic-biological characteristics, the combination of which would make it possible to get a varied selection material.

Appropriateness of the use of diallel crosses schemes was shown first at the 28 hybrid combinations (8 initial varieties), followed by 10 combinations (5 initial varieties). Results intraspecific crosses conducted in years with large time intervals (1973 and 2009) and a manifestation of unequal meteorological factors indicate that conditions are favorable for obtaining a sufficient amount of high-varieties, viable seeds of spring wheat. The hybrid grains set grain on average in 1973 on 28 combinations was 15.4%, in 2009 on 10 combinations – 18.7%.

Comparative evaluation of hybrid and parental forms in the field was performed at sowing units with an area of power plants each 10×20 cm in the first stage (F1) and families at sowing (seeds per a plant) – at later stages (F2, F3, etc.).

To expand the genetic diversity of varieties and hybrids of spring wheat studied the effectiveness of the method of experimental mutagenesis. Seeds of spring and winter wheat were treated with gamma- rays (three doses) and chemical mutagen phosphemid (2 concentrations). To determine the patterns of mutation process used to study the traditional scheme rules plant responses to mutagens in terms of sensitivity in M1 and M2 in mutability.

To improve the efficiency of research, seed production and rapid evaluation of breeding material in a controlled environment setting of experiments carried out in a climatic chamber MLR- 351.

Analysis of soft spring wheat varieties that have undergone the complex breeding and valuable features on seven state testing plots of the Tyumen region: Tobol, Ishimsk, Yalutorov, Aromashev, Omutin, Nizne-Tavdin, Berdyuzh in this state conducted based on the reports of the State Commission of the Russian Federation for Testing and Protection of Selection Achievements for the period from 1980 to 2005.

24.3 RESULTS AND DISCUSSION

Any variety is not only a means to improve productivity, but also is a factor, without which it is impossible to realize the achievements of science and technology. In agricultural production a variety acts as the biological system that cannot be replaced by anything [19]. In Russia, all varieties tested state strain testing, which shall be determined on the basis of the territorial boundaries of their cultivation. Overall management responsibility of the State Commission for Variety Testing crops under the Ministry of Agriculture of the Russian Federation. In use come best varieties and hybrids best adapted to local conditions.

Analysis of the 357 varieties of soft spring wheat grown on 7 state testing plots located in different soil-climatic zones of the Tyumen region, revealed significant differences between varieties for breeding and valuable attributes. The varieties tested long period of time is shown in (Table 24.1).

For the duration of the growing season varieties studied in accordance with the International Classification of the Council for Mutual Economic Assistance (CMEA) genus *Triticum L.* divided into 4 groups: early maturing, middle, medium, late-ripening. Largest middle and late-ripening varieties prevailed. In conditions of low mean temperature and excessive moisture, an increase in terms of ripening grain.

TABLE 24.1 Analysis of Some Varieties of Spring Wheat Breeding and the Manifestation of Symptoms in Different Environmental Conditions (According to the Average Testing Stations of Tyumen Region)

Variety	Growing season, days			Height of one plant, cm			Resistance to lodging, score			Yields, t/ha			Weight of 1000 grain, g		
	1*	2**	3***	1	2	3	1	2	3	1	2	3	1	2	3
Ilinskaya (n=8)	82.7	83.7	98.8	92.8	93.0	104.2	4,0	4.6	3.5	3.1	3.5	3.3	39.6	41.5	42.5
Kazahstanskaya 10 (n=8)	85.8	81.8	101.8	86.2	87.2	96.6	4.5	4.9	4.3	3.2	3.4	3.5	40.6	41.1	44.3
Lutescens 70 (n=10)	80.3	80.5	94.4	78.9	77.9	93.1	4.4	4.7	4.2	2.9	2.7	3.1	35.4	35.3	38.7
Novosibirskaya 67 (n=13)	84.9	78.5	85.5	84.4	86.0	94.7	4.2	4.6	4.2	3.4	2.9	3.9	34.6	32.9	32.9
Omskaya 12 (n=11)	79.3	87.0	87.0	64.9	72.0	73.6	4.5	5.0	4.5	3.1	1.9	1.9	33.9	25.2	25.2
Omskaya 20 (n=9)	83.7	82.4	98.5	87.5	88.6	101.5	4.3	4.9	4.0	3.1	3.1	3.6	36.3	36.3	38.1
Rang (n=10)	88.8	86.0	86.0	76.1	83.5	83.5	4.8	4.8	4.8	3.5	1.9	1.9	35.5	30.2	30.2
Skala (n=12)	79.0	74.5	83.0	71.6	61.5	83.0	4.6	4.2	4.5	2.5	1.5	2.3	33.6	31.6	33.2
Tulunskaya 12 (n=9)	76.4	78.1	88.7	79.6	81.5	93.2	4.8	4.8	4.8	2.5	2.4	3.0	32.0	32.1	34.3
Tuymenskaya ranniaya (n=9)	77.8	75.0	86.0	71.4	79.0	85.0	4.5	4.5	2.0	2.4	2.4	2.9	32.2	35.9	34.2
Tuymenskaya 80 (n=14)	80.6	80.4	95.2	83.7	83.2	97.8	4.5	4.7	4.5	3.0	2.9	3.0	38.2	38.7	41.0

* 1, the average value of the characteristic for the entire study period; ** 2 – average value of the characteristic in dry years; *** 3 – average value of the flag during the setback; n – the number of years of testing varieties.

The particular qualities of spring wheat were manifested in the variability of plant height in different growing conditions. Plant height at 82% of the studied cultivars ranged from 66 to 95 cm of plant resistance to lodging ranged from low (up 2.9 points) to high (4–5). Win varieties resistant to lodging was 70% of the total studied.

Yields 277 varieties (80%) of the 357 studied was 3–4 t/ha. High-yield group (more than 5 t/ha) included 22 varieties, or 6% of the total. Dominated varieties with 1000 grain weight 31–38 g (54%) and 39–46 g (33%).

Defeat loose smut was observed in 170 varieties (degree of damage 0.02–1%), 10 varieties – 1–2%, 5 varieties – 2–3%, variety 1–4%, 2 varieties (Krasa 2 and the Turinskaya) – 4.6–4.9%. The lowest resistance to loose smut characterized variety Tarasovskaya 2, the degree damage that was 7.2%.

On the basis of a comprehensive assessment during the study period was zoned for soil-climatic zones of the Tyumen region 65 varieties, including 3 varieties – ripening, 17 – mid, 33 – middle, 12 – mid-late-ripening varieties.

During the growing season crops, especially in the temperate and high latitudes are often affected by a number of adverse influences. To survive in the rapidly changing environmental conditions during plant evolution has developed nonspecific complex responses.

In a series of experiments showed that the selection of spring wheat varieties with high ecological plasticity effective laboratory methods to conduct a rapid assessment of plant resistance to adverse factors (reduced temperature, salinity, pathogens infestation) with further study in the field.

In cold conditions and prolonged summers, due to low soil temperatures may be the loss of the ability of seeds to germinate or further destruction of plants. Seed germination of spring wheat in the laboratory at low temperature growth was retarded both roots and shoots, indicating that the perception of external influences plant organism as a whole, rather than the local response of the individual body. There was a sharp decline in seed vigor, which averages to 19 varieties studied was 6.6% [20]. Similar results were obtained in studies of Reimers and Illy [21].

In dry conditions (e.g., 1998) was shown the advantage of large grains. Average field germination of the samples with a mass of 1000 grains above 40 g was 59.0%, and samples with small grain only 48.8%. The relationship between grain size and field of seed germination was significant, as evidenced by the high correlation coefficient ($r = 0.70$).

The study of the response of spring wheat hybrids obtained by crossing on diallel scheme and initial varieties on the action of environmental factors in terms of germination and survival of plants. In the first generation (F_1, 2010), the group of hybrids (20%) was characterized by low germination of seeds that we associate with a smaller margin of nutrients in the small seeds. The inhibition of growth can be caused by trauma to the underdevelopment and endosperm hybrid grains, is a provider of nutrients to the embryo.

However, in subsequent generations (F_2, 2011 and F_3, 2012) hybrid seeds, due to the increased ability to germinate, provide full amicable shoots. Observations on the growth and development of plants at different stages of ontogeny with the calculation of biological stability index revealed a significant advantage over the original hybrid forms varieties for adaptation properties. When comparing the mean values of daily increment starting plants and hybrid forms revealed differences. So in the tillering stage marked the most active plant growth hybrids compared to the parental forms. While in phase earing this figure was significantly higher in the initial samples. Upon the occurrence of lactic ripeness hybrid forms of continued growth in height more active than the parent varieties.

Theoretical basis of plant immunity to infectious diseases developed by Vavilov [22]. He identified the patterns of distribution of immunity in cultivated plants and their congeners, developed the doctrine of genetics stability and determined the direction of selection.

The factors limiting crop seed quality include fungal infection pathogens. Phytopathological analysis of four varieties of seeds (Tyumenskaya 80, Mir 11, Comet, Saratovskaya 57) showed that the overall infection was 6, 25–38 and 50%. Allocated among the most frequent fungi *Alternaria tenuis*, *Helmintosporium sativum*, *Pennicilium* sp. and fungi of the genus *Fusarium*. According to the index prevalence of illness (PI) affected seedling varieties Comet, Tyumenskaya 80, Mir 11 are related to moderately susceptible (PI = 28.15–30,45%, 21.59–28,43%, 20.40–29.50%, respectively), varieties Saratovskaya 57 – to low susceptible (PI = 12.48–18.48%).

Seed germination on infectious backgrounds revealed mixed reaction varieties to pathogens. The presence of the fungus in the substrate *Helmintosporium sp.* induced growth inhibition of primary root system and some stimulation of above-ground organs. On infectious background comprising pathogen *Fusarium sp.*, an increase of the linear dimensions of the germinal roots in lag values characterizing shoots.

On the natural infectious background of spring wheat plants according to our data the most frequently affected with powdery mildew, leaf rust spotting.

Powdery mildew (*Erysiphe graminis* DC). Pathogen fungus *Erysiphe graminis* DC belongs to the class *Ascomycetes*, order *Erysiphales*. The disease affects the leaves, leaf sheaths, stems, manifested in the form of a whitish pubescence plaque, which later becomes mealy appearance, gradually turning into dense mycelia pads.

Brown leaf rust (*Puccinia recondita* Rob. Ex Desm f. Sp.). Pathogen *Puccinia* recondita Rob. ex Desm f. sp. *tritici* belongs to the class Basidiomycetes, order Uredinales, family Pucciniaceae. On the lamina and leaf sheaths develop brown uredopustuly, and later – with a glossy black shade teleytopustulies.

Dark brown spot (*Helminthosporium sativum* Pammel, C.M. King et Bakke). Pathogen named *Cochliobolus sativus* (Ito et Kurib.) Drechsler ex Dastur, anamorph Bipolaris sorokiniana (Sacc.) Shoemaker (synonym – *Helminthosporium sativum* Pammel, C.M. King et Bakke) belongs to the class Hyphomycetes, family Dematiaceae.

In studying the collection fund of spring wheat on the example of 99 samples revealed patterns of geographic manifestations stability samples of spring wheat to the agents of fungal diseases. No infection or weak degree registered with samples of the Tyumen region, China and Mexico.

In the analysis of intraspecific diversity of the collection of spring wheat found that complex disease resistance showed samples linked to varieties *lutescens* (Alef) Mansf., *Graecum* (Koern.) Mansf., *Eritrospermum* Korn. To mildew less susceptible neural stunted form rust and spotting – beardless shape with an average height of plants (Table 24.2)

TABLE 24.2 Origin and Morphological Characteristics of Spring Wheat Samples With a High Resistance to Disease (Percentage of Samples of n = 99)

Sign	Powdery mildew, %		Mildew, %		Spotted, %	
Origin	Mexico	55.6	Tyumen area	17.5	Tyumen area	40.5
	Tyumen area	13.9	China	12.6	China	20.6
	Others	30.5	Mexico	12.6	Mexico	9.5
			Argentina	15.5	Others	29.4
			Others	38.8		
Species	*lutescens*	43.7	*lutescens*	48.3	*lutescens*	63.6
	graecum	33.3	*eritrospermum*	21.1	*eritrospermum*	13.6
	eritrospermum	18.0	*ferrugeneum*	16.5	*graecum*	9.2
	Others	5.0	*graecum*	7.2	Others	13.6
			Others	6.9		
The presence of spines	awned	66.7	beardless	52.8	beardless	68.2
	beardless	33.3	awned	47.2	awned	31.8
Height of one plant	undersized	85.7	medium tall	54.1	medium tall	68.2
	(51–80 см)	9.5	(81–110 см)	43.1	(81–110 см)	31.8
	medium tall	4.8	low tall	2.8	low tall	0
	(81–110 см)		(51–80 см)		(51–80 см)	
	dwarf		dwarf		dwarf	
	(<50 см)		(<50 см)		(<50 см)	

Predicted the emergence of new races of pathogens and perhaps even new diseases, which require protection from new resistance genes that are in short supply. This situation causes to pay special attention to preserving well-tested resistance genes and conduct new searches. Increase the genetic diversity of wheat to mitigate the environmental vulnerability, possibly based on the mobilization of biological diversity of this culture by engaging in material selection process of different ecological and geographical origin.

One of the priorities spring wheat is to create a raw material with a high ecological plasticity using modern methods, one of which hybridization.

Hybrids in the first generation (F1) are characterized less pronounced susceptibility to powdery mildew. When dividing the sample into groups of stability revealed that 10% of hybrid forms signs of injury were found. Among the initial varieties large proportion (40%) had very low resistance to powdery mildews fungi predominated among hybrids combination with moderate losses.

In relation to the spotting (*Alternaria* sp., *Helmintosporium* sp.) 50% of hybrid forms showed a high level of stability. While at the initial varieties, this group accounted for only 20%. Among the varieties dominated groups with very low and medium resistance, including hybrids, they accounted for 10% and 20%, respectively.

According to studies in 2010, 2011 and 2012 in terms of resistance to powdery mildew stood hybrid forms ♀Hybrid × ♂Lutescens 70 and SKENT 3. Very high resistance to hybrid forms had spotted ♀SKENT 3 and ♂SKENT 1; ♀Hybrid × Lutescens 70♂; ♀Hybrid × ♂SKENT 3 and ♀Cara × ♂SKENT 3, the initial variety was Lutescens 70.

Dedicated us samples of wheat immune to a certain extent are unique due to simultaneous resistance to the most harmful pathogens – pathogens of fungal diseases and can be used in breeding as potential donors in developing new competitive edge in the global market of varieties.

One of the major abiotic factors, leading to a decrease in productivity of wheat is soil salinity. In the study of the response of 18 varieties of spring wheat on salt stress (solution NaCL) revealed mixed reaction. Most sensitive to the effects of adverse factors showed Omsk 24 and Satu whose seed germination in experimental variants were 40 and 38%, respectively, lower than in the control. Two varieties Lutescens 70 and Erythrospermum not observed negative influence on the germination, germination as compared to the control was higher than 4% and 18%, respectively.

Given the fact that these data may not be sufficiently informative for further development of the seedlings was investigated several morphometric variability parameters (Table 24.3). The average number of germinal roots was 4.8 units in the control and 2.7 units in the experience. Length of the roots of the control seedlings ranged from 9.3 to 10.8 cm in seedlings exposed to stress, from 1.0 to 2.3 cm Significant differences were controlling and experienced options to escape length (9.4 cm – control, 0.8 cm – experience). Ratio of the length of roots and shoots in standard conditions was close to unity in the experience of this ratio from 1.00 to 7.67, which may indicate significant changes in the growth processes.

TABLE 24.3 Effect of Salinity on the Primary Symptoms of the Root System and the Aerial Sphere Spring Wheat (Average Data of Samples, n = 18)

Sign	Control $X \pm m_x$	Experiment $X \pm m_x$	Deviation from control, %
Root:			
number, pieces	4.7±0.22*	4.3±0.16	−8.5
length, см	10.0±0.34*	8.1±0.38	−19.0
weight, g	4.7±0.50	4.0±0.41	−14.9
Shoot:			
length, см	16.0±0.42	13.0±0.40	−18.8
weight, g	5.2±0.50	3.7±0.37	−28.8
Leaf:			
number, pieces	2.0±0.04*	1.7±0.04	−15.0
length, см	11.6±0.49*	8.7±0.41	−25.0
weight, g	0.3±0.01*	0.2±0.01	−33.3

* The differences between the control and experimental variants are authentic.

The scoring rankings with all the studied traits allowed to distinguish varieties from high resistance to salinity. These data confirm a high reliability integrated assessment varieties for salt tolerance. The most suitable varieties for saline soils are Lutescens 70, Tyumenskaya 80, Omskaya 20, Zyryanovka, Skorospelka, Satu, combine salt tolerance with other economic characteristics.

Experimental mutagenesis is considered as an extension method of genetic diversity of spring wheat. With the help of artificially induced mutations can significantly extend the limits of a species or population genetic variability of plants and thereby increase the effectiveness of selection events [23].

Initial studies on the application of experimental mutagenesis, carried out in 1972–1976 in the North Trans-Ural region. On cultivars and hybrids of soft spring wheat during that period were relevant, timely, and have both theoretical and practical significance. Experimental mutagenesis attributed to the relatively new techniques, much attention was paid to the current tendency of sharing sources and mutational variability combination to create new varieties. This became the basis for the first time in the region to put the work to determine the effectiveness of radiation mutagenesis at 8 and 28 varieties of hybrids obtained by crossing an incomplete diallel scheme [24].

It was shown that for breeding purposes for complex forms with improved properties to effectively apply ionizing radiation on intervarietal hybrids. Spring wheat hybrids surpass parent varieties for radioresistance in M1 and M2 is found in greater frequency and varied spectrum of mutations induced by gamma radiation. Induced mutational variability increases the likelihood of hybrids forms with transgressive expression productivity traits. The reason for the enhanced stability of hybrids in comparison with the initial variety is, apparently, as a hybrid and heterozygosity level of heterosis and heterozygosity induced by gamma radiation, which substantially change adaptive, including radioprotective properties of plants.

Found that genotypic characteristics of varieties and their hybrids on radiosensitivity and mutability relative yield breeding and valuable forms can be set only by studying their response to several doses of mutagenic factors (Table 24.4).

TABLE 24.4 Mutability of Spring Wheat Depending on the Dose of Gamma Radiation

Dose X-rays	Variety			Hybrid		
	Number of families, pieces		% changed	Number of families, pieces		% changed
	Analyzed	With changes		Analyzed	With changes	
Control	1200	0	0	3617	40	1.1
1000	1116	216	19.4	3367	456	13.5
3000	1008	174	17.3	3417	660	19.3
5000	1100	130	11.8	3377	690	20.4
10000	1078	97	8.9	-	-	-

Received a large variety of mutant forms, including features characteristic species *Triticum* L.: T. *macha*, T. *sphaerococcum*, T. *spelta*, T. *compactum*, etc.

Manifestation of heterosis in first-generation hybrids (F1) – one of the performance criteria crosses. In 1919 I.M. Fremen first observed heterosis in wheat plant height, leaf width, maturity ear formation. The magnitude of heterosis in wheat hybrids of the first generation (F_1) can vary widely [25]. The obtained data on the inheritance of productivity first generation hybrids, heterosis and dominance indices indicate that wheat possible intermediate inheritance, deviations from the additive scheme in the direction of one of the parents, or superiority over both parental forms (strongly expressed positive or negative heterosis).

Some hybrid populations that have changed under the influence of gamma radiation elements productivity retain these figures in the second and third generations. This indicates that under the influence of mutagenic factors in intervarietal hybrids arises induced variability, which in some cases provides inheritance productivity traits.

From the standpoint of genotypic nature of quantitative traits is important to study the action of the mutagen on the total and specific combining ability. Mathematical models proposed by Griffing [26] allow to consider how the general and specific capacity, and the effects caused by environmental factors. Analysis of variance of combining ability effects on such elements as productivity number of grains per ear and grain weight per plant showed that the variability under the influence of gamma rays on the ACS and SCS was significant. Varieties and hybrids (Saratovskaya 29, Strela, Leningradka, Novosibirskaya 67, which along with high both for both types of combining ability have high variance.

Dedicated us highly productive, with a shorter growing season, nondrowning and high quality grain mutants further used as a starting material in the breeding of spring wheat.

Based on these results now begun research on varieties and hybrids of spring wheat on the effectiveness of alkylating compounds of phosphemid (phosphemidum, synonym phosphazin) – di-(etilene imid)-pyrimidyl-2-amidophosphoric acid [27, 28]. The main objective of this work is to increase genetic diversity and create a collection of mutant forms of wheat using mutagenic seed treatment. Mutational variability is studied on seedlings in the lab and on the plants – in the field.

Mutation in combination with hybridization provides heterogeneity of the plant material and allows you to maintain it for a long time. According to Schmalhausen [29] "mobilization reserve of genetic variation " is an indisputable value for use in extreme environmental conditions. It is no accidents in the center of modern projects are genetic variation and genetic diversity.

24.4 CONCLUSIONS

The importance to use of different samples from the world collection of Institute of Plant Industry N.I. Vavilov as well hybrid and mutant forms in genetic and breeding studies, due to the need to introduce a new culture of dominant genes that determine, as a rule, ecological plasticity, high homeostasis, resistance to unfavorable climate, disease, etc. Using these genes, we can solve the most important task of the selection process associated with genetic adaptation of varieties to regional conditions, which significantly eliminates the dependence of yield changes in climatic and other conditions.

Using recombination and mutation variability effectively to isolate and create sources of genetic resistance to biotic and abiotic stresses.

In a changing climate and increasing anthropogenic pressure is expected to increase the number of diseases caused by warming. In this regard, particular values are samples of a spring-sown field with resistance genes. Their identification is possible on the basis of the mechanisms underlying intrinsic interactions, implementation of screening of the gene pool of spring wheat for resistance to phytopathogenic fungi – pathogens of powdery mildew, rust, spotting various etiologies.

When selecting a plastic spring wheat varieties in the south of the Tyumen region effective laboratory methods to conduct a rapid assessment of plant resistance to adverse factors (reduced temperature, salinity, pathogens infection) using provocative and infectious backgrounds. This approach allows us not only to know the peculiarities of the biomass of seedlings, but also to predict the rate of reaction of the plant organism to changing environmental conditions in the later periods of growth and development. Integrated assessment reduced the time for the selection of forms with high adaptive properties.

KEYWORDS

- **experimental mutagenesis**
- **gamma rays**
- **hybrid**
- **phosphemid**
- **phytopathogenic fungi**

REFERENCES

1. Alexanian, S. M. (2007). Gene Banks Interaction Strategy in a Globalizing World, Bulletin of Applied Botany, Genetics and Breeding, *164*, 11–33 (in Russian).
2. Modern Methods and International Experience Conservation of Genetic Diversity of Wild Plants (for Example of Wild Fruit), Collective Monograph, Almaty (2011) 188p (in Russian).
3. The United Nations Conventionon Biological Diversity, the International Secretariat, Geneva, Executive Center (1992) http://www.un.org/ru/documents/decl_conv/conventions/biodiv.shtml.
4. Hammer, K. (2004). Resolving to Challenge Posed by Agro Biodiversity and PGR an Attempt, Journal of Agriculture and Rural Development in Tropics and Subtropics, Beiheft, *76*, 184p.
5. Tyutyuma, N. V. (2009). Theoretical and Applied Aspects of the Study of the Gene Pool Breeding Value of Cereal Crops in the Arid Conditions of the Lower Volga Region, The Thesis Abstract on Competition of the PhD of Agricultural Sciences, Astrakhan, 44p (in Russian).
6. Zamolodchikov, D. G. (2011). Ratings Changes Climatogenic Diversity of Tree Species According to the Forest Fund Advances in Modern Biology, *131(4)*, 382–392 (in Russian).
7. Cherkashenina, E. F. (1972). Agroclimatic Resources of the Tyumen Region (Southern Part), Leningrad, Gidrometeoizdat, 151p (in Russian).
8. Ivanenko, A. S. (2008). Agroclimatic Conditions of the Tyumen Region, Ivanenko, A. S., Kulyasova, O. A., Tyumen State Agricultural Academy, 206p (in Russian).
9. Gradchaninova, O. D. (1987). Methodological Guidance on the Study of the World Collection of Wheat, Gradchaninova, O. D., Filotenko, A. A. & Rudenko, M. I. Leningrad, VIR, 28p (in Russian).
10. International Classification of CMEA (1984) Genus *Triticum L.*, Leningrad, 84p.
11. Naumova, N. A. (1970). Analyse is of Seeds on Fungal and Bacterial Infection, Leningrad Publishing Hous Kolos, 32p (in Russian).
12. Evaluation of Crop Resistance to Disease in Siberia (1981) Methodological Guidance, Novosibirsk, 48p (in Russian).
13. Mihajlina, N. I. (1983). Comparative Evaluation of Methods for Determining the Severity of Root Rot of Spring Wheat, Agricultural Biology, *4*, 95
14. Methodological Guidance on the Study of Stability of Cereal Grasses to the Agents of Fungal Diseases for Conditions Chernozem Zone of the RSFSR (1977) Leningrad, All-Russian Research Institute of Protection of Plants (VIZR), 60p (in Russian).

15. Zrazhevskaya, T. G. (1979). Determination of Resistance of wheat Plants to Common Root Rot, Mycology and Phytopathology, *13(3)*, 58 (in Russian).
16. Udovenko, G. V. (1977). Salt Tolerance of Crop Plants, Udovenko, G. V., Proceedings of Agricultural Sciences, Leningrad, Publishing House Kolos, 215p (in Russian).
17. Lukyanenko, P. P. (1973). A New Method for Artificial Pollination of wheat, Selected Works, Moscow, Publishing House Kolos, 52–54 (in Russian).
18. Dorofeev, V. F. (1990). Flowering, Pollination and Hybridization of Plants, Dorofeev, V. F., Laptev, Y. P., & Cherkashina, I. M. Moscow, Agropromizdat, 144p (in Russian).
19. Efremov, V. V. (1997). Change Varietal Composition of Agrotcenoz of Winter Field, Anniversary Edition for the 75th Anniversary Krasnodar State Agrarian University, Efremov, V. V., Aistova, Y. T., & Terpugova, N. I. Krasnodar Agro Ecological Monitoring in Agriculture Krasnodar Territory, 468p (in Russian).
20. Bome, Ya A. (2012). Investigation of the Gene Pool *Triticum Aestivum* L on Plant Response to Low Temperatures, Bome, Ya A., & Bome, N. A. Natural and Engineering Sciences, Publishibg Sputnik, *1(57)*, 117–121 (in Russian).
21. Reimers, F. E. (1978). Seed Germination and temperature, Reymes, F. E. & Illy, I. E. Novosibirsk, Nauka, 168p (in Russian).
22. Vavilov, N. I. (1964). Problem of Immunity of Cultural crops, Selected Works, *4,* Moscow-Leningrad, 450p (in Russian).
23. Rapoport, I. A. (1980). Chemical Mutagenesis, Rapoport, I. A., Shigaeva, M. H., & Ahmatullina, N. B. Alma-Ata, Nauka, 317p (in Russian).
24. Bome, N. A. (1980). Study of Mutational Variability of Hybrids of Spring Wheat Induced by Gamma Radiation and its Selective Use, the Thesis Abstract on Competition of the Candidate of Agricultural Sciences, Leningrad, Vavilov, N. I., All-Russian Research Institute of Plant Breeding, 21p (in Russian).
25. Fedin, M. L. (1967). About Heterosis in Wheat, Fedin, M. L., Agricultural Biology, *2(6),* 887–889 (in Russian).
26. Griffing B. (1956). Concept of General and Specific Combining Ability in Relation to Diallel Crossing System, Australia J. Biology Science, *9,* 463–493.
27. Chernov, V. A. (1964). Cyto Toxic Agents in the Chemotherapy of Malignant Tumors, Moscow, Medicine, 320p (in Russian).
28. Weisfeld, L. I. (1965). Cyto Genetic Effects of Phosphasin on Human and Mouse Cells in Tissue Cultures, Genetics, *4,* 85–92 (in Russian).
29. Shmalgausen, I. I. (1968). Factors of Evolution (the Theory of Stabilizing Selection) Ed Second Revised and Enlarged, Moscow, Nauka, 451p (in Russian).

CHAPTER 25

THE PECULIARITIES OF SEEDLINGS FORMATION OF *TRITICUM AESTIVUM* L. UNDER SALINIZATION CONDITIONS

ANNA A. BELOZEROVA[1], NINA A. BOME[1], ALEXANDER Y. BOME[2], and SVETLANA M. MOSTOVSHCHIKOVA[1]

[1]Tyumen State University, d.10, Semakov St., Tyumen, 625003, Russia; E-mail: anna-bel@bk.ru; bomena@mail.ru

[2]Tyumen basing point of the N.I. Vavilov, d. 42-44, Bol'shaya Morskaya St., St. Petersburg, 190000, Russia; E-mail: genbank.d@gmail.com

CONTENTS

ABSTRACT

The results of laboratory assessments of spring mild wheat for the salinization during the sprouting of seeds on salt solutions are listed. The energy of sprouting, seeds germination, length and mass of sprouts, amount, length and mass of roots, square of leaves were taken into account. It was set that substratum salting leads to considerable reduction in the majority of studied features. In accordance with the complex of signs more salt tolerant samples were discovered. Comparison of several methods of plants salt tolerance laboratory diagnostics allowed to uncover disadvantages along with positive sides. With the aim of sorts adaptation mechanisms activation towards the stressful factor the influence of para-aminobenzoic acid (PABA) on the vegetative organism on the background of the salinization was studied. It was shown that the preliminary processing of the seeds with PABA solutions increases defensive features of plants on the first stages of ontogenesis. The nature of the response to the PABA in different concentrations has a graded specificity. Optimal concentration of the PABA solution (0.01%) was set when the length of the root system and sprouts of the wheat as well as their raw mass melted down at a lesser degree in all versions. The application of the PABA solution in the concentration of 0.05% was less effective.

25.1 INTRODUCTION

The problem of plants salt tolerance is one of the most relevant in the plant growing. The necessity of its comprehensive study is defined by the presence of big squares with saline soils in many countries posing a considerable hinder for growth, development and increase in harvest of agricultural plants [1]. Saline soils are met predominantly in the southern arid areas, often in spots among the nonsaline soils [2]. Surplus content of mineral salts is characteristic of 25% of soils on the planet. One third of the soils exposed to irrigation should be added [3]. The areas of the saline lands have a tendency towards permanent and considerable increase as a result of processes of secondary salinization, which causes a lot of damage to many fields of plant growing and restricts the extension of areas for different agricultural crops in arid districts of Russia yearly [4–6]. In accordance with the information given by Sabolch in the monograph "Simulation of the salinization and "solonetzization" processes of the soils" (cited in Ref. [7]) total area of saline soils in the world accounts for more than 950 mln ha.

According to the data of the State report "About the condition of the environment of the Russian Federation in 1994" (cited in Ref. [8]), from 186 mln ha of the agricultural lands in Russia about 60 mln ha are exposed to erosion, 40 mln ha are presented by saline and solonetz complexes, 26 mln ha are water saturated and swamped.

From the total area of the West-Siberian economic region (224.9 mln ha) saline soils take up 15.5 mln ha (7% from the total area). Within the precincts of the agricultural lands of the economic region saline soils take up 17.6% (5.2 mln ha) including among the arable lands – 8.4% (1.5 mln ha), according to the data for the year 1996. In the solonetz complexes the soils account for 21.5% (6.4 mln ha) on the lands of agricultural holdings and 16.1% (2.9 mln ha) – on the arable land. On the whole, saline soils predominate in the southern part of the West-Siberian plain. Saline soils or soils with characters of the former salinization are met in the northern regions of the West-Siberian plain as well including the coastal zone of the northern seas. These soils are not included in the fund of lands used in the agriculture. That is why they were not taken into account during the calculation of the saline soils squares.

Salinity is the most important peculiarity of the forest steppe soils of the Tyumen District. As a whole, frequent change of diverse soils on the limited area is characteristic of the Tyumen District. Zonal chernozemic soils include meadow and soddy solonets, meadow, meadow-marsh saline lands. The soils of river valley flood plains are distinguished by considerable complexity. Soddy-podzolic, chernozemic, meadow-chernozemic and gray forest soils are the main basis of agriculture and economy of the district [9]. Saline soils of the Tyumen District take up the square of 286.6 thsd ha within the precincts of agricultural lands and 66.5 thsd ha on the arable lands, 564.1 thsd ha of agricultural lands are occupied with solonetz complexes and 18.7 thsd ha of the arable lands of the district [7].

One of the ways of reduction of the negative impact on plants is the selection of the varieties, which are tolerant to increased composition of the salt in the soil.

For the diagnostics of the plants tolerance towards the salinization of the soil a whole number of methods are applied. They all have a necessary condition in the basis – the creation of the provoking background of salinization. It is connected with the fact that the level of plants salt tolerance is consolidated genetically and it manifests only when this extreme factor operates [10].

The existing methods and ways of plants salt tolerance diagnostics are divided into two groups – direct and indirect methods. In direct methods the definition of biological indicators underlies – harvest, elements of productivity, morphophysiological character. Direct methods possessing higher precision are more prolonged and laborious at the same time. Indirect methods are based on the calculation of the plants response to salinization, which is defined due to the measuring of a number of physiological and biochemical processes registered with appropriate devices. The advantage of these methods is the changes setting possibility for a small period of time; given that devices are technically advanced, higher performance is provided compared to the direct evaluation methods.

Many explorers suggest conducting an evaluation of the salt tolerance on different indicators accompanying the germination of the seeds. In different formulations of these assessment ways different technical modifications are offered [11], varying in the way of seeds germination, concentrations of salt solution and elements of account (the velocity of seeds swelling in saline solutions, energy of germination, germination power, length and mass of seedlings and so on). The evaluation of salt tolerance at plants on the germination of the seeds gives the possibility to compare both closely related groups (varieties and species of one crop) and biologically distant (different crops).

Gurova, and Berezina [12] for the evaluation of salt tolerance of spring wheat and barley suggest using luminescent methods. Among the authors a strong linear correlation is set between the slowed fluorescence of the plant tissue and inhibition of growth processes at the salinization ($r = 0.80–0.96$).

Polonski, and Surin [13] group the existing diagnostics in the following way: 1) methods based on the definition of seeds ability to germination in saline solutions [14]; 2) techniques taking into account growth indicators of the seedlings: the length and the roots surface [15], the total mass of 5-day seedlings [16]; 3) biophysical approaches basing on the measuring of bioelectrical potentials amplitude [17]; 4) biochemical methods in which the activity of superoxide dismutase ferment is defined [18]; 5) vegetative-field methods [14], which are the most prolonged but at the same time more precise.

The analysis of the generalization of results from literary sources gives the ground to suppose that the study of the peculiarities of the early ontogenesis with detailed analysis of the initial root system and over-ground seedlings organs basing on the complex of quantitative characters can be of value for the selection of highly tolerant to the salinization forms.

25.2 MATERIALS AND METHODOLOGY

Experimental part of the work is conducted on the chair of botany, biotechnology and landscape architecture of the Tyumen Biology Institute of the State University. 28 varieties of spring mild wheat served as an object of the study (*Triticum aestivum* L.) (Amurskaya 90, Aviada, Dvulineynaya, Eritrospermum, Fora, Grecum 114, Ikar, Ilyinskaya, Irgina, Iya, Kazahstanskaya 10, Kometa, Latona, Lutescense 70, Mir 11, Novosibirskaya 15, Novosibirskaya 31, Omskaya 10, Omskaya 20, Omskaya 24, Saratovskaya 57, Satu, SCENT 1, SCENT 3, GDS 11, Skorospelka, Tyumenskaya 80, Zyryanovka) which are attributed to the varieties *erytrhrospermum, ferrugineum, graecum, lutescens, milturum, and pirothrix*.

The evaluation of the spring mild wheat on salt tolerance was conducted with the usage of two methodical approaches: (1) seeds green sprouting in saline solutions in Petri dishes; (2) seeds green sprouting in salt sand in vegetative vessels.

Petri dishes with filter paper were preliminary sterilized in a dry heat oven at $t = 175°C$ for 3 h. The seeds were disinfected with 1% $KMnO_4$ solution during 5 min. The paper on the bottom of the dish was moistened (in 7 mL) with distilled water (control) and salt solution (experimental variants). The chloride salinization was created with the use of 1.40% NaCl solution, in the sulfate salinization 2.78% solution of Na_2SO_4 was added, in the carbonate salinization – 2.06% solution of Na_2CO_3. The given concentrations of salts correspond to the osmotic pressure of 1 MPa. In each dish 50 seeds in triple repetition for each variant were put.

Para-aminobenzonic acid is a genetically active substance, which is able to enhance plasticity and adaptive properties of plants [19]. That is why the treatment of the spring mild wheat seeds with PABA was included in the experiment in the conditions of chloride salinization.

Seeds were germinated in sterile Petri dishes on distilled water (control 1, 2) and on the provoking background with 1.40% solution of NaCl (experimental variants 1–4). In control 1 and experiment 4 dry seeds were used, in control 2 and experiment 3 the seeds were aged in the water; in experiment 1 and 2 the seeds were treated with PABA solutions in the concentration 0.01 and 0.05%. The period of seeds treatment is 12 h. Sample size is 20 seeds in triple repetition for each variant.

The dishes with the seeds were placed into a thermostat (TCO-1/80 СПУ) с t=21–24°C. The volume of the sample for the calculation of the morphometric parameters accounted for 30 plants from each variant. On the third day of the experiment the calculation of the germination energy of the seeds was conducted. On the seventh day the following parameters were taken into account: laboratory germination power of the seeds, the length of seedlings and roots, amount of roots, wet and dry mass of seedlings and embryonic roots.

At the second methodical approach calcined sand was used as a substratum dewy up to 60–70% of complete water-absorbing capacity with which the vegetative vessels from inert material were filled. The salinization took place after the appearance of seedlings inserting NaCl solution up to the level of the osmotic pressure 0.7 MPa or at once after the apportionment of the seeds with NaCl, Na_2SO_4 and Na_2CO_3 solutions with the osmatic pressure 1 MPa. In the control the sand was exposed to the moistening with distilled water.

The seeds of each variety were uniformly apportioned on the dewy sand in 50–100 pcs. (the repetition of the experiment is quadrupled). Then the seeds were scattered with a 2 cm thick even layer of sand. The vessels were closed with the glass for the reduction of the evaporation. The germination of the seeds took place under the conditions of artificial illumination on special racks at the temperature of 22–24°C or in a universal capsule for the simulation of the climatic conditions of the environment MLR-351 at t = 24–26°C and 50% humidity. The control weighing of vessel was performed daily and then added to the vessels amounts of water that has evaporated during the day. Before the beginning of the experiment the mass of the vessels, sand of given humidity and seeds was defined.

On the tenth day of the experiment in each variant the germination power of the seeds, length of the seedlings and roots, the amount of the roots, length, width and square of the first leave, wet and dry mass of seedlings and embryonic roots were taken into account.

The calculation of the main statistic parameters was conducted according to the standard techniques [20]; the programs for the statistic data processing Statistica 6.0 and STATAN were also used.

25.3 RESULTS AND DISCUSSION

When evaluating the initial material in standard laboratory methods among the salt tolerance criteria are such indicators as germination energy, laboratory germination power of the seeds. However, in the scientific literature it is remarked that the change of seeds germination indicators during the salinization often correlates weakly with the extent of salinization of the plants. In this connection it is recommended to define the amount of embryonic roots, the length of seedlings and roots, intensity of the seedlings growth in the time [21] except for the germination power and energy of plants germination so that to receive more trustworthy and impartial results of the research. Between the growth intensity of the seedlings on the saline substratum and extent of the salt tolerance of the varieties a coefficient of the correlation turned out to be 0.62–0.83.

Laboratory germination power of the seeds in standard conditions and on the saline substratum.

The study of the plants adaptation to salinization shows that this process goes irregularly in the duration of the ontogenesis [22]. As it is set in the works of Abutalibov and Kovalskaya (cited in Ref. [23]) as well as Alderov and Shichmuradov [24], the biggest sensitivity to salts plants show in the phase of seedling, during the budding and bloom.

According to the generalized results of Udovenko's and Goncharova's [25] experiments, negative impact of the increased concentration of salts tells on the early stages of the plant life. It shows in the delay of seed swelling at a high level of salinization (15–20 atm) seeds do not turn up.

When salting the soil with sodium chloride the drop in the magnitude of its osmotic potential takes place which leads to the decrease in the intension of water absorption by the roots [26]. The osmotic pressure of the roots increases which indicates unspecific reaction of the roots in response to the salinization [27].

In our experiment of the plants growth on the saline substratum the considerable influence of NaCl on the indicators of laboratory germination power of plants was considerable. At the same time the reaction of the sorts on the salinization was ambiguous. In standard conditions (during germination on calcined sand) high germination ability was characteristic of the experimental seed samples.

Such varieties as Omskaya 24 and Satu showed the most sensitivity to unfavorable factor impact. Their germination power at experimental variants was by 40.0 and 38.0% accordingly lower than in the control. In the varieties Amurskaya 90, GDS 11, Grecum 114 this magnitude accounted for 20.0–22.0%, which also testifies about their not high tolerance to salinization.

The two varieties Lutescense 70 and Eritrospetrum did not have the negative influence on the seed germination of the studied factor; on the contrary, the increase of the sideling by 4.0 and 18.0% was marked. Though in other variants the reduction of the germination power prevailed, it was less considerable than at above-mentioned varieties.

On average among the varieties 35.1 seeds germinated which accounted for 70.1% on the background of the salinization 29.4 pcs. (57.6%) – by 12.5% less (Fig. 25.1).

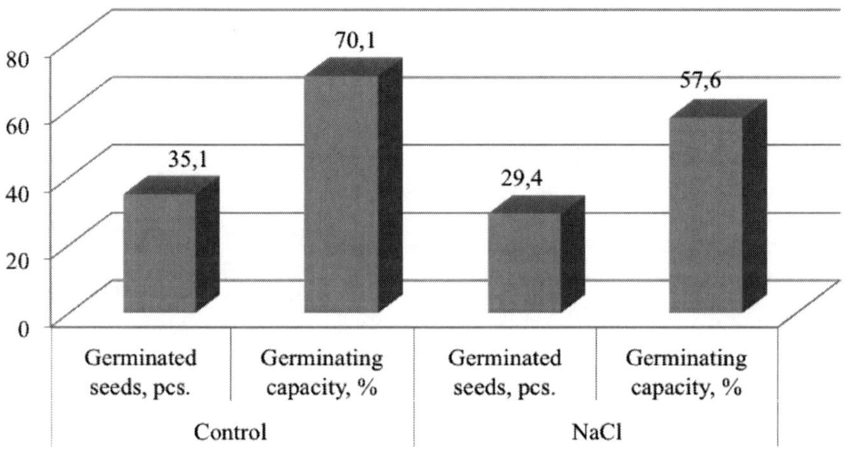

FIGURE 25.1 The germination of the wheat seeds on the saline control background (18 varieties on average).

In this way the germination power allows giving preliminary indirect evaluation of the spring wheat on the sensitivity to salinization. On the ground of this indicator the experimental samples were divided into three groups: tolerant, middle-tolerant and sensitive to salinization. The difference between the control and experimental variants on the germination served as a criterion during the distribution: 1st group 10–20%; 2nd group – 21–30%; 3rd group – 31% and more.

In the first group of the most tolerant varieties 10 of them were included: Fora, Iya, Kometa, Lutescence, Omskaya 10, Omskaya 20, Saratovskaya 57, Skorospelka, Tyumenskaya 80, which showed a relatively high germination power of the seeds both in the control and experimental variants.

Five varieties are attributed to the group with the average tolerance at which the reduction of laboratory germination power is marked by 20–25% on the saline substratum in comparison with the control: Amurskaya 90, Dvulineynaya, GDS 11, Zyryanovka, and Mir 11.

More sensitive to the impact of the saline solution varieties were included in the third group of which the maximal reduction of the laboratory germination power in the experimental variants is characteristic. There were three such varieties: Grecum 114, Omskaya 24, Satu. The biggest gap between the given data of control and experimental variants (40%) is marked at the variety Omskaya 24. In the standard conditions 88% of seeds germinated at this variety, but on the substratum with salinization this indicator was reduced up to 48%. At Grecum 114 the reduction of the laboratory germination accounted for 32%, but at the variety Satu – 38% (in the control the germination power of the seeds was equal to 50% and 90% accordingly).

The changeability of the seedling morphometric parameters in the conditions of the salt stress.

However, you cannot make a conclusion about the norm of the reaction only by the indicators characteristic of the seeds ability to germination, because these data are not informative in the relation of the seedlings development. In this connection the calculation of the quantitative characters of the initial root system and over-ground organs. In our experiments took place; at the same time the amount of the embryonic roots, the length of the roots and seedlings as well as their mass were defined.

Essential differences have been set at a number of the experimental and control variants on the amount of the embryonic roots, their length and mass.

At the varieties Amurskaya 90, Grecum 114, and Omskaya 24 the growth in the sand with the addition of NaCl solution led to the trustworthy reduction in all the studied characters of the root system. So, in the variety Amurskaya 90 in the experimental variant with respect to the control the amount of the embryonic roots accounted for 84% and their length – 70%, mass – 50%.

Of the varieties Grecum 114 and Omskaya 24 a stronger oppression of the initial embryonic roots especially their length is characteristic. When the seeds germinating, Grecum 114 formed the roots the length of which accounted for 9.1 sm on usual substratum on the day of account, at Omskaya 24 this indicator was equal to 10.9 sm. The salinization contributed to the reduction in the character at the both varieties and its magnitude was equal to 5.1 sm (Grecum 114) and 5.6 sm (Omskaya 24).

The response to the created background of the Omskaya 10 variety is of interest. On the amount of embryonic roots and their mass according to the variants of the experience trustworthy differences were not discovered while the salinization led to essential increase in the roots length by 1.3 sm (23.2%).

The average meaning of the given character in all the varieties in control accounted for 10.0±0.34 sm, but in experiment – 8.1±0.38. Correspondingly in the experimental and control variants the amount of the embryonic roots was 4.7±0.22 and 4.0±0.41 g (Table 25.1).

TABLE 25.1 The Influence of the Salinization on the Characters of the Primary Root System and Over-ground Sphere of Spring Wheat Seedlings (18 Varieties on Average)

Characters	Control $X \pm m_x$	Experience $X \pm m_x$	The deviation from the control, %
Embryonic roots			
Number, pcs.	4.7±0.22*	4.3±0.16	-8.5
Length, sm	10.0±0.34*	8.1±0.38	-19.0
Mass, g	4.7±0.50	4.0±0.41	-14.9
Seedling			
Length, sm	16.0±0.42*	13.0±0.40	-18.8
Mass, g	5.2±0.50*	3.7±0.37	-28.8
Leaves			
Number, pcs.	2.0±0.04*	1.7±0.04	-15.0
Length, sm	11.6±0.49*	8.7±0.41	-25.0
Width, sm	0.3±0.01*	0.2±0.01	-33.3

*the differences with the control are statistically true at $P<0.05$.

The study of the character changeability of the over-ground organs showed considerable intervarietal diversity in the length of the seedling, amount of leaves, lineal parameters (length and width) of the leave blade.

Analogically to the characters of the root system, vegetative organs underwent the growth oppression in some cases, which showed itself in the trustworthy reduction in characters, in others the depression was absent or was insignificant. So, the studied parameters were on the level of control; sometimes they even exceeded it.

On average at all the samples high sensitivity to the salinization in the development of seedlings is marked. Compared to the control, the reduction of the seedling length accounted for 18.8%, mass of the seedling – 28.8%, the amount of the leaves – 15.0%, length of the leaf blade – 25.0%, width – 33.3% (the differences with the control statistically trustworthy).

For the normal passing of the growth and development processes of the spring wheat the square of the assimilative surface, magnitude of which defines the efficiency of photosynthesis to a considerable extent. In this connection the study of both lineal sizes of the leave blade and the square of the whole surface in different environmental conditions is of interest.

In our experiment considerable differences in the square of leaves as at the samples of the control variant and between variants also were set. The best results were received at the variety Tyumenskaya 80. The square of the leave blade in the experimental variant accounted for 2.28 cm^2, which is essentially higher than the average meaning of this character in all samples included in the study, – 1.56 cm^2.

The square of the leave blade and the amount of leaves on a plant are determinable when forming the assimilating surface of both one plant and plants growing on some square unit. High results on this character at plants grown on the sand with salinization were discovered at the varieties Fora – 5.23 cm^2, Lutescense 70–4.93 cm^2, Satu – 6.07 cm^2, Skorospelka – 6.48 cm^2, Tyumenskaya 80–7.23 cm^2.

The average magnitude on the square of the leave blade to all the samples accounted for 2.60 cm^2 in the control and 1.56 cm^2 in the experiment, but the square of all the leaves on the plant 5.20 cm^2 and 2.81 cm^2 conformably.

For the distribution of the wheat varieties into groups on salinization tolerance point ranging on morphological characters was held. The maximum amount of points corresponded to the number of the varieties and was equal to 18. Each variety received a number of points on a concrete character corresponding to the value of the character. The variety with the lowest value received one point, with the highest – 18 points. The rest of the varieties were situated in the order of character value increase. According to the total amount of the points collected on the estimated characters, the variety acquired a corresponding range.

According to the manifestation of the characters, characterizing the root system and over-ground organs of the plants in the earlier ontogenesis into the group of high tolerance to salinization varieties Lutescense 70, Omskaya 20, Satu, Skorospelka, Tyumenskaya 80, Zyryanovka are attributed. The amount of points at them changed from 111 up to 90 and was maximum compared to other varieties.

At the varieties Amurskaya 90, Dvulineynaya, Eritrosperum, Grecum 114, Kometa, Mir 11, Omskaya 24 high sensitivity to the stress factor is marked which became apparent in the substantial reduction in the morphological characters when growing on the saline substratum.

Juxtaposition of the sample distribution groups according to the quantitative characters with the distribution in accordance with seeds germination. The varieties Lutescense 70, Omskaya 20, Skorospelka, and Tyumenskaya 80 both with the ability of the seeds germination and with the further formation of biomass in the conditions of the salinization showed the highest results, which allowed attributing them to the first group of tolerance.

The reduction of the indicators on all characters (the third group) was steadily characteristic of the varieties Grecum 114 and Omskaya 24.

The similar regularity was not observed at two varieties such as Kometa, Satu. So, at the variety Kometa the amount of germinated seeds on the saline background decreased slightly compared with the control, which has become the ground for its ascription to the group of the tolerant. However, the plants of this variety showed the characters of burden and depression, which was expressed by the reduction in the parameters of the root system and seedlings. The variety Satu, on the contrary, protuberated among the others on the development of the root system and over-ground mass, but it sharply reduced the germination power of the seeds under the influence of the salinization.

The given data indicate the specificity of the response of the spring wheat to the stress factor as well as approve the opinion about the practicality of the evaluation not only on the germination power but those characters typical of the growth and development of plants.

The complex evaluation of the spring wheat according to the seed germination and appearance of quantitative characters in the early ontogenesis on the substratum with NaCl allowed distributing them into three groups of the tolerance: high, middle, and weak. Such varieties as Lutescense 70, Omskaya 20, Satu, Skorospelka, Tyumenskaya 80, and Zyryanovka are attributed to the group of salt tolerant.

In the 50s of the XX century, Stroganov turned his attention to the difference of the lesion symptoms caused at plants by chloride and sulfate salinizations. On this basis the conception of the toxic ion effect was put forward according, to which different ions entail destruction of the metabolism leading to specific toxic effects. The researches conducted with the use of isosmotic solutions of salts on tomatoes showed that during the salinization of the nourishing ground with Na_2SO_4 and NaCl anatomic-physiological changes go unequally. The results of these experiments show that at isosmotic concentrations Na_2SO_4 depresses the growth and development of the plants, than NaCl. But at the same time, in the presence of NaCl the leaves of the plants acquire higher succulence than at Na_2SO_4. The regularity of stronger toxic action of Na_2SO_4 salts in comparison with NaCl is set for guayule, flax and others. For some plants NaCl is less noxious salt compared to Na_2SO_4 [28]. Genkel [29] remarks that chloride salinization is stronger than sulfate one comparing the types of the salinization (at low levels – 5–7 atm). It depresses the growth and reduces productivity of the wheat, which is connected with high toxicity of chlorine for the cells. Self-increase of the osmotic pressure has the decisive meaning at a stronger salinization and the differences disappear.

The evaluation of the wheat variety tolerance to different types of the salinization with the use of different methodical approaches.

Aimed at study of different types of salinization influence on the changeability of a number of the root system characters and seedlings of 10 varieties of the spring wheat we germinated the seeds in the Petri dishes on filtrated paper and in vegetative vessels on calcined sand. In the experimental variants they used NaCl. Na_2SO_4 and Na_2CO_3 with osmotic pressure 1 MPa; in control distilled water was added.

It is known that during the salinization of the soil in plants big amounts of ions ballast salts (ashes) accumulate, the intension of the physiological-biochemical processes changes. At that the germination power of the seeds and plants growing intension considerably decreases [11, 28].

The reaction of the varieties on salinization on seeds germination in the Petri dishes and vegetative vessels.

In the result of our researches it is set that in the conditions of the salinization in Petri dishes and vegetative vessels considerable reduction of the laboratory germination power has been observed: on the background with NaCl by 25.3–37.9%, and on the background with Na_2SO_4 by 47.7–58.4% in comparison with the control (on average for the sorts). In the conditions with the carbonate salinization seeds did not grow in the Petri dishes completely, 1% of the germinated seeds are registered in the vegetative vessels, which indicates its strong negative impact (Fig. 25.2).

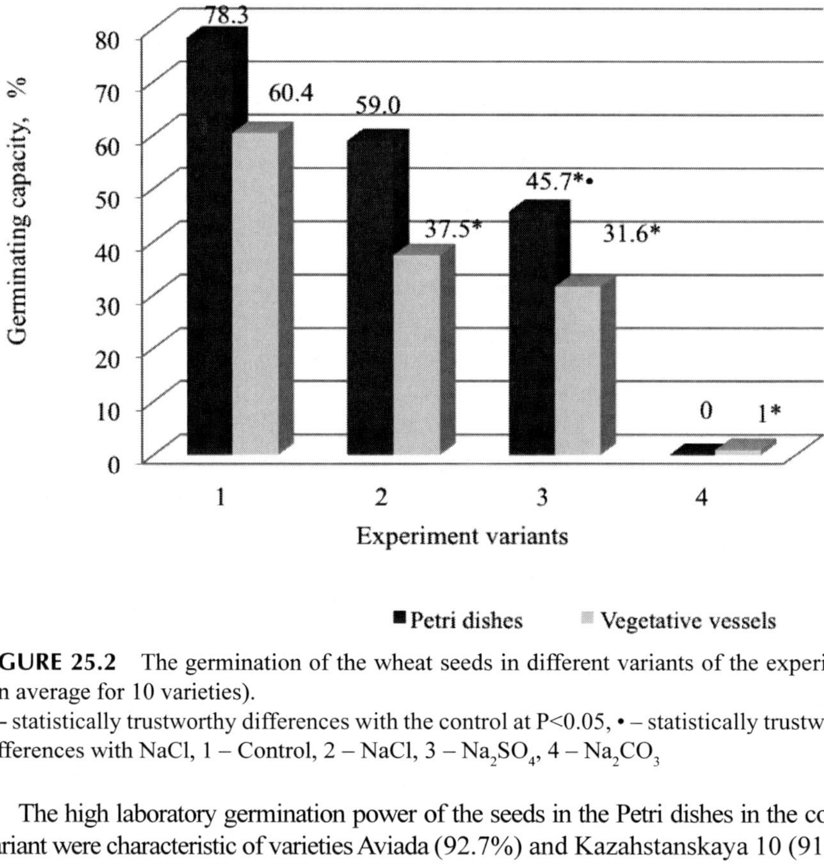

FIGURE 25.2 The germination of the wheat seeds in different variants of the experiment (on average for 10 varieties).
* – statistically trustworthy differences with the control at $P<0.05$, • – statistically trustworthy differences with NaCl, 1 – Control, 2 – NaCl, 3 – Na_2SO_4, 4 – Na_2CO_3

The high laboratory germination power of the seeds in the Petri dishes in the control variant were characteristic of varieties Aviada (92.7%) and Kazahstanskaya 10 (91.3%) on the background with NaCl – SCENT 3 and Kazahstanskaya 10 having 80.7% of germinated seeds; on the background with Na_2SO_4 varieties Aviada and Tyumenskaya 80–63.3% and 60.0% accordingly. The lowest interest of the germinated seeds is marked at SCENT 1 (control – 48.0%, background with NaCl – 21.3%, background with Na_2SO_4–18.7%).

In the vegetative vessels varieties with the biggest laboratory germination power were singled out: in control – Lutescense 70 (81.7%) and Aviada (79.7%), in the experience with NaCl – Lutescense 70 (71.3%) and Novosibirskaya 15 (51.0%) on the background with Na_2SO_4 – Lutescense 70 (50.3%) and Novosibirskaya 15

(41.7%) as well. The considerable reduction of the germinated seeds interest in the conditions of the chloride and sulfate salinization with the relation to the control at the Aviada variety is marked (by 76.5% and 77.8% correspondingly), but less reduction of the germination power on the background with NaCl marked at the variety Lutescense 70 (12.7%), on the background with Na_2SO_4 – at Latona (by 26.4%) and Tyumenskaya 80 (by 28.5%).

The peculiarities of the seedling biomass formation in different experimental conditions.

Change of the lineal parameters of the growth processes reflects the salt tolerance more trustworthy, than the evaluation of the indicators of seeds germination, which is connected with the general nature of plants resistance. The suppression of the cell division and especially expansion during the influence of unfavorable conditions leads to the delay of the plant growth, which is considered as the defensive reaction. Sustainable forms adapt to stress quicker and revive their growth. The extent of the salt tolerance is defined by the magnitude of the chosen parameters deviation from the control under the influence of the salinization.

In our experiment the salinization led to depression of the most morphometric parameters (Table 25.2). Seedlings have especially high percentage of the character reduction in the Petri dishes. Ambiguous reaction of the varieties to salinization was marked on the amount of the roots. Both reduction and increase in the given indicator independently of the type of the salinization and way of evaluation was observed. On average the reduction of the character in experimental variants by 24.5–36.7% is marked in the varieties during the comparison with the control in the Petri dishes when in vegetative vessels by 4.2–6.3%. The variety SCENT 3 in all the variants in the both experiments formed the biggest number of roots. During the comparison of two salinization types it was set that the sulfate salinization led to considerable reduction in the number of the roots in Petri dishes; in vegetative vessels no differences among the varieties on average were discovered.

TABLE 25.2 The Influence of Different Salinization Types on the Appearance of Morphometric Parameters of Wheat Seedlings

Characters	Control		NaCl		Na_2SO_4	
	$X \pm m_x$	CV,%	$X \pm m_x$	CV,%	$X \pm m_x$	CV,%
Petri dishes (7 varieties on average)						
Number of roots, pcs.	4.9±0.07	19.78	3.7±0.07*	27.74	3.1±0.06*•	29.90
Length of roots, mm	116.9±3.15	38.98	23.5±0.77*	47.20	7.7±0.35*•	65.26
Length of seedlings, mm	84.2±1.94	33.99	4.8±0.36*	109.84	2.2±0.05*•	36.52

TABLE 25.2 *(Continued)*

Vegetative vessels (10 varieties on average)						
Number of roots, pcs.	4.8±0.04	15.58	5.1±0.04*	14.62	5.0±0.05	14.92
Length of roots, mm	116.3±2.45	36.19	45.8±0.68*	25.08	42.4±0.63*•	25.08
Length of seedlings, mm	168.0±2.72	27.80	69.8±1.82*	44.03	65.5 ±1.81*	46.69
Length of a leaf, mm	133.9±2.45	36.36	41.5±1.70*	69.25	37.5±1.62*	72.55
Width of a leaf, mm	2.9±0.03	15.91	2.5±0.02*	15.86	2.4±0.03*	20.65
Square of a leaf, mm^2	335.3±7.24	36.97	89.5±3.84*	72.50	81.2±3.61*	74.18

* – statistically trustworthy differences with the control at $P<0.05$, • – statistically trustworthy differences with NaCl.

The depression of the roots was observed at the time of the evaluation of the varieties in Petri dishes according to the indicators of their length. On the background without the salinization the length of the initial roots varied from 88.7±3.53 mm (SCENT 3) up to 150.7±9.12 mm (Latona), in average value on the varieties 116.9±3.15 mm. During the germination of the plants on salt solutions the length of the embryonic roots decreased to7.7±0.35 mm (in the variant with Na_2SO_4) – 23.5±0.77 mm (in the variant with NaCl). The varieties Kazahstanskaya 10 and Latona have the least deviation from the control (by 39.0–76, 7%) according to the given indicator in the variants with the salinization. The considerable depression of the roots on the background with NaCl was registered at the variety Tyumenskaya 80, on the background with Na_2SO_4 – at the variety SCENT 1.

In the vegetative experiment the length of the primary root system in the control changed from 32.0±3.33 mm (SCENT 1) to 153.4±4.00 mm (Ikar) (the average on the varieties 116.3±2.45 mm). In the experimental variants according to the given factor the differences between the various types of salinization were less expressed compared to the experiment in Petri dishes. The primary root system of the seedlings on the background with NaCl reduced its length by 60.6%, on the sulfate background – by 63.5% in stressful conditions.

At the time of seeds germination in Petri dishes the seedlings formed shorter sprouts compared to the results of the vegetative experiment. In the control the length of the sprouts accounted for 84.3±1.94 mm on average in the first experiment and 168.0±2.72 in the vegetative vessels. Sulphate salinization in Petri dishes led to the decrease in the sprouts compared to chloride one the length of which on average accounted for correspondingly 2.2±0.05 and 4.8±0.36 mm on the varieties. In the vegetative vessels the given indicator in experimental variants changed correspondingly from 65.5 ±1.81 to 69.8±1.82 mm.

The vegetative experiment allowed us comparing the influence of different salinization types on the morphometric parameters of the first real leaf at different varieties, which is impossible while conducting an experiment in the Petri dishes. The seedlings of experimental variants had short and narrow leaf blades. In the control variant the length of a leaf varied from 36.9 (SCENT 1) to 164.4 mm (Novosibirskaya 15). Novosibirskaya 15 (3.4 mm) and SCENT 1 (2.5 mm) with maximum and minimum values of the given character differed from each other by the width of a leaf. As a consequence, on the square of the leaf blade the Novosibirskaya 15 variety (466.73 mm^2) exceeded other varieties, but SCENT 1 considerably yielded (77.03 mm^2).

With the maximum length of the sprouts in the experiment with sodium chloride the varieties Kazahtanskaya 10 (108.3±4.6 mm) and Lutescense 70 (88.5±4.54 mm) protuberated. The variety Aviada (32.3±1.57 mm) had the minimum length. On the whole all the studied varieties at the substratum with NaCl had the considerable decrease in the sprout length (by 58.5%) compared to the control.

In the study of the morphometric parameters of the seedlings in the conditions of the sulfate salinization it is set that the Kazahstanskaya 10 variety was characteristic of more developed sprouts (the length of a sprout – 102.0 mm, the length of a leaf – 68.4 mm, the square of a leaf – 131.36 mm^2). The considerable depression of the over-ground sphere is characteristic of the varieties Aviada, SCENT 1 (the length of a sprout – 31.6–41.6 mm, the length of a leaf – 12.4–19.1 mm, the width of the leaf – 1.5–2.1 mm, the square of a leaf blade – 21.74–23.92 mm^2). On all the characters the experimental seedlings were inferior to control.

The analysis of the dry and wet biomass showed that in stressful conditions the reduction of the dry mass of the embryonic roots, dry and wet mass of the sprouts takes place independently of the salinization types and the used methodological evaluation. On the contrary, the dry mass of the roots in a number of cases increases, which we connect with the damaged root system and ion salts accumulation.

In the structure of wet and dry seedlings biomass differences on the correlation of distinct elements are discovered. The part of the roots in the structure of the common seedling biomass increased by 9.3% and accounted for 70.8%. Correspondingly, the part of the over-ground sphere reduced from 38.5% to 29.2%. It can be explained by the fact that a sprout has water-excess in the tissues compared to the root system and loses more mass during the drying than roots (Fig. 25.3).

The essential change in the correlation between the part of the root system and sprouts under the influence of sodium chloride is marked. In the structure of the dry biomass of the explored varieties the over-ground sphere in the experiment accounted for 5.4% from the common mass of the seedlings, roots correspondingly — 94.6%, which points at the considerable depression of the sprouts under the influence of NaCl. In the correlation of the dry biomass elements the similar tendency is traced – the part of sprouts in the experimental variant accounted for 4.4%, the part of the roots – 95.6%.

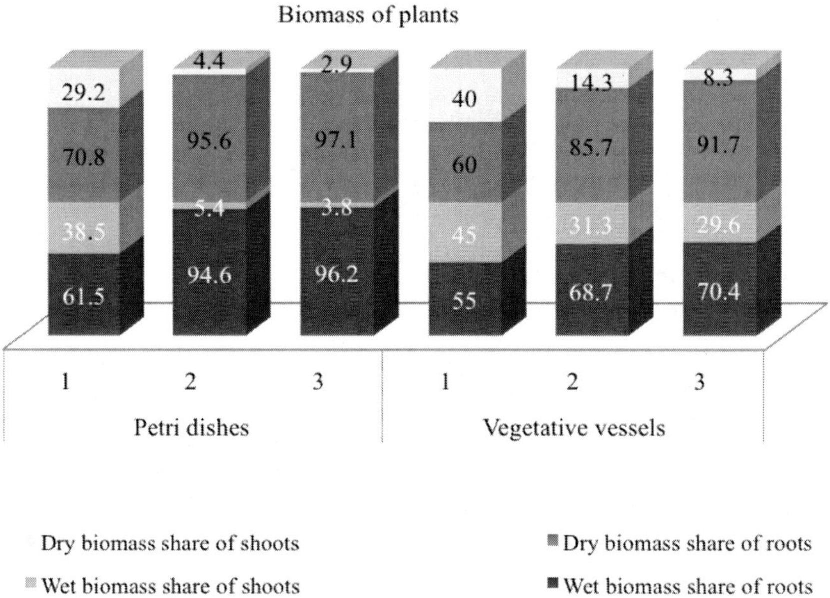

FIGURE 25.3 The ratio of the root system and shoots in wet and dry biomass of seedlings, %. (1 – Control, 2 – NaCl, 3 – Na_2SO_4).

Similar results were received in the experiment with sulfate salinization. In the structure of the wet and dry seedlings biomass roots predominated, the part of which accounted for 96.2 and 97.1% correspondingly.

In the vegetative experiment the analysis of the seedlings dry biomass components of the explored components showed that to the share of the over-ground sphere in the experiment with NaCl fell up to 31.3% from the general mass, in the experiment with Na_2SO_4–29.6%, when in the control 45.0%; to the share of the roots 68.7, 70.4 and 55.0% correspondingly. In the correlation of the dry mass elements the similar tendency was traced – the part of the sprouts in the experimental variants accounted for 14.3 and 8.3%, in the control – 40.0%, the part of the roots 85.7, 91.7 and 60.0% correspondingly.

With the aim of salt tolerant species revelation score ranging was conducted based on the percentage account of the study characters reduction in the relation of control. According to the results of the research, on the backgrounds with NaCl and Na_2SO_4 in Petri dishes the sorts Ikar, Kazahstanskaya 10, Latona, Lutescense 70, SCENT 3 and Tyumenskaya 80 are singled out which were depressed to a less extent compared to the control on the majority of the studied characters. The variety SCENT 1 turned out to be the most sensitive to chloride and sulfate salinization.

In the vegetative vessels the varieties Kazahstanskaya 10, SCENT 1, Tyumenskaya 80 are attributed to the most tolerant to the chloride salinization, to sulfate – Kazahstanskaya 10, Latona, Lutescense 70. The biggest sensitivity to the operation of NaCl and Na_2SO_4 showed the variety Aviada.

Therefore, at the uncovered general regularity of the seedlings growth processes depression under the influence of the salts the varieties varied from each other considerably on the manifestation of the characters. The spring wheat variety Kazahstanskaya 10 independently of the type of salinization and way of evaluation owed maximum salt tolerance. Moreover, according to the point ranging on the complex of indicators the varieties with minimum reduction of the characters in the stressful conditions – Lutescense 70, SCENT 3, Tyumenskaya 80. To the number of the worst in the first experiment the variety SCENT 1 is attributed, in the second – Aviada.

The evaluation of the PABA influence on the plants salinization.

For the enhancement of the adaptive properties and productivity of the cultivated plant species different biologically active compounds are being used providing the regulation of the growth processes and stimulation of the defensive mechanisms at the present time. The specific properties of low concentrations of PABA as a factor, which is able to activate a wide range of useful for the organism biological processes on the phenotypic level were open by I. A. Rapoport on *Drosophila* in 1939 [30]. PABA is an ecologically clean substance, vitamin, antioxidant taking part in the creation of the ecologically clean production [31]. PABA is used during the correction of the characters development at animals and plants, weakened by unfavorable environmental conditions, for the increase in harvest of crops [32–38].

Garnizonenko, and Beletskiy [39], Bome, and Govoruchina [40] and other authors [41, 42] mark that salt tolerance of plants can be raised with the help of physiologically active substances (para-aminobenzoic acid, weak solutions of the boric acid, glycine betaine, kinetin and benzylaminopurine).

For the evaluation of the para-aminobenzoic acid impact on the germination of the seeds and morphometric parameters of the seedlings germination of 10 wheat varieties in the condition of the chloride salinization in Petri dishes the seeds were treated with the PABA solution in different concentrations. The experiment included the following variants: (1) control 1 (dry seeds, placed into water); (2) control 2 (the seeds 12 h seasoned in the water) (3) experiment 1 (the seeds seasoned in the 0.01% PABA, then placed into the solution of NaCl); (4) experiment 2 (the seeds seasoned in в 0.05% PABA, then placed into the solution of NaCl, experiment 3 (the seeds seasoned in the water, then placed into the solution of NaCl, experiment 4 (dry seeds placed into the solution of NaCl).

During the growth of the seeds on saline substratum a considerable difference from the control variants was marked, according to the indicators of laboratory germination power of the seeds. The percentage of the germinated seeds varied on the provocative background from 23.3 up to 100 in the standard conditions from 80 to 100. High laboratory germination power (more than 80%) was characteristic of the seeds of the varieties Lutescense 70, SCENT 1, SCENT 3. Low germination ability (23.3–61.7% in different variants of the experiment) distinguished the varieties Irgina and Kazahstanskaya 10. On average on the varieties with 0.01% (experiment 1) and preliminary soaking of the seeds in water (experiment 3) of statistically trustworthy differences were not noticed. The solution of PABA in the concentration 0.05% led to considerable reduction in the germination power of the seeds (to 63.5%). Analogical figure was observed in the experiment 4, in which the percentage of all the germinated seeds accounted for 68.8, while in the control variants the given indicator was equal to 87.4 and 86.7%.

In the conditions of the chloride salinization the increase in the amount of the embryonic roots in all the experimental variants in the majority of the varieties took place, the biggest number of which SCENT 3 had (5.7±0.11 pcs. – experiment 2), their smallest number is recorded at the variety Ikar (3.5±0.22 pcs. – experiment 4). On the whole the seedlings on the provocative background exceeded the control variants on that character by 6.7–19.0%. On average the number of the roots varied from 4.2 in the control 2 up to 5.0 pcs. in experiment 1 (Table 25.3).

TABLE 25.3 The Variability of the Characters in the Root System and Sprouts of Spring Wheat in Different Variants of the Experiment (On Average at 10 Varieties)

Variant	The number of roots, pcs.	Length of roots, mm	Length of a sprout, mm
Control 1	4.5±0.22	122.5±3.58	94.9±4.71
Control 2	4.2±0.18	120.9±4.90	89.8±6.46
Experiment 1	5.0±0.14•	32.2±3.66*•	32.4±4.61*•
± to control, %	+11.1 +19.0	-73.7 −73.4	-65.9 −63.9
Experiment 2	4.9±0.20•	28.4±3.40*•	28.6±4.97*•
± to control, %	+8.9 +16.7	-76.8 −76.5	-69.9 −68.2
Experiment 3	4.8±0.20•	33.6±3.82*•	32.2±5.28*•
± to control, %	+6.7 +14.3	-72.6 −72.2	-66.1 −64.1
Experiment 4	4.9±0.20•	28.8±3.21*•	25.0±4.50*•
± to control, %	+8.9 +16.7	-76.5 −76.2	-73.7 −72.2

Notice: The number of roots: top row - deviation from the control 1, %, the bottom line - deviation from the control 2, %; statistically trustworthy differences P<0,05: * – with control 1; • – with control 2.

The changeability of the morphometric parameters of the primary root system and over-ground organs at the salinization can characterize the extent of the sustainability of the population, its ability to resist unfavorable factors.

On the background with NaCl of all the studied varieties weakly developed root system and sprouts were characteristic. The primary root system considerably decreased in stressful conditions (by 54.2–88.3%) if to consider the experimental seedlings. On the length of the embryonic roots the variety Lutescense 70 was distinguished which had the most developed root system in all the experimental variants. The least length of the roots was marked at the variety Novosibirskaya 31 both in the control conditions and on the background of the salinization.

Eight varieties (Ikar, Irgina, Lutescense 70, Novosibirskaya 15, Novosibirskaya 31, SCENT 1, SCENT 3, Tyumenskaya 80) reduced their length of the root system during the treatment of the seeds with 0.01% solution of the PABA at least compared to the experiment 4. At the time of the preliminary seeds treatment with water in the experiment 3 the reduction of the given parameter at the seven mentioned above varieties was marked (with the exception of Ikar). The varieties SCENT 1 and SCENT 3 turned out to be more responsive to the treatment of the seeds with PABA; the percentage of the reduction of the character with the relation to the controls accounted for from 64.6 up to 69.0 in the experiment 1 (0.01% PABA) and from 71.4 up to 75.2% in the experiment 4 (on the background with NaCl). The treatment of the PABA seeds in the concentration 0.05% led to the reduction of the root system at the varieties Kazahstanskaya 10 and Novosibirskaya 15. On average on the varieties in the experimental variants varied from 28.4±3.21 mm up to 33.6±3.82 mm, in the control these indicators were higher (122.5±3.58 mm in control 1 and 120.9±4.90 mm in control 2).

Chloride salinization led to considerable reduction in the sprout length at all the varieties (by 28.5–95.8%). The varieties changed the given indicator on average from 25.0±4.50 mm in experiment 4 to 32.4±4.61 in the experiment 3. On the length of the sprouts in experimental versions the variety SCENT 1 was distinguished which had the biggest indicators. Weakly developed over-ground organs were characteristic of the varieties Aviada, Ikar and Irgina. At the same time compared to the control variants the plants of the variety Lutescense 70 were depressed to a less extent (by 26.9–41.9%), which says about their higher tolerance to the salinization. Under the influence of PABA the percentage of the characters reduction varied from 28.5 to 90.0 in the experiment 1 and from 41.6 to 95.8 in the experiment 4. The preliminary treatment of the seeds with PABA solutions and water led to the increase in he given parameter at the varieties Aviada, Irgina, Lutescense 70, SCENT 1, SCENT 3, Tyumenskaya 80.

In the stressful conditions the reduction in the wet mass of the root system and sprouts was observed. At that in comparison with the control variants at the variety Lutescense 70 reduced the mass of the root system noticeably (by 21.4–42.9%), the rest of the varieties – the mass of the sprouts (by 44.4–95.0%). During the treatment of the seeds with PABA the reduction of the character was less visible at the majority of the varieties. The majority of the varieties had less considerable character reduction during the treatment of the seeds with PABA.

The varieties Aviada, Irgina, Kazahstanskaya 10, Lutescense 70, Novosibirskaya 15, Novosibirskaya 31, SCENT 1, Tyumenskaya 80 increased their dry mass of the roots by 50.0–300.0% in the experimental variants compared to controls. It may be connected with the excess ingress of salt ions to the root system in the result of its lesion. Ambiguous data on the dry sprouts mass were received. The variety SCENT 3 has both the increase and the reduction in the given indicator in different variants of the experiment. The variety Lutescense 70 increased the dry mass of the sprouts but the rest of the varieties reduced it. In the majority of the cases no essential differences between the experimental variants on the dry mass of the roots and sprouts under the influence of PABA were revealed.

Biomass of a plant along with the primary root system is defined to a great extent by the development of the over-ground organs in the early ontogenesis and their tolerance to unfavorable environment factors.

In the structure of the wet and dry biomass on the background with the salinization in standard conditions roots predominated (Fig. 25.4). On average on the share of wet root system mass fell 57.9% (control 2) to 80.0% (experiment 4), after the drying – from 11.8% (experiment 2. 4) to 37.5% (control 2).

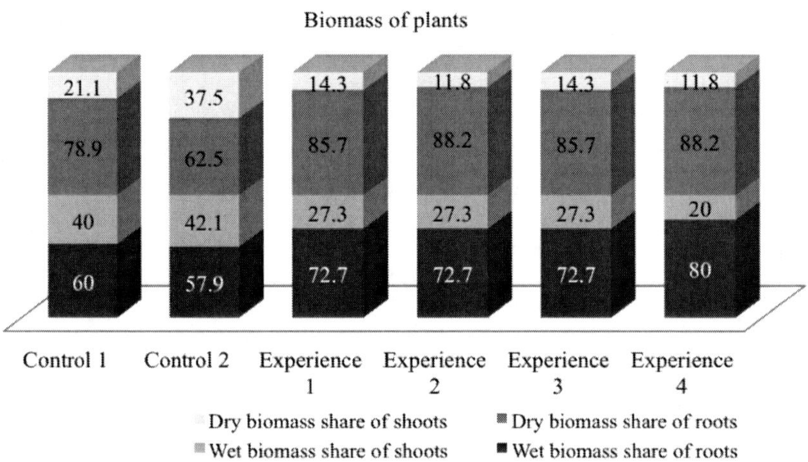

FIGURE 25.4 The correlation of the root system and sprouts in the wet and dry biomass of the seedlings, %. Control 1 – dry seeds; control 2 – seeds seasoned in the water 12 h; experiment 1 – seeds seasoned in the 0.01% PABA 12 h then placed into NaCl solution; experiment 2 – seeds seasoned in 0.05% PABA 12 h then placed into NaCl solution; experiment 3 – seeds seasoned in the water 12 h and then placed into NaCl, experiment 4 – dry seeds, placed into the NaCl solution.

During the point ranging holding with account of the percentage of the studied characters reduction with the relation to the control 1 it is set that the varieties Lutescense 70 and SCENT 1collected the biggest amount of points steadily in all the experimental variants. In the experiment 3 (preliminary treatment of the seeds with water) and experiment 4 (with the use of dry seeds) the varieties Ikar and Irgina collected minimal amount of points. In respect of control 2 (with the preliminary treatment of the seeds with water) only the variety Lutescense 70 is attributed to the best from the above mentioned, to the worst – Ikar and Irgina. The given varieties reacted positively to the seeds treatment with PABA in the concentration of 0.01%, with 0.05% of PABA. It had a positive influence on the variety Ikar and did not have any impact on the variety Irgina, which had only minimal points in the experiment 2.

It follows from the above said that the chloride salinization has a negative influence on the seedlings, causing the reduction of such parameters as the length of the roots and sprouts, wet mass of roots and sprouts, dry mass of sprouts. The number of the embryonic roots and their dry mass serve as an exception, which increases concerning their control variants.

The preliminary treatment of the seeds with water and PABA solution in the conditions of the salinization influenced positively on the manifestation of the morphometric parameters of the sprouts. Under the influence of 0.01% PABA the length of the root system and sprouts as well as their dry mass decreased to the less extent on exposure of the stress in all the variants. The PABA solution in the concentration 0.05% in a number of cases enhances the adaptive characters of the seedlings, but less effective.

25.4 CONCLUSIONS

At the time of the comparison of the sulfate and chloride salinization it is set that maximum sensitivity of the spring wheat varieties to the impact of the stress factor shows itself on the stages of seed germination and seedling development (on the 7-th day), later on (on the 10-th day) the differences between chloride and sulfate salinization are less expressed.

In the result of the detailed evaluation of spring wheat varieties on the complex of characters it is set that all the types of salinization render a negative impact causing the reduction of morphometric parameters of the seedlings. At the same time the carbonate salinization is the most baneful, but sulfate one operates stronger than chloride one on the first stages of plant ontogenesis.

During the usage of two diagnostics methods no equal results in the relation of salt tolerance at the studied varieties of the spring wheat are received. The conduction of the experiment in Petri dishes is different in its less continuation and laboriousness, which allows carrying out the evaluation of large material scope. The growth of plants in vegetative vessels gives the possibility of bigger amount of parameters account, which increases the precision of the experiment as far as we are concerned.

Variety-specific reaction to the treatment of the seeds in different concentration of PABA was discovered. The PABA effect was more considerable for varieties with their low salt tolerance (Ikar, Irgina). The optimal concentration of PABA, providing a more favorable manifestation of the characters on the phenotypic level, accounted for 0.01%.

KEYWORDS

- **ontogenesis**
- **para-aminobenzoic acid**
- **salt tolerance**
- **spring mild wheat**

REFERENCES

1. Boiko, L. A. (1969). Physiology of the Root System in the Conditions of Salinization, Boiko, L. A. Leningrad, Nauka, 164p (in Russian).
2. Kovda, V. A. (1946). The Origin and Saline Regime of Soils, Kovda, V. A. Moscow-Leningrad, Publishing House of AS USSR, *1*, 568p (in Russian).
3. Chirikova, T. V. (2002). Physiological Foundations of Plant Resistance, Chirikova, T. V., Street Petersburg, Publishing House of Street Petersburg University, 244p (in Russian).
4. Udovenko, G. V. (1977). Salt Tolerance of Cultural Plants, Moscow, Kolos, 150p (in Russian).
5. Kovrigina, L. V. (2006). Comparative Evaluation of Salt Tolerance of Barley Varieties in the Laboratory Conditions, Kovrigina, L. V., Zaushintseva, A. V., & Petunkina, L. O., Proceedings on Applied Botanics, Genetics and Selection, *162,* 44–48 (in Russian).
6. Shahov, A. A. (1956). Salt Tolerance of Plants, Shahov, A. A. Moscow, Publishing House of AS USSR 552p (in Russian).
7. Pankova, E. I. (2006). Saline Soils of Russia, Pankova, E. I., Vorobyova, L. A., Gadgiev, I. M. et al., Moscow, Akademkniga, 854p (in Russian).
8. Dobrovolski, G. V. (1998). Soils, Encyclopedia of the Russian Nature, Dobrovolski, G. V., Sheremet, B. V., Aphanasyeva, T. V. & Palechek, L. A., Moscow, ABF Publishing House, 368p (in Russian).
9. Bakulin, V. V. (1996). Geography of the Tyumen District, Bakulin, V. V. & Kozin, V. V. Ekaterinburg, Middle-ural publishing House, 240p (in Russian).
10. Udovenko, G. V. (1988). Diagnostics of the Plant Tolerances to Stress ful Impact (Methodical Guidance), Udovenko, G. V., Leningrad, VIR, 228p (in Russian).
11. Udovenko, G. V. (1970). The Diagnostics Method of Plant Resistance, Udovenko Leningrad, G. V., Gidrometeoizdat, 74p (in Russian).
12. Gurova, T. A. (2001). Assessment of Salinity Tolerance Varieties of Spring Wheat and Barley Fluorescent Methods, Gurova, T. A., & Berezin, Yu V. Methods and Technical Tools of Studying Physical Processes in Agriculture, Proceedings of Siberian Physical-Technical Institute, Novosibirsk, 18–26 (in Russian).

13. Polonskiy, V. I. (2003). Assessment of Crop Plants for Resistance to Adverse Environmental Factors, Monograph, Polonsky, V. I. & Surin, N. A. Novosibirsk, Russian Academy of Agricultural Sciences, Siberian Branch, 128p (in Russian).
14. Methodological Guidance for Determining the Salt Tolerance of Barley (1980). Leningrad, VIR, 14p (in Russian).
15. Semushina, L. A. (1976). Morphological and Physiological Changes in the Roots of Wheat and Barley under Salinity, Proceedings of Applied Botany, Genetics and Breeding, *57(2),* 107–110 (in Russian).
16. Jurba, T. P. (1984). Rapid Method for Estimating Endurance Samples of Rice to Salinity, Jurba, T. P., & Vorobyov, N. V., Breeding and Seed, *11*, 22–24 (in Russian).
17. Udovenko, G. V. (1977). Using Bioelectrical Plant Response in Determining their Salt Tolerance, Udovenko, G. V. & Semushina, L. A. Journal of Agricultural Science, *4,* 124–125 (in Russian).
18. Beletsky, Yu D. (1987). A Method for Selecting Salt-Tolerant Sunflower Forms, Beletsky, Yu D., Sizov, L. I., Karnauhova, T. B., & Korobov, L. N. USSR Author's Certificate, 1337005, Bull, *34* (in Russian).
19. Rapoport, I. A. (1989). PABA Action in Connection with the Genetic Structure, Rapoport, I. A., Chemical Mutagens and Para-Aminobenzoic Acid in Increasing the Yield of Crops, Moscow, Nauka, 3–37 (in Russian).
20. Lakin, G. F., & Biometrics, G. F. (1988). Lakin, Moscow Graduate School, 294p (in Russian).
21. Udovenko, G. V. (1976). Features of Different Methods of Evaluation of Salt Tolerance, Udovenko, G. V., Semushina, L. A., & Sinelnikova, V. N., Methods for Assessing the Resistance of Plants to Adverse Environmental Conditions, Leningrad, Kolos, 228–238 (in Russian).
22. Zhuchenko, A. L. (1988). Adaptive Capacity of Cultivated Plants (Ecological and Genetic Basis), Zhuchenko, A. L., Kishinev, Shtiintsa, 768p (in Russian).
23. Henkel, P. A. (1970). Basic way of Exploring the Salt Tolerance of Plants, Henkel, P. A., Agricultural Biology, *2*, 292 302 (in Russian).
24. Aldarov, A. A. (2007). Salt Tolerance Wheat Species, Aldarov, A. A. & Shikhmuradov, A. S. Genetic Resources of Cultivated Plants in XXI Century, Abstracts of II Vavilov International Conference on 26–30 November (2007), Street Peterburg, 400–401 (in Russian).
25. Udovenko, G. V. (1982). Impact of Extreme Environmental Conditions on the Structure of Agricultural Crop Plants, Udovenko, G. V. & Goncharova, E. A. Leningrad, Gidrometeoizdat, 215p (in Russian).
26. Houchi, R. (1988). Effects of NaCl and Na_2SO_4 on Water Relations of Plantago Maritime, L., Morant-Avice, A., Ferard, G., & Coudret, A. Biologia Plantarum, *6,* 457–460.
27. Ahmadeeva, A. K. (1973). The Role of Primary and Secondary Roots of Barley in the Absorption of Salts from the Soil, Ahmadeeva, A. K., Agricultural Biology, *8(7),* 927–928 (in Russian).
28. Polevoy, V. V. (1989). Plant Physiology, Field, V. V. Moscow, Graduate School, 464p.
29. Henkel, P. A. (1954). Salt Tolerance of Plants and Ways to Improve the Directional, Henkel, P. A. XII Timiryazevskiy Reading, Moscow, Academy of Science of the USSR, 84p (in Russian).
30. Stroeva, O. G. (2001). Phenotypic Activation, a New Scientific Trend Established, Rapoport, I. A., & Stroeva, O. G. Iosif Abramovich Rapoport, Scientist, Warrior, Citizen, Essays, Memoirs, and Materials, Moscow, Science, 178–182 (in Russian).
31. Eiges, N. S. (2013). The Historical Role of Joseph Abramovich Rapoport in Genetics, Further Research using the Chemical Mutagen, Eiges, N. S. Vavilov Journal of Genetics and Breeding, *17(1),* 162–172 (in Russian).

32. Bome, N. A. (2007). Resistance of Crop Plants to Environmental Stress, Monograph, N. A., Bome, A. Y., & Bome, N. A., Belozerova, A. A, Tyumen, Publisher Tyumen State University, 192p. (in Russian).
33. Eiges, N. S. (1993). Laws Action Para-Amino Benzoic Acid in Cereal, Eiges, N. S., & Weisfeld, L. I. Chemical Mutagenes is and Objectives of Agricultural Production, Moscow, Science, 191–198 (in Russian).
34. Novikova, P. N. (2011). Efficiency of Use of PABA in Improving Drought Resistance of Spring Wheat, Novikova, P. N. & Belozerova, A. A. (2011). Tobolsk Research Proceedings of the VIII All-Russian Scientific-Practical Conference, Tobolsk, Poligrafisdat P, 54–55 (in Russian).
35. Cherkashina, I. A. (2012). Increased Productivity and Adaptive Properties of *Triticum aestivum* L with a Physiologically Active Substances, Cherkashina, I. A., & Bome, N. A. (2012) Tobolsk Research Materials IX All-Russian Scientific and Practical Conference, Tobolsk, 9–10 November (2012), Tyumen, Tyumen Publishing House, 168–171 (in Russian).
36. Mostovshchikova, S.M. (2013). Assessing the impact of para-amino benzoic acid on morphometric parameters of wheat plants (*Triticum aestivum* L.) under conditions of chloride salinity / Mostovshchikova, S.M. & BelozerovaA.A. Successes contemporary science, 8, 20–21
37. Bekuzarova, S. A., Bome, N. A., & Weisfeld, L. I. (2012). Activation of Cereal Seeds by Para-Aminobenzoic Acid, Journal of Information, Intelligence and Knowledge, *4(4),* 317–324.
38. Bekuzarova, S. A., Bome, N. A., Weisfeld, L. I., & Luschenko, G. V. (2012). Activation by Para-Aminobenzoic Acid of Sowing Properties of Seeds of Winter Grain Crops and Forage Cereals, Bekuzarova, S. A., Bome, N. A., Weisfeld, L. I. & Luschenko, G. V. Polymers Research Journal, *7(1),* 1–8.
39. Garnizonenko, T. S. (1989). Effect of PABA on the Yield of Barley Plants Grown under Saline Conditions, Garnizonenko, T. S., & Beletsky, Y. D. Chemical Mutagens and Para-Aminobenzoic Acid in Increasing the Productivity of Agricultural Plants, Moscow, Nauka, 126–129 (in Russian).
40. Bome, N. A. (1998). Efficiency of Para-Aminobenzoic Acid on the Ontogeny of Plants under Stress, Bome, N. A. & Govorukhina, A. A. Bulletin of the Tyumen State University, *2*, 176–182 (in Russian).
41. Genckel, P. A. (1975). Plant Physiology, Henkel, P. A., Moscow, Education, 335p (in Russian).
42. Iqbal, M. (2006). Wheat Seed Priming in Relation to Salt Tolerance, Growth, Yield and Levels of Free Salicylic Acid and Polyamines, Iqbal, M., & Ashraf, M. Annual Botancy Fennici, *43(4)*, 250–259 http://www.sekj.org/PDF/anbf43/anbf43–250.pdf.

CHAPTER 26

ADAPTIVE REACTION OF *TRITICUM AESTIVUM* L. ON SOIL AND CLIMATIC ENVIRONMENTAL FACTORS UNDER IMPACT OF PHYSIOLOGICALLY ACTIVE COMPOUNDS

IRINA A. DUDAREVA[1], and NINA A. BOME[2]

[1]Tobolsk complex scientific station, Ural Division of the Russian Academy of Sciences, d. 15, Yuriy Osipov St., Tyumen area, Tobolsk, 626152, Russia; E-mail: sovet29@rambler.ru

[2]Tyumen State University, d. 1, Semakova St., Tyumen, 625003, Russia; E-mail: bomena@mail.ru

CONTENTS

ABSTRACT

The results of studies of the cultivated residual-carbonate soddy small podzolic soil of the subtaiga area of the Tyumen region by morphological features, physical properties and content of chemical elements. There is determined the positive effect of physiologically active compounds on the chemical composition of spring wheat plants. There are identified the peculiarities of distribution of chemical elements in the vegetative organs of plants and grain. There is noted the positive correlation between the content of Cu, Mn, Pb, Sr and pigments in leaves of plants. There are established essential differences between the varieties of wheat at the rate of biological uptake of elements from the soil.

26.1 INTRODUCTION

Growth and development of any vegetable organism are determined by a complex of environmental factors specific to a particular region [1]. Conditions of Siberia are characterized by a significant length of Siberia territory in the meridional direction, as well as difficult terrain with a wide variety of soil types, their different physicochemical properties, highly variable climatic conditions. Factors limiting the livelihoods of cultural Phytocoenosis in northern latitudes are heterogeneity of weather and lack of mineral nutrients in the soil. In this aspect, there is actual the application of physiologically active compounds (FAC) on soft spring wheat plants for minimizing the action of stress factors and activation of biochemical processes in complex environmental conditions.

The study of the chemical elements Cu, Mn, Pb, Sr in the composition of the soil and plants is significant. According to chemical nomenclature these elements are particularly necessary for normal growth and development of the plant organism physiological processes. Furthermore, they can belong to the group of heavy metals if there is large content of them in the soil, and to the group of macro and microelements if the concentrations are optimal [2, 3]. The availability of these elements to plants of spring wheat in the conditions of the northern territories, and the impact on their amount in vegetative organs during the action of FAC have been studied not enough.

The objectives of this study were: determination of physical and chemical properties, the gross contents and forms of chemical elements (Cu, Mn, Pb, Sr) in the cultivated residual-carbonate soddy small podzolic soil of the Tobolsk district; the chemical composition of the vegetative mass and grain of spring wheat plants; coefficients of biological absorption; the study of the pigment complex in the cells of flag leaves in heading stage.

26.2 MATERIALS AND METHODS

Field studies were carried out in 2009–2011 at the experimental plot of Small Zorkaltsev village, the Tobolsk district, Tyumen region, which is located in accordance with agro-climatic zoning in the subtaiga zone – 58°25'00" N, 68°24'00" east longitude. Field studies packing were carried out according to the method developed by Dospehov [4]. Total area of the plots – 1 m^2, four-order repeatability, the number of seeds sown per plot – 650, row spacing – 15 cm, seed depth – 5–6 cm.

Data on meteorological conditions during the years of study and the average annual values were obtained from the "United hydro meteorological stations" of the Tobolsk region.

Type of the soil – cultivated residual-carbonate soddy small podzolic soil on ancient alluvial deposits with a wavy surface due to plowing furrows. No signs of erosion has been revealed, profile character – simple, unstrained. Textural and structural heterogeneity of the profile were evidenced by – interlayers of heavy (to the horizon B) and light (to the horizon C) granulometric composition. No signs of gleyzation, carbonate content – weak. Profile's power is of 110 cm/t. Soil profile structure shall be as follows: Ap (0–38 cm), E (38–48 cm), U (48–76 cm), Bh, f, al (76–93 cm), C (93–110 cm).

Laboratory studies were carried out on the basis of an accredited laboratory "Ecotoxicology" of Tobolsk complex scientific station. Selection of soil samples for physicochemical studies, determination of chemical elements was carried out in accordance with GOST requirements, adopted in the Russian Federation.

In order to determine the moisture content and physical properties of soil, the samples were taken with the help of soil sampling tube in May (15.05.2009–2011), June (15.06.2009, 24.06.2010, 26.06.2011), July (17.07.2009–2011), August (15.08.2009–2011), September (12.09.2009, 19.09.2010, 17.09.2011), during a relatively dry period of the month – on the days without precipitation. Samples of soil for chemical analysis were selected twice each year during the growing season (June 15 and August 15 – combined sample). Sampling was carried out at 30 locations at the plow-depth of 30 cm; the experimental plot was divided lengthwise into three zones, 10 samples-each.

Determined: pH of the aqueous extract (18-channel instrument Anion-7050), the dry residue, the amount of anions, mg·eq – Cl^-, SO_4^{2-}, HCO_3^-, cations, mg·eq – Mg^{2+}, Ca^{2+}, Na^++K^+ (calculation method), nutrients, mg/kg – NH_4^+, NO_2^-, NO_3^-, $H_2PO_4^-$ and HPO_4, humus, % – according to methods developed by I.V. Tyurin and modified by V.N. Simakov. Soil density, solid phase density, porosity, aeration or air supply – were determined according to methods developed by Kaczynski.

Gross and fractional content of chemical elements in the soil was determined by atomic emission methods on inductively coupled plasma spectrometer OPTIMA-7000 DV (PerkinElmer) and repeated twice. Sequential fractionation of chemical elements was performed using the Sposito method.

We have studied the effect of presowing treatment of spring wheat varieties, such as – Annette, Irene, Ricks, Icarus with physiologically active compounds (8 h exposure) on the elemental composition of the vegetative mass and grain, biological absorption coefficients, pigment flag leaf complex. Experimental compounds concentration was as follows: zircon (a mixture of hydroxycinnamic acids in alcohol – 0.1 mg/cm^3) – 4.98×10^{-5} g/cm^3, and appin (24-epibrassinolide) – 1.25×10^{-5} g/cm^3.

Vegetable samples were ashed in a mixture of ultrapure concentrated HNO_3 (acid cleaning BSB-939-IR) and concentrated H_2O_2 in microwave decomposition pressure system Speedwave MWS-2 (made in Germany BERGHOF Products + Instruments Gmb H), using individually selected mode.

During assessing the content of chemical elements in the soil the maximum permissible concentration (PDK) [5] and normalization scale of elements concentrations in soils [6] were used. Conditional landmark World clarke soils [7] were taken as a background. Calculation and evaluation of chemical concentrations in soil (Kc) was made according to methods developed by Ozersky [8], the coefficient of biological absorption of chemical elements from the soil (PCU) – was made according to Perelman [9].

Content of photosynthetic pigments in the flag leaves cells of spring wheat during the heading stage was determined with a spectrophotometer UNICO-1200, using the electronic program Hlorof. The results were processed statistically.

26.3 RESULTS AND DISCUSSION

Meteorological conditions during the period of studies were characterized by high variability, both over years and during the growing season. In general, they reflected the main features of the cold climate zones of the subtaiga area of the Tyumen region: lack or surplus of moisture, the frequency of high summer temperatures with sharp declines and fluctuations even during one day.

Average daily air temperature ranged from 10.8 to 17.2°C in 2009, from 10.4 to 17.5°C – in 2010, from 10.9 to 18.0°C – in 2011. The maximum average daily temperature recorded in 2011 was 18.0°C (June) and the minimum – 10.9°C (May). May 2009, 2010 and 2011 was particularly hot and differed from normal at 1.1°C. The average daily temperature in August and September 2009 and 2011 exceeded the average annual values at 0.7 and 2.5°C, respectively. The active temperatures amount was as follows: in 2009 – 1977.4°C, in 2010 – 1855.4°C, in 2011 – 1925.3°C (the average excess over normal 219°C).

The amount of precipitation during the growing season in 2009 (311.3 mm) and 2011 (358.2 mm) was close to normal. Vegetation period in 2010 can be characterized as arid, because the precipitation amount was 221.9 mm, which is below normal at 73.1 mm. Distribution of rainfall during the growing season was uneven, and dry periods observed during the ontogeny of plants were as follows: 2010 – July (19.9 mm), 2011 – May (8.5 mm).

May of 2009–2011 is characterized by low soil moisture. Critically low moisture content in the soil was noted in 2011 (3.4%). In June, the soil moisture content was as follows: 11.7% – in 2009, 13.6% – in 2010, 39.4% – in 2011. Accordingly, in 2011 the soil moisture content was most favorable for amicable and full germination, plant growth and development. June 2009 and 2010 was critical to the soil moisture content, which affected the germination and survival of plants.

Water consumption is especially important during the phase of leaf-tube formation and heading stage, that is, through the period of reproductive organs formation (July). 2010 was the most difficult for wheat plants, with its dry season, low rainfall and high temperatures, the humidity level was – 6.8%. August of 2011 was also less favorable for the crop due to the soil moisture content at a level of 9.0%. This is called the milk phase of grains when wheat consumes 20–30% of the total moisture in the growing season. Warm air temperatures characterize September, moderately favorable rainfall during the growing season and optimum soil moisture. All this facts are contributed to the full maturation of the grain in the ear.

Average data for the years of soil density research were as follows: 0.188±0.0001 cm^3 (May) – 0.899±0.055 g/cm^3 (June). Such substrate density is favorable for the growth and development of spring wheat within the area. Dense phase of the soil is on the following level: 1.018±0.000 g/cm^3 (August) – 1.117±0.000 g/cm^3 (July). The soil constitution is crumbly and loose, it is well–structure, but structural units are poorly cemented with each other. The data indicates that the soil contents mineral inclusions and organic substances in sufficient quantities.

Porosity and soil aeration were at a high level. Overall performance has been close to 100% of the total porosity and aeration. Maximum porosity value was fixed in May – 91.2% minimum value was fixed in July – 69.9%. The aeration ranged from 36.3% (July) to 88.4% (May). Obviously, these results are related to the amount of moisture in the soil and precipitation, which has occurred during this period.

The soil of slightly alkaline type is determined by the following factors: residual carbonates, calcium and magnesium ions in the cations composition, low content of hydrocarbons, long organic fertilizers. Dry residue – an indicator of soil salinity – is 0.35%. Nitrogen is excreted in three forms: ammonium (9.30±3.42 mg/kg), nitrate (8.87±1.79 mg/kg), nitrite (8.77±2.31 mg/kg) – the average content in the soil was fixed during 2009–2011. Phosphorus is found in large quantities and is available for plants mobile forms $H_2PO_4^-$ and HPO_4^- (268.7±98.65 mg/kg). Humus content in the soil is low: 1.9% – in 2009, 1.5% – in 2010 and 1.8% – in 2011.

Uptake and transport of chemicals in the soil-plant system and in plant bodies largely depends on a number of factors; environmental pH, calcium and phosphorus content are very important, as well as the ratio of some chemical elements in the soil solution. The forms in which they are presented in the soil, that is, the components solubility degree, and the way they are linked to solid and mineral phase of soil, mostly cause the elements shortfall in plants. There is often a deficiency of elements in soils with high pH values, soils containing carbonates and soils with low organic matter content [10].

According to the results of statistical data processing and annual dynamics analysis, it has been revealed that the number of related forms for Cu in the soil remains relatively constant and is within experimental error. There has been a slight increase in the number of Sr; the maximum was recorded in 2011. There has been a decrease of Mn related forms – 68.57–59.22 mg/kg and Pb – 8.77–5.63 mg/kg (Table 26.1).

TABLE 26.1 Contents of Cu, Mn, Pb, Sr in Residual-Carbonate Soddy Small Podzolic Soil of the Tobolsk Region (2009–2011)

Elements	Gross contents, mg/kg	Connected forms, mg/kg	Mobile forms, mg/kg
Cu	0.99±0.19	0.57±0.0001	0.42±0.0001
Mn	96.65±1.97	64.98±2.90	31.67±2.70
Pb	25.71±1.64	6.74±1.01	18.97±0.0001
Sr	1.86±0.12	1.04±0.000	0.85±0.0001

The mobile forms of elements (capable of migration and actively participating in biochemical processes in the soil-plant) have been defined and separated into fractions (exchangeable, organic, carbonate, oxide, residual, water-soluble) in order to identify the role of these chemicals in the plant organism.

Dominant factions of the lead are organic and exchangeable – 28.7% and 26.9%, respectively (of the total number of mobile forms). The content of water-soluble forms reaches 20.4%. Lead is associated with various soil components (mineral components, hydroxides and oxides, colloids) and organic substance in almost equal proportions and forms stable complexes with these substances. These factional compounds play major role in plant nutrition. Water-soluble forms are the most mobile and thus potentially more transportable, since they transfer surface and ground waters and can be easily involved in biogeochemical migration, available for plants and participate in their nutrition process. Exchangeable and water-soluble forms are direct nutrition reserve of plants. The highest strontium content in the soil revealed in organic (27.1%) and exchangeable (21.2%) fractions. The predominant manganese proportion is presented in form of stable complexes with organic matter (60.6%) and only 16.7% referrers to exchangeable form. The copper content in soil associated with carbonates and bicarbonates has been revealed in one faction – carbonate (100%).

It has been established that Mn, Pb, Sr elements can be found in varying amounts in all fractions under study in bound and movable forms, representing its gross structure. Most of the elements revealed are in organic, exchangeable and water-soluble forms, being a direct source of plants nutrition. Nature and form of the elements migration capacity are determined by the properties of the element itself, its reactivity, connections pattern and migration conditions – temperature, humidity, pH, anion-cation ratio, biogenic compounds and organic matter in the soil.

As measured by the chemicals concentration in soil, it has been revealed that compared to the conventional world Clarke, Pb can be characterized by high content in the soil, evidencing its accumulation, but not above the PDK level. Chemical concentration coefficient (K_c) has reached – 2.62. Element high concentration may respond to geochemical characteristics of parent rocks. There is a clark deficiency for Mn, Sr, Cu and their critically low content in the soil (K_s = 0.11, 0.006, 0.04, respectively) suggests the elements removal (this may be due to soil leaching regime).

Content of these elements in the soil is very important for plants, especially during the cultural phytococnosis. The chlorophyll content in the plant is increased under the influence of copper, enhancing the photosynthesis. Manganese, as well as copper, plays an important role in the redox reactions of the plant. It is a part of enzymes, which take part in these processes. Element are involved in the photosynthesis reaction, respiration, carbohydrate and protein metabolism. Strontium is also very important for metabolic processes, formation and growth of chloroplasts and is a substitute for calcium. Lead, in very small quantities, is also important for the plant body as a trace element and its role is not fully understood.

The ratio of elements content in the plants ash to their content in soil has been used to assess the plants extraction degree of chemical elements from soil and accumulation of the biomass. Preplanting zircon and appin treatment had a positive impact on the elements content in plants and in most cases helped to increase the biological absorption coefficients. Perhaps, this is due to an increase of the endogenous level of phytohormones in plant and biochemical processes strengthening, which contributed to extraction of elements from the soil solution (Table 26.2).

The content of Mn in Annette variety has increased in plants vegetative mass and grains during zircon and appin treatment. In contrast to the controls – Pb has been detected. Cu content under the decreased by 19.5% influence of these specimens. Sr content has increased after appin treatment of seeds.

Icarus variety has demonstrated and increase in Cu and Mn content in experimental variants of plants compared to the controls. Strontium has been revealed in grains – in its minimum amount and in the vegetal mass – in its maximum amount.

TABLE 26.2 The Content of Chemical Elements in Soft Spring Wheat and Biological Absorption Coefficients (PCU)

Grade	Experience option	Mn		Cu		Sr		Pb	
		mg/kg	PCU	mg/kg	PCU	mg/kg	PCU	mg/kg	PCU
Annette	Control	2.78±0.05	0.03	1.99±0.00	2.58	2.61±0.02	1.39	0.00	0.00
		0.44±0.01	0.01	2.90±0.03	3.76	0.10±0.01	0.05	0.00	0.00
	Zircon	3.29±0.00	0.03	1.60±0.00	2.08	2.56±0.00	1.36	0.20±	0.01
		0.51±0.00	0.01	2.25±0.01	2.92	0.08±0.00	0.04	0.00	0.00
								0.00	
	Appin	3.40±0.01	0.04	1.61±0.00	2.08	2.72±0.02	1.45	0.19±	0.01
		1.58±0.00	0.02	2.26±0.00	2.94	0.08±0.01	0.04	0.05	0.00
								0.00	

Icarus	Control	4.61±0.01	0.05	1.98±0.02	2.57	3.29±0.03	1.75	0.00	0.00
		2.68±0.01	0.03	2.26±0.03	2.94	0.19±0.01	0.10	0.00	0.00
	Zircon	4.89±0.03	0.05	1.99±0.00	2.58	3.41±0.03	1.81	0.00	0.00
		3.37±0.05	0.04	2.31±0.01	2.95	0.15±0.00	0.08	0.00	0.00
	Appin	5.12±0.03	0.05	2.36±0.02	3.06	4.23±0.04	2.25	0.00	0.00
		5.46±0.03	0.06	2.27±0.02	2.95	0.16±0.00	0.09	0.00	0.00
Iren	Control	2.65±0.02	0.03	1.95±0.02	2.53	3.47±0.01	1.84	0.00	0.00
		1.93±0.01	0.02	2.30±0.01	3.00	0.12±0.01	0.07	0.00	0.00
	Zircon	2.43±0.00	0.03	1.96±0.00	2.55	3.41±0.03	1.81	0.00	0.00
		1.94±0.00	0.02	2.30±0.00	2.99	0.13±0.03	0.07	0.00	0.00
	Appin	2.29±0.01	0.02	1.97±0.00	2.56	3.46±0.01	1.84	0.00	0.00
		2.46±0.01	0.03	2.57±0.00	3.33	0.15±0.00	0.08	0.00	0.00
Ricks	Control	3.93±0.05	0.04	1.97±0.00	2.56	3.04±0.01	1.62	0.00	0.00
		2.53±0.02	0.03	2.26±0.02	2.93	0.09±0.01	0.05	0.00	0.00
	Zircon	3.05±0.01	0.35	1.93±0.00	2.51	3.14±0.00	1.67	0.00	0.00
		2.63±0.00	0.03	2.28±0.04	2.96	0.10±0.00	0.05	0.00	0.00
	Appin	3.40±0.05	0.04	1.96±0.00	2.54	3.06±0.00	1.63	0.00	0.00
		3.13±0.02	0.03	3.26±0.01	4.23	0.07±0.00	0.04	0.00	0.00

Note: the numerator – in vegetative mass (culm + leaf), the denominator – the grain; no excess over the MPC values has been revealed, statistically significant results.

The Iren variety turned to be the most responsive to zircon and appin treatment of seeds. Significant increase has been obtained in vegetative parts of the plant and grain Cu. The increase of Mn content in the grain has been observed during the appin treatments (difference with controls – 2.78 mg/kg). Conent of Sr has been increased to 0.13 mg/kg (zircon) and to 0.15 mg/kg (appin) under the control value – 0.12 mg/kg.

Ricks variety reaction on growth regulators characterized by increase of Cu content in plants relative to controls in the experimental variants and Sr – in zircon options (0.10 and 3.14 mg/kg, respectively).

Chemical elements from the soil solution under the FAS influence have been absorbed with different selective ability. Copper, manganese and strontium, according to the CBI, are relatively easily absorbed by the plant organs, whereas lead is poorly absorbed by the plants. Biological absorption coefficient in the Sr soil-plant system refers to the strong accumulation elements in vegetative organs and secondary capture in grain, Cu – is an energetic accumulation element; Mn and Pb – are the weak grip elements.

Biochemical adaptations are reflected in the physiological processes of the plant organism [11]. Photosynthesis – is the process of formation of organic matter from inorganic compounds, which plays an important role in plants energy transformation and adaptive responses to rapidly changing environment.

Photosynthetic pigments (*a, b* chlorophylls and carotenoids) are main pigments of photosynthesis, which ensure absorption and storage of solar energy. Quantitative content, changing their ration within leaves – are important and sensitive indicators of physiological state of plants and photosynthetic apparatus, directional selectivity of adaptive responses when plants are exposed to stress conditions [12]. Decrease of leaves chlorophyll content leads to the proportion increase of b chlorophyll auxiliary pigments or carotenoids, which may be regarded as an adaptive response of the assimilation apparatus [13, 14].

In this aspect, the studies shall be focused on the pigment complex in flag plant leaves, as it is a long-functional complex, which occurs during the whole growing season and is the main donor that supplies assimilates to the ear and the final product – the seeds [15]. This is confirmed by a positive correlation between the total content of photosynthetic pigments in the flag leaf cells and content of chemical elements in the vegetative mass of plants: copper – r = 0.76; manganese – r = 0.63; strontium – r = 0.12.

The ratio of photosynthetic pigments in the flag leaf cells of the experimental is as follows: a chlorophyll (Chl. a) > carotenoids > *b* chlorophyll (Chl. b). It is represented in Fig. 26.1, as exemplified by controls, since under the FAS impact this feature has not been changed.

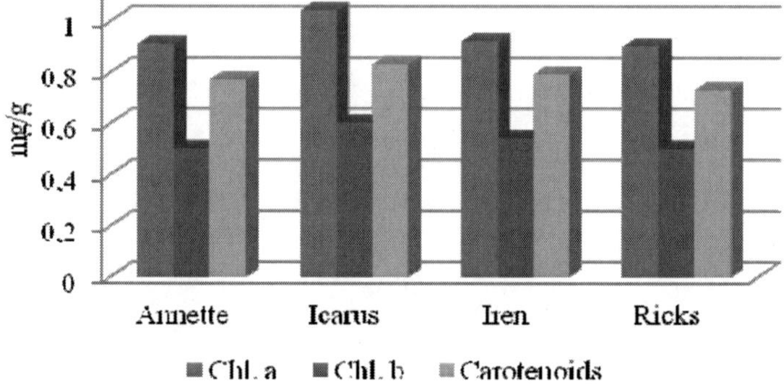

FIGURE 26.1 The ratio of photosynthetic pigments in the flag leaves cells of spring wheat in controls (mg/g).

Significant increase obtained in the prototypes with respect to the pigments content under the influence of zircon, on average, reaches 49.1% and appin increase within the range of 22.1% compared to the controls. Data analysis revealed differentiation of content and ratios of photosynthetic pigments in flag plant leaves, depending on the used adjuster (Table 26.3).

The average ratio of *a* and *b* chlorophylls increased as follows: 60.3% zircon impact and 34.6% – appin impact, respectively, compared to the controls. Quantity content of carotenoids in the plant leaves has grown during the experiment from 19.2 to 52.6%. No significant differences in intervarietal reaction was observed.

TABLE 26.3 Content and the Ratio of Photosynthetic Pigments in Flag Leaves of Spring Wheat During Its Heading Stage Under the FAS Preliminary Treatment (mg/g)

Grade	Experience option	A+B	C	A+B+C	(A+B): C	(A: B)
Annette	Control	1.42	0.77	2.19	1.84	1.76
	Zircon	2.16	1.16	3.32	1.86	2.38
	Appin	1.78	0.93	2.71	1.91	1.97
Icarus	Control	1.65	0.83	2.48	1.99	1.73
	Zircon	2.18	1.16	3.34	1.88	2.39
	Appin	1.92	0.95	2.87	2.02	1.85
Iren	Control	1.46	0.79	2.25	1.85	1.70
	Zircon	2.19	1.25	3.44	1.75	2.13
	Appin	2.00	1.00	3.00	2.00	2.51
Ricks	Control	1.40	0.73	2.13	1.23	0.80
	Zircon	2.20	1.18	3.38	1.86	2.33
	Appin	1.61	0.84	2.45	1.92	1.56

Note: (A+B) – *a* chlorophyll + *b* chlorophyll; (C) – carotenoids, (A+B+C) – chlorophylls + carotenoids; (A+B): C – the ratio "chlorophylls: carotenoids," (A:B) – the ratio of "*a* chlorophyll: *b* chlorophyll;" statistically significant results.

We have reordered the accumulation of photosynthetic pigments in the flag leaf during zircon processing within the range from 0.86 mg/g (Icarus) to 1.25 mg/g (Ricks), appin – 0.32 mg/g (Ricks) to 0.75 mg/g (Iren). Taking into account this fact, we can conclude that there is a specific response within the variety of physiologically active compounds influence. As was previously noted, the Ricks varieties had the most responsive effect on zircon, which was evidenced by the maximum total accumulation of pigments in the flag leaves cells of the varieties under study. Presowing impact of appin has contributed to a minimum increase of the chlorophylls and carotenoids concentration compared to the controls in this variety class. It is clearly that, in this case a drug concentration its physiological effects on plants shall be important. The specificity of impact caused by FAS of different chemical compounds classes shall be considered during the experiments.

During assessing the ratio "chlorophylls: carotenoids" it was noted that the appin impact manifested itself in greater extent (6.5% – the difference under zircon treatment experience), 13.3% – the difference under the controls. The maximum values of the varieties against "*a* chlorophyll: *b* chlorophyll" ratio has also been received under zircon treatment options. The exception was noted in Iren varieties, similar values within the range of – 2.13 and 2.51 mg/g (zircon and appin) have been obtain during the experiments.

The impact of physiologically active compounds on varieties did not alter the "ratio of pigments" in the flag leafs compared to controls, however it has only increased their concentration in the leaves.

In general, the concentration of zircon within the range of 4.98×10^{-5} g/cm^3 helped to maximize the carotenoids and *a*, *b* chlorophylls content and in the soft spring wheat. It can be concluded that daily impact of FAS on plants is capable of minimizing the impact of critical soil and climatic conditions, causing adaptive reactions, which are expressed by increase of carotenoids and *a, b* chlorophylls content.

It is known that the main function of chlorophyll is to absorb and transmit the energy. The study shows that the average content of a chlorophyll in the control and experimental variants of wheat reaches 67.0% and is a part of the light-harvesting complexes (SSC), acting as an antenna, which transmits the energy to the photosystem reaction center. The most active energy transmission flows from long wave *a* chlorophyll, which is the main component of the reaction centers; *b* chlorophyll – is an auxiliary pigment, with an absorption maximum at a short wavelength. Carotenoids absorb the light at those wavelengths, that are least absorbed by chlorophyll, on average, 36.0% of such pigments are included in the light harvesting complexes.

The effect of physiologically active compounds used in the experiment, has contributed to the a chlorophyll preservation during the vegetation period by increasing and maintaining the certain level of carotenoids. The treatment method used allows to keep the pool carotenoids in line with the growth of the *a* chlorophyll pool, thereby increasing its protective function and the *b* chlorophyll pool growth compared to the control variant, due to the lack of photo-oxidation of green pigment. Therefore, the carotenoids have performed the light-shielding function along with the light-harvesting function; the light-shielding function protects the chlorophylls from photo-oxidation by reactive oxygen species. This is confirmed by the links between photosynthetic pigments with normally distributed traits. Differences in responses between the varieties have been observed. Mostly, it is strong positive significant correlation (Fig. 26.2).

Analysis of variance revealed the exposure of influence factors to the variability of the pigment amount in plant leaves. It has been revealed that the percentage impact on the variability of cells chlorophyll content is 3.8%, the share of FAS – 96.2%. The varieties affect the variability of carotenoids amount within the variation share of 5.6%, FAS – 94.4%. Accumulation of flag-leaf pigments occurs mainly due to the effect of compounds (96.9%). The increase in pigments accumulation is determined by the profiled reaction within a given cultural variety, only by 3.1%.

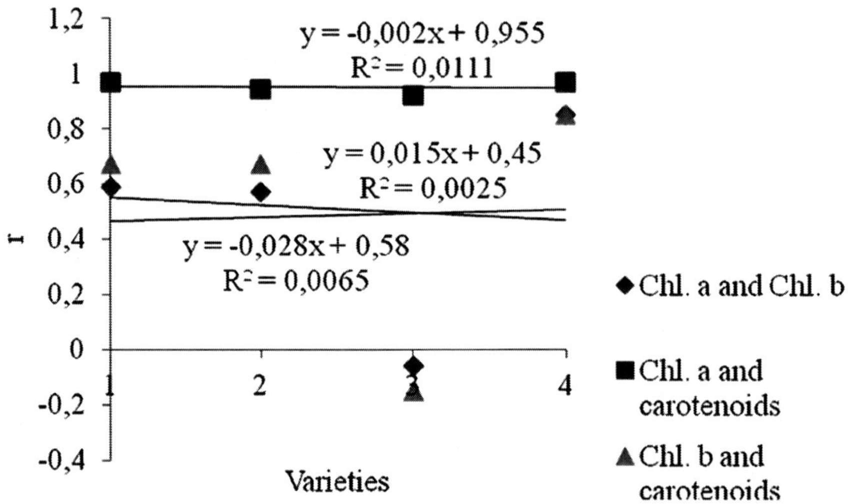

FIGURE 26.2 Spearman correlation between photosynthetic pigments in the flag leaf cells during the heading stage of soft spring wheat. Varieties: 1 – Annette; 2 – Icarus; 3 – Iren; 4 –Ricks. Chl. a – *a* chlorophyll, Chl. *b* – *b* chlorophyll.

Thus, the use of growth regulators can influence the chemical composition of soft spring wheat and the final product – grain, by enhancing and strengthening the internal regulation mechanisms of selective absorption of chemical elements from the soil solution depending of the environment, as well as morphological and physiological processes of the plant organism.

26.4 CONCLUSIONS

1. There is determined that cultivated residual-carbonate soddy small podzolic soil on ancient alluvial deposits of the subtaiga area of the Tyumen region refers to basic-type soils and its cationic and anionic composition, dry residues, density and porosity creates conditions for spring wheat growth on a favorable level.
2. During some plants phenological phases the soil moisture content reaches its critical level: germination and tillering in 2009 and 2010 (11.7 and 13.6%, respectively), stem elongation and earing in 2010 (6.8%), milky ripeness of grains in 2011 (9.0%).
3. Pigment composition of spring-wheat flag leaves cells, the chemical composition of the vegetative organs and grain varies significantly under the

influence of physiologically active substances, taking into account their concentration, methods of application and variety-specific reaction to their impact.

KEYWORDS

- **appin**
- **chemical elements**
- **photosynthetic pigments**
- **soil**
- **wheat**
- **zircon**

REFERENCES

1. Sudachkova, N. E. (2009). Adaptive Responses of Scots Pine to the Impact of Adverse Abiotic Factors on the Rhizosphere, Sudachkova, N. E., Milyutina, I. L., & Romanova, L. I., Ecology, *6*, 411–416. (in Russian)
2. Voronchihina, E. A. (2002). The Basic of Landscape Hemoecology, Voronchihina, E. A., Larionova, 148p. (in Russian)
3. Kaygorodov, R. V. (2010). Plants Resistance to Chemical Contamination, Study Guide, Kaygorodov, R. V., Perm State University, 134p. (in Russian)
4. Dospekhov, B. A. (1979). Methods of the Field Experiments (Basis of Statistical Processing of the Research Results), Moscow, Kolos, 416p. (in Russian)
5. Kabata-Pendias, A. (1989). Trace Elements in Soils and Plants, Kabata-Pendias, A., & Pendias, H., Moscow, Mir, 439p. (in Russian)
6. Pokatilov, Y. G. (1993). Biogeochemistry of Components of the Biosphere and Biomedical Problems (Environmental Problems in Biosphere Chemistry and Public Health), Pokatilov, Y. G., Novosibirsk, Nauka, 168p. (in Russian)
7. Malyuga, D. P. (1963). Biogeochemical Method of Prospecting for Ore Deposits, Malyuga, D. P., Leningrad, AS USSR Publisher, 264. (in Russian)
8. Ozersky, A. Y. (2008). Fundamentals of Environmental Chemistry, Study Guide, Ozersky, A. Y., Krasnoyarsk, Siberian Federal University, 316p. (in Russian)
9. Perelman, A. I. (1999). Relationship Doctrine of Biogeochemical Provinces and Landscapes Geochemistry, Perelman, A. I., Problems of Biogeochemistry and Geochemical Ecology, Moscow, Nauka, *23*, 115–133. (in Russian)
10. Litvinovich, A. V. (2011). The Effect of Long-Term Application of Phosphate Fertilizers and Meliorantov on Stable Strontium Accumulation in Soils and Plants, Litvinovich, A. V., Nebolsina, Z. P., Yakovleva, L. V., & Vitkovskaja, S. E., Agrochemicals, *1,* 35–41. (in Russian)
11. Dudareva (Cherkashina), I. A. (2013). Chemical Elements Content in the Soil-Plant System under the Influence of Biologically Active Substances, Dudareva (Cherkashina), I. A., Bome, N. A., Sciences Statement of Belgorod State University, Natural Sciences Series, *10(153)*, 23P, 116–124. (in Russian)

12. Golovko, T. K. (2010). Pigment Complex of Natural Flora Plants of the European North-East, Golovko, T. K., Dalke, I. V., Dimova, O. V., Zakhozhiy, I. G., & Tabalenkova, G. N. Proceedings of Komi Scientific Center of Ural Branch of RASNo.1, 39–46. (in Russian)
13. Zotnikova, A. P. (2006). Ecophysiological Response of Siberian Cedar Leaf Apparatus on Climate Change, Zotnikova, A. P., Bender, O. G., & Rudnik, T. I., The Atmospheric and Oceanic Optics, *19(11)*, 969–972. (in Russian)
14. Astafurova, T. P. (2011). The Effect of Nanoparticles of Titanium Dioxide and Aluminum Oxide on Morphometric Parameters of Plant Beans, Astafurova, T. P., Morgalev, Y. N., Zotnikova, A. P., Verhoturova, G. S., Mikhailova, S. I., Burenina, A. A., Zaytseva, T. A., Postovalova, V. A., Tsytsareva, L. K., & Borovikova, G. V. Bulletin of the Tomsk State University, Biology, *1(13),* 113–121. (in Russian)
15. Bome, N. A. (2012). Improvement of the Seed Viability and Resistance of *Triticum aestivum* L Plants to a Changing Environment, Cherkashina, I. A., & Bome, N. A. Natural Sciences and Engineering, *2*, 108–112. (in Russian)

CHAPTER 27

LIPID PEROXIDATION IN PLANTS OF COASTLINE OF PONDS OF URBANIZED TERRITORIES

ALEXEY L. KORNILOV and GALINA A. PETUKHOVA

Tyumen State University, 625003, Pirogova st., 3, Tyumen, Russian Federation;
E-mail: lelik_tgu@mail.ru

CONTENTS

ABSTRACT

The content of the products of lipid peroxidation in the plants of coastline of some ponds in Tyumen city was studied. The anthropogenic effect on the content of primary and secondary products of lipid peroxidation in the plants of coastline of ponds in Tyumen city was detected. Under the long-term anthropogenic pollution of urban ponds and their coastlines the content of products of lipid peroxidation in the plants and ponds under study did not exit the values observed in plants from the pond with low anthropogenic stress [Yuzhny (Southern) pond]. Two plant species (broadleaf plantain and broadleaf cattail) of coastlines of two ponds Kristalnye rodniki (Kristal springs) pond and Lesnoy (Forest) pond], from which the first one has the highest pollution level among ponds under study] presented the exclusion. Broadleaf plantain among waterside plants and broadleaf cattail among waterside-water plants showed to be more sensitive, as just in these plants the concentrations of primary products of lipid peroxidation were elevated.

27.1 INTRODUCTION

Urban environment is exposed to the effects of multiple pollutants and anthropogenic factors. In the present time the central issue of ecological security is man-caused pollution, which affects the state of ecosystems of water objects. As plants are immovable and often are situated at the coastline of rivers or ponds, they are the most suitable for the monitoring of the state of coastal and water zones of urban water objects.

Plants are often affected by unfavorable environmental factors, which evoke the accumulation of active forms of oxygen in plant tissues and activate oxidation processes (oxidation stress). The level of the products of lipid peroxidation (LP) is one of indicators of cell membranes damage [1, 2].

In the urban environment first of all the biochemical and physiological parameters are subjected to changes. The level of plant damage depends mainly on two factors – the concentration of toxic substance and the duration of the effect [3].

The formation of primary and intermediate products of LP (diene conjugates and malonic aldehyde, respectively) is considered as one of unfavorable consequences of lipid peroxidation as a result of the action of oxigene radicals and further rupture of polyene acids. This aldehyde, in its turn, creates Shiff bases with amides of proteins. This leads to the formation of insoluble lipid-protein complexes, which are sometimes called "wear-and-tear pigments" (lipofuscins).

The effect of unfavorable factors, in particular soil pollutants, is displayed in plants by the change of concentration of products of primary lipid oxidation – diene conjugates, and of products of secondary lipid oxidation – Shiff bases.

The aim of the study is the analysis of the anthropogenic effect on the concentration of primary and final products of lipid peroxidation in the coastline plants of ponds of Tyumen city.

27.2 MATERIAL AND METHODS

The concentrations of diene conjugates and Shiff bases were determined by spectrophotometer using standard methods [4].

27.2.1 THE DETERMINATION OF DIENE CONJUGATES

In order to find the concentration of diene conjugates we added 4 mL of the mixture of heptane and isopropanol (1:1) to 0.2 mL of plant extract and shook it during 10–15 min in the laboratory shaker. Then we added in the test-tube 1 mL of HCl (pH 2) and 2 mL of heptanes, shook it intensively and, after desilting and exfoliation of the mixture into phases (what takes 20–25 min) we collected the upper, heptanes layer, which was then used for the detection of diene conjugates according to the absorption of light with wave length 233 nm. As a control we applied a sample containing 0.2 mL of water instead of plant extract and submitted to all above described treatments.

The calculation of the concentration of products of lipid peroxidation was performed in conditional units according to the equation:

D_{233} per 1 mL of solution $= (D_{233} \times V_{he}) / V_{pe} = (D_{233} \times 4)/0.2$

where D_{233} is the optical density of the sample for 233 nm, in relative units per mL (ml^{-1}); $V_{he} = 4$ mL is final volume of heptane extract; $V_{pe} = 0.2$ mL – volume of plant extract.

27.2.2 THE DETERMINATION OF SHIFF BASES

In order to measure the concentration of Shiff bases we placed the ground sample of plant material (P, in g) with a small amount of Na_2SO_4 in the test-tube, added 4 mL of the mixture of heptane and isopropanol (1:1) and shook 10 min. Then we added 1 mL of HCl solution (pH 2.0) and 2 mL of pure heptane. Then we shook it again and waited 20 min for desilting. After it we collected the upper layer, measured its volume (V, mL) and optical density of sample (D_s) and pure heptane (D_h) under wave length of 365 nm on the spectrophotometer using "control" sample, which was prepared the same way as others, but instead of plant extract we added the same volume of water. The concentration of Shiff bases (C, conditional units per 1 mg of lipid (mg^{-1})) was calculated according to the equation:

$$C = (D_s - D_h),$$

where C is the concentration of Shiff bases; D_s is the optical density of sample; D_h is the optical density of pure heptane.

Measurements were carried out in eight replicates for each type of plants from one body of pools.

27.3 RESULTS AND DISCUSSION

According to the hydrochemical analysis according to the level of mineralization all ponds under study could be attributed to sweet water pools with low mineralization (Lesnoy and Yuzhny ponds, Krugloye lake), medium mineralization (Utiny pond, Alebashevo and Obrochnoye lakes) and high mineralization Kristalnye Rodniki pond. The excess of maximum permissible concentrations (MPC) for biogenic substances was found in the ponds Kristalnye Rodniki (ammonium nitrogen 10.5 times, phosphates 5.6 times), Lesnoy (ammonium nitrogen 1.1 times), Utiny (ammonium nitrogen 2.1 times, nitrites 1.5 times), lake Obrochnoye (ammonium nitrogen 2.1 times, nitrites 5.5 times, phosphates 2.9 times), Krugloye (nitrites 4 times, phosphates 1.8 times), Alebashevo (phosphates 2.1 times). According to the content of organic matter the following objects stand out: lakes Obrochnoye (the indicator of 5-days biological oxygen consumption (BOC_5) exceeds MPC 2.7 times), Alebashevo (BOC_5 exceeds MPC 1.8 times), pond Kristalnye Rodniki (BOC_5 exceeds MPC 1.8 times). The concentration of oil products exceeds MPC in the ponds Kristalnye Rodniki (3.6 times), Utiny (1.6 times) and lakes Alebashevo and Obrochnoye (1.2 times). Our analysis allowed to estimate the index of aquatic pollution (IAP [5]) and range the water objects under study according to IAP in descending order: Kristalnye Rodniki pond, Obrochnoye, Alebashevo and Krugloye lakes, Lesnoy, Utiny and Yuzhny ponds.

We analyzed 10 hydrochemical indicators, namely ammonium nitrogen, nitrate nitrogen, nitrite nitrogen, phosphates, BOC_5, calcium, suspended substances, Na+K, sum of ions, oil products, which exceeded MPC (Table 27.1). The integrated indicator of chemical pollution (ICP_{10} [6]) was calculated, which allowed to range pools in descendent order of pollution: Kristalnye Rodniki pond; Obrochnoye, Alebashevo, Krugloye lakes; Lesnoy, Utiny and Yuzhny ponds. According to the results of calculation of pollution indices it is possible to detect the most polluted pools –Kristalnye Rodniki pond, Obrochnoye, Alebashevo, Krugloye lates, Lesnoy and Utiny ponds.

TABLE 27.1 Classifications of Pools Under Study According to Pollution Level

Pool	ICP_{10}	IAP
Lake Alebashevo	6.7	Polluted
Lake Krugloye	6.3	Very polluted
Lake Obrochnoye	6.4	Polluted
Pond Kristalnye Rodniki	7.2	Very polluted
Pond Lesnoy	6.3	Polluted
Pond Utiny	6.1	Polluted
Pond Yuzhny	6.1	Very polluted

Note: ICP_{10} is integrated indicator of chemical pollution of water (10 indicators); IAP is the index of aquatic pollution (based on 6 indicators).

The most pure from hydrochemical point of view was Yuzhny pond. In the water samples from these pond low concentrations of biogenic elements, organic and suspended substances, oil products. Therefore this pond was taken as reference object (control pool).

Traffic is the source of pollution in all pools under study except Krugloye lake, where traffic intensity is low. The highest traffic intensity was observed in the streets Alebashevskaya, 50 let VLKSM and Dambovskaya close to Alebashevo lake, Utiny and Kristalnye Rodniki ponds, Obrochnoye lake, respectively. Railway (ponds Kristalnye Rodniki and Utiny) and water transport (lake Krugloye) are also presented.

Among stationary industrial sources of pollution are cogeneration plant TETs-1 situated near lake Obrochnoye (which is used as its cooling pool and is exposed to atmospheric pollution), and cogeneration plant TETs-2 situated near Lesnoy pond, which is watered open pit and is exposed to atmospheric pollution. Waste water from shore zone, formed by water from melted snow and precipitation, fall into all pools. Waste water from sewer system falls into lake Alebashevo [7, 8] (volume is unknown). All pools are used for recreation and fishing except Kristalnye Rodniki pond. The Yuzhny pond is situated in court zone and has improved recreation zone along its whole perimeter.

There are construction works at the ponds Kristalnye Rodniki, Utiny and at the lake Obrochnoye, and construction waste and wastewater fall on the shore and into the water. At the shore of Alebashevo sand harvesting is performed.

All pools are in similar climatic conditions. Shore territories of most of pools (except Krugloye and Obrochnoye lakes) consist from natural soils and landscapes modified by human activity. Ponds are quarries (ponds Kristalnye Rodniki, Lesnoy, Utiny) and one pond is former river-bed of an internal pool (pond Yuzhny).

The study of the state of pools (4 ponds and 3 lakes) was conducted using frequent plants: broadleaf plantain (*Plantago major* L.) and white clover (*Trifolium repens* L.) in the shore zone (1–2 meters from summer water minimum of pool, and round fruit rush (*Juncus compressus* Jacq.), katniss (*Sagittaria sagittifolia* L.), broadleaf cattail (*Typha latifolia* L.) for waterside-water zone (pools swash territory). The sampling of plant material was performed in July under the same weather conditions (partly cloudy, 24–26°C, wind sw. 1–2 m/s).

Broadleaf plantain from shore zone of Kristalnye Rodniki contains more diene conjugates (Fig. 27.1) ($P<0.05$), than broadleaf plantain × 3 from shore zone Yuzhny pond. The high concentration of primary products in broadleaf plantain from shore zone of Kristalnye Rodniki pond is probably the reaction on unfavorable factors (anthropogenic pollution), which strongly affect broadleaf plantain of shore zone.

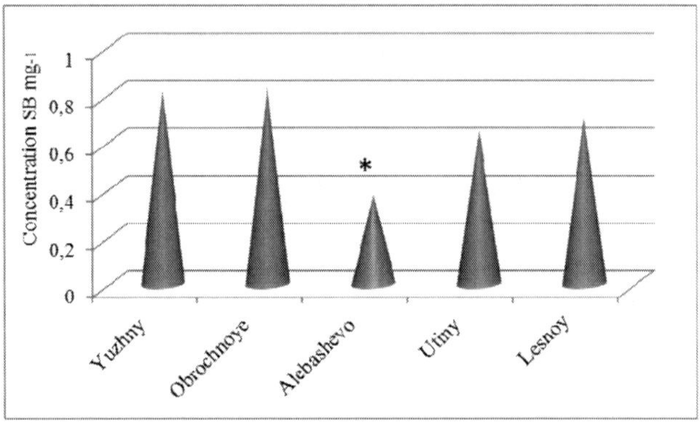

FIGURE 27.1 Content of diene conjugates in the plants of broadleaf plantain from objects under study, relative units per mL of solution.
Note: * – statistically significant differences of obtained data with control values of diene conjugates concentration in the plants of shoreline of Yuzhny pond (reference object) (*P*<0.05).

The concentration of diene conjugates in broadleaf plantain from shore zone of other pools under study was lower (P<0.05) relatively to the concentration in the plants from shore zone of Yuzhny pond. The decrease of concentration of primary products of lipid peroxidation relatively to reference object is, probably, the indicator of high activity of protective systems (ascorbic acid, catalase, superoxide dismutase) of plant cell and adaptation to the complex of pollutants [9].

The analysis of Shiff bases concentration (Fig. 27.2) in broadleaf plantain from shore zone of pools under study showed, that in all samples it was lower (P<0.05), than in broadleaf plantain from shore line of Yuzhny pond.

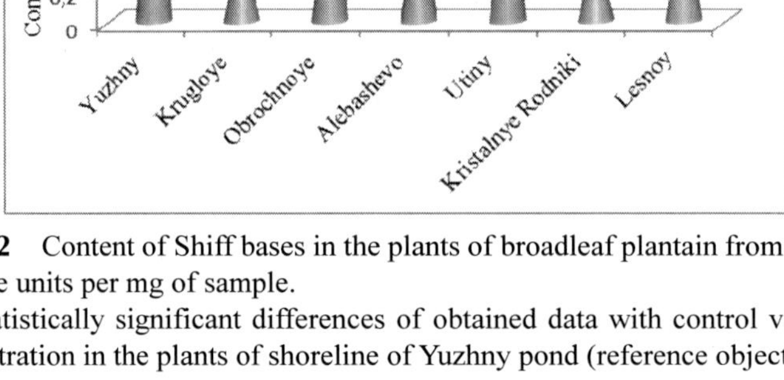

FIGURE 27.2 Content of Shiff bases in the plants of broadleaf plantain from objects under study, relative units per mg of sample.
Note: * – statistically significant differences of obtained data with control values of Shiff bases concentration in the plants of shoreline of Yuzhny pond (reference object) (*P*<0.05).

Thus, cell protection systems and biochemical processes of protective nature (neutralization of intermediate LP products and their involvement in biochemical cycles) in broadleaf plantain from shore line of pools under study increased plant resistance to unfavorable environmental factors (man-caused pollution).

Content of diene conjugates in white clovere (Fig. 27.3) from shore zones of all pools under study was lower (P<0.05), than in white clover from shore zone of Yuzhny pond, except white clover from shore zones of Kristalnye Rodniki pond and Alebashevo lake.

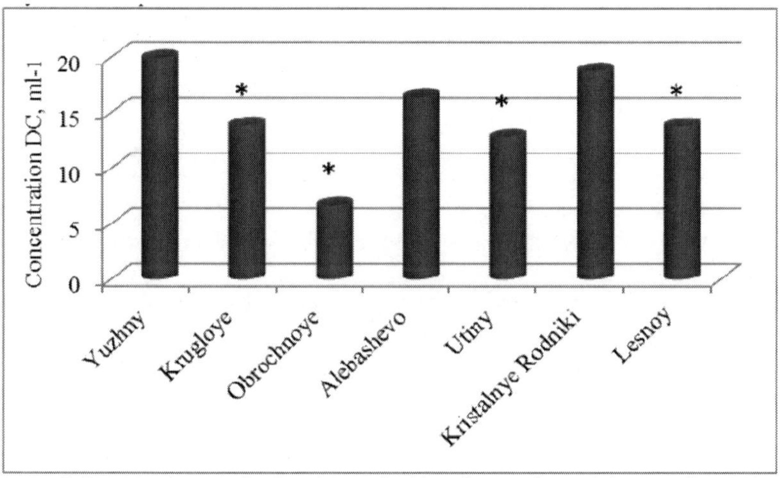

FIGURE 27.3 Content of diene conjugates in the plants of white clovere from objects under study, relative units per mL of solution.
Note: * – statistically significant differences of obtained data with control values of diene conjugates concentration in the plants of shoreline of Yuzhny pond (reference object) (P<0.05).

Figure 27.4 shows that the content of Shiff bases in white clover from shore zones of all pools under study was lower (P<0.05) relatively to white clover from shore zone of Yuzhny pond.

Thus, coastal plants in urban conditions are permanently exposed to pollution (air and soil pollution) and are stressed during many years. Occasionally, during spring or summer flooding and under summer precipitation the structure of pollution of shore zone changes (the attenuation of currently presented pollutants and additional input of substances cumulated in the snow cover of pool territory occur), which is also a stress factor. This leads to the activation of protective systems and, consequently, the resistance of organism to anthropogenic influence increases. Probably, a selection of resistance individuals takes place under the effect of long-term anthropogenic stress, which leads to the plants adaptation [10]. Broadleaf plantain from shore zone of Kristalnye Rodniki pond presented exclusion, because the content of diene conjugates in its cells (22.35 mL^{-1}) was above control (19.93 mL^{-1}). Probably, broadleaf plantain was recently settled in this plant community and is not adapted to the given type of pollution of Kristalnye Rodniki pond territory.

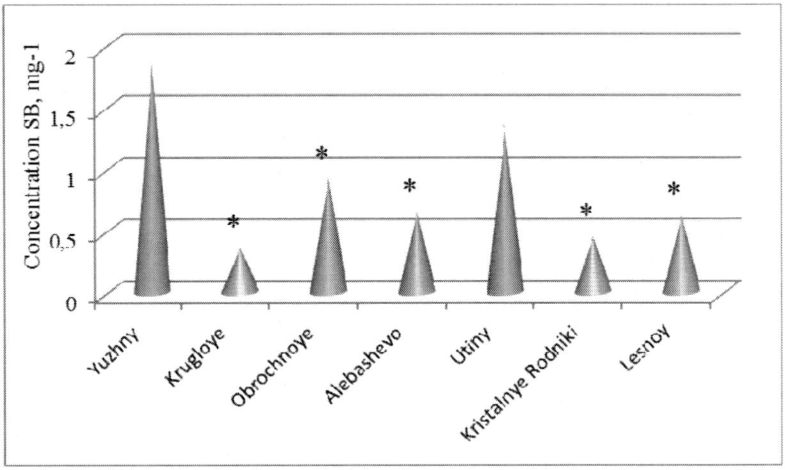

FIGURE 27.4 Content of Shiff bases in the plants of white clovere from objects under study, relative units per mg of sample.
Note: * – statistically significant differences of obtained data with control values of Shiff bases concentration in the plants of shoreline of Yuzhny pond (reference object) ($P<0.05$).

The neutralization of products of lipid peroxidation in white clover is going more effectively and more rapidly than in broadleaf plantain from the same ecotopes, which is probably related to such peculiarity of legumes as symbiosis with tuber microorganisms. Production of higher amount of proteins in legumes allows to fix more intermediate LP product with formation of Shiff bases. Final LP are more cumulated in white clover, than in broadleaf plantain. Probably the increased fixation of intermediate LP products (namely of malonic aldehyde) in white clover is evoked by higher proteins production and formation of Shiff bases. The concentration of diene conjugates in broadleaf plantain of shore zones of most of pools under study is higher than in white clover from similar territories, excluding ponds Yuzhny (19.92 mL^{-1} in broadleaf plantain and 19.95 mL^{-1} in white clover) and Lesnoy (8.6 mL^{-1} in broadleaf plantain and 13.8 mL^{-1} in white clover).

Broadleaf plantain reacts more intensively to anthropogenic influence by the formation of higher concentrations of primary LP products (diene conjugates), than white clover. Therefore, the use of broadleaf plantain as test-object of anthropogenic effect is more sounded.

The concentration of diene conjugates in roundfruitrush in shore Zone of Alebashevo Lake is lower (P<0.05) than in roundfruitrush from shore zone of Yuzhny pond (Fig. 27.5). In other cases the difference was insignificant. In spite of high level and different type of pollution of pools under study, the level of adaptation of roundfruitrush to man-caused pollution is high. In spite of high anthropogenic pressing, the rush from waterside-water zone of Alebashevo lakes has low content of diene conjugates, what is probably related both with ecotope conditions (big pool, overwatered substrate of shoreline), as well as with the type of pollution (organic, oil products – main pollutants lake Alebashevo).

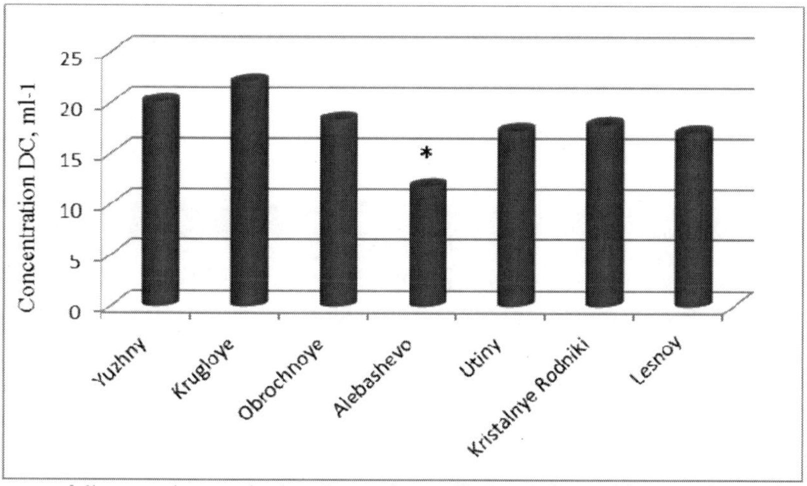

FIGURE 27.5 Content of diene conjugates in the plants of roundfruitrush from objects under study, relative units per mL of solution.
Note: * – statistically significant differences of obtained data with control values of diene conjugates concentration in the plants of shoreline of Yuzhny pond (reference object) (P<0.05).

Content of Shiff bases in roundfruitrush (Fig. 27.6) from waterside-water zone of Kristalnye Rodniki, Lesnoy, Utiny ponds, Alebashevo and Krugloye lakes was lower (P<0.05), than in roundfruitrush from waterside-water zone of Yuzhny pond. Low concentration of both, diene conjugates, as well as Shiff bases is, probably, related with the achieved adaptation of organisms to the complex of unfavorable factors. The increase of hydroperoxide level contributes to the activation of antioxidant systems, which, in their turn, contribute to the slowing of senescence processes [11]. It is necessary to notice very low content of Shiff bases (0.18 mg^{-1}) in roundfruitrush from waterside-water zone of Kristalnye Rodniki pond. The maintenance of so low values plant needs to spend considerable energy resources.

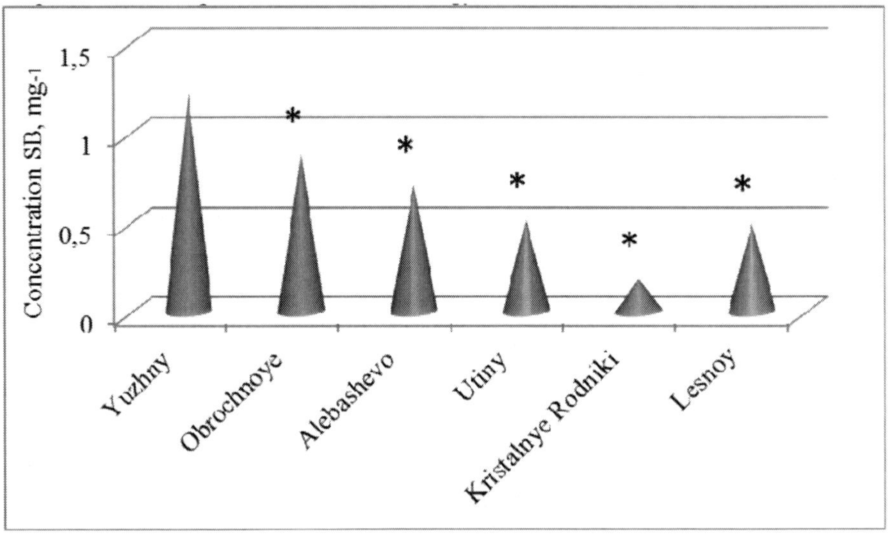

FIGURE 27.6 Content of Shiff bases in the plants of roundfruitrush from objects under study, relative units per mg of sample.
Note: * – statistically significant differences of obtained data with control values of Shiff bases concentration in the plants of shoreline of Yuzhny pond (reference object) (*P*<0.05).

Beside of high concentration of Shiff bases (1.2 mg^{-1}) in roundfruitrush from waterside-water zone of Yuzhny pond, high concentrations were detected in roundfruitrush from waterside-water zones of Obrochnoye (0.9 mg^{-1}) and Krugloye (0.8 mg^{-1}) lakes. Probably, namely these pools have the most favorable ecological conditions (ICP_{10} index less than that of the most dirty pond) for this type of plants.

The analysis of our data (Fig. 27.7) shows that katniss from waterside-water zone of Lesnoy pond, Obrochnoye, Alebashevo lakes and Kristalnye Rodniki pond contains less diene conjugates ($P<0.05$), than katniss from Yuzhny pond. In katniss from waterside-water zones of pools under study protective systems compensate the negative effect of complex man-caused load and natural conditions. It is known [12] that under normal conditions a certain level of lipid peroxidation induced by the formation of active forms of oxygen is permanently presented in cells.

When comparing the concentration of diene conjugates in the arrowhead of the studied reservoirs can be identified of pond Kristalnye Rodniki and lake Alebashevo. In katniss from waterside-water zones of these pools the highest content of diene conjugates was observed what indicates higher pollution level of their coastal and water environment relatively to other pools under study. Hydrochemical analysis showed indeed that Kristalnye Rodniki pond and Alebashevo lake are the most polluted among pools under study.

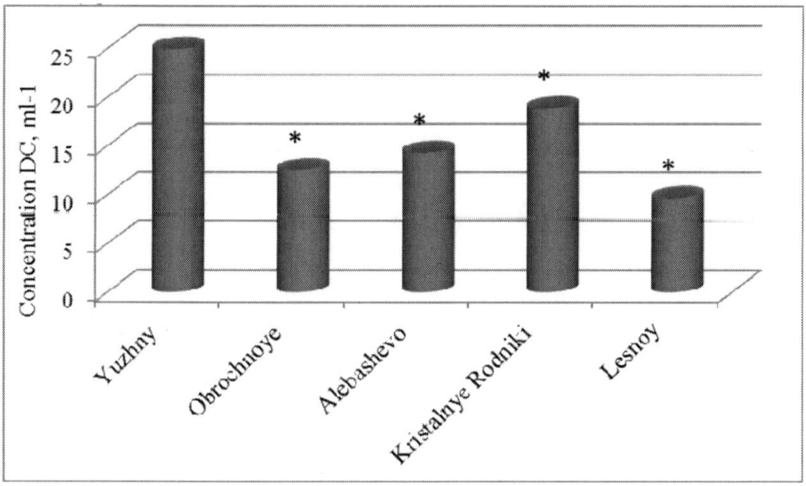

FIGURE 27.7 Content of diene conjugates in the plants of katniss from objects under study, relative units per mL of solution.
Note: * – statistically significant differences of obtained data with control values of diene conjugates concentration in the plants of shoreline of Yuzhny pond (reference object) ($P<0.05$).

Content of Shiff bases (Fig. 27.8) in katniss from waterside-water zones of Alebashevo lake, Lesnoy and Kristalnye Rodniki ponds was lower ($P<0.05$) than in katniss from waterside-water zone of Yuzhny pond. No significant difference in Shiff bases concentration in katniss from waterside-water zone of Obrochnoye lake ($P>0.05$) with katniss from waterside-water zone of Yuzhny pond. Content of peroxidation products in katniss from waterside-water zones of pools under study remains within limits of control in spite of anthropogenic load.

Apparently, LP process and antioxidant protective system (AOP) reflect adaptive abilities of plants, their adaptation to the ecotope conditions. The relationship between LP and AOP allows to evaluate the resistance of organisms to extreme effects and to forecast their state. The LP induction is a nonspecific universal reciprocal reaction to stressing effect of environmental factors. In the studies [13, 14] the intensification of LP under the effect of benzene hydrocarbons and metals was shown.

The increase of diene conjugates in broadleaf cattail (Fig. 27.9) from waterside-water zone of Lesnoy pond ($P<0.05$) relatively to their concentration in broadleaf cattail from waterside-water zone of Yuzhny pond was observed. The increase of primary products of lipid peroxidation (diene conjugates) in broadleaf cattail from waterside-water zone of Lesnoy pond is apparently evoked by the same reasons as in broadleaf plantain from shore zone of Kristalnye Rodniki pond. Probably, the elevated content of biogens, organic pollution and algal bloom in Lesnoy pond evoke the increase of primary LP products in broadleaf cattail.

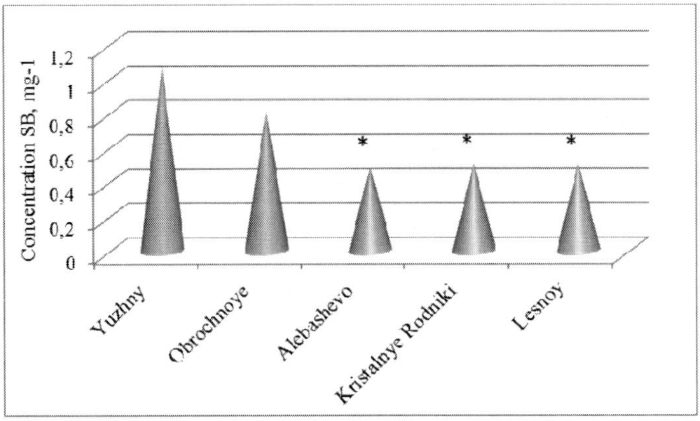

FIGURE 27.8 Content of Shiff bases in the plants of katniss from objects under study, relative units per mg of sample.
Note: * – statistically significant differences of obtained data with control values of Shiff bases concentration in the plants of shoreline of Yuzhny pond (reference object) (P<0.05).

Beside of high concentration of diene conjugates (26.44 mL^{-1}) in broadleaf cattail from waterside-water zone of Lesnoy pond, high content of diene conjugates (23.78 mL^{-1}) was also observed in broadleaf cattail from waterside-water zone of Alebashevo lake, which is from hydrochemical point of view one of the most polluted pools under study.

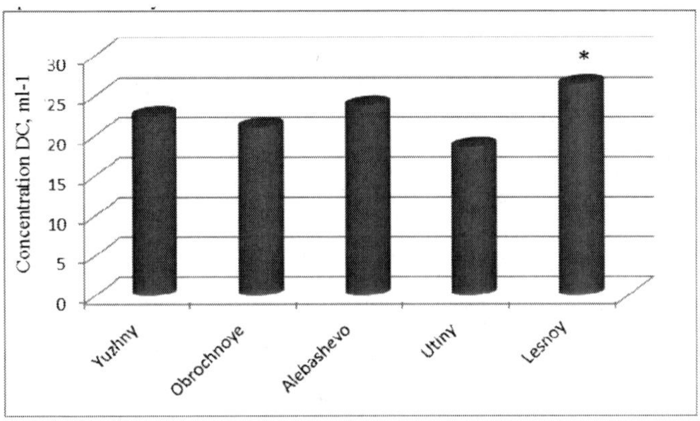

FIGURE 27.9 Content of diene conjugates in the plants of broadleaf cattail from objects under study, relative units per mL of solution.
Note: * – statistically significant differences of obtained data with control values of diene conjugates concentration in the plants of shoreline of Yuzhny pond (reference object) (P<0.05).

At the last stage of the transformation products of peroxidation the content of Shiff bases in broadleaf cattail (Fig. 27.10) from waterside-water zone of Alebashevo lake is less (P<0.05), than in broadleaf cattail from waterside-water zone of Yuzhny pond. There is no significant difference in the indicators analyzed among other pools under study. Consequently, the decrease of concentration of these compounds is probably related to the high level of antioxidant protection in response to complex unfavorable factors and it does not lead to formation of large amount of final products of lipid peroxidation [15]. The highest concentration of Schiff bases was observed in broadleaf cattail from waterside-water zones of Yuzhny pond (0.81 mg^{-1}) and Obrochnoye lake (0.84 mg^{-1}).

The decrease of content of LP products can be also explained by insufficient amount of substrate [membranes and "lipid storage" of organism (spare substances)] for the development of LP reactions, because it was used for energy needs of organism. It is known from the literature about close relation between LP and the activity of ferments of antioxidant system [16].

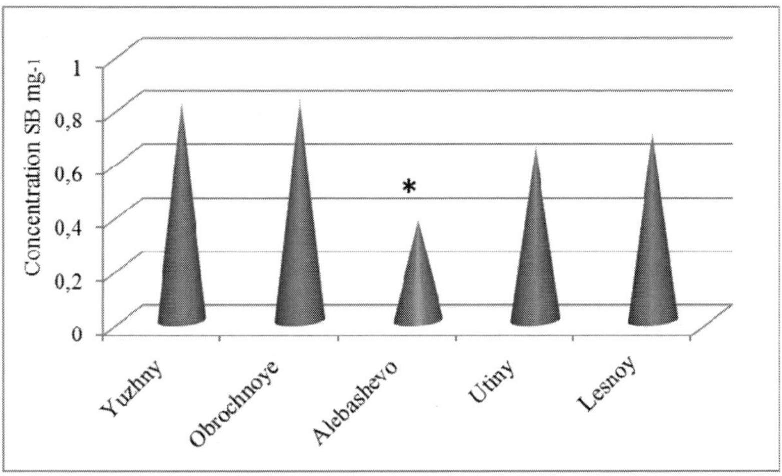

FIGURE 27.10 Content of Shiff bases in the plants of broadleaf cattail from objects under study, relative units per mg of sample.
Note: * – statistically significant differences of obtained data with control values of Shiff bases concentration in the plants of shoreline of Yuzhny pond (reference object) ($P<0.05$).

All appearances, territory study periodically in flooded condition is, and therefore in plants in waterside-water zone have more wet substrate, and sometimes they are in the aquatic environment. Consequently plants absorb substances both, from the substrate on which they grow, as well as from water and air. They have more dynamic conditions and higher absorbing capacity. Therefore, plants of these areas are more susceptible to sudden changes than the coastal vegetation. Consequently plants from this zone are more exposed to sharp changes than the vegetation of shore zone.

Pools, as relief depressions, accumulate a considerable amount of various substances, including substances of anthropogenic origin. Therefore, plants of shore and waterside-water zones are exposed to the most stressed ecological conditions. The highest burden falls on the coastal and aquatic plant communities. The need to neutralize high concentrations of diene conjugates was more successful in plants of waterside-water zones of Alebashevo and Obrochnoe lakes and Lesnoy pond. Probably plants of waterside-water zones of these ponds adapted to the incoming pollution and other anthropogenic factors that cause the growth of hydroperoxides in plant cells.

Broadleaf cattail from waterside-water zone of Lesnoy pond presented exclusion, as its content of diene conjugates (26.4 mL^{-1}) was above control value (22.47 mL^{-1}). Probably, the settling of broadleaf cattail in this phytocenosis was relatively recently and hence it did not adapt yet to the unfavorable conditions of this pool (algal bloom in the pond, elevated content of biogens, etc.).

Among waterside-water plants katniss has morphological features of *Alismataceae, namely* several types of leaves of different ecological confinement, shortened rhizome with tuber-shaped bulges containing starch. These features likely help to maintain low content of primary and final products of lipid peroxidation relatively to control – 4.88 mL^{-1} of diene conjugates and 1.07 mg^{-1} of Shiff bases. These morphological features of katniss do not play a significant role with respect to other species of plants of waterside-water zone as the concentration of diene conjugates in katniss from waterside-water zones of most of pools under study, except Lesnoy pond (9.6 mL^{-1}) and Obrochnoe lake (12.5 mL^{-1}), was higher than in waterside-water plants growing in similar territories.

Protection systems of roundfruitrush from waterside-water zone of zone of Alebashevo lake function actively (catalase, ascorbic acid, carotenoids, flavonoids other antioxidants) so it has low content of diene conjugates.

Conjugated diene content in the studied waterside-water plants from the Alebashevo, Obrochnoe lake and Lesnoy pond, and content of Schiff bases in plants from waterside-water zone of Kristalnye Rodniki and Lesnoy ponds and Alebashevo lake was less than in other pools under study.

27.4 CONCLUSIONS

1. Under chronic stress, in particular under anthropogenic influence the plants of shore and waterside-water zones of pools under study adapted and increased their resistance to the complex of unfavorable factors. The balance of lipid peroxidation and antioxidant activity reflects the adaptive capabilities of living systems, their adaptation to environmental conditions.

2. The content of lipid peroxidation products in the most plants from the shore and waterside-water zones of pools under study is reduced, which probably results from the selection of resistant genotypes under anthropogenic impact on aquatic objects.
3. Under strong and complex anthropogenic impact broadleaf plantain from shore zone of Kristalnye Rodniki pond and cattail from waterside-water zone of Lesnoy pond the content of primary products of lipid peroxidation (conjugated diene) increases.
4. Further deterioration of environmental situation and ongoing development of water pollution will lead to the oppression of the vegetation cover and to the change of species composition of coastal vegetation, to the more intense algae blooming and reduction of fisheries function.

KEYWORDS

- **broadleaf cattail**
- **broadleaf plantain**
- **diene conjugates**
- **katniss**
- **Shiff bases**
- **white clover**

REFERENCES

1. Dremuk, I. A. (1956). Lipid Peroxidation in Barley Germs under the Joint Effect of Low Temperature Stress and Hydration, Minsk, Institute of Biophysics and Cell Engineering, National Academy of Sciences of Belarus, 119p (in Russian).
2. Chirkova, T. V. (2002). Physiological bases of Plant Resistance, Street Petersburg State University, 244p (in Russian).
3. Kurbatova, A. S., Bashkin, V. N., & Kasimov, N. S. (2004). Urban Ecology, Moscow Scientific World Publishing, 624p (in Russian).
4. Shvedova, A. A. & Polyansky, N. B. (1992). Method of Determining of Lipid Peroxidation Final Products in the Tissues, Fluorescent Schiff Bases, Collection of Scientific Papers Study of Synthetic and Natural Antioxidants *In vitro* and In VIVO [Burlakova, E. B. Ed] Moscow Nauka (*Science*), 72–73 (in Russian).
5. Guidance Document of Comprehensive Assessment of Surface Water Pollution on Hydro Chemical Indicators, Rostov-on-Don (2002) Hydro Chemical Institute of Roshydromet, 24p (in Russian).
6. Criteria for the Assessment of the Environmental Situation to Identify the Ecological Emergency Zones and Zones of Ecological Disaster (1992) Moscow, Manual of Ministry of Natural Resources of Russian Federation Dated November 30, 51p (in Russian).
7. Agarkov, A. (2013). Is Alebashevo a Lake Valley? Tyumen News, *12(5689)*, 3–4 (in Russian).

8. Isachenko-Bome, E. A., Mikhailova, L. V., Bondar, V. O., & Chernyak, M. A. (2000). Status of the Alebashevo Lake Ecosystem, Abstracts of the Regional Scientific-Practical Conference "Environment", Tyumen, Tyumen State University, 166–168 (in Russian).
9. Kornilov, A. L., & Petukhova, G. A. (2013). Ascorbic Acid Content in Plants from Coastal Zone of Pools of Tyumen City, International Journal of Applied and Basic Research, *11*, 10.
10. Fazlieva, E. R., & Kiseleva, I. S. (2012). Antioxidant Activity of *Melilotusalbus* and *Trifolium Medium* Leaves from Technogenically Disturbed Habitats of the Middle Ural under Effect of Copper, Plant Physiology, *3,* 365–375 (in Russian).
11. Vadov, D. L., Brilkina, A. A., & Veselov, A. P. (2008). Activity of Antioxidant Enzymes and the Content of Lipid Peroxidation Products under the Effect of Short-Term Salinization on Plants Differing in Abscisic Acid Content, Bulletin of Lobachevsky, N. I. Nizhni Novgorod State University, Biology, *1*, 73–76 (in Russian).
12. Baraboi, V. A. (1991). Stress Mechanisms and Lipid Peroxidation, Modern Biology Advances, *6*, 923–932 (in Russian).
13. Symons, P. E. (1973). Behavior of Young Atlantic salmon (*Salmosalar*) Exposed to Force-Fed Fenithrothion, on Organophosphate Insecticidae, J. Fisheries Res. Board of Canada, *30*, 651–655.
14. Walsh, P. J., Bergman, H. L., & Wood, C. M. (1993). Effect of Ammonia on Survival, Swimming and Activities of Enzymes of Nitrogen Metabolism in the Lake Magaditilapa, Exp. Biology, *180*, 323–327.
15. Petukhova, G. A. (2007). Ecological and Genetic Consequences of Oil Pollution on Organisms: the Thesis of Doctor of Biological Sciences, 03.02.08, Petukhova, G. A., Tyumen, 390p (in Russian).
16. Saprin, A. N., & Kalinina, E. V. (1999). Oxidative Stress and its Role in the Mechanisms of Apoptos is and the Development of Pathological Processes, Biological Chemistry Advances, *39*, 289–326 (in Russian).

CHAPTER 28

PROSPECTIVE SPECIES OF ELMS (*ULMUS L.*) FOR ARID ZONES OF KHAKASIA

GALINA N. GORDEEVA

Khakas Scientific Research Institute of Agrarian Problems of Russian Academia of Agriculture Science, Abakan, p/o box 709, Republic of Khakasia, 655019, Russia; E-mail: gordeeva.gal2011@yandex.ru

CONTENTS

ABSTRACT

The article deals with the collection of elms of different background tested in the steppe zone of Khakasia. It focuses on their biological and morphometric peculiarities with the aim to determine the reason preventing their wide spread and usage in the region.

28.1 INTRODUCTION

In Khakas flora there are no representatives of genus *Ulmus* L. However, one of the species of this genus plays an efficient role in forming cultivated lands in the steppe zone of Khakasia. 2-year-old seedlings of *Ulmus L.* were brought from Ulan-Ude in 1960. This species was tested at the State Agriculture Experimental Station (at present the Khakas Scientific Research Institute of Agrarian Problems) where a forest strip was created to protect the water canal. Some time later the forest strip served as a mother plantation for seed harvesting and further reproduction of the species. In the dryland conditions the height of *U. pumila* does not exceed 3.5–4 meters. At present this species has adapted and it is widely spread out with the help of self-seeding, forming wooded meadows on the derelict lands. Like a poplar, an elm grows fast that is why it is used for creating a protective forest strip. Moreover, having a sufficient drought resistance and unpretentiousness during its cultivating, elm is actively used for settlement forestry as well as for a biological recultivation of mine tips. There it actively reseeds itself not requiring any imported soil nutrients and creating almost impassable overgrowths, elm groves in the lower and facing away from the wind areas of mine tips.

About 11 species of this genus of elms are being studied among the collection of woody plants of the Institute. They differ in crown forms, leaves and height. These species have been being tested in the extreme conditions of Khakasia. Their variety let us expand the plant assortment with the view to melioration and forestry. However, this collection is spread slowly and at present it has not gone in most cases outside the arboretum.

The goal of the present research is to analyze the possibility of a wide spreading of this valuable culture in Khakasia.

28.2 MATERIALS AND METHODOLOGY

In the collection of woody plants of the Institute 11 species of elm genus that differ in the forms of a crown, leaves and branches, tree height are being studied. These species have already been being tested in the extreme climates of Khakasia for a long time.

The plant testing station is situated on the second terrace above flood-plain of Abakan river, 13 km from the capital of Khakasia, Abakan. The climate of the steppe zone of the republic, where the collection is situated, is extremely continental with huge daily temperature swings during transitional seasons (spring, autumn). Strong parching winds up to 25–30 m/s are typical of spring, autumn is long-lasting, warm and dry. Winters are freezing (an average temperature of January – 21°C), February is characterized by winds with the air temperature – 38–40°C. Winter precipitation is little, the snow line in the arboretum reaches 15–17 sm, and consequently, there is a deep freezing of the soil – up to 3 m. Soils are dark chestnut, calcareous with unfavorable physicochemical qualities for plant growing and alkaline soil reaction. Annual precipitation is 298–300 мм, 60% of which falls during summer-time – the end of July – August [1].

During the plant studying phonological observations according to the methods of Luchnik were made [2], winter resistance and prospectivity were observed according to the methods of Lapin [3, 4].

28.3 RESULTS AND DISCUSSION

In the collection of the arboretum of the Institute studied species of *Ulmus* have different background and number from one to 4 samples of each species: *Ulmus laciniata* (Trarv.) Mayr, *U. pumila* L., *U. parvifolia* Jacg., *U. macrocarpa* Hance, *U. japonica* (Rehd.) Sarg., *U. scabra* Mill., *U. americana* L., *U. fulva* Michx, *U. laevis* Pall., *U. celtidae* (Rogow) Litw., *U. carpinifolia* Rupp. ex Suckow. Their ecological belonging, occupied range, some biometrical parameters under the conditions of the arboretum are set out in Table 28.1. All species of *Ulmus* were taken mostly as seeds, gathered during the expeditions or ordered by post from other regions of our country. The exceptions are *U. pumila* and *U. japonica* that were brought as live seedlings.

TABLE 28.1 Bio-morphological Characteristic of Species of Genus *Ulmus* in the Arboretum of the Khakas Scientific Research Institute of Agrarian Problems

Species	**Occupied range**	***Ecological group**	**Age, years**	****Height, m**		**Diameter of crown, m**	*****Length of leaf, sm**		**Diameter of trunk, sm**	**Crown form**
				in nature	**In nursery garden**		**In nature**	**In nursery garden**		
U. laciniata	Primorsky Krai, Sakhalin, China, Korea	M	31	Till 16	6	3.5–4	8–16	13.9±0,6	11,2	spreading

TABLE 28.1 *(Continued)*

U. parvifolia	Far East, Mongolia, Korea, Japan	Mx	26	Till 12	9	3. 5–5	2–5	9.0±0,2	9,2	spreading
U. macrocarpa	Far East (southern part), China, Korea	Mx	30	Till 8	6	3.3	7–9	8.7±0,2	7.2	narrow
U. pumila	Far East, Western Tien Shan, Mongolia, China	X	51	Till 16	13	8.5	3–7	6.5±0,1	25–30	widely spreading
U. japonica	Asia, Japan	M	35	25	9	3.5	8	8.7±0,3	5.5	widely spreading
U. americana	North America	Mx	25	20–30	9,5	3.5	5–10 (15)	10.6±0,3	8.5	spreading
U. fulva	North America	Mx	32	Till 22	15	5.5	10–20	9.6±0,3	10.6	widely spreading
U. laevis	Europe, Caucasus	Mx	40	Till 35	13	8	6–12	5.6±0,2	18–20	widely spreading
U. celtidae	Europe	M	42	15	8	8	6–12	11.2±0,4	18–20	widely spreading
U. carpinifolia	Europe, Central Asia, Caucasus	M	7	Till 25	5	3	12	8.9±0,2	6.3	widely spreading
U. scabra	Europe, Crimea, Caucasus	M	37	Till 30	6	5	8–16	7.3±0,2	12	spreading

* – the height of the plants are listed according to "Trees and bushes" [2]; ** – M – mesophyte, X – xerophyte, Mx – mesoxerophyte, according to N.I. Lichovid [5]; *** – indices of leaf length in natural conditions are listed according to "Trees and bushes" [2]; indices of leaf length in nature of *U. japonica, U. carpinifolia* are listed according to [7, 8].

The main limited factors during the testing of new plant species under the conditions of the steppe zone of the region are winter resistance and water supply. It is known that under a change of growing conditions, being adapted, they can change their habit.

Also new growing conditions can have an impact on winter resistance and reproductive sphere of the studied plants. Tested in the arboretum elms are grown during irrigation but even under these conditions the most representatives of this genus are below the height that is typical in the open environment (see Table 28.1). In addition to the above, dark chest-nut often saline soils have a great meaning, they influence negatively on the plant growth and development as well as parching winds, especially during spring-time.

According to ecological belonging the most of the tested elm species are mesophytes (45%) and mesoxerophytes (45%), and only one species (*U. pumila*) refers to the typical xerophytes, this helps to use it widely in the region. Leave sizes of the most of elm species equal the sizes of ones in the open environment. Only leaves of *U. parvifolia* are far larger under the conditions of introduction.

The species under consideration of genus *Ulmus* undergo all the phases of the growth and development under the conditions of the nursery (Table 28.2).

TABLE 28.2 Periods of Growth and Development Phases, Winter Resistance and Prospectivity of the Species of Genus *Ulmus* in the Arboretum of the Khakas Scientific Research Institute of Agrarian Problems

Species	Beginning of bud development, vegetative/reproductive by month	Blossom		Seed maturing	End of abscission period	*Winter resistance, point	*Prospectivity, point
		Beginning	Ending				
U. laciniata	11.05/16.04	29.04	10.05	14.06	12.10	I-II	II
U. parvifolia	18.05/9.04	25.04	6.05	15.06	12.10	I-II	II
U. macrocarpa	16.05/9.04	26.04	16.05	12.06	12.10	I-II	II
U. pumila	11.05/4.04	23.04	11.05	9.06	10.10	II-III	II
U. japonica	2.05/11.04	23.04	10.05	18.06	15.10	I-II	II
U. Americana	26.05/8.04	24.04	7.05	12.06	1.10	II-III	II
U. fulva	16.05/11.04	23.04	13.05	12.06	15.10	II-III	III
U. laevis	12.05/14.04	28.04	11.05	11.06	10.10	I-II	II
U. celtidae	23.05/16.04	24.04	16.05	16.06	15.10	III-IV	IV
U. carpinifolia	14.05/29.04	27.04	16.05	17.06	8.10	I-II	II
U. scabra	16.05/9.04	24.04	13.05	16.06	12.10	I-II	III

* – winter resistance and prospectivity data are listed according to Lichovid [5].

According to the beginning of growth, periods of blossom and seed maturing elms refer to early spring plants. It may be pointed out almost the same dates of their beginning of different elm species with the difference in several days. The length of the vegetation period reaches by an average of 176.3±3.06 days.

According to the analysis results of winter resistance of the studied species it was found out that 63.6% have I – II points, i.e. they are winter hardy enough, 27% have II × III points. It is related to the fast growth of elms, especially in their young age (annual growth is up to 0.6 m). The growth of the majority of the presented species continues till September, the shoots don't manage to become woody and annually they are frozen. As the shoots get older, their growth periods are shifted to earlier ones (I – II decades of July) that allow them to ripe and resist successfully winter time. The most of the studied plants are several tens years old. Four species – *U. pumila, U. fulva, U. celtidae, U. scabra* have the signs of aging in the form of dead basal branches, trunk aging and trunk dieback (*U. fulva*).

With the aim to guard the collection, further reproduction and to get prospective elm species accepted into production the quality of seed products was studied. Seeds of the listed above species are set annually, but their quality and quantity differ. While checking a field germination it was determined: *Ulmus pumila* has 35–60% of germination depending on the year conditions, *U. macrocarpa* – 46,2%, *U. japonica* – 37%, *U. laciniata* – 30%, *U. laevis* – 25–30%, *U. carpinifolia* – 20–25%, seeds of other elm species turned out to be imperfect. The main reason of their imperfection may be atmospheric drought that often happens during the seed blossom, formation and maturing (May–June). Besides, *Ulmus laevis* has a splendid blossom in the conditions of Khakasia and its seed production happens rarely. *U. celtidae* bears blossom and fruit on the very top of the crown, thus getting under the negative influence of the abiotic factors (drought).

One of the characteristics of species adaptation to his new conditions is the presence of self-seeding. According to this characteristic the studied species are divided into two groups. 54% species, whose self-seeding is observed, were referred to the first group – *U. pumila, U. macrocarpa, U. japonica, U. laevis, U. laciniata*. The second group unites 45% species that have no self-seeding: *U. parvifolia, U. americana, U. fulva, U. celtidae, U. scabra, U. carpinifolia*. There are not large forest strips with *U. laevis* at the introduction station. The small amounts of its self-seeding are observed not far from them, which under the nonirrigated conditions, being frozen for the first years, reaches safely 4–4.5 m high, having a beautiful widely spreading crown form.

Regarding the research results of *Ulmus* collection in the arboretum of the Institute their prospectivity was determined (see Table 28.2). While defining the prospectivity an integral assessment of the state of plants is held according to seven characteristics: shoot ripening degree, winter resistance, habit conservation under the new growing conditions, shoot-forming capacity, the presence of continuous height growth in case of plant damage, reproductive ability, practicable reproduction methods. The dependence of the prospectivity of the studying species on their occupied range in the open environment is not found.

28.4 CONCLUSIONS

The results of assessment show that species of genus *Ulmus* L. – *U. fulva, U. celtidae, U. scabra* have III × IV points of prospectivity. 72% of the studying species of the collection have II points of prospectivity – *Ulmus laciniata, U. parvifolia, U. macrocarpa, U. japonica, U. americana, U. laevis, U. carpinifolia*, this goes to prove their sufficient stability under the conditions of the introduction station. However, the bad quality of seeds keeps down their spreading out with the view to phytomelioration and forestry to the different districts of the republic of Khakasia.

KEYWORDS

- **agro-landscape**
- **arboretum**
- **introduction**
- **steppe zone**

REFERENCES

1. Krasnoyarsk Krai Tuva (1961). Autonomous Region Agro Climatic Guide Publishing Editor Bakhtin, N. P. Leningrad, Hydrometeoizdat, 288p (in Russian).
2. Trees and Bushes of USSR (1951) Moscow-Leningrad, AN USSR Publishing, *2*, 578p (in Russian).
3. Lapin, P. I. (1967). Seasonal Rhythm of Development of Plants and its Importance for Introduction, Lapin, P. I., Bulletin Main Botanical Garden AS USSR, *65,* 13–18 (in Russian).
4. Lapin, P. I. (1973). Evaluation of Promising Introduktsii Woody Plants According to Visacial Observations, Lapin, P. I. & Sidneva, S. V. Experience Introduction of Woody Plants (Collection Scientific Papers), Moscow, GBS, 7–67 (in Russian).
5. Likhovid, N. I. (2007). The Introduction of Woody Plants under Arid Conditions of the South of Central Siberia, Abakan, 288p (in Russian).
6. Luchnik, Z. I. (1964). Method for Studying the Alien Trees and Shrubs, Questions Decorative Horticulture, Barnaul, Altai Publishing House, 6–22 (in Russian).
7. Japanese Elm. URL: http://www.ag.ndsu.edu/trees/handbook/th-3-115.pdf.
8. Smooth-Leaved Elm. URL: http://www.prostranstvo.biz/lekarstvennye-rasteniya/vyaz-grabolistnyy.html.

PART VIII

THEORETICAL ASPECTS AND THE ANALYSIS OF BIODIVERSITY

CHAPTER 29

ANALYSIS OF CORRELATION STRUCTURE IN BILATERAL TRAITS

ALEKSEY B. TRUBYANOV[1*], GEORGY YU. SOFRONOV[2], and NIKOLAY V. GLOTOV[1]

[1]Mari State University, 1, Lenin Square, Yoshkar-Ola, Republic of Mari El, 424000, Russia

[2]Macquarie University, Sydney NSW, 2109, Australia

*E-mail: a.b.trubyanov@gmail.com

CONTENTS

ABSTRACT

When studying asymmetry of bilateral traits, it is very important to take into account their correlation structure. The assumption for quadrivariate normal distribution of traits considers possible models of equality of four correlation coefficients between traits using different criteria and approaches on the basis of two large lamina samples (N_1=500, N_2=521) of drooping birch (*Betula pendula* Roth.). This study has shown that different traits give evidence of different models of correlation structure.

29.1 INTRODUCTION

When studying morphological traits of biological objects a researcher faces an important biological phenomenon – symmetry of organism structure. Weyl defines different types of symmetry [1]; in this work we turn our attention only to bilateral symmetry – left/right symmetry. By virtue of different factors (such as heterogeneous living environment) perfect symmetry does not exist in nature, that is why the study of perfect symmetry deviation – different types of asymmetry – is an important task both from basic and applied point of views. Van Valen distinguishes fluctuating, directive asymmetry and antisymmetry [2]. After publication of Astaurov's work [3] the phenomenon of fluctuating asymmetry has become the subject of much research. One of the research courses for this phenomenon was an investigation of its quantitative estimation methods. The works of Palmer and Strobeck [4–6] contain an extensive summary. Zakharov drew attention to the influence of correlation coefficient between different parts of one trait (left/right correlation) in estimating fluctuating asymmetry [7]. In our work [8] we have suggested a new quantitative estimation index for fluctuating asymmetry – coefficient of variation – correlation $CVR = CV \cdot (1-\rho^2)$. This index based on the probability model presents left/right correlation in an explicit form as one of its components. This index has theoretical rather than practical importance as it is calculated on the basis of just one trait. Considering the symmetry of the object as a whole, usually a set of traits is used in practice. This raises the question of correlation structure between traits of the object with bilateral symmetry. This aspect is rather topical from several points of view. Firstly, it may be an additional argument when choosing traits. Secondly, according to pilot studies, it may play a significant role in estimating degree of asymmetry of bilateral objects. Thirdly, this question is important by itself, from the point of view of morphological structure of such objects. It was shown that the correlation coefficient between traits influenced fluctuating asymmetry assigning the variation range of left/right correlation (Figs. 29.1–29.6), that is, with the increase of correlation value between the traits range of possible values of left/right correlations diminished becoming a speck when correlation coefficient between traits was 1. Therefore, it becomes essential to study the correlation structure between the traits [9].

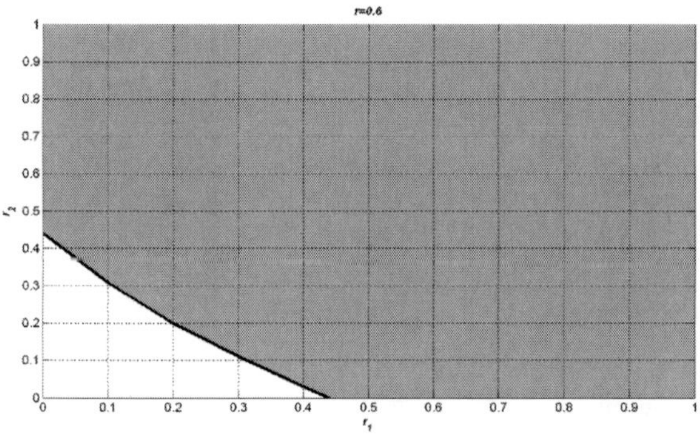

FIGURE 29.1 The range of variation of left/right correlation for two traits in case when correlation between traits is ≤0.5 (along abscissa axis left/right correlation is for the first trait, along ordinate axis it is for the second one).

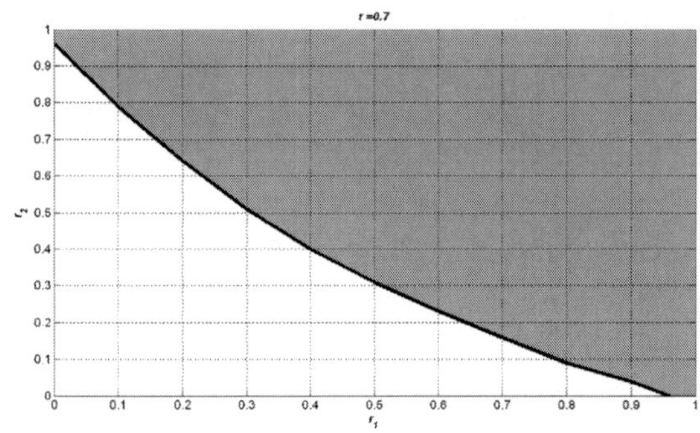

FIGURE 29.2 The range of variation of left/right correlation for two traits in case when correlation between traits is 0.6 (along abscissa axis left/right correlation is for the first trait, along ordinate axis it is for the second one).

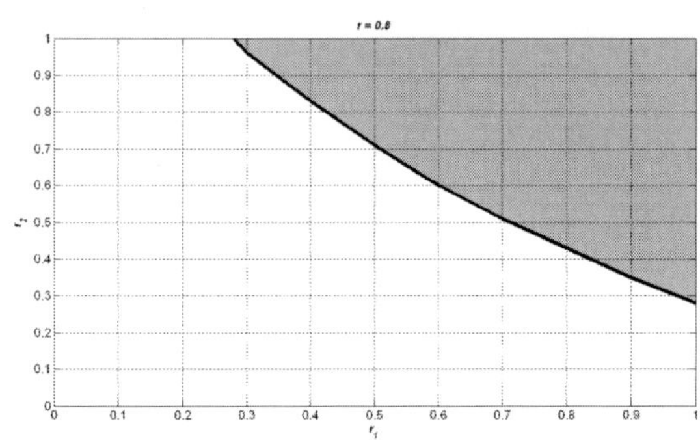

FIGURE 29.3 The range of variation of left/right correlation for two traits in case when correlation between traits is 0.7 (along abscissa axis left/right correlation is for the first trait, along ordinate axis it is for the second one).

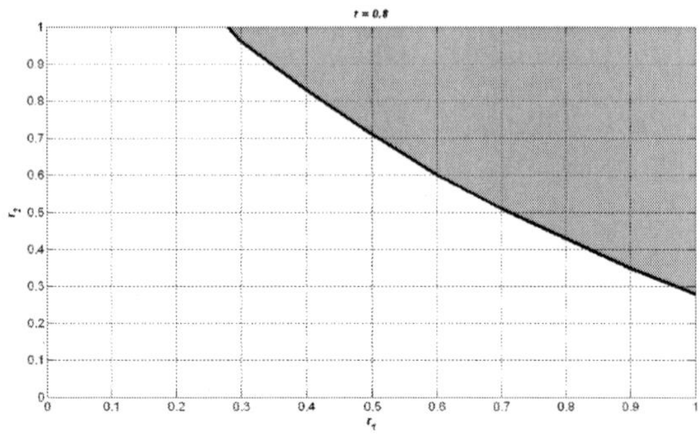

FIGURE 29.4 The range of variation of left/right correlation for two traits in case when correlation between traits is 0.8 (along abscissa axis left/right correlation is for the first trait, along ordinate axis it is for the second one).

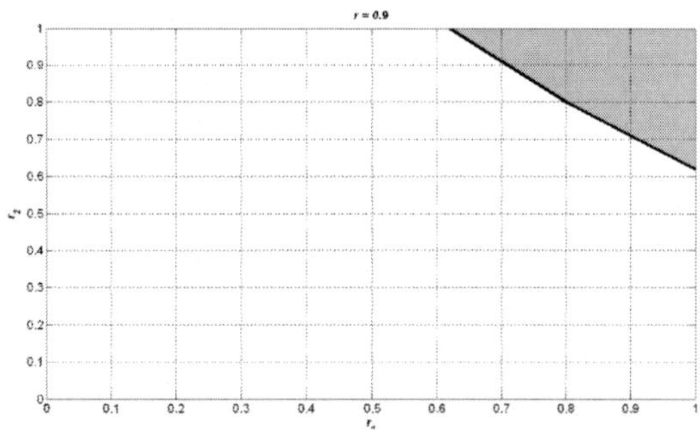

FIGURE 29.5 The range of variation of left/right correlation for two traits in case when correlation between traits is 0.9 (along abscissa axis left/right correlation is for the first trait, along ordinate axis it is for the second one).

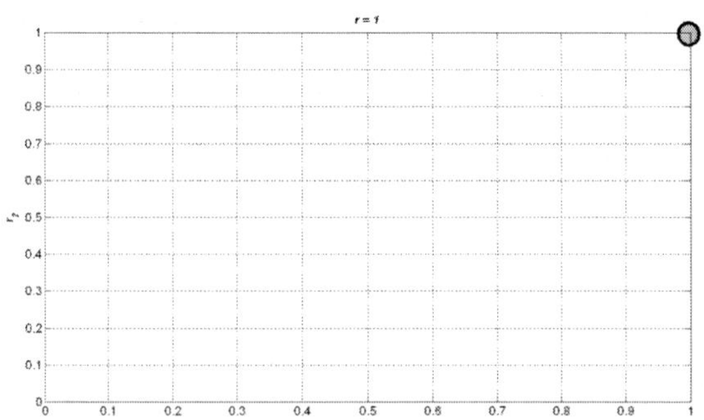

FIGURE 29.6 The range of variation of left/right correlation for two traits in case when correlation between traits is 1.0 (along abscissa axis left/right correlation is for the first trait, along ordinate axis it is for the second one).

In this chapter, we assume that data come from a multivariate normal distribution. In the case of two traits the correlation matrix contains two left/right correlation coefficients and 4 correlation coefficients between the parts of different traits (Fig. 29.7).

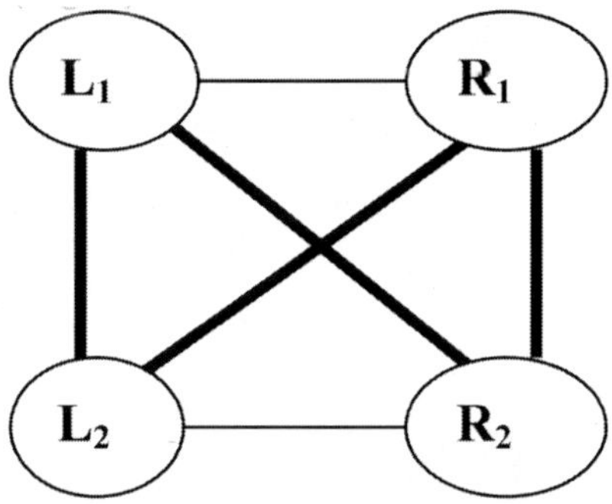

FIGURE 29.7 Correlation structure of two traits of the object with bilateral symmetry: L_1, R_1 are the values of the first trait on the left and on the right; L_2, R_2 are values of the second trait on the left and on the right.

Fine lines signify correlations between left and right parts for a separate trait, heavy lines – correlations between traits. If we assume that the traits have normal distribution, Pearson linear correlation coefficients may act as a quantitative measure of these connections. In this case there are four such correlation coefficients. The easiest estimation method for communication between traits is averaging out of these correlation coefficients or calculation of correlation coefficient between averaged value of the first trait ($(R_1 + L_1)/2$) and the second one ($(R_2 + L_2)/2$). However, this may lead to the loss of considerable part of information.

It is possible to hypothesize regarding the equality of these correlation coefficients in different ways. It is worth noting that the correlations are dependent.

29.2 MATERIALS AND METHODOLOGY

The research was carried out on the basis of materials of two independent lamina samples (V1 and V2) of drooping birch *Betula pendula* Roth. The material for the first sample was collected in the section of Nizhny Novgorod City on the other side of the Volga River in July August of 2004 in five localities (habit areas), in the quantity of 10 leaves from a separate tree. The total sample size is 500 leaves. For each lamina the values of five traits were measured on the left and on the right (Fig. 29.8). All measurements were conducted straight on the collected leaves using the protractor and the ruler. In this work we have used only first four traits.

The material for the second sample was collected in July of 2007 along the roadway of the 40th Anniversary October Revolution Street, the City of Nizhny Novgorod. The total sample size was 521 laminas. The measurements were conducted on the scanned material using electronic ruler.

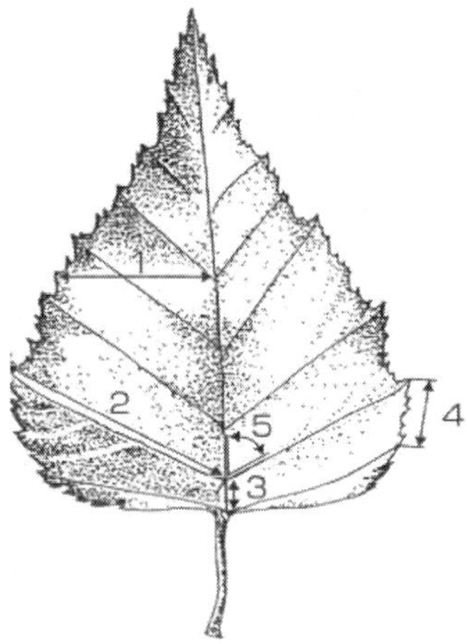

FIGURE 29.8 Measurement scheme for the leaf of drooping birch [10].

The collection methods of the first and the second samples were different. The material of the first sample was collected according to the standard method [10]: we took 10 leaves developed enough from each tree, 10 trees per each locality. The material of the second sample was collected in the following way: we cut one of the branches of each tree from every exposition at a height maximum available for the hand shears and collected all laminas; laminas collected from one branch were enumerated and then 20 laminas, which were chosen using a random number generator, were scanned.

Both samples were submitted to examination for the presence of outliers. On the basis of Rosner's test [11] 2 observations were removed from sample V1, 5 observations – from sample V2. In sample V1 the most of the observations are presented within the accuracy of integral values, however, some values (generally closer to the average ones) are presented within the accuracy of 0.5. Such values are rounded to the integral values.

The difference of traits distribution in sample V1 is statistically significant in comparison with the normal distribution (Table 29.1).

TABLE 29.1 Testing For Normality of Traits in Sample V1

Tests	L_1	R_1	L_2	R_2	L_3	R_3	L_4	R_4
χ^2 test	0.12	2.2×10^{-9}	2.4×10^{-4}	3.2×10^{-5}	8.3×10^{-12}	4.4×10^{-11}	1.2×10^{-4}	2.7×10^{-5}
Bera-Jarque test	0.67	0.38	0.04	0.08	1.3×10^{-7}	0.01	0.27	0.90
Kolmogorov–Smirnov test	0.01	0.01	3.5×10^{-4}	3.8×10^{-3}	5.9×10^{-11}	3.9×10^{-10}	1.5×10^{-7}	4.0×10^{-6}
Shapiro–Wilk test	2.0×10^{-14}	5.1×10^{-14}	6.7×10^{-14}	1.7×10^{-15}	3.8×10^{-13}	4.3×10^{-14}	1.8×10^{-14}	7.3×10^{-14}

In sample V2–5 distributions of 8 do not differ significantly from the normal ones (Table 29.2).

TABLE 29.2 Testing For Normality of Traits in Sample V2

Tests	L1	R1	L2	R2	L3	R3	L4	R4
χ^2 test	0.17	2.3×10^{-4}	0.62	0.31	0.21	0.88	0.06	6.9×10^{-4}
Bera-Jarque test	0.22	0.03	0.20	0.20	0.23	0.26	0.02	0.01
Kolmogorov–Smirnov test	0.59	0.11	0.72	0.58	0.53	0.59	0.23	0.08
Shapiro–Wilk test	0.08	1.8×10^{-3}	0.11	0.01	0.06	0.17	4.9×10^{-3}	1.0×10^{-4}

Parameters of sample V1 are presented in Tables 29.3–29.4. All correlation coefficients differ significantly from zero value ($P<10^{-2}$).

TABLE 29.3 The Means and the Variances of Sample V1

Parameters	L1	R1	L2	R2	L3	R3	L4	R4
Mean	22.11	22.36	36.06	35.97	5.22	5.22	13.30	12.57
Variance	10.49	9.99	22.23	21.65	2.53	2.54	4.00	3.75

TABLE 29.4 The Correlation Matrix of Sample V1

	L1	R1	L2	R2	L3	R3	L4	R4
L1	1	0.74	0.80	0.71	0.26	0.27	0.52	0.46
R1	0.74	1	0.71	0.77	0.21	0.16	0.44	0.52
L2	0.80	0.71	1	0.87	0.22	0.26	0.60	0.45
R2	0.71	0.77	0.87	1	0.28	0.23	0.49	0.54

TABLE 29.4 *(Continued)*

	L1	R1	L2	R2	L3	R3	L4	R4
L3	0.26	0.21	0.22	0.28	1	0.60	0.10	0.18
R3	0.27	0.16	0.26	0.23	0.60	1	0.17	0.14
L4	0.52	0.44	0.60	0.49	0.10	0.17	1	0.55
R4	0.46	0.52	0.45	0.54	0.18	0.14	0.55	1

Parameters of sample V2 are presented in Tables 29.5–29.6. All correlation coefficients differ significantly from zero value ($P<10^{-3}$).

TABLE 29.5 The Means and the Variances of Sample V2

Parameters	L1	R1	L2	R2	L3	R3	L4	R4
Mean	16.62	16.41	26.52	26.40	3.45	3.56	8.92	8.92
Variance	19.72	20.48	48.90	50.21	1.91	2.01	5.99	6.69

TABLE 29.6 The Correlation Matrix of Sample V2

	L1	R1	L2	R2	L3	R3	L4	R4
L1	1	0.93	0.95	0.93	0.14	0.14	0.85	0.83
R1	0.93	1	0.93	0.95	0.16	0.15	0.84	0.85
L2	0.95	0.93	1	0.97	0.16	0.18	0.87	0.83
R2	0.93	0.95	0.97	1	0.19	0.16	0.83	0.87
L3	0.14	0.16	0.16	0.19	1	0.72	0.11	0.14
R3	0.14	0.15	0.18	0.16	0.72	1	0.11	0.13
L4	0.85	0.84	0.87	0.83	0.11	0.11	1	0.84
R4	0.83	0.85	0.83	0.87	0.14	0.13	0.84	1

Hotelling [12] was the first who suggested the criterion for comparison the dependent correlation coefficients; he constructed an asymptotic criterion for comparison of 2 correlation coefficients of a three-dimensional normal distribution. Later on Williams [13] improved his result having increased the power of this criterion. Lawley [14] generalized Anderson's result [15] having built the criterion to compare nondiagonal elements of the correlation matrix. For these hypotheses Aitkin et al. [16] constructed the test based on the likelihood ratio. Dunn and Clark [17, 18] suggested several criteria to compare two dependent correlations; they calculated asymptotic correlations between *z*-values (values of correlation coefficients after Fisher's *z*-transformation), compared the powers of these criteria with already known Hotelling's [12] and Williams's [13] criteria. Steiger [19] suggested the solution for a more general problem – comparison of more than two dependent correlation coefficients. To solve this problem he considered two criteria that differ by estimation method of the unknown parameters: X_1 criterion uses the maximum likelihood estimates, and X_2 criterion is based on the generalized least squares method. It was shown that these estimates are equivalent [20].

In this chapter, we will use the following criteria and methods: the classical likelihood ratio test [21], the Akaike information criterion (AIC) and the Bayesian information criterion (BIC) [22–23], simulation-based criteria (model-based and nonparametric) and Steiger's X_2 criterion [19], as well as the estimation of correlation structure using partial correlation coefficients [24].

The likelihood ratio test is calculated as $-2ln\Lambda = -2ln\frac{\max_{\theta\in\Theta_0} L(\theta)}{\max_{\theta\in\Theta} L(\theta)}$, where the numerator is the maximum of likelihood function under the null hypothesis (Θ_0 parameter set), and the denominator is under the alternative, that is, the case of the difference of all 14 parameters of a 4-dimensional normal distribution (Θ parameter set). This statistic is approximately distributed as χ^2 -distribution with $(\nu - \nu_0)$ degrees of freedom, where ν is Θ dimension, and ν_0 is Θ_0 dimension.

The information criteria AIC and BIC can be written as follows

$$AIC = 2k - 2ln(L),$$

$$BIC = kln(N) - 2ln(L),$$

where k is the number of the model parameters to be estimated, L is the maximum value of likelihood function of the model, N is the sample size.

The proposed simulation criteria perform decision-making on the basis of the distribution of the differences of the compared correlation coefficients. In the first case (model-based simulation or parametric bootstrap), the specified distribution is constructed by repeated sampling from the multivariate normal distribution whose parameters are estimated from the sample. In the second case, the distribution is built by repeated sample replication by means of simple random sampling with replacement (nonparametric bootstrap analog).

The statistic of the Steiger's criterion has the following form:

$$X_2 = (N-3)[z(r) - z(\hat{\rho}_{GLS})]'S_{LS}^{-1}[(r) - z(\hat{\rho}_{GLS})],$$

where N is the sample size, $z(.)$ –is the Fisher's z-transformation, r is the vector-column of the sample correlation coefficients, $\hat{\rho}_{GLS}$ is the vector-column of the estimated correlation coefficients by the generalized least squares method under the null hypothesis, S_{LS}^{-1} is the estimate of the inverse covariance matrix between correlation coefficients received by the least squares method. The X_2 statistic has asymptotical χ^2 -distribution with $(k–q)$ degrees of freedom, where k is the number of upper-diagonal elements of the correlation matrix, that is, $k = (m^2 - m)/2$ for the matrix $m\times m$, q is the number of different correlation coefficients under the null hypothesis.

The variances of the traits for samples V1 and V2 differ about twofold (see Tables 29.3 and 29.5). The reason for this is different methods of material collection. In the first sample the objects were selected according to methodological recommendations [10], i.e. laminas were selected within quite small range of sizes, and in the second sample laminas were selected at random without considering their size. This led to the fact that the correlation structures of different samples are incomparably different: the traits in sample V2 are stronger correlated (see Tables 29.4 and 29.6). To compare the data of different samples we use partial correlation coefficients [24], where we choose lamina width as a conditional variable; it is calculated as the sum of the values of trait 1 on the left and on the right (denoted by *d*):

$$\rho_{L_iL_j\cdot d} = \frac{\rho_{L_iL_j}-\rho_{L_id}\rho_{L_jd}}{\sqrt{1-\rho_{L_id}^2}\sqrt{1-\rho_{L_jd}^2}},\quad \rho_{L_iR_j\cdot d} = \frac{\rho_{L_iR_j}-\rho_{L_id}\rho_{R_jd}}{\sqrt{1-\rho_{L_id}^2}\sqrt{1-\rho_{R_jd}^2}},\quad \rho_{R_iR_j\cdot d} = \frac{\rho_{R_iR_j}-\rho_{R_id}\rho_{R_jd}}{\sqrt{1-\rho_{R_id}^2}\sqrt{1-\rho_{R_jd}^2}},\quad i,j=\overline{2,4}$$

29.3 RESULTS AND DISCUSSION

For the sake of convenience, we will denote the model when all of the four correlation coefficients between the traits are equal as 1111. If there are two different correlation coefficients, then the following models may be possible: 1112, 1121, 1122, 1211, 1212, 1221, and 1222. In case of three different correlation coefficients, the models are as follows: 1123, 1213, 1223, 1231, 1232, and 1233. Finally, if all four correlation coefficients are different, there is only one model 1234. For further details see Table 29.7.

In both samples for all pairs of traits, the likelihood ratio test accepted several different models. However, except for the pair of traits 1–2 in sample V2, in all cases model 1221 is not rejected at the 1% significance level (P = 0.02–0.94). The results for traits 1–2 in sample V1 are presented in Table 29.7. Model 1221 can be interpreted as a model in which all equal correlations between traits on one side of the lamina (left or right) differ from equal correlations between the values of the trait on different sides of the lamina.

TABLE 29.7 The Results of the Likelihood Ratio Test

Models	Interpretation	χ^2 test	Degree of freedom	P-value
1111	$\rho_{L_1L_2} = \rho_{L_1R_2} = \rho_{R_1L_2} = \rho_{R_1R_2}$	115.48	3	0.00
1112	$\rho_{L_1L_2} = \rho_{L_1R_2} = \rho_{R_1L_2}$	60.70	2	0.00
1121	$\rho_{L_1L_2} = \rho_{L_1R_2} = \rho_{R_1R_2}$	52.81	2	0.00
1122	$\rho_{L_1L_2} = \rho_{L_1R_2}\ \rho_{R_1L_2} = \rho_{R_1R_2}$	115.37	2	0.00

1211	$\rho_{L_1L_2} = \rho_{R_1L_2} = \rho_{R_1R_2}$	34.58	2	0.00
1212	$\rho_{L_1L_2} = \rho_{R_1L_2}\ \rho_{L_1R_2} = \rho_{R_1R_2}$	113.97	2	0.00
1221	$\rho_{L_1L_2} = \rho_{R_1R_2}\ \rho_{L_1R_2} = \rho_{R_1L_2}$	2.14	2	0.34
1222	$\rho_{L_1R_2} = \rho_{R_1L_2} = \rho_{R_1R_2}$	32.72	2	0.00
1123	$\rho_{L_1L_2} = \rho_{L_1R_2}$	4.82	1	0.00
1213	$\rho_{L_1L_2} = \rho_{R_1L_2}$	20.82	1	0.00
1223	$\rho_{L_1R_2} = \rho_{R_1L_2}$	0.12	1	0.72
1231	$\rho_{L_1L_2} = \rho_{R_1R_2}$	1.27	1	0.25
1232	$\rho_{L_1R_2} = \rho_{R_1R_2}$	14.55	1	0.00
1233	$\rho_{R_1L_2} = \rho_{R_1R_2}$	22.83	1	0.00
1234	All correlation coefficients are different	-	-	-

The values of the information criteria for this model are also small (see Table 29.8), which means that model 1221 is very plausible.

TABLE 29.8 The Results of the Information Criteria, AIC and BIC

Traits	Models	Rank AIC	Rank BIC
Sample V1			
	1111	13	13
1–2	1221	2	1
	1234	4	4
	1111	11	3
1–3	1221	10	6
	1234	6	14
	1111	13	13
1–4	1221	1	1
	1234	4	10
	1111	13	13
2–3	1221	1	1
	1234	6	11

	1111	14	13
2–4	1221	4	1
	1234	3	5
	1111	13	3
3–4	1221	1	1
	1234	9	15
Sample V2			
	1111	14	14
1–2	1221	4	4
	1234	2	2
	1111	10	1
1–3	1221	2	3
	1234	13	15
	1111	14	12
1–4	1221	6	2
	1234	5	8
	1111	13	13
2–3	1221	1	1
	1234	5	9
	1111	13	13
2–4	1221	1	1
	1234	4	4
	1111	6	1
3–4	1221	7	4
	1234	12	15

To test model 1221, simulation-based criteria (model-based and nonparametric) were also used (see Table 29.9).

TABLE 29.9 The Results of Simulation Study For Testing Various Hypotheses About Equality of the Correlation Coefficients (in bold – the corresponding hypothesis is rejected)

Traits	Hypotheses	Model-based simulation		Nonparametric simulation	
		2.5%-quantile	97.5%-quantile	2.5%-quantile	97.5%-quantile
Sample V1					
1–2	r1=r4	–0.0146	0.0559	–0.0214	0.0618
	r2=r3	–0.0566	0.0403	–0.0599	0.0426

1–3	r1=r4	0.0033	0.1841	0.0005	0.1817
	r2=r3	–0.0322	0.1436	–0.0322	0.1418
1–4	r1=r4	–0.0637	0.0866	–0.0716	0.0966
	r2=r3	–0.0895	0.0718	–0.0985	0.0799
2–3	r1=r4	–0.0923	0.0731	–0.0972	0.0767
	r2=r3	–0.0989	0.0601	–0.1005	0.0635
2–4	r1=r4	0.0088	0.1407	–0.0019	0.1517
	r2=r3	–0.1220	0.0317	–0.1279	0.0324
3–4	r1=r4	–0.1265	0.0710	–0.1242	0.0704
	r2=r3	–0.0949	0.1030	–0.0872	0.0950
Sample V2					
1–2	r1=r4	–0.0270	–0.0083	–0.0631	0.0077
	r2=r3	–0.0329	–0.0060	–0.0683	0.0109
1–3	r1=r4	–0.1085	0.0291	–0.1210	0.0383
	r2=r3	–0.0466	0.0900	–0.0554	0.1035
1–4	r1=r4	–0.0281	0.0191	–0.0382	0.0251
	r2=r3	–0.0551	0.0010	–0.0653	0.0077
2–3	r1=r4	–0.0955	0.0370	–0.1100	0.0455
	r2=r3	–0.0217	0.1040	–0.0403	0.1326
2–4	r1=r4	–0.0147	0.0234	–0.0193	0.0268
	r2=r3	–0.0306	0.0210	–0.0341	0.0258
3–4	r1=r4	–0.1383	0.0120	–0.1488	0.0208
	r2=r3	–0.0776	0.0718	–0.1082	0.0909

The model-based simulation criterion based on the multivariate normal distribution model rejects the hypothesis for pairs of traits 1–3 and 2–4 in the first sample as well as for pair of traits 1–2 in the second sample. The nonparametric simulation criterion is more conservative and rejects model 1221 only for pair of traits 1–3 in the first sample.

Using the Steiger's criterion, we tested models 1111 and 1221. The results are shown in Table 29.10.

TABLE 29.10 The Results of the Steiger's Criterion of Testing Models 1111 and 1221

Traits	Model 1221			Model 1111		
	χ^2	df	P	χ^2	df	P
Sample V1						
1–2	2.12	2	0.35	104.72	3	$<10^{-12}$

1–3	6.99	2	0.03	10.14	3	0.02
1–4	0.12	2	0.94	25.04	3	1.5×10^{-5}
2–3	0.62	2	0.73	25.53	3	1.2×10^{-5}
2–4	5.21	2	0.07	97.59	3	$<10^{-12}$
3–4	0.29	2	0.86	7.96	3	0.05
Sample V2						
1–2	15.55	2	4.2×10^{-4}	130.05	3	$<10^{-12}$
1–3	1.17	2	0.56	5.15	3	0.16
1–4	4.96	2	0.08	29.34	3	1.9×10^{-6}
2–3	1.73	2	0.42	58.83	3	1.0×10^{-12}
2–4	0.20	2	0.90	139.96	3	$<10^{-12}$
3–4	2.96	2	0.23	4.76	3	0.19

Except for a pair of traits 1–2 in the second sample, model 1221 is accepted everywhere. Model 1111, that is, the model of equality of all four correlation coefficients is rejected in most cases. Tables 29.11 and 29.12 show the partial correlation coefficients.

TABLE 29.11 The Partial Correlation Coefficients for Sample V1 (in bold – the corresponding hypothesis is rejected)

	L2	R2	L3	R3	L4	R4
L2	-					
R2	0.65	-				
L3	0.02	0.14	-			
R3	0.16	0.07	0.58	-		
L4	0.37	0.15	-0.03	0.06	-	
R4	0.06	0.24	0.06	0.02	0.39	-

TABLE 29.12 The Partial Correlation Coefficients for Sample V2 (in bold – the corresponding hypothesis is rejected)

	L2	R2	L3	R3	L4	R4
L2	–					
R2	0.73	–				
L3	0.09	0.16	–			
R3	0.20	0.09	0.70	–		
L4	0.33	0.12	–0.03	0.01	–	
R4	0.15	0.36	0.06	0.05	0.41	–

The analysis of the partial correlation coefficients shows that a large proportion of the correlations between traits are caused by the dependence on the size of lamina. Deliverance from this dependence leads to a situation that is very similar for both samples: traits 2 and 4 are uncorrelated, and the correlations between traits 2 and 3 are significant only when the traits located on the same side of lamina (left or right), while the correlation between the left and right sides of the same trait remains sufficiently high. Based on the values of the partial correlation coefficients, model 1221 also seems the most plausible, however, this requires a rigorous statistical analysis and this is the direction of our further research.

29.4 CONCLUSIONS

When studying asymmetry of bilateral traits one cannot but take into account their correlation structure. Based on the results received we may see that when estimating correlation between traits we cannot use the averaged values of these correlation coefficients as their equality model is one of the worst ones. It is confirmed by all criteria examined in this work. The most credible is the model in which the equal correlations between traits on one part of the lamina (left or right) differ from the equal correlations between trait values on different parts of the lamina. The examined correlation structure depends significantly on the chosen range of lamina sizes, which is struck off using partial correlation coefficients to a wide extent, at the same time communication between left and right parts of one trait stays strong enough. Based on the estimates of partial correlation coefficients we may suppose that the same model is the most probable for the correlated traits, but some traits appear to be noncorrelated.

The investigation of the influence of correlation structure between traits to the integrated indexes of asymmetry of bilateral objects seems to be the important direction of further research as for a set of traits. The selection of traits to study fluctuating asymmetry depending on their correlation is also of interest.

In order to understand the scope of generalization of the received results, it is necessary to expand the analysis conducted in this work to other objects and other traits different in structure of fluctuating asymmetry, for example, to bilateral traits of rodents skull [25].

The elaboration of this range of problems may allow understanding the mechanisms of the most important biological problem at least partially. Schmalhausen formulated it as follows: an organism as a whole in individual and historical development [26].

ACKNOWLEDGEMENT

The reported study was partially supported by RFBR, research project No. 12–04–01251-a. The authors are grateful to Doctor of Biological Sciences, Head of the

Chair of Ecology at the Lobachevsky State University of Nizhni Novgorod D.B. Gelashvili and Candidate of Biological Sciences, Senior Research Assistant A.A. Nizhegorodtsev for the material provided. We express gratitude to the lecturer of the Chair of Cross-Cultural Communication E.E. Fliginskikh and Candidate of Pedagogical Sciences, Head of the Chair of Cross-Cultural Communication at Mari State University S.L. Yakovleva for the translation of this article to the English language.

KEYWORDS

- ***Betula pendula* Roth.**
- **bilateral symmetry**
- **dependent correlation coefficient**
- **fluctuating asymmetry**

REFERENCES

1. Weyl, H. (1952). Symmetry New Jersey Princeton University Press, 170p.
2. Van Valen, L. A. (1962). Study of Fluctuating Asymmetry, Evolution, *16(2),* 125–142.
3. Astaurov, B. L. (1927). Researches of Hereditary Change of Galter at *Drosophila Melanogaster,* Journal of Experimental Biology Series A, *3(1/2)*, 1–61 (in Russian).
4. Palmer, A. R., & Strobeck, C. (1986). Fluctuating Asymmetry, Measurement, Analysis, Patterns, Annual Revision Ecology System, *17*, 391–421.
5. Palmer, A. R., & Strobeck, C. (1992). Fluctuating Asymmetry as a Measure of Developmental Stability, Implications of Non-Normal Distribution and Power of Statistical Tests, Acta Zool Fennica, *191,* 57–72.
6. Palmer, A. R., & Strobeck, C. (2003). Fluctuating Asymmetry Analysis Revisited, Developmental Instability, Causes and Consequences, Polak, M. ed. Oxford, Oxford University Press, 279–319.
7. Zakharov, V. M. (1987). Asymmetry of Animals, Moskow, Nauka, 216p (in Russian).
8. Trubyanov, A. B., & Glotov, N. V. (2010). Fluctuating Asymmetry, Trait Variation and the Left-Right Correlation, Doklady Biological Science, *431(1),* 103–105.
9. Trubyanov, A. B., Sofronov, Yu G., & Glotov, N. V. (2013). Correlation Structure of Bilateral Traits, Voskresenskaya, O. L., Zhukova, L. A. (Eds.) Principles and Methods of Biodiversity Conservation, Proceedings of the Fifth International Research Conference, Yoshkar-Ola, Mari State University, *2,* 208–211 (in Russian) (http://marsu.ru/science/conferenc/files/(2013)/12/SBORNIK_chast_%202.pdf).
10. Methodical Recommendations about Performance of an Assessment of Quality of the Environment about a Condition of Living Beings (2003) (an Assessment of Stability of Development of Live Organisms on Level of Asymmetry of Morphological Structures), Entered 16.10.03, *460,* 24p (in Russian).
11. Rosner, B. (1983). Percentage Points for a Generalized ESD Many-Outlier Procedure, Techno Metrics, *25(2),* 165–172.

12. Hotelling, H. (1940). The Selection of Varieties for use in Prediction with Some Comments on the General Problem of Nuisance Parameters, Annals of Mathematical Statistics, *11(3)*, 271–283.
13. Williams, E. J. (1959). The Comparison of Regression Variables, Journal of the Royal Statistical Society, Series B, *21*, 396–399.
14. Lawley, D. N. (1963). On Testing a Set of Correlation Coefficients for Equality, the Annals of Mathematical Statistics, *34(1),* 149–151.
15. Anderson, T. W. (1963). Asymptotic Theory for Principal Component Analysis, the Annals of Mathematical Statistics, *34,* 122–148.
16. Aitkin, M. A., Nelson, W. C., & Reinfurt, K. H. (1968). Tests for Correlation Matrices, Biometrika, *55,* 327–334.
17. Dunn, O. J., & Clark, V. (1969). Correlation Coefficients Measured on the Same Individuals, Journal of the American Statistical Association, *64(325)*, 366–377.
18. Dunn, O. J., & Clark, V. (1971). Comparison of Tests of the Equality of Dependent Correlation Coefficients, Journal of the American Statistical Association, *66(336),* 904–908.
19. Steiger, J. H. (1980). Tests for Comparing Elements of a Correlation Matrix, Psychological Bulletin, *87*, 245–251.
20. Browne, M. W. (1977). The Analysis of Patterned Correlation Matrices by Generalized Least Squares, British Journal of Mathematical and Statistical Psychology, *30*, 113–124.
21. Johnson, R. A., & Wichern, D. W. (2007). Applied Multivariate Statistical Analysis (Sixth Ed), New Jersey, Prentice Hall, 216–220, 456–459.
22. Akaike, H. (1973). Information Theory and an Extension of the Maximum Likelihood Principle, in Procedure 2nd Int. Symp., Information Theory, Petrov, B. N. & Csaki, F. Ed. Budapest, Akademia Kiado, 267–281.
23. Schwarz, G. (1978). Estimating the Dimension of a Model, Annual Statistics, *6*, 461–464
24. Afifi, A. A., & Azen, S. P. (1979). Statistical Analysis, A Computer Oriented Approach, Second Edition, New York, Academic Press, 442p.
25. Trubyanov, A. B. (2010). Analysis of Fluctuating Asymmetry, PhD dis. Nizhni Novgorod, Lobachevsky State University of Nizhni Novgorod, 138p (in Russian).
26. Schmalhausen, I. I. (1938). Organisms as Whole in Individual and Historical Development, Moscow Leningrad, Academy of Science USSR, 144p (in Russian).

CHAPTER 30

THE ANALYSIS OF ONTOGENETIC SPECTRUM OF HETEROGENEOUS POPULATION

NIKOLAY V. GLOTOV[1*], GEORGY YU. SOFRONOV[2], SERGEY M. IVANOV[1], YULIA G. SUETINA[1], LYUDMILA V. PROKOPYEVA[1], and ALEXEY A. TEPLYKH[3]

[1]Mari State University, 1 Lenin Square, Yoshkar-Ola, Republic of Mari El, 424000, Russia

[2]Department of Statistics, Macquarie University, Sydney, New South Wales, 2109, Australia

[3]Russian Center for Forest Protection, 83 Komsomolskaya Street, Yoshkar-Ola, Republic of Mari El, 424004, Russia

*E-mail: nvglotov@inbox.ru

CONTENTS

ABSTRACT

The distribution of discrete ontogenetic states of individuals is usually spatially and temporally different within a population. If a sample from the population sample consists of several subsamples, the comparison of their ontogenetic spectra reveals heterogeneity of samples, i.e. different subsamples cannot be described by the same polynomial distribution. Therefore, the comparison of the samples using the aggregate data is not correct and tends to result in false inferences of biological importance. The paper proposes three methods for comparison of ontogenetic spectra of heterogeneous samples: a randomized variant of ANOVA, principal components analysis and ordinal regression analysis. The following approaches are exemplified in natural populations of cowberry *Vaccinium vitis-idaea* L. and epiphytic lichens *Hypogymnia physodes* (L.) Nyl. and *Pseudevernia furfuracea* (L.) Zopf.

30.1 INTRODUCTION

Each individual of any living organism is characterized by its age: a chronological one measured in time units and a biological one defined on the basis of different morphological, physiological, biochemical, etc. characteristics of an organism. The notion of a biological age is used in population biology of plants and lichens especially if the assessment of a chronological age is impossible [1–3]. An individual development (ontogeny) of plant and lichens has successive ontogenetic periods – latent, pregenerative, generative and postgenerative ones defined on the basis of the scope of morphological markers. Each of these periods has successive discrete ontogenetic states (see Table 30.1). The virginal ontogenetic state of some plants and lichens are divided into v_1 and v_2 on account of their peculiar formation of morphological structures [4, 5].

Thus, a continuous ontogenetic process of plants and lichens is described as a set of successive ontogenetic states, and an ontogenetic state is a qualitative (not a quantitative) marker of an individual. Note that since successive ontogenetic states are arranged in time, then, inevitably, chronological and biological ages are strongly correlated, with individuals of the same ontogenetic state of a different chronological age.

Based on logistical growth curve of an individual during ontogenesis, A. Uranov introduced a weighting coefficient (numerical characteristic) of each ontogenetic state – k_i (Table 30.1).

The whole set of population individuals enables to build an ontogenetic spectrum of population. Though, it is necessary to specify that a notion of a population have two meanings in biological studies. On the one hand, population is a theoretical notion, an elementary microevolutional unit [6]. On the other hand, field researchers use a working term of population when describing demographic indicators of a group of certain individuals and definitive characters in different discontinuous habitats. Therewith, Soviet/Russian botanical studies use the term of coenopopulation – a set of individuals of a certain plant within a single phytocoenosis [3, 5, 7]. Several types of spectra can be distinguished for different plants (Fig. 30.1). The analysis of age spectra elides seeds and germs because of fluctuations in their possible outbreak and accidental mass mortality. When a virginile ontogenetic state bisects, weight coefficients are calculated on the assumption of equal intervals on im-g_1 segment and make up accordingly v_1 – 0.0884, v_2 – 0.1589 [8].

TABLE 30.1 The Discrete Description of Plant Ontogenesis [2]

Period	Number of ontogenetic state i, its name and symbol	Weight coefficient of ontogenetic state k_i
Latent	1. Seeds, sm	0.0025
Pregenerative	2. Seedling, p	0.0067
	3. Juvenile, j	0.0180
	4. Immature, im	0.0474
	5. Virginal, v	0.1192
Generative	6. Young generative, g_1	0.2700
	7. Mature generative, g_2	0.5000
	8. Old generative, g_3	0.7310
Postgenerative	9. Subsenile, ss	0.8808
	10. Senile, s	0.9529
	11. Subcadaveric, sc	0.9819

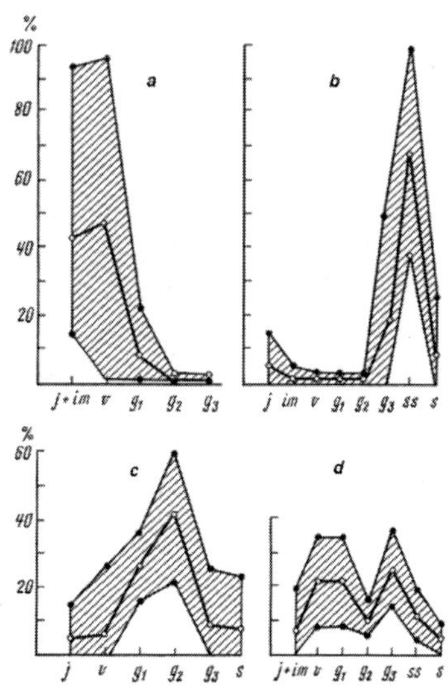

FIGURE 30.1 The types of basic coenopopulation spectra (average scores); ±3σ is shaded, a – left hand spectrum, b – right hand spectrum, c – symmetric single peak, d – double-peak [7].

A. Uranov [2] introduced an *average age* Δ that became a standard parameter of an ontogenetic state of a population (coenopopulation)

$$\Delta = \frac{\sum_{i=1}^{11} k_i n_i}{\sum_{i=1}^{11} n_i},$$

n_i – the number of individuals in an ontogenetic state with *i*-number, k_i – a weight coefficient of an ontogenetic state with *i*-number.

However, different characteristics of an ontogenetic spectrum are used. Thus, N. Glotov [9] proposed to calculate *a modified recovery index* I_1:

$$I_1 = \frac{\sum_{i=3}^{5} n_i}{\sum_{i=3}^{8} n_i},$$

here a part of pregenerative individuals among pregenerative and generative ones characterizing a left part of a spectrum (Figure 1). *An aging index* I_2 [9], a part of postgenerative individuals in a population characterizes the right part of a spectrum:

$$I_2 = \frac{\sum_{i=9}^{11} n_i}{\sum_{i=3}^{11} n_i}.$$

Methodology of data collection to characterize an ontogenetic spectrum of a population consists of samples that include several subsamples collected in different parts of a population, in different years, etc. For instance, in case of herbaceous plants all plants within one square meter quadrate are collected (considered), with the ontogenetic state of each individual being defined; data on all quadrates within a transect are summed, the ontogenetic spectrum of a population (coenopopulation) is found, frequencies (percentage) of individuals of each ontogenetic state and parameters characterizing the ontogenetic spectrum of a population are calculated. Thus, a quadrant is here a measurement unit. For example, in the study of epiphytic lichens, individuals (thalli) of all ontogenetic states separately on each tree are considered, with data within a habitat or a certain phorophyte (substratum) being summed. Thus, a single tree is here a measurement unit.

However, the aggregation of subsamples (quadrants for plants, trees for lichens) and, therefore, the study of the aggregated sample (population) are only eligible if the sample is homogeneous – the distributions of different subsamples within the sample are not significantly different, i.e. can be described by the same polynomial distribution [10].

The chi-square criterion is a standard method to compare ontogenetic spectra. It is a asymptotic criterion, and it use for analyzing contingency tables is eligible only if expected values are big enough. The conditions of correct use of chi-square criterion are standard: minimal anticipated value is not less than 5 or an average observable in a cell of a contingency table is not less than 5. Sokal and Roplf [10] propose to lower the requirement to the minimal expected abundance with growing degrees of freedom. Simonoff-Tsai criterion [11] is often used to determine the possibility to use chi-square criterion. Observed abundance of adjacent classes is often combined to get bigger expected values. If chi-square criterion is still inapplicable, the homogeneity analysis of RxC contingency tables uses an exact test that is a generalized Fisher's exact test for 2×2 contingency tables [12].

The present study focuses on proposing statistical methods to analyze ontogenetic structure of a population adequate to the field data collection methods taking into account heterogeneity of plant and epiphytic lichen populations in space (or in time).

30.2 MATERIALS AND METHODOLOGY

To solve this problem we propose the following three methods.

Method 1 – Estimation and comparison of the parameters Δ, I_1 and I_2, ontogenetic spectra using randomization analysis of variance (ANOVA). The values of the parameters are calculated for each subsample. In this case the assumptions of one-way ANOVA hold: levels of a factor are the k samples, which are represented by n_i (i=1, 2, –, k) subsamples. Each subsample is characterized by the value of the parameter Δ_{ij} (j=1, 2, –, n_i, the number of subsamples in the ith sample). It can similarly be done for the parameters I_1 and I_2.

Using the F-test and Scheffé's method (in order to account for multiple comparisons) in one-way ANOVA, we can compare the sample mean values of the parameter Δ ($I_1 \times I_2$, respectively). Let us denote this fixed-effect model as Model I. Using the random-effect ANOVA model (denoted by Model II), we can estimate the effect of the factor s_a^2 / s_t^2, where s_a^2 is the variance between samples, s_t^2 is the total variance. However, under ANOVA model the following assumptions should be met: equality of the variances within all groups and the residuals are normality distributed [13]. Since these requirements are often violated, various randomized procedures in ANOVA have been increasingly used in recent years. In this chapter, we use the randomized variant of ANOVA [14].

In Model I ANOVA, we carry out random permutation of the values of the parameter of subsamples between the samples without repetitions. After this randomization we calculate the value of F-test taking into account the sizes of the subsamples. Under these permutations, the subsample weight is moved together with the value of its parameter, while the total weight of the sample may vary. We repeat this procedure N=10,000 times, then we obtain the distribution of F-values (F_i). For the randomization procedure, the distribution of F-values is not important; the values are only used as quantities characterizing the differences between the samples. Then we find F-value for the initial data (F_{exp}) and count the number of F-values that greater than or equal to the F_{exp} (that is, $F_i \geq F_{exp}$). Finally, we determine the proportion of the values such that $F_i \geq F_{exp}$ assuming that the F_{exp} has been obtained using the randomization procedure. If the results are significant, then Scheffé's method for multiple comparisons can be carried out in a similar way.

The randomized variant of Model II ANOVA is based on the bootstrap method [15]. The standard bootstrap procedure is carried out within each sample, that is, a new sample is constructed using resampling with repetitions from the initial sample. Then the effect of between-sample variability for the obtained randomized data is calculated. This procedure is repeated N=10,000 times, as a result we obtain the distribution of the effect. The median of this distribution is used as an estimate of the effect, $\alpha\times100\%$-confidence interval is obtained using $(1-\alpha)/2$- and $(1+\alpha)/2$-quantiles (for example, $\alpha = 0.95$). When calculating the effect, the procedure of weighting the size of subsamples is also used.

Method 2 – Application of principal component analysis [16]. The initial data are the distribution of frequencies of individuals belonging to each ontogenetic state. We apply the Fisher φ-transformation for the obtained frequencies of ontogenetic states such that subsamples are treated as observations and variables (markers) are j-values of the frequencies of individuals from the same ontogenetic state. Then we obtain the correlation matrix taking into account the weights that were defined in Method 1. It can be shown that the correlation matrix will remain the same if the weights are proportionally increasing.

As a result of principal component analysis, we obtain new variables, the components, which are linear combinations of the original variables. The sum of frequencies of individuals from different ontogenetic states for each subsample is equal to 1. Therefore, the correlation matrix based on the frequencies of ontogenetic states will be singular and the variance of the last component (that is, the smallest eigenvalue of the matrix) must be 0 [16]. Formally speaking, the matrix should not be singular after the Fisher φ-transformation and, therefore, the smallest eigenvalue of the matrix is not equal to 0. However, the variances of the last components are usually very small and, taking into consideration dependence between frequencies of individuals from different ontogenetic states, it is likely that the variance of the last component under the Fisher φ-transformation will be very small. Then for each of the obtained components, we carry out the randomized variant of ANOVA taking into account the weights of the subsamples.

Method 3 – Application of cumulative link mixed models [12], also known as ordinal regression models, which belong to generalized linear models. The proposed model can be described as follows:

$$P(Y_i < j) = g(\theta_j - u(\text{Sample}) - v(\text{Subsample})), \quad i=1, \ldots, n; \quad j=1, \ldots, J-1;$$

$$u(Sample) \sim N(0, \sigma_a^2), \quad v(Subsample) \sim N(0, \sigma_e^2),$$

where n is the total number of individuals in all samples, J is the number of ontogenetic states, j is the ontogenetic state. The random variable Y_j represents the ontogenetic state of the ith individual. Then $P(Y_i \leq j)$ is the probability that the ontogenetic state of the ith individual is less than or equal to j. We take the sample and the subsample effects to be random, and assume that they have normal distributions $N(0, \sigma_a^2)$ and $N(0, \sigma_e^2)$, respectively. The g is a link function usually logit or probit (although it is possible to use other functions), that is, $g(y) = \exp(y)/(1+\exp(y))$ or $g(y) = \Phi(y)$ (the cumulative normal distribution function).

This model can be interpreted in the following way. Each individual has a certain age ξ, which increases over time such that $-\infty$ indicates the birth of the individual whereas $+\infty$ represents the death of the individual. Thus, the jth ontogenetic state is a time interval and θ_j is the value of the age ξ such that $\theta_{j-1} < \xi \leq \theta_j$. In this case the g is the cumulative distribution function of ξ. The θ_j, j =1, …, J–1, are known as threshold parameters or cut-points. The link function g and the values of θ_j are the same for all samples and subsamples. Distributions for different subsamples differ only by a horizontal shift of the function g, this shift is the sum of the effects in our model.

In order to test that the sample random effect is significant, we propose to use the following two nested models:

$$M_{full}:\ P(Y < j) = g\left(\theta_j - u(Sample) - v(Subsample)\right),$$

$$M_{cut}:\ P(Y < j) = g\left(\theta_j - v(Subsample)\right).$$

We use the likelihood ratio test to test the null hypothesis H_0: $\sigma_a^2 = 0$. The estimates of σ_a^2 and σ_e^2 can be used in order to assess the contribution of each predictor (Sample or Subsample).

The data were analyzed using statistical software R. We implemented the algorithms based on Methods 1 and 2, R-package "ordinal" was applied in order to analyze the data using Method 3.

Suggested methods are tested on the data from natural populations of cowberry (*Vaccinium vitis-idaea* L.) and epiphytic lichens *Hypogymnia physodes* (L.) Nyl. and *Pseudevernia furfuracea* (L.) Zopf on the territory of the Republic of Mari El.

The study of 10 coenopopulations of cowberry was carried out within the State Nature Reserve "Bolshaya Kokshaga" (№№ 1, 3–6) and in the neighborhood of the Ismentsy settlement of the Zvenigovsky district (№№ i1-i5). Coenopopulation No.1 is situated in a pine tree forest recovering from ground fire in 1995, coenopopulations No. 3, i1, i4 and i5 – in a pine tree forest with cowberries, coenopopulations No. 4, 5 and 6 – in a pine tree forest with green mosses and cowberries, coenopopulation No. i2 – in a pine tree forest with firs and cowberries, coenopopulation No. i3 – in a pine tree forest with carex, heath and cowberries. The age of forest in the habitats of different coenopopulations is 55–85 years, its normality amounts to

0.5–1. Geobotanical descriptions were processed according to D. Tsyganov's ecological scales [17, 18]. Different habitats have 12.68–13.60 scores (with 11.5% of its range for cowberry) according to Hd-scale (soil moistening), 4.55–5.56 (14.4%) according to Tr-scale (soil wealth), 4.27–5.32 (21.0%) according to Nt-scale (nitrogen abundance), 5.28–6.55 (21.2%) according to Rc-scale (soil acidity), 4.29–4.84 (6.9%) according to Lc-scale (illuminance) and 3.32–5.31 (39.8%) according to fH-scale (soil moisture variability). Habitats are different in their floristic composition: Jaccard's coefficient of community differs from 9.4% to 43.8%. Each coenopopulation had 10–30 quadrants (1 sq.m.) observed, the number of partial cowberry bushes within a quadrant is 5–778, the number of partial cowberry bushes in different coenopopulations is 1498–7117. The diagnostics of partial bushes is described in [19]. The spectrum of ontogenetic states involves the range im-sc, juvenile individuals cannot be found as partial bushes are of vegetative genesis. Some, possibly tens of partial bushes within a single quadrant belong to the same cowberry individuals [20].

The study of *Hypogymnia physodes* was carried out in two habitats within the Starozhilskoye forestry in Medvedevsky district on the thalli (up to 3 m high) of little-leaved linden (*Tilia cordata* Mill.), Siberian fir (*Abies sibirica* L.) and common pine (*Pinus sylvestris* L.). The first habitat in a flood lime-tree forest with pines, ostrich ferns and lilies-of-the-valley in a high-water bed of the Bolshaya Kokshaga river gave two samples: on 23 little-leaved lindens (the number of thalli on a tree is 42–785, the total number of thalli is 3652) and on 7 Siberian firs (the number of thalli on a tree is 4–1229, the total number of thalli is 3562). The age of forest stand is 70–80 years, the length of linden girth on the height of 1.3 m is 0.81 m, the length of fir girth is 0.92 m. The second habitat is located in a woodland park of pine tree forest with fescue and bent grass in the settlement of Starozhilsk: the sample of 16 *Hypogymnia physodes* was taken (the number of thalli on a tree is 221–691, the total number of thalli is 3562).

The second habitat is located in a woodland park of pine tree forest with fescue and bent grass in the settlement of Starozhilsk: the sample of 16 common pines was taken (the number of thalli on a tree is 221–691, the total number of thalli is 7651). The age of forest stand is 75 years, the length of linden girth is 1.04 m. The pine tree forest receives strong man's impact – stocking. Ontogenetic states of *Hypogymnia physodes* are described in [21].

Data on *Pseudevernia furfuracea* were collected in 8 habitats. In pine tree forest with green mosses and *Betula pendula* Roth. within the State Nature Reserve "Bolshaya Kokshaga," 2 samples were collected: sample No. 2a on a pine tree (the age of trees is 65 years, the length of girth is 0.6 m) and sample No. 2b on a birch tree (the age of trees is 45 years, the length of girth is 0.46 m). In the Starozhilskoye forestry in the Medvedevsky district, three samples were collected: sample No. 3 on a pine tree in a pine tree forest with blueberry (the age of trees is 70 years, the length of girth is 0.54 m); sample No. 4 on a pine tree in a low bush and bog moss pine tree forest (the age of trees is 110 years, the length of girth is 0.63 m); sample No. 5 on a birch tree in birch tree forest with carex, calla and bog moss (the age of trees is 50–60 years, the length of girth is 0.55 m). Four samples were taken from pine trees within the Kerebelyakskoye forestry of the National Park "Mariy Chodra": on the nearly located areas of blueberry pine tree forest sample No.6 was taken (in a meso-rise) and sample No.7 (in a mesofall), the age of trees is 60 years, the length of girth is 0.63 m. Sample No.4 includes 67 trees, other samples have 10–20 trees each. The number of thalli on a tree differs from some to a hundred and a half, only in a sample No.8 the number of thalli differs from 64 to 238 and sample No.9 has 93 to 359 thalli on a tree. Ontogenetic states of *Pseudevernia furfuracea* are described in [22].

Mass data collection in nature populations of lichens enables to construct ontogenetic spectra from v_1 ontogenetic state, since thalli of initial ontogenetic states are particulate and specific belonging of im individuals is sometimes difficult to determine. Note that the frequency of g_3, ss and s individuals in nature populations *Pseudevernia furfuracea* is insignificant (about 1 percent), since they are likely to remain on a tree trunk [23, 24].

30.3 RESULTS AND DISCUSSION

Coenopopulations of cowberry. For all coenopopulations, the ontogenetic distributions of subsamples within each coenopopulation differ significantly ($p < 10^{-15}$). This means that statistical analysis of the aggregated sample for coenopopulation would not be correct.

Using Method 1, we showed that the differences in all three parameters for 10 coenopopulations are statistically significant: p=0.0001 for Δ, p= 0.0001 for I_1, p=0.0012 for I_2. Fig. 30.2 shows the results of pairwise comparison of all parameters of the coenopopulations. It can be seen that 20 pairs of Δ (out of possible 45 pairs) are different at the 5% significance level, with the values of Δ for different coenopopulations are located fairly uniform. For the parameter I_1, 28 differences are significant at the 5% significance level and the values for coenopopulations №№ 1, 3, 6 are clearly shifted to the left. For the parameter I_2, only 8 differences are significant at the 5% significance level and the values of I_2 are located fairly evenly. A more detailed analysis of the values of the parameters for ontogenetic spectra of coenopopulations as well as comparison of these results with the characteristics of habitat is the objective of a special study. Note that if we were to do pairwise comparisons of ontogenetic spectra using the aggregated data for each coenopopulation, then we would get (even taking into account the Bonferroni correction) that all 45 comparisons are significantly different ($p < 10^{-12}$). The impacts of differences between the samples are 0.254 for Δ (95% confidence interval is 0.180–0.337), 0.374 for I_1 (0.306–0.443), 0.103 for I_2 (0.040–0.183).

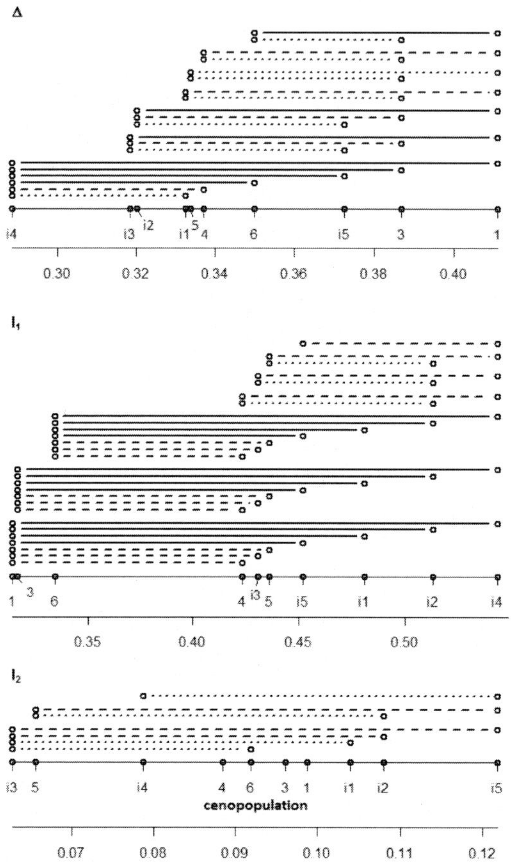

FIGURE 30.2 Significance of differences between the parameters of coenopopulations of cowberry, solid line – $p < 0.001$, dashed line – $p < 0.01$, dotted line – $p < 0.05$.

The axis (bottom line) is the scale of the values.

The results of applying principal component analysis (Method 2) are shown in Table 30.2. It can be seen that all principal components are significantly different, while the effect of variability between coenopopulations is large enough only for the first four principal components. Given that the principal components are independent (by definition), we calculated the average factor effect taking into account the weight of the corresponding principal component:

$$0.445\times0.454 + 0.247\times0.349 + 0.164\times0.395 + 0.081\times0.214 + 0.028\times0.087 + 0.019\times0.144 + 0.017\times0.226 = 0.379$$

TABLE 30.2 The Results of ANOVA of the Principal Components For Coenopopulations of Cowberry

Principal components	PC1	PC2	PC3	PC4	PC5	PC6	PC7
Contribution of the principal component in the total variance	0.445	0.247	0.164	0.081	0.028	0.019	0.017
Cumulative contribution	0.445	0.692	0.855	0.937	0.965	0.983	1.000
Difference between coenopopulations, p	0.0001	0.0001	0.0001	0.0001	0.0273	0.0001	0.0001
The effect of variability between coenopopulations (95% confidence interval)	0.454 (0.392–0.516)	0.349 (0.245–0.455)	0.395 (0.323–0.468)	0.214 (0.142–0.302)	0.087 (0.038–0.171)	0.144 (0.076–0.234)	0.226 (0.152–0.312)

Using ordinal regression models (Method 3), we have demonstrated that the differences between coenopopulations are highly significant both for the probit and for logit models ($p = 10^{-9}$). The effects of variability between coenopopulations are quite close to each other: 0.213 for the probit model and 0.244 for the logit model. The meaning of these effects is clear from Fig. 30.3. Here the coenopopulations are ordered by ascending value of the effect of coenopopulation, the effects for separate quadrants within this coenopopulations are also arranged in ascending order. It can be seen that the range of variability of the coenopopulation is much smaller than the range of variability of the effect of individual quadrants.

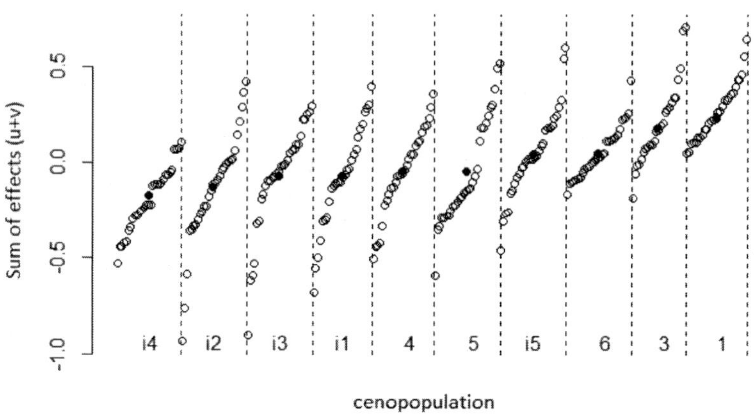

FIGURE 30.3 The results of ordinal regression analysis for ontogenetic spectra of coenopopulations of cowberry (probit model). The filled circles are the effects for the coenopopulations, the blank circles are the effects for individual quadrants.

Thus, all three methods, which take into account the heterogeneity of coenopopulations of cowberries, identify differences between ontogenetic spectra of the coenopopulations. All of the methods also reveal that the effects of differences in variability of the coenopopulations are close to each other. It can be seen that the variability between coenopopulations is relatively low (0.103–0.379), that is, variability between ontogenetic spectrum is mainly concentrated between sample quadrants within the coenopopulation. This may be due to individual characteristics of cowberry or the fact that the study area (territory of the Republic of Mari El) is the southern limit of the individual' area so that the described feature is typical for populations on the border of the area.

Hypogymnia physodes on different substrata. This example shows the need for careful preliminary examination of the experimental data as well as for an accurate statistical analysis. The sample ontogenetic spectra for all three phorophytes are heterogeneous ($p < 10^{-15}$). Method 1 detected the differences in all three parameters: p=0.0007 for Δ, p=0.0025 for I_1, p=0.0002 for I_2. The 95% confidence intervals for the effects between samples with different phorophytes are 0.080–0.719 for D, 0.067–0.707 for I_1 and 0.168–0.539 for I_2. Sufficiently large 95% confidence interval for I_2 may be explained by the fact that we have only three samples and the number of thalli on three firs (in three subsamples) is very small: 4, 13 and 16. The confidence intervals of the effects of between-sample differences for Δ and I_1 do not have much meaning since they cover 63.9% and 73.0% of all possible values, respectively. In order to understand this situation, we plotted the histograms of the distributions of the effects of differences between the samples (Fig. 30.4). It can easily be seen that the distributions have at least two modes. This indicates that the ontogenetic spectra of some individual trees may have systematic biases. Such bias was identified for linden trees: with high irregular variability of spectra, tree 22 appears to be an outlier, it is characterized by much greater frequency of thalli v_1, the highest frequency of thalli v_2 among all the other trees and, therefore, lower frequencies of thalli of all other ontogenetic states (Fig. 30.5). It seems reasonable to exclude linden tree 22 from the analysis. The distinctive feature of ontogenetic spectrum of thalli *H. physodes* on linden tree 22 may be explained by its position: the tree is located on the edge of a forest at the cliff of high bank of the river Bolshaya Kokshaga.

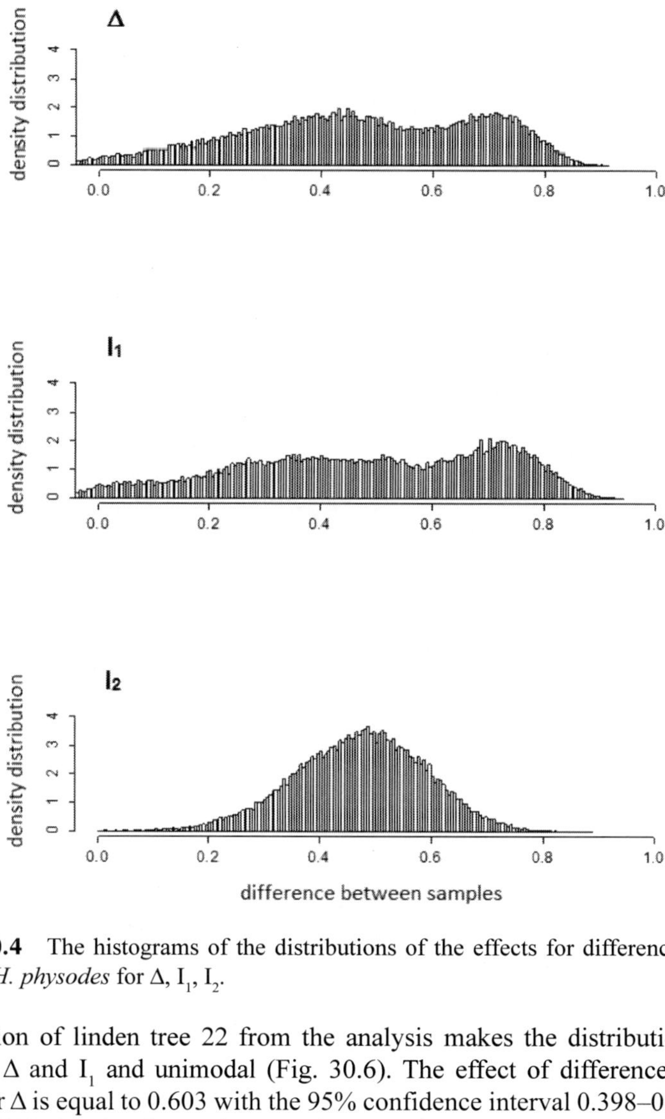

FIGURE 30.4 The histograms of the distributions of the effects for differences between samples of *H. physodes* for Δ, I_1, I_2.

Exclusion of linden tree 22 from the analysis makes the distributions of the effects for Δ and I_1 and unimodal (Fig. 30.6). The effect of differences between samples for Δ is equal to 0.603 with the 95% confidence interval 0.398–0.756, for I_1 – 0.704 (0.479–0.819), almost the same for I_2 – 0.365 (0.167–0.573). Multiple comparisons of the parameters demonstrated that Δ significantly differs for pairs linden-fir, linden-pine ($p < 0.001$), whereas the parameter does not differ for pair fir-pine ($p = 0.11$). The parameter I_1 is different for the same pairs ($p < 0.001$) and for pair fir-pine at 5% significance level ($p = 0.014$). The parameter I_2 is significantly different for pairs linden-pine ($p = 0.0002$) and for linden-fir at 5% significance level ($p = 0.048$) and it does not differ for pair fir-pine ($p = 0.80$). If we ignored the heterogeneity of the samples, the ontogenetic spectra of *H. physodes* on all substrates would be significantly different ($p < 10^{-15}$).

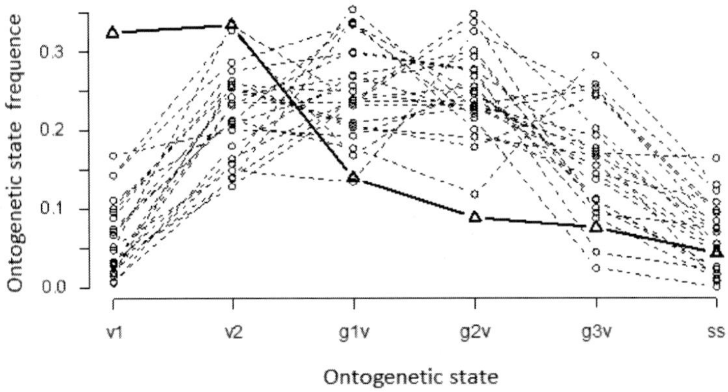

FIGURE 30.5 Ontogenetic spectra of *H. physodes* on linden trees, ontogenetic spectrum of tree 22 is highlighted by solid line.

The results of principal component analysis are shown in Table 30.3. The average effect of between-sample variability of ontogenetic spectra with the weight of the corresponding principal component is equal to 0.512.

TABLE 30.3 The Results of ANOVA of the Principal Components For Samples of *H. physodes*

Principal components	PC1	PC2	PC3	PC4	PC5
Contribution of the principal component in the total variance	0.6789	0.1288	0.1189	0.0584	0.0150
Cumulative contribution	0.6789	0.8077	0.9266	0.9850	1.0000
Difference between coenopopulations, p	0.0001	0.0636	0.0076	0.7098	0.9658
The effect of variability between coenopopulations (95% confidence interval)	0.6876 (0.4147–0.8160)	0.2020 (0.0481-0.4324)	0.1279 (0.0285–0.3231)	0.0304 (0.0009–0.1557)	0.0422 (0.0015–0.2211)

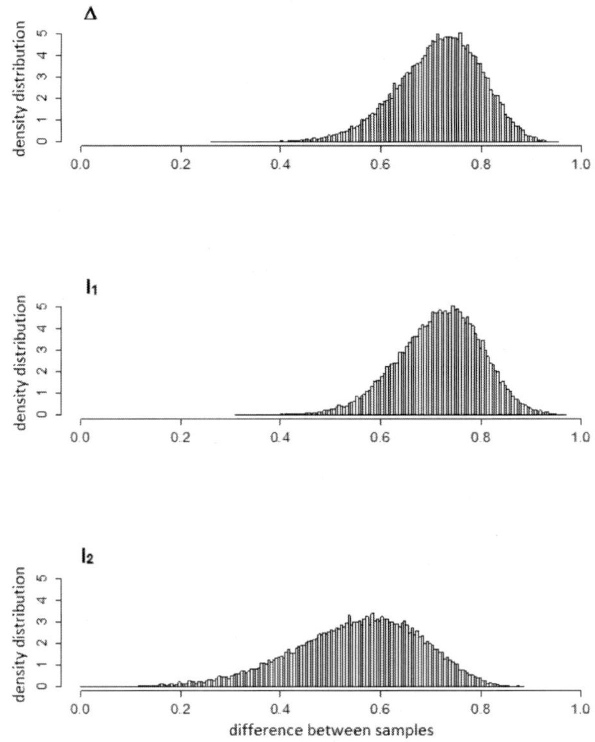

FIGURE 30.6 The histograms of the distributions of the effects of differences between samples of *H. physodes* for Δ, I_1, I_2, linden tree 22 is excluded.

The ordinal regression analysis yields similar results. Ontogenetic spectra are significantly different ($p < 10^{-6}$), the effect of between-sample differences is 0.634 for the probit model and 0.641 for the logit model.

Thus, analyzing ontogenetic spectra of *H. physodes*, all three methods produce consistent results. Unlike for coenopopulations of cowberry, the effect of between-sample variability for *H. physodes* is much higher.

Pseudevernia furfuracea in different habitats and on different substrates. We discovered that ontogenetic spectra of samples 8 and 9 were heterogeneous ($p < 0.003$), whereas spectra of sample 2a were homogenous ($p = 0.1256$). Using Method 1, we demonstrated that the differences between the sample parameters Δ and I_1 were significant ($p = 0.0001$). Pairwise comparisons of Δ show that only one pair (out of 36 possible pairs) is significant the level of 0.0001, two pairs at the level of 0.001 and 7 pairs at the level of 0.05 (Fig. 30.7). Only 5 paired differences of I_1 are significant at the level of 0.05 (Fig. 30.7). If we performed incorrect pairwise comparison of aggregated ontogenetic spectra of 9 samples, then we would make the conclusion about the significance of 28 differences (out of possible 36) at the level << 0.001, four at the level of 0.01, one at the level of 0.05, and only 3 differences would not be statistically significant. The effects of differences between the samples are 0.339 for Δ (0.181–0.514) and 0.204 for I_1 (0.087–0.351).

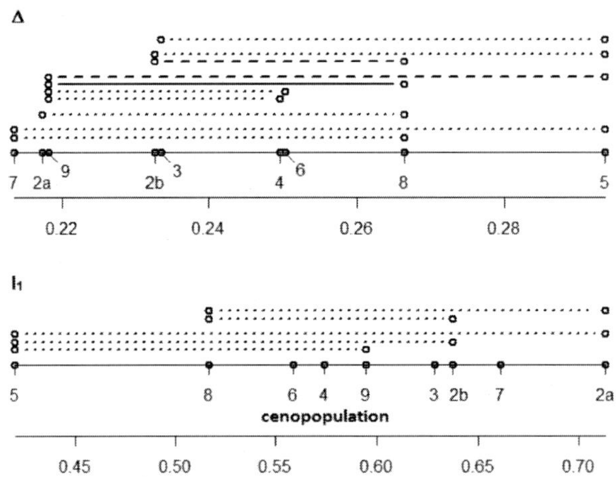

FIGURE 30.7 The significance of differences between the parameters of samples of *P. furfuracea*, solid line – $p < 0.001$, dashed line – $p < 0.01$, dotted line – $p < 0.05$. The axis (bottom line) is the scale of the values.

The results of principal component analysis (Method 2) are shown in Table 30.4. The average effect of between-sample variability of ontogenetic spectra with the weight of the corresponding principal component is equal to 0.293.

TABLE 30.4 The Results of ANOVA of the Principal Components For Samples of *P. furfuracea*

Principal components	PC1	PC2	PC3
Contribution of the principal component in the total variance	0.4232	0.3729	0.2040
Cumulative contribution	0.4232	0.7961	1.0000
Difference between coenopopulations, p	0.0003	0.0001	0.0101
The effect of variability between coenopopulations (95% confidence interval)	0.2297 (0.1080–0.3675)	0.4432 (0.2502–0.6218)	0.1487 (0.0497–0.2862)

Figure 30.8 shows the positions of all subsamples on the plane of the first and the second principal components. As an example, for four samples we encircled the regions that cover the subsamples belonging to the same sample. It can easily be seen that both the size and the position of these regions are very distinct. This type of examination should become the objective of a special study.

The ordinal regression analysis reveals significant differences between ontogenetic spectra of different samples ($<10^{-4}$). The analysis yields similar results for both models; the effect of between-sample differences is 0.404 for probit model and 0.395 for the logit model.

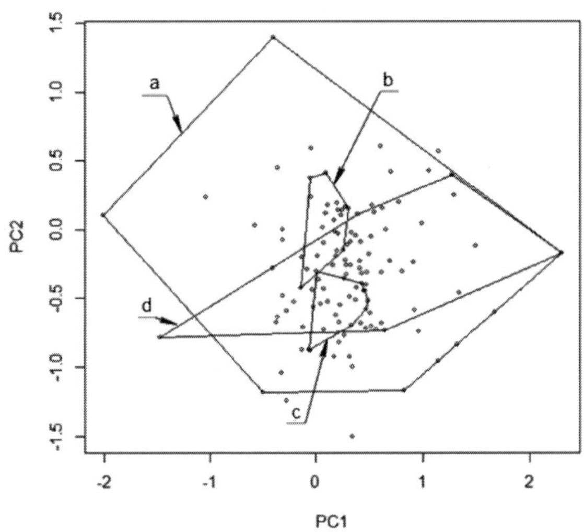

FIGURE 30.8 The positions of all subsamples (thalli of *P. furfuracea*, collected from an individual tree) on the plane of the first two principal components, PC1 and PC2. The regions covering the subsamples of the same sample are encircled: a – sample 4, b – sample 9, c – sample 8, d – sample 7.

30.4 CONCLUSIONS

The proposed methods allow correct comparisons of ontogenetic spectra of heterogeneous samples from populations of plants and epiphytic lichens and assess the effect of between-sample differences of ontogenetic spectra in the total variability. Note that principal components analysis ignores the order of ontogenetic states unlike ordinal regression and in full the comparison of Uranov's age coefficients whose calculation uses age coefficients of ontogenetic states.

It is important that the testing of these methods on data from nature populations of cowberry and epiphytic lichens raises new questions in population studies, for example, Is small cowberry between-population variability connected with border location of populations under study? Does between-population variability correlate with the differences in ecological conditions of different habitats? Does within-population variability correlate with various conditions within a population?

ACKNOWLEDGMENT

This study was partially supported by the Russian Foundation for Basic Research (RFBR), research project No. 12–04–01251-a. The authors are grateful to M.V. Frolova for the translation of the paper into English.

KEYWORDS

- **compare of parameters of ontogenetic spectra**
- **heterogeneous samples**
- ***Hypogymnia physodes***
- **lichens**
- **ontogenetic spectrum of population**
- **ontogenetic state**
- **ordinal regression**
- **plants**
- **principal components analysis**
- ***Pseudevernia furfuracea***
- ***Vaccinium vitis-idaea***

REFERENCES

1. Rabotnov, T. A. (1950). The Life Cycle of Perennial Herbaceous Plants in Meadow Coenoses, Proseeding of Komarov Botanical Institute of the Academy of Sciences of the USSR, Series 3 (Geobotany), *6,* 70–204 (in Russian).
2. Uranov, A. A. (1975). The age Spectrum of a Phytocoeno Population as a Function of Time and Energy Wave Processes, Biological Sciences, *2,* 7–33 (in Russian).
3. The Population Structure of Vegetation (1985) Handbooks of Vegetation Science, 3 [White, J. Ed], 669p.
4. Suetina, Yu G. (2001). Ontogeny and Population Structure of *Xanthoria Parietina* (L) Th Fr. under Different Ecological Conditions, Russian Journal of Ecology, *32(3)*, 185–190.
5. Markov, M. V. (2012). Population Biology of Plants, Moscow, KMK Scientific Press Limited, 386p (in Russian).
6. Timofeeff-Ressovsky, N. W., Jablokov, A. V., & Glotov, N. V. (1977). Grundriss Der Populations Lehre Jena VEBG, Fisher Verlag, 266p (In German).

7. Zaugolnova, L. B., Zhukova, L. A., Komarov, A. S. et al. (1988). Coeno Populations of Plants (Essays of Population Biology), Moscow, Nauka, 184p (in Russian).
8. Suetina, Yu V., Glotov, N. V., Milyutina, L. N., & Kshnyasev, I. A. (2005). Population Research of *Xanthoria Parietina* (L) Th. Fr with Varying Degrees of Urban Pollution, Bulletin of Lobachevsky State University of Nizhni Novgorod, Series Biology, *1(9)*, 197–205 (in Russian).
9. Glotov, N. V. (1998). About the Estimation of the Parameters of the Age Structure of Plant Population, Life Populations in Heterogeneous Environment Part 1 Yoshkar-Ola Periodika Mari E. L. 146–149 (in Russian).
10. Sokal, R. R., & Rohlf, F. J. (1995). Biometry, New York, H. W. Freeman and Company, 891p.
11. Simonoff, J. S. & Tsai, C. L. (1991). Higher Order Effects in Log-Linear and Log-Nonlinear Models for Contingency Tables with Ordered Categories, Journal of the Royal Statistical Society, Series C (Applied Statistics), *40(3)*, 449–458.
12. Agresti, A. (2002). Categorical Data Analyse is, John Wiley & Sons, 710p.
13. Scheffe, H. (1999). The Analyse is of Variance, John Wiley & Sons, 477p.
14. Manly, B. (2006). Randomization, Bootstrap and Monte Carlo Methods in Biology, Chapman and Hall, 480p.
15. Efron, B., & Tibshirani, R. J. (1993). An Introduction to the Bootstrap, New York, Chapman & Hall, 456p.
16. Jolliffe, I. T. (2002). Principal Components Analysis, New York, Berl in, Heidelberg, Springer-Verlag, 489p.
17. Tsyganov, D. N. (1983). Phyto Indication of Ecological Regimes in the Coniferous-Broad Leaved Forest Sub Zone, 196p (in Russian).
18. Zaugolnova, L. B., & Khanina, L. G. (1996). Experience in Development and Use of Database in Phytocenology of Forests, Russian Forest Sciences, *1*, 76–83 (in Russian).
19. Prokopyeva, L. V., Zhukova, L. A., & Glotov, N. V. (2000). Ontogenesis of Cowberry (*Vaccinium Vitis-Idaea,* L) Ontogenetic Atlas of Medicinal Plants, Yoshkar-Ola, Mari State University, 39–46 (in Russian).
20. Glotov, N. V., Semerikov, V. L., & Prokopyeva, L. V. (2008). Research on Genetic Structure of Cowberry Population (*Vaccinium Vitis-Idaea,* L) in the Reserve, Scientific Papers of the State Nature Reserve "Bolshaya Kokshaga", *3*, Yoshkar-Ola, Mari State Technical University, 110–130 URL http: //b-kokshaga.ru/content/files/docs/v3.zip (in Russian).
21. Suetina, Yu G., & Glotov, N. V. (2014). Variability of Characters in Ontogenes is of Epiphytic Lichens (*Hypogymnia Physodes,* L) Nylon Russian Journal of Developmental Biology, *45(3).*
22. Suetina, Yu G., Teplyh, A. A., & Bogdanov, G. A. (2007). Foliaceous Form of the Lichen (*Pseudevernia Furfuracea,* L) Zopf Scientific Papers of the State Nature Reserve "Bolshaya Kokshaga", *2*, Yoshkar-Ola, Mari State Technical University, 230–234. URL: http://b-kokshaga.ru/content/files/docs/v2.zip (in Russian).
23. Suetina, Yu V., & Teplykh, A. A. (2006). Age-Vitality Structure of Population Epiphytic Lichen (*Pseudevernia Furfuracea,* L) Zopf in Lichen-Moss Pine Forest, Principles and Methods of Biodiversity Conservation, Yoshkar-Ola, Mari State University, 286–288 (in Russian).
24. Teplykh, A. A. (2008). Spatial and Age-Vitality Population Structure of Lichen (*Pseudevernia Furfuracea,* L) Zopf in Upland Bog, Scientific Papers of the Nature Reserve "Bolshaya Kokshaga", *3*, Yoshkar-Ola, Mari State Technical University, 143–158. URL: http://b-kokshaga.ru/content/files/docs/v3.zip (in Russian).

CHAPTER 31

PRINCIPLES AND CHALLENGES IN THE Development of Parametrical Systems of Biological Taxa

ZELEEV RAVIL M.

Kazan (Volga region) Federal University, d. 18, Kazan, Kremlevskaya St., 420018, Russia, Tatarstan region; E-mail: zeleewy@rambler.ru

CONTENTS

ABSTRACT

The article examines the causes of nonconformity of the hierarchical model of assessment of biodiversity to the needs of practical systematics. The necessary groundings and approaches to create multidimensional parametric systems are presented. The system of terms that allows making the correct description of taxa in accordance with the rules of parametric systematic is given. The algorithm of transformation of the hierarchical system of biological taxon into parametric form is developed. The variants of algorithm implementation are illustrated on a number of examples of individual groups of organisms of different taxonomic ranks.

31.1 INTRODUCTION

Biological diversity is so great and specific that it requires a special type of thinking to perceive it integrally. It is difficult to use discrete concepts here, you have to widely use metaphors, as the variety of biodiversity implementations, especially for high rank taxa, is easier to perceive in images than in formula. There is even the notion of "taxonomic continuum" [1], reflecting the gradual nature of the manifestation change of the individual features and their specific values for different taxa in the multidimensional taxonomic space. At the same level of quality the taxa themselves (as you know, their selection procedure is purely intuitive) are perceived as a kind of "cloud" in the multidimensional feature space with a distinctive core consisting from the most typical representatives and the periphery, where used features are combined in a different way. It is here where the so-called taxa *incertae sedis* (Latin for "uncertain situation") violating the hierarchy harmony are concentrated. Apparently, they play the key role in understanding the structure of the entire taxonomic space for specific taxonomic groups of organisms. Paleontological data analysis indicates changes in the position in this space – both the taxa and their nuclei, which evidentially reflects changes of enclosing coenotic context. In everyday awareness it is difficult to imagine other variants of biological systematics, except the hierarchical subordination of lower taxa to higher ones. This is due to the historical circumstances, primarily, to the decisive contribution to modern systematics of its creator, C. Linnaeus. Thanks to E. Haeckel, and in virtue of typological similarities, hierarchical approach is caught in Darwinism, and then in the synthetic theory of evolution. After creation of cladistics by Hennig [2], modern taxonomy is impossible to imagine without this methodology. That's why today biologists in any situation requiring a systematization of objects under study, regardless of their nature, primarily use hierarchical approach [3], which seems to many the most immanent in both the nature and human thought. As the main strategy in this case similarity is used as the criterion of congeniality (assuming that the reemergence of feature is unlikely to happen), and the system is identified with the phylogeny.

31.2 THE REASONS OF DECLINE FROM HIERARCHY PRINCIPLES

Meanwhile, there is a surprising constancy of classification systems based on different principles [4–11]. In many cases, the nature of the material itself, classified in accordance with the principles of the hierarchy [12–14] and many others, indicates combinative character of syndromes of classified objects traits. This implies the use of a parametric approach as the most adequate to the objectives of rational description of biodiversity. The reasons for the inevitable deviations from the principles of the hierarchy are associated with the obvious disadvantages: many cases of parallelism, the ambiguity of location signs used in the hierarchy, the emergence of alternative hierarchical taxonomic trees when the taxonomic weight of used features is changed. The latter circumstance is due to the inclusion of features, which by their psychological novelty, attributes the increased importance. A well known example is the new megasystematics [15], based on the molecular and genetic characteristics, where, in particular, the taxon of molting animals (Ecdysozoa) is formed among the invertebrates, and including groups previously attributed to completely unrelated organisms. The former similarities are now interpreted as evidence of not congeniality, but parallelisms or convergences. Because of the nature of this criterion the above group Ecdysozoa should also include, in particular, molting snakes and other reptiles, cortex losing trees and a number of others, similar in this respect, but completely unrelated organisms, which, of course, is absurd. It is obvious that any new taxonomically significant features (today – it's mostly molecular and genetics), when used in a hierarchical manner, cause the revolutionary and unpredictable change in the entire system of the taxon and its relations with neighboring groups of the same rank. These genus alternative systems of the same set of objects, if they are based on different attributes and criteria, which leads to clutter of systematics and complicates the possibility of its use in practice [10, 11, 16, 17]. The described situation is further exacerbated by the subjectivity of the procedure of description and classification of biodiversity [11, 18] and installations for discreteness or continuity of changes in the properties of classified objects.

31.3 THE PECULIARITIES OF TAXA *INCERTAE SEDIS*

The taxa *incertae sedis* are one of the inevitable consequences of the use of hierarchical principals in systematics and is "not formalized" in this kind of classification of vestige. They are present in all taxonomic ranks from kingdom and type to kind. Thus, among the kingdoms the combinatory features of *incertae sedis* are mostly presented in fungi [6, 19, 20], the well-known example of *incertae sedis* from protozoan – Microsporidia, among the types of histic animals Onychophora, Pentastomida and Tardigrada, Myzostomids, Acoela, Gnathostomulida and many others are found, among the class of arthropods – Pantopoda, among the insects – Strepsiptera, Mantophasmatodea, Grylloblattida and others. [15, 21]. These and some groups of Ctenophora, that has convergent similarity with flatworms [22], and Ctenophora themselves [23]. It is not easy to tell the real number of such taxa, but specialists on determine groups of organisms can indicate them without mistake in accordance with their combinatory character of features mix, which form usual correlations (syndromes) in "good" taxa. That's why the parametric method of description despite its famous form redundancy [24], for taxa *incertae sedis* (due to its combinatory character of features connection) gives advantage in its fullness and predictive value, in comparison with the traditional for modern systematics hierarchy, and helps to understand more deeply the evolutionary regularities of morphogenesis [7]. There is also hope for perspective future of parametric systems used to describe any taxa, at least because of the opportunity of simultaneous use of complex of different features that almost makes systematics multidimensional. This implies the search of ways of rationalization of existing systematics. The solution of this task requires the elaboration of principles of parametric systems construction, finding of the subordinate categories and defining of their discretion criteria. Earlier we [25–27] proposed the approaches for this that were based on our own interpretation of the term "biosystem" and subsequent ideas.

The proposed interpretation of the definition of the term "biosystem" [27] is based on structural approach [10, 28], principals of systemacity, and also on identification of nature of system forming factor. Biosystem is any system the behavior of which is determined by laws of biological processes with its characteristic maintenance and extension of its own specific biological circulation (as immanent way of life of energy connection). All the diversity of biosystems (organism, population, ecosystem, and also possibly biological taxa) is a result of the processes of their integration within the three ***functional modules***: metabolic (synthesis of ecosystems from various (heterogeneous) organisms), reproductive (synthesis of population from homogeneous organisms) and locomotory (integration of ecosystems and/or populations into organisms of higher structural level). In short, the idea of functional modules leads to the following. The ***metabolic module*** is initial providing the very fact of existence of organism – structure which exists only in the flow of substance and energy. Besides the trophic functions, the additional opportunities providing the stability of metabolism are developed within its limits. Higher animals develop excretion, breath, blood circulation and also resources hoarding and their ability to mobilize. Some other forms of metabolism regulation like biorhythms, diapause, and anabiosis might be placed here. In case of an excess of resources the size of the body inevitably increases, that is considered to be thermodynamically favorable strategy, but it leads to poor interaction with the environment due to the scale effect. The way out of this situation is a bisection that returns the original order in the relationship with the environment. Thus the ***reproductive module*** usually manifesting itself under excess of resources is formed. In case of their lack there is a need in the third – ***locomotor module*** that allows the organism to replace the spent medium either moving into it or pumping it through itself. We used the indicated modules in the form of the corresponding vectors as the coordinate axes, in which it is possible to graphically reflect the place of any biosystem in form of a point, a three-dimensional range or track. Each of these axes (vectors) is a series of refrains moving with respect to which, depending on the specific environmental conditions (potentially – in all possible directions) biosystems experience the historical (or ontogenetic) transformation that we had observed.

We also used the notions of functional modules to rationalize the description of biodiversity. In particular, the vector of locomotor module allowed separating more intricate organisms (e.g. animal) from less integrated options of biosystems. In the received taxonomic space the separate group of organisms of any taxonomic rank can be arranged in accordance with the values of used features selected for the most complete taxa distinguishing. In the case of transition to other ranks, such as the need to consider the structure of subordinate taxa, taxonomic space axes content varies according to the specifics of these taxa. The taxon rank changes are supposed to happen as it develops historically. This reduces the dimension of describing its system – from a multidimensional variant in evolutionarily young taxa to the linear form in relict groups often represented by a single species. Form of the system of the present taxon for a specific period of its existence, if it is made without substantial "abuse of facts" can be a criterion of its age. The combinative principle of features correlating manifests itself in case of taxa of specific rank. Further, at the level of group the usual hierarchy becomes comfortable to more extent that seems the degenusation of combinative system (decrease of dimension of system describing it). By this analogy we obtain a linear version of the system of taxa of the level of types and kingdoms that is a further dimension to the degenusation of the unit. And finally, the zero dimension of describing system means the extinction of the taxon. The possibility of analysis of all realized diversity of forms in which the definite eras are likely to be presented by a bit different coordinates of diversity appears when the taxa is totally extinct. The semantic of these differences may also be the subject of study.

31.4 THE LIST OF TERMS FOR PARAMETRIC TAXONOMY

The list of terms used in traditional (phylogenetic) taxonomy should not be used in parametric systematic, which may provide new explanatory and predictive capabilities, in order to avoid confusion. Using the well-known experience in creating terminology, "to clear the decks" [12], we believe it is possible to use as a counterpart the periodic table, parametric by its form, by D.I. Mendeleev, in which the concepts period and group have their serial numbers and filled with the corresponding elements are used as basic references. Reducing the number of all possible measurements to three (the most convenient and clearly evident variant) used vectors of functional modules, we can mark the area formed by the vectors of reproductive and metabolic modules as a ***period***, and the vertical columns of similar morphs of different organizational levels associated with the motion along the vector of locomotor module – by term ***group***. Thus, these three vectors, forming a potential taxonomic space, which may to greater or lesser extent, be filled during phylogeny by some taxa. The degree and order of filling of the taxonomic space of these taxa (due to differences in their adaptation capabilities) will not be the same. As an obvious consequence we get penetration into a single cell potential taxonomic space of different phylogenetic branches of trees or redevelopment of this cell by branches of one tree. Hence, the prevalence of well-known cases of isomorphism and convergence on a wide range of features is used in the taxonomy. It is inevitable to introduce new terms for more productive combining of traditional and parametric approaches. We will accept the concept of ***taxon,*** that has a certain rank in the system, used in phylogenetic taxonomy and presumably monophyletic (holophyletic taxon) origin. Then it is necessary to designate similar forms trapped in the same cell of taxonomic space independently (para- and polyphyletic taxa). By the analogy with the periodic table by D.I. Mendeleev the following notions will be correct: ***bioisotope*** (unrelated biosystems that are in a single cell taxonomic space), ***bioisomere*** (relatively close biosystems in a series of successive transformations within the period or group in this taxonomic space), and also ***bioantimere*** (biosystems emerged as alternative variants of implementation of evolutionary solutions).

There are various embodiments of the taxonomic space. The ***real taxonomic space*** specific to the present taxon differs from the ***potential taxonomic space*** – by the specifics of vectors filling of functional modules by definite features used in the system of this taxon. As previously indicated, the transition from a higher rank on the lower level of the hierarchy, for example, when considering the content of individual families within the enclosing order, there is a change in the content of vectors of functional modules from the order level to the level of families. This kind of "jumps" in the contents reflect the phenomenon of quantization of evolution. The subject of study here may the semantics of changing of these features, which apparently reflect paleoecologic context. In addition to jumps in the ranks, originating (roughly speaking) the levels of kingdoms, types, classes, orders, families up to the species, the intermediate (subordinate) ranks designated by prefixes: over, sub, infra-, etc. are known. These categories can be specified within the limits of the taxonomic space enclosing the ranks. Three-dimensional parametric version of the system assumes the possibility of simultaneous attributing taxa of lower rank (in our example – families of one unit) to multiple subordinate ranks (e.g. infra- or suborders), if they stand on the grounds of different functional modules. Such property is not absolutely possible within a phylogenetic approach, but, in our opinion, may be useful for a better understanding of the biodiversity formation laws.

As it was mentioned previously, three-dimensional system is preferred because of its convenience for perception. However, the most immanent nature, apparently, are of a very high dimension. But the real multidimensional characteristic of biosystems makes it virtually impossible to their successful perception, so it is necessary to select the features that are relevant for the classification of specific taxa. Specifically, they determine the specifics of this taxonomic space.

Along with the indicated concepts ***group*** and ***period***, we consider necessary to designate as a variant form of the taxon – ***range*** (three-dimensional figure, which occupies part of the taxonomic space), ***sector*** (line, axis of period; there may be two versions of sectors, depending on the specifics of the vector: the reproductive or metabolic), ***segment*** (single cell of taxonomic space). Different taxa of traditional taxonomy can fit in one of the designated forms. For evolutionarily young taxa we can expect form of a group or range within the taxonomic space, and it is sufficient segment only for the extinct ones. The more complete correlation ranks of traditional phylogenetic and parametric (nomogenetic in fact) taxonomy requires additional work. Existing tradition of taxonomy created obvious difficulties in making fuller use of such an approach. First of all, they are determined by the selection of features focused on the unique and one-time occurrence (accidental and not the essential features are used). The cases of parallelism in features occurrence objectively make it difficult to use a hierarchical system, that's why they are to avoid in the preparation of identification keys. To overcome this difficulty it is necessary to use the most comprehensive diagnoses in order to easily localize the position of the studied taxon in three-dimensional taxonomic space.

31.5 GENERAL CHARACTERISTICS AND CREATION ALGORITHM OF TAXON PARAMETRIC SYSTEM

In genus, the proposed option of taxon parametric system, taking into account the terms outlined above [27], is seen as a series of nested multidimensional space features, each of which relates to a corresponding rank of subordinate taxa. In addition, the considered taxon should also be included into the system enclosing a taxon of a higher rank with its set of features as the coordinates of the taxonomic space. Here are some guidelines in the construction of parametric systems, which further eliminate the traditional problems with alternative systems of taxon in their hierarchical form.

In order to simplify the perception of the parametric system of taxon, as previously noted, it is recommended not to use more than three significant features (spatial coordinates that relate to any function module), providing the satisfying level of distribution of subordinate groups of taxon in the taxonomic space to be formed. For a more complete distribution of the features it is necessary to use the features, the values of which are variable at most taxa of given rank (as a rule, such features most often used in conventional determinants). The only condition is accurate values of selected characteristics for all subordinate taxa, regardless of whether they are used in traditional identification keys or not. The features which have the quantitative expression (competent distribution of qualitative characteristics in accordance with their natural gradation changes often require an acceptable level of intuition) are more useful to construct parametric systems of taxon. Some experience of studying the properties of parametric systems that we obtained in the course of their creation for individual taxa allows us to talk about universal principles, rules and even algorithm f such works. At the current stage of development of the topic the algorithm is as follows:

1. Collect the information on synonyms and alternative systems of classification of the present taxon, create a universal operating system of taxon as detailed as possible, including all its subordinate ranks. Create the most complete identification keys for all subordinate taxon groups, especially for groups *incertae sedis.*
2. Analyze the characteristics used in the systematics of taxon and its distribution by functional modules, build the refrains identified within the character states, as well as search for possible links (functional, ontogenetic and phylogenetic) between different taxonomically significant features. Distribute the features in accordance with the taxonomic rank.
3. Create a worksheet on the states of actual characteristics for all analyzed subordinate groups of the present taxon according to their ranks.
4. Select each of the ranks of the three most differentiating features to construct a three-dimensional taxon system with the location of characteristic values in accordance with the most likely sequence of evolutionary transformations.
5. Designate the selected variant of taxonomic space ranges of subordinate groups in accordance with the identified values in their features.
6. Analyze the received configuration of realized taxon states: – identify the areas of potential taxonomic space, search vacant sites in it and link them to the previously known subordinate taxa; – identify areas of "evolutionary prohibitions" and search for possible reasons of their existence.
7. Identify plesiomorphic and different versions for a given taxon of apomorphic groups, search for convergence zones and their environmental (and possibly paleoecological) sense – for further reconstruction of the evolutionary scenario of the taxon; examine of the need and possibilities for revising the existing system of taxon.

The separate functional stages are shown at an indicated point of this algorithm: 1–3 – preparatory, 4 and 5 – taxon parametric system construction, 6 and 7 – analysis of the resulting biodiversity. If the progress of these items, because of the impossibility occurs not in full compliance with the indicated sequence, the further progress on these points becomes less successful.

It seems that the implementation of the proposed algorithm is limited only by the degree of mastery of the subject-matter of taxa. List of groups to which we have been made to transfer their systems in parametric form is still quite modest. Besides, due to the peculiarities of prevailing interests of the author, in one way or another, these taxa are included into the category of articulated animals, belong to different taxonomic ranks, tend to have a category *incertae sedis*, and in varying degrees are "parametrically mastered." Nevertheless, the experience obtained in this work is of particular interest, since each of these taxa takes place in the corresponding node of the system of articulated animals that lets you get closer to an understanding of the entire system of articulated animals in genus. We used the term "articulated animals" not in the traditional taxonomic (genealogy), but structural and morphological (ecomorphological) sense. Originally we conceptualized the parametric principles of systematics in the process of building of our own version of the system of ecomorphs (living forms) by modifying of hierarchical system of ecomorphs by Yu.G. Aleev [12] into the parametric form [29] and then we transferred them to the real taxa. Here are some examples of the realization of the above-mentioned algorithm with the designation of some of the emerging issues of genus biology.

31.6 THE EXAMPLES OF THE PROPOSED ALGORITHM REALIZATION

31.6.1 STREPSIPTERANS

At the moment we parametrically assimilated to the greater extent the system of families and genus of order of strepsipteran insects (Strepsiptera) – almost to the stage 7 [30,31]. Strepsipterans (*Strepsiptera*, Kirby, 1813) – a small (about 600 species) order of highly experiences insects-endoparasites with an expressed sexual dimorphism, hypermetamorphosis in the life cycle and unclear systematic position [32]. Paleontological evidence does not reveal their congeniality with other orders because fossil strepsipterans differ from more recent species by more plesiomorphic state for the same unique set of features of this order. The system of strepsipterans finally couldn't withstand. It is associated with a secretive way of life, creating difficulties in finding and identifying them, especially females (for many species they are unknown). In accordance with the requirements of the third paragraph of the above-mentioned algorithm, we have drawn up a work table (Table 31.1) of values of taxonomically important characters for some of the genus cited in the literature. The parametric system of this order we propose is based on three key traits of male adults, the values of which are arranged in the order of their phylogenetic changes: the number of tibeotarsus segments (from 5 to 2), the number of tentacula segments (from 8 to 4), the distribution of the lateral processes on the segments of the flabellum tenacula (from 5 up to one). All three are used in our characteristic manifestation of locomotor module that is explained by peculiarities of biology of Strepsiptera males: the absence of functioning mouthparts requires perfect sense organs and locomotion to reach the success in search for females during released several hours of activity.

TABLE 31.1 Values of Quantitative Traits of Males of Some Strepsipterans Genus

Family	Genus	Traits: 1	2	3	Family	Genus	Traits: 1	2	3
Protoxenidae	*Protoxenos*	8	5	5	Myrmecolacidae	*Myrmecolax*	7	1	4
Cretostylopidae	*Cretostylops*	8	5	5		*Coenocholax*	6	1	4
Bahiaxenidae	*Bahiaxenos*	8	5	5		*Stylops*	6	1	4
Mengeidae	*Mengea*	7	2	5	Stylopidae	*Hylechthrus*	5	1	4
Mengenillidae	*Mengenilla*	6	3	5		*Halictoxenos*	4	1	4
	Eoxenos	6	2	5	Xenidae	*Xenos*	4	1	4
Bohartillidae	*Bohartilla*	7	3	4		*Halictophagus*	7	4	3
Corioxenidae	*Blissoxenos*	6	3	4	Halictophagidae	*Dipterophagus*	6	3	3
	Dundoxenos	5	2	5		*Driozocera*	4	1	3
	Triozocera	7	2	4	Elenchidae	*Elenchus*	5	1	2

Notes: trait 1 – number of tentacula segments; trait 2 –grade of reduction of the lateral processes on the segments of the flabellum tentacula; trait 3 – number of tibeotarsus segments.

The above-mentioned features are used as the three axes of the taxonomic space to designate portions corresponding to the place of individual families and the constituent strepsipterans genus. For more details on locations of individual taxa in the system of genus it is also possible to include in the refrains of additional features. Thus, within the family Corioxenidae, in parallel with the reduction of the number of tibeotarsus segments, there is a reduction of the number of tarsal segments and their complete further disappearance. In a number of families with apomorphic states of this trait it is observed the appearance of microvilli on the lower surfaces of the remaining tarsal segments in order to be fixed on the surface, which reaches the limit within the family of Elenchidae with their biarticulated and lacking claws tarsus. It is possible to consider isomers reduction variants of lateral processes on different tentacula segments (e.g., in the genus *Bohartilla*) in order to further specify the system of strepsipterans.

The resulting three-dimensional system of strepsipterans the "assimilated" area by the known families, beginning with plesiomorphic characteristic values continues to the center of taxonomic space, wedging out there by the zone of evolutionary young genus with apomorphic states of features. Since the shape of the resulting system of this unit is difficult to capture graphically, we made its volume model revealing a number of vacancies for realization for genus still undiscovered by science and suggesting that they belong to the now well-known families and make predictions of further morphological evolution of this order. Tabular option of strepsipterans parametric system is shown in Fig. 31.1, where the third spatial dimension (perpendicular to the plane of the paper) is shown in numerals in brackets referring to the number of tarsal segments.

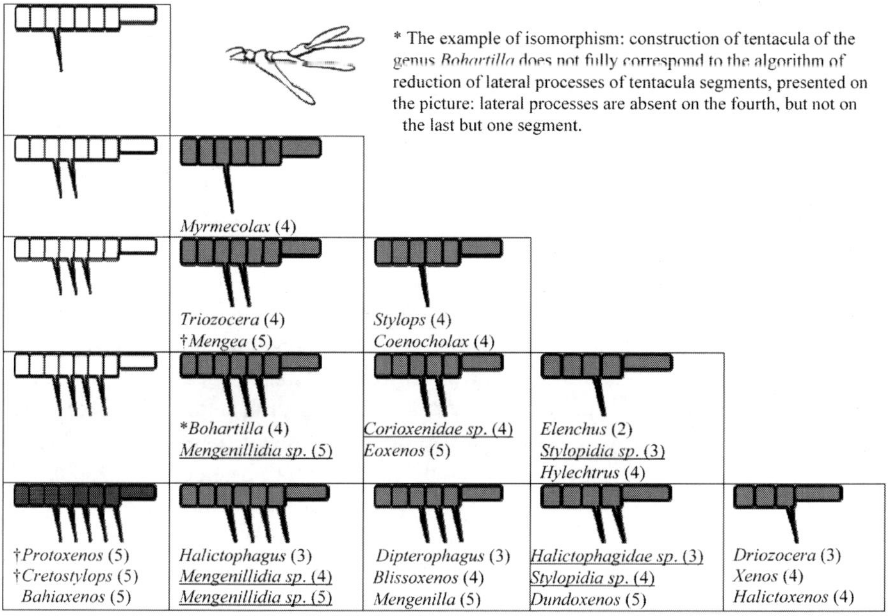

FIGURE 31.1 Tentacula form of some genus of *Strepsiptera* males in three-dimensional space of traits (terminal segments are situated on the right, and head case is relatively on the left); horizontal rows from left to right – the reduction sequence of the number of tentacula segments; vertical columns from the bottom upwards – the process of reduction of lateral processes of tentacula segments; – numerical values in brackets in the names of families – the number of tarsal segments. † – extinct genus.

The names of some of the predicted and taxa undescribed by science are underlined. Their taxonomic rank is reflected in the names endings corresponding to family or suborder. The contour color filling of tentacula: dark – plesiomorphic state of tentacula, gray – more recent genus, white – the area of "evolutional prohibitions," a consequence of simultaneous reduction of values of both used traits of tentacula

The structure of the tentacula in the figure is reduced to a scheme, which allows determining the position of the segments with the lateral processes. It is noteworthy that in this variant of taxonomic space of strepsipterans the genus of one family forms rather compact ranges. This allows carrying the vacancies that are not occupied by known forms, to the known families. The numerous cases of multiple inclusions of often unrelated groups of strepsipterans into one cell of taxonomic space are also to be noticed. Thus, in the lower left cell corresponding to the most plesiomorphic states of used features, occurred to be three different kinds, two of which (*Protoxenos* and *Cretostylops*) have already become extinct. In one of the cells there were different genuses of families: *Stylops* (Stylopidae) with *Coenocholax* (Myrmecolacidae), in another cell – *Halictoxenos* (Stylopidae) with *Xenos* (Xenidae). This feature is a testament to the independent acquisition of similar forms of representatives of different families, assimilating due to their specific nature and peculiarities of coenosis script the taxonomic space in different directions.

Within the parametric approach to the systematic the question of about the direction for further growth of genus diversity is adequate to ask. The taxa that are closer to the upper right edge of Fig. 31.1: representatives of families of *Elenchidae, Xenidae*, and partly *Halictophagidae* and *Stylopidae* – are considered the most evolutionarily developed. The characteristic for them are not only minimal values of used traits, but also related to their supposed evolutionary status – the values of many other features not used in our order, as well as some advanced features of their biology. The limits of the further evolution of the order within a given taxonomic space should be considered extreme states of the reduction of used traits: reduction in the number of tarsal segments to one, and for the tentacula segments, apparently, the possible limit of the reduction has already been achieved – 4 segments. Such a set is due to functional features of the tentacula itself: first two segments (scapus and pedicellus), unlike segments of flabellum, retain mobility, and none of the known insects has processes, as well as the terminal tentacula segment. The presence of lateral processes determines the specificity of the tentacula of males of strepsipterans. When moving their tentacula their fan may fold and unfold that probably provides high sensitivity to pheromones of females and prevents the development of addiction to them. If the tentacula has less than 4 segments, the required effect of fan-shape becomes geometrically impossible and tentacula externally acquires the rope form that does not correspond to the genus habit of the males of the group.

Unfortunately, the accomplishment of the 1st paragraph of the above mentioned algorithm is not fully completed, since for some of the known genus the values of traits we use are not found in the available literature, so there may be additions and clarifications to the localization of the subordinate taxa. So, in the rank of family the genus Callipharixenos has been recently isolated from the family Halictophagidae on the basis of some traits of females, while the males have not yet been found in nature [33]. Therefore, it seems impossible to localize the genus and justify the need to single out on its basis the individual family.

The subsequent improvement of the strepsipterans system, as it is supposed, will be connected with the confirmation of the predictions made earlier in the sequence of filling of vacant cells of the potential diversity of forms. To complete the construction of the order system, as an example, it should be created a parametric system of at least one kind, for example, related to the studied genus *Stylops*, represented by a sufficiently large number of species. In accordance with the characteristics of parametric systematics for its detail should be used other than in the genus traits' system differentiating diversity of well-known species of this genus in more details. It may be, in particular, peculiarities of proportions of individual segments of the tentacula [34]. The obtained parametric version of this order, despite its preliminary nature, provides opportunities for further improvement by specifying of values of used traits for genus and species ranks, as well as – using an additional quantitative traits with significant diversity of their values. The main advantage of this approach is a combination of alternative possibilities (and therefore unacceptable) of hierarchical variant of classification ways and predictability of the system (taxonomic forecast). As a significant obstacle to the creation of parametric strepsipterans system and similar systems in genus should be considered the deeply established tradition of publishing of incomplete diagnoses in taxonomy which don't present the comprehensive data on all traits used and for all taxa examined.

The indicated question in addition to the aspects discussed here poses problems beyond the topic, for example, the ontological essence of biological taxa at all. The interest caused by the position of the strepsipterans order in the system of insects is connected with the possibility of construction of the entire system of insects in accordance with the principles of parametric systematics. There is no clarity at the moment in the nature of congeniality of strepsipterans with other orders, at least four versions of this relationship are indicated: part of beetles order (Coleoptera), their sister group, a sister group of the two-winged order (Diptera) and a sister group to all the insects with complete metamorphosis (Holometabola) [2, 33, 35–40]. The construction of a parametric system of this order allows us to undertake similar work attempts for such key order giants as Coleoptera and Diptera, without which it is impossible to understand the nature of the congeniality/similarity of these three groups, and to start using the parametric system of the whole class of Insecta.

The same approach was taken in respect of the order Diptera which is brought together with stresipterans in the latest version of insects system [40]. The success here is much more modest due to huge scope and diverse set of traits, known for Diptera, at the level of the first two paragraphs of the previous algorithm. In the future, the use of the proposed approach can remove the contradiction between the alternative systems of classification existing today based on different traits of suborder rank – for imago (the number of tentacula segments) and immature stage (specific structure of the pupa). The use of characteristics such as morphology of larvae (structural features of the head case, mouthparts, tracheal system, etc.) [41], the features of the reproductive biology, etc. seem to be perspective. In the preliminary version of the parametric classification these traits form independent axes that allow you to see areas of "taxonomic space" that are not occupied by any known forms and very "popular." The degree of filling of these "popular" areas by specific families, genus and species, does not seem so much an indicator of their congeniality as a parallel formation of similar combinations of characteristics. Combinative correlations of features in separate groups are noticed even in existing determinants of this order [42]. There are similar examples from other groups of traits: ontogenic peculiarities, ecology, trophism and etc. Numerous cases of convergence of some traits of some subordinate taxa [43, 44], is apparently due to repeated getting into the same cell of "taxonomic space" of the different branches of the phylogenetic "tree." In genus, it is too anticipating to demonstrate the shape of the parametric system of such a large and diverse order as Diptera. Currently, the list of such characteristics is specified for the subsequent analysis and use in accordance with the indicated algorithm.

31.6.2 SEA SPIDERS

Another taxon convenient for our purposes is a class of sea spiders (*Pantopoda*, Gerstaecker, 1863), numbering according to various estimates about 1–1.5 thousand species, and also traditionally related to taxa category of *incertae sedis*. The idea of place and structure of this class is important to build a system of all types of arthropods, as in this class features of crustaceans combine with some Chelicerata species and Early Cambrian groups, but within the hierarchical approach the place of Pantopoda is still not clearly defined [8,45]. Earlier in the beginning of last century, Shimkevich [46] presented the pantopoda system in the form of tables, where the number of segments of the first three pairs of limbs sets the quantitative limits for the determination of individual families. The values used for the construction of the parametric features found in the literature [5] are presented in Table 31.2.

When placing the values of the number of segments of each of these limbs in the three space coordinates, the known families of Pantopoda occupy strictly defined, almost not intersecting areas. Within all the taxonomic space they fill basically only two out of six edges. In the inner parts of this space, which is, apparently, a zone of "evolutionary prohibitions" a small number of genus classified as *incertae sedis* appeared. The phenomenon of "evolutionary prohibitions," as in other cases, has probably paleoecological nature. The location of characteristic values in accordance with the logic of phylogenetic (reducible to most arthropods to a reduction initially large number of metameric units) allows us to see some regularity in the resulting parametric pantopoda system (Fig. 31.2). Thus in the sphere of plesiomorphic state the families are found having a "brutal" habitus and tending in adult state to living exclusively at the bottom, so unable to float in the water unlikely more evolutionary advanced rather than the more evolutionarily advanced "gracile" forms besides having advanced features of biology [45]. It is also indicative that in Pantopoda a unique feature of all arthropods was discovered, it is the multiplicity of genital openings located not on the trunk, but on the second segment of the walking legs [45, 47].

In most plesiomorphic state of this character genital openings are on all the limbs of both sexes, as we move towards the "apomorphic" pole of the taxonomic space it also becomes noticeable the reduction of genital openings starting with the front pair of walking legs, moreover in males, this trend is more obvious. So, members of the Colossendeidae family, as well as the genus *Decalopoda* and *Dodecalopoda*, attributable in accordance with the latest classifications to the same family, this trait is fully expressed, while in genus *Rhynchothorax* (Tanystylidae) and *Pycnogonum* (Pycnogonidae) the genital openings in both sexes are expressed only in the last pair of limbs.

In the preliminary version of the parametric system to construct this class we relied on the system of V.M. Shimkevich in which all these genus of Pantopoda are provided with the exact values of trait we used. The significant problem for the improvement of the parametric system of Pantopoda is absence of genus accepted classification, numerous cases of changes in the volume and even the names of individual families, the nature of their relationship with each other, which results in an abundance of options to form a super family and groups [48, 49]. To further promotion by the stages of the indicated algorithm to build the final version of the system of Pantopoda it is necessary to analyze the differences in the available alternative classifications, to reveal similarity, to compare the list of families by genus in different systems.

Several rich diverse forms of bloodlines with severe signs of variations used, also, require updating of the value of these characteristics to separate genus. Such a procedure has already been done when dealing with twisted-winged insects and allowed, as shown above, to achieve volumetric visualization, which has some heuristic properties. On the other hand, the analysis of individual cells fullness of pantopodas parameter space with specific genus points to two polar categories of genus there are genus, holding a number of cells, and there are cells in which there are several genus. This fact highlights the need for careful selection of features that better differentiate Pantopoda taxa according to their ranks and makes to think again about the ontological essence of this phenomenon as a biological taxon and its rank. Certain complexities to build a system of this class are also instances of intraspecific variation, sexual dimorphism and ontogenetic changes in traits [5, 47], as well as numerous instances of convergence characteristics in unrelated forms.

TABLE 31.2 Values of Quantitative Traits of Males of Some Genus of Pantopoda (According to Shimkevich, 1930 with Additions)

Family	Genus	Traits:			Family	Genus	Traits:		
		1	2	3			1	2	3
Nymphonidae	*Nymphon*	2	5	10	Paraphyopodidae	*Decalopoda*	3	10–9	10
	Chaetonymphon	2	5	10		*Dodecalopoda*			
	Pentanymphon	2	5	10	Ammotheidae	*Ammothea*	2	10–6	10
	Boreonymphon	2	5	10		*Ammothella*	3	9–8	10
Ammotheidae?	*Paranymphon*	2	7	10		*Ascorhynchus*	3–2	10	10
Incertae sedis	*Platychelus*	2	5	10		*Nymphopsis*	3	10–9	10
Pallenidae	*Pallene*	2	0	10		*Eurycyde*	3	10	10
	Pallenopsis	3–2	1	10		*Trigaeus*	1	6	9
	Pseudopallene	2	0	10		*Fragilia*	3	6	10
	Austropallene	2	0	10		*Boehmia*	2	7	10
	Parapallene	2	0	10	Incertae sedis	*Alcynous*	3	9	9
	Metapallene	2	0	10		*Paralcinous*	3	7	8
	Cordylochele	2	0	10		*Pasithoe*	0	8	9
	Hannonia	2	0	10		*Endeis*	0	7	9
	Oropallene	2	4	10	Colossendeidae	*Colossendeis*	0	10–9	10
	Propallene	2	2	10		*Rhopalorhynchus*	0	10–9	10
	Pallenella	3	0	10		*Lecythorhynchus*	0	9	10
	Neopallene	2	0	10		*Pipetta*	0	8	10

Incertae sedis	*Phanodemus*	2	3	10	Oorhynchidae	*Oorhynchus*	2	9	10
Phoxichilidiidae	*Anoplodactylus*	2	0	9–6		*Scipiolus*	2	9	10
	Halosoma	2	0	7–5		*Cilungulus*	2	9–8	10
	Enoplodactylus	2	0	7	Tanystylidae	*Tanystylum*	1	7–4	10
	Phoxichilidium	2	0	5		*Rhynchothorax*	0	5	10
	Phoxichilopsis	1	0	7		*Discoarachne*	0	5	10
Incertae sedis	*Oomerus*	2	1	6		*Austroraptus*	0	6	10
	Pephredo	2	3	6	Incertae sedis	*Austrodecus*	0	6	10
Phoxichilidae	*Phoxichilus*	0	0	7	Pycnogonidae	*Pycnogonum*	0	0	9
						Pentapycnon	0	0	9

Notes: trait 1 – number of segments of the 1st pair of limbs (chelophore); trait 2; – number of segments of the 2nd pair of limbs (palp); trait 3 – number of segments of the 3rd pair of limbs (ovigery).

Individual taxa incertae sedis located are situated next to the groups, the characteristic values of which are close to them.

The above-mentioned feature of pantopods is unique among all known arthropods – the multiplicity of genital pores on the second articles of the limbs, raises a number of questions about the patterns of evolution of arthropods morphogenesis, in particular, about the specifics of the formation of limbs, reflecting their fractal nature, and also puts this class in particular place in the arthropods system.

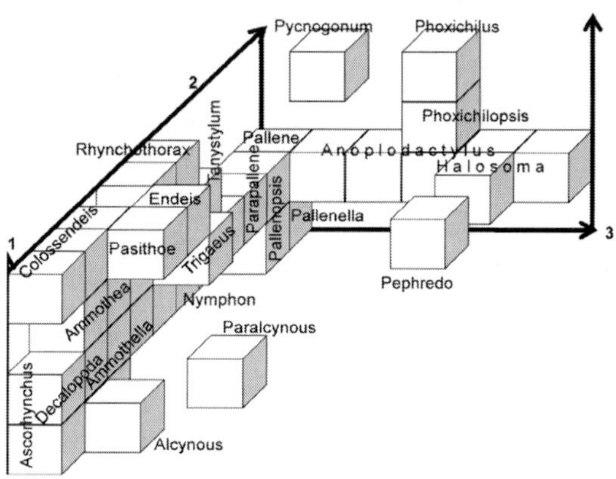

FIGURE 31.2 Parametric system of Pantopoda indicating places of some of the genus listed in the Table 31.2 (the traits used are indicated by the same numerals as in Table 31.2).

31.6.3 TRILOBITES (TRILOBITA, WALCH, 1771)

Even more promising in terms of the possibility of constructing a parametric systems are trilobites. For more than 300 million years of its evolution this class seems to have exhausted the potential of shaping in the conditions of actual terrestrial scenarios, so it is convenient to explore this group of animals for the specifics of the distribution characteristics for members of specific time periods. Currently they are from 1.5 to 5000 species on various sources, vary greatly in size, proportions, reconstructed biology features, and time and areas of existence, etc. Classification of trilobites also in flux: in systems of different authors, there are differences in grades, composition, quantity and the names of some subordinate taxa [48–51]. In particular, according to one of the latest guidelines for zoology [15] one of the Early Cambrian trilobite groups – Agnostida is derived from the class and included in one of the groups of crustaceans. In its morphology and reconstructed biology trilobites combine some signs of chelicerates and crustaceans, although quite differently than pantopods. Unlike recent forms for which in the preparation of their systems leading traits today are almost exclusively molecular genetics, for fully extinct trilobites features of their morphology is only remain relevant and allow you to create visual signs of their ranks variability of their skeleton. This fact allows the use of methods of parametric systematics to be no less successful than the above taxa.

In the first approximation there is most likely for building a parametric system of trilobites' class, the use of traits with expressed variability in various unrelated groups. These symptoms may be – the number of segments of the trunk, from two in *Agnostus*, to 44 in Olenellidae. Another sign – overall dimensions of the body (isolated groups by size- up to 1 cm, medium – 25–40 mm and macroform – more than 80 mm). Finally – the ratio of the size of head and tail carapaces: there are micropygous forms (tail carapace is much smaller than brain carapace), geteropygous (the differences are less significant), isopygous (carapaces are almost equal), and macropygous (tail carapace is even slightly larger than the head carapace). The latter feature, in our opinion, reflects different strategies in the growth rate. Pygidium as well as the head carapace, especially in iso- and macropygous, formed by the confluence of the telson with the front segments, reducing their number in the composition of the body, so the size of the pygidium are inversely proportional to the number of body segments [52]. In accordance with the laws of metameric animal growth, the formation of new body segments occurs in the division zone just before the tail carapace, so macriygous and iso- forms have been able to produce large parts immediately, unlike mikropygous in which the youngest segments are very small. By virtue of growth regularity of the pygidium in the trilobites evolution, for historically later forms the increase of the iso- and macropygous forms is noticeable amid a genus increase in body size. It is noteworthy that similar proportions are noticeable to Early Cambrian Agnostida that reached izopygous state and the minimum number of body segments at microsized rank already. This implies a long evolution for oligomerization of their trunk in the Precambrian. This regularity can be a manifestation of the periodic law in the system of trilobites.

31.6.4 ONYCHOPHORA AND OTHER LOBOPODIANS

Finally, we should mention a small (about 200 recent species of 48 genus and 2 families) equatorial – Southern Hemispherical (Gondvanian) velvet worms (*Onychophora*, Grube, 1859) type that is the subject of constant debate of taxonomists since their discovery. Previously, they were placed somewhere between annelids and arthropods [53]. Today they are formally included together with tardigrades into a group of Lobopodia, included with arthropods into super type Panarthropoda [15], but a number of signs convergently arizing in other types, make such a classification arbitrary to a certain extent. We believe that within the hierarchical system Onychophora place cannot be determined, and requires the use of parametric approaches, but due to low availability of specialized literature, the implementation of the above algorithm in genus does not go beyond the first or the second paragraph. As an indication for the construction of the parametric system of this taxon, promising may be the following: the number of pairs of limbs, body size ranges, the degree of development of the reproductive system – from egg-laying – up viviparous forms with the placenta. The feature space can be used for the application of values of the above-mentioned tardigrades and pentastomids, whose place on the basis of new molecular-genetics taxonomy is indicated today among one of the groups of crustaceans [15]. Meanwhile, the structure of their larvae makes possible consideration of pentastomids under typical Lobopodians. Variant of taxonomic space for the three lobopod groups presented in Table 31.3.

TABLE 31.3 Values of Quantitative Traits of Certain Groups of Lobopod Animals (According to Ref. [15] with Additions)

Taxon	Approximate number of species	Number of pairs of segments	Reproductive system development	Average size (mm)
Pentastomida (tongue worms)	100	2	Oviparous	0.1 (larva)
Tardigrada	800	4	Oviparous	0.05–1.2
Onychophora of Peripatopsidae family	100	13–25	Oviparous and ovoviviparous	15–20
Onychophora of Peripatidae family	100	22–43	Ovoviviiparous and viviparous	up to 200

Furthermore, the structural features of insect larvaes, in particular larvae of Lepidoptera, have the majority of external signs of lobopody. This is especially true of their limbs. So, this category of animals cannot be seen as holophyletic taxon, but as a life form or constructive type of organism that does not stop to consider it as an object parameter systematics.

31.7 CONCLUSION

From a brief review of the matter under consideration the character of development of its taxonomic space is obviously unique (which is not surprising because of the uniqueness of evolutionary scenarios), and, apparently, has an indicator value to determine the extent and nature of specialization, evolutionary age of taxon, or any other of its properties. Continuation of this theme involves both the solution of these issues simultaneously, and expanding the list of taxa to produce a multilevel parametric system of articulated animals.

This work was funded by the subsidy of the Russian Government to support the Program of competitive growth of Kazan Federal University.

KEYWORDS

- **articulated animals**
- **biodiversity**
- **functional modules**
- **taxon incertae sedis**
- **taxonomic space**
- **taxonomy**

REFERENCES

1. Zarenkov, N. A. (1976). Lectures on the Theory of Systematics, Moscow, and Publishing House Moscow state University, 140p (in Russian).
2. Kluge, N. Y. (2000). Principles of Systematics of Living Organisms and the General System of Insect with Classification of Apterygota and Palaeoptera, Saint-Petersburg, Publishing House "Lan", 366p (in Russian).
3. Skvortsov, A. K. (1992). Mechanisms of Organic Evolution and Progress of Knowledge, Nature, *(7),* 3–10 (in Russian).
4. Belousov, L. V. (1975). Parametric System of Hydroids of the Caphora and Possible Genetic Regulations of Species Differences, Journal of General Biology, *36(5),* 654–663 (in Russian).
5. Dogiel, V. A. (1951). Class of Geniculata (Pantopoda), Manual on Zoology, *3*, Invertebrata Edited by Zenkevich, L. A. Moscow, State Publishing House "Soviet Science", 45–106 (in Russian).

6. Zavarzin, G. A. (1979). Space of Logical Possibilities in a Variety of Bacteria and their Phylogeny, Nature, *6,* 9–19 (in Russian).
7. Kovalenko, E. E., & Popov, Yu I. (1997). A New Approach to the Analysis of Variability, Journal of General Biology*, 58(1)* (in Russian).
8. Pavlov, Ya A. (2000). Periodic System of Arthropoda, Moscow, Russian Federation Research Institute of Fishery and Oceanography, 186p (in Russian).
9. Popov, Yu I. (2008). Periodic Systems and Periodic Laws in Biology, Saint-Petersburg, Moscow KMK, 223p.
10. Chaykovsky, Yu V. (2008). Active Connected World, Experience of the Theory of Evolution of Life, Moscow, KMK, 726p (in Russian).
11. Chaykovsky, Yu V. (2010). Diatropics, Evolution and Taxonomy, to Anniversary of Meyen, Collected Works, Moscow, KMK, 407p (in Russian).
12. Aleev, Yu G. (1986). Ecomorphology, Kiev, Scientific Thinking, 423p (in Russian).
13. Marfenin, N. N. (1993). Phenomenon of Colonialism, Moscow, Publishing House Moscow State University, 239p (in Russian).
14. Sharova, Kh I. (1981). Life forms of Ground Beetles (Coleoptera, Carabidae), Moscow, Science, 360p (in Russian).
15. Ruppert, E. E. et al. (2008). Zoology of Invertebrates, Functional and Evolutionary Aspects, Manual for Higher Education Studentsin 4 Volumes,*3* Arthropoda Moscow Publishing Centre "Academy", 496p (in Russian).
16. Lubishchev, A. A. (1982). Problems of Forms, Evolutions and Systematics of Organisms Collected Works, Moscow, Sciences, 278p (in Russian).
17. Meyen, S. V. (1978). The Basic Aspects of Typology of Organisms, Journal of General Biology, *39(4)*, 495–507 (in Russian).
18. Mikhaylov, K. G. (1995). Typology and Practical Systematics, Theoretical Problems of Ecology and Evolution (Lubishchev's Second Reading), Tolyatti, Inter-Volga, 40–46 (in Russian).
19. Bekker, Z. E. (1975). About Converged Similarities of Fungi with Some Groups of Animals, Journal of General Biology, *36(5)*, 670–687 (in Russian).
20. Zmitrovich, I. V. (2006). Vegetable Epiphenomena and their Ecomorphological Essence, Vestnik of Ecology, Forest and Landscape, *7,* 3–28 (in Russian).
21. Chaykovsky, Yu V. (2003). Evolution, Series 22 "Cenology Investigations", Moscow, Centre of Systemic Research, Institute of Science and Technical History of Russian Academy of Science, 472p (in Russian).
22. Vorontsov, N. N. (2004). The Development of Evolutionary Ideas in Biology, Moscow, KMK, 266p (in Russian).
23. Joseph Ryan, F., Kevin Pang, Christine Schnitzler, E., Anh-Dao Nguyen, R., Travis Moreland, David Simmons, K., Bernard Koch, J., Warren Francis, R., & Paul Havlak (2013). NISC Comparative Sequencing Program, Stephen Smith,A., Nicholas Putnam,H., StevenHaddock, H. D., Casey Dunn, W., Tyra Wolfsberg, G., James Mullikin, C., Mark Martindale, Q., & Andreas Baxevanis, D. The Genome of the Ctenophore *Mnemiopsis Leidyi* and its Implications for Cell Type Evolution, *Science*, 13 December *342,* 1336, DOI: 10.1126, Science,1242592 (in English).
24. Rasnitsyn, A. P. (2005). Phylogeny and Taxonomy, selected Works on Evolutionary Biology, Moscow, KMK, 84–89 (in Russian).
25. Zeleev, R. M. (2011). Opportunities of Approaches of Ecomorphology and Traditional Systematics in Distinguishing of Features of Plants and Animals, Modern Problems of Popularized Ecology, Geological Botany, Systematics and Floriculture, Proceedings of the International Scientific Conference Devoted to the 110th Anniversary of Uranov, A. A. in 2 volumes, *1,* Kostroma, Kostroma State University, Nekrasov, N. A., 275–280 (in Russian).

26. Zeleev, R. M. (2011). Proportionality as a Characteristic of the Development of Biological Systems, Option of Biological Axioms, Scientific Proceeding of Kazan University, *153,* Series Natural Science, 3–21 (in Russian).
27. Zeleev, R. M. (2012). Option of Biological Axiom and its Ability to Describe Biodiversity, Scientific Proceeding of Kazan University, *154*, Series Natural Science, 8–25 (in Russian).
28. Meyen, S. V. (1978). The General Aspects of Organisms Typology, Journal of General Biology, *39(4),* 495–507 (in Russian).
29. Zeleev, R. M. (2007). The Variant of Construction of the System of Life Forms Organisms, Modern Problems of Evolution, Collection of Articles, XXI Lubishchev's Reading, Ulyanovsk, Ulyanovsk State Pedagogical University, 77–90 (in Russian).
30. Zeleev, R. M. (2013). Parametric Systematics, Problems and Prospects, Modern Problems of Evolution and Ecology, Collection of the International Conference, XXVII Lubishchev's Reading, Ulyanovsk, Ulyanovsk State Pedagogical University, 45–52 (in Russian).
31. Zeleev, R. M., & Safin, R. A. (2013). Parametric System of Stresipterans Insects (Insecta, Strepsiptera), Scientifics Proceeding of Kazan University, 155, Series Natural Science, 221–238 (in Russian).
32. Medvedev, L. M. (1969). Strepsiptera Order, Determinator of Insects of the European Part of the USSR Edited by Bey-Biyenko, Ya.G., *2,* Coloptera and Stresipterans, Moscow-Leningrad, Science, 641–645 (in Russian).
33. Pohl, H. & Beutel, R. G. (2005). The Phylogeny of Strepsiptera (Hexapodas), Cladistics, *21*, 328–374 (inEnglish).
34. Kogan, M., & Poinar, G. J. (2010). Systematics Morphology and Physiology New Fossil Stylops (Strepsiptera, Stylopidae) from Dominican Amber, Neotropical Entomology, *39(2),* 227–234 (inEnglish).
35. Bey-Biyenko, Ya G. (1962). About the General Classification of Insects, Entomological Survey, *41(1)*, 6–21 (in Russian).
36. Boerner, C. (1904). Zur Systematik der Hexapoda (Strepsiptera), Zoology Anz., *27*, 511–533 (in German).
37. Crowson, R. A. (1981). The Biology of the Coleoptera, London, Academic Press, 802pp (in English).
38. Kristensen, N. P. (1991). Phylogeny of Extant Hexapods, 125–140 in the Insects of Australia, A. Textbook for Students and Research Workers, 2nd edition (Naumann, I. D., Cornell Carne, P. B., Lawrence, J. F., Neilson, E. S., Spradberry, J. P., Taylor, R. W., Whitten, M. J.,& Littlejohn, M. J.,eds) CSIRO, Melbourne University Press, Melbourne, Ithaca, New York (in English).
39. Kristensen, N. P. (1995). Forty Years insect Phylogenetics Systematic, Zoologische, Beiträge, N. F., *36,* 83–124 (in English).
40. Whiting, M. F., Carpenter, J. C., Wheeler, Q. D., & Wheeler, W. C. (1997). The Strepsiptera Problem, Phylogeny of the Holometabolous Insect Orders Inferred from 18S and 28S Ribosomal DNA Sequences and Morphology, Systematic Biology, *46*, 1–68 (in English).
41. Krivosheeva, N. P. (1969). Ontogeny and Evolutions of Diptera Insects, Moscow Science, 291p (in Russian).
42. Narchuk, E. P. (2003). Determinant of Diptera Families (Insecta, Diptera) of Russia and Adjacent Countries (with a Brief Overview of the Families of the World Fauna), Zoological Institute of Russian Academy of Science, *294*, Saint-Petersburg, 250p (in Russian).
43. Kovalev, V. G. (1987). Classification of Diptera in the Light of Paleontological Data, Diptera, Taxonomy, Morphology, Ecology, Collection of Scientific Work, Leningrad, Zoological Institute of Academy of Science of USSR, 40–48 (in Russian).

44. Krivosheeva, N. P., & Zaytsev, A. I. (1989). Phylogeny and Evolutionary Ecology of Diptera, Results of Science and Technology, Series Entomology, *9*, Moscow, all-Union Institute of Scientific and Technical Information, 162p (in Russian).
45. Bogomolova, E. V., & Malakhov, V. V. (2006). Sea Spiders, Nature, *8,* 37–45 (in Russian).
46. Shimkevich, V. M. (1929). Manygeniculate (Pantopoda), Leningrad, 555p (in Russian).
47. Raysky, A. K. (2011). About Forming of Sea Spiders (Arthropoda Pycnogonida) by Maintaining of Juvenile Traits, Modern Problems of Evolutionary Morphology of Animals, (Proceedings of the II All-Russian Conference with International Participation), "Modern Problems of Evolutionary Animal Morphology" the 105th Anniversary of the Birth of Academician, Ivanov, A. V. October (2011) 17–19, Saint-Petersburg, Zoological Institute of Russian Academy of Science, 382p, 296–299 (in Russian).
48. http://www.gbif.org/species/109541246 (Accessed 28 February 2014).
49. http://www.gbif.org/species/109764619 (Accessed 01 march 2014).
50. Fortey, R. A. (2001). Trilobite Systematics, the Last 75years, Journal Paleontology, *75(6),* 1141–1151 (in English).
51. Hughes, N. (2007). The Evolution of Trilobite Body Patterning (2007) annual Revision Earth Planet Science, *35*, 401–434 (in English).
52. Chernysheva, N. E. (1960). Class of Trilobita (Trilobites), Fundamentals of Paleontology, Handbook for Paleontologists and Geologists of the USSR, Arthropoda, Trilobitomorpha and Crustaceans, Moscow Governmental Scientific and Technical Publishing House of Literature on Geology and Mineral Resources Protection, 17–54 (in Russian).
53. Kuznetsov, Ya N. (1951). Class of Primary Uniramia (Protracheata Sive Onychophora), Manual on Zoology, *3*, Part 2: Invertebrates, Moscow, State Publishing House, "Soviet Science", 107–123 (in Russian).

PART IX

NATURAL DISASTERS AND WAYS TO MINIMIZE THEIR HARM

CHAPTER 32

LANDSCAPE CONDITIONS OF DEVELOPMENT OF NATURAL DISASTERS IN THE REPUBLIC OF NORTH OSSETIA-ALANIA

FATIMA M. KHACAJEVA and VADIM A. TOMAJEV

K.L. Khetagurov North-Ossetian State University, Vladikavkaz, d. 46, Vatutin St., Republic of North Ossetia-Alania, 362025, Russia, факс: 8 (867 2) 545108; E-mail: tomaevvadim@mail.ru,.

CONTENTS

ABSTRACT

At the turn of two centuries natural disasters have acquired unprecedented popularity, large scale, power-intensity and destructive force all over the planet. Modern activation of dangerous processes on the mountainous territories makes the fragile ecosystems and unique mountain landscapes rather vulnerable.

More than a thousand dangerous processes centers have been recorded in the mountains of North Ossetia. Dozens of them have been ascribed to the category of catastrophes by their power. A high degree of affection by mud torrents, landslides and avalanches is conditioned by the availability of a number of favorable natural and anthropogenic factors here.

Modern activation is due to the climatic anomalies: 0.6 °C rise of the average annual temperature for the last several dozens of years and the fall of the maximum amount of precipitation. The apotheosis was the fall of a three-month norm during two days in 2002 that resulted in a mass manifestation of dangerous processes, the catastrophe of Kolka glacier among them.

Having high energetic and natural resources potential, mountain systems are becoming a significant component of the modern nature use. A growth of anthropogenic load also leads to the activation of dangerous processes. The analysis of complex landscape conditions of dangerous processes development enables us to appreciate the danger and to work out measures for the prevention and lowering the risk of catastrophes.

32.1 INTRODUCTION

Risk of accidents covered all the natural environment and the Earth system. The change of millenniums delineated total catastrophically also in all spheres of public life, which remains steadily in the first decade of the XXI century. The planet is in a phase of high-speed and energy-intensive processes of evolutionary development. Along with this man-made systems provide additional press on geophysical environment and biosphere in general. As a result of accumulation of technogenic energy and its imposing on energy frame ground is breaking geographical regularities [1].

Geosynclinal region most affected by geomagnetic disturbances and reconstruction of climate. For the last two decades have intensified dangerous slope processes, landslides, mudflows and avalanches. Their power is acquired destructive, often destroying nature [2]. In mountain areas of the Republic of North Ossetia-Alania registered over 1000 foci of hazardous natural and anthropogenic processes, dozens of them attributed to the category of disasters [3].

Modern natural factors of the development of landslides, mudflows and avalanches are intensification of geological-geomorphological processes and climatic anomalies. Anthropogenic factor is interpreted as an additional factor in the development of the disaster.

All natural processes and factors in mountain areas have vertical zonality, therefore, the geographic spread of the dangerous natural processes has a clear correlation with the absolute heights. The largest landslides happen in mountainous regions – south zone Jurassic shale depression with numerous tectonic faults, overlap, folded easily decaying shales. The unloading material mudflows and avalanches also made the midlands in the form of cones outstations. If we consider the anthropogenic factor, the residential zone, with its economic complex are nutrigenie hollow and valley middle, partially mountains and foothills [1, 4, 5].

In general, the development of dangerous natural processes is the result of a complex of the natural landscape conditions and anthropogenic factors.

32.2 LANDSLIDE PROCESSES

Formation and development of one or another type of landslide determined by the specific landscape conditions. When analyzing data, field studies were shaped "landscape situation" as a condition of the occurrence of structural and deluvial landslides. Different types of landslides have different landscape characteristics and are confined to certain geosistema.

In the course of our field research found that structural landslides extrusion are confined mainly to the slopes, composed of indigenous breeds of high density of granite, granitoids, crystalline slates, sandstones of the Paleozoic. For the formation of landslides on the bedrock you need an active action of subsoil waters, as one of main factors, with a rate of 0.05–0.08 liters per second and the most powerful layer of surface sediments (aluvia, diluvija, moraines, fluvioglacial) with a capacity of 3–8 m and more on the slopes steeper than 45°. The exposure distribution of the structural landslides, the most extensive. This suggests that landslides this type less dependent on the origin of meteorological conditions. The role of vegetation in education structural landslides extrusion is insignificant as the main deforming horizon is much deeper than the possible penetration of roots of woody vegetation. For this reason, countersigned landslide strata role roots as a deterrent occurrence of landslide factor is missing. However, if the surface-sliding slope covered with woody vegetation with high mass, the load vegetation may lead to activation of landslide processes.

Structural landslides extrusion develops mainly in alpine, subalpine meadows and has the biggest capacity of 10–20 million cubic meters and more.

Geosystem, in which they develop structural landslides extrusion is very steep (45–60°) slopes of the western and northern exposures with alpine and subalpine grass-forb meadows on low-powered mountain-meadow soils on granites, granitoids and Paleozoic sandstones.

Structural sliding landslides occur mainly on the slopes, folded shales, argillite, siltstone Jurassic characterized by smaller density, than granite.

For the formation of landslides on easily breakable loose Jurassic rocks enough action of subsoil waters with a rate of 0.02–0.05 liters per second, much less than for the formation of landslides extrusion. Under the action of groundwater easily destroyed shales, mudstones and siltstones formed moist layer fragmental small material bedrock, which is an active engine of landslide masses and acting as a master of deforming the horizon. Landslides slip formed when the thickness of sedimentary deposits of alluvium and diluvia not less than 1–3 meters with capture indigenous rocks on the mountain slopes with a gradient of 30–45°.

The exposure distribution of landslides slip is more limited than landslides extrusion. As landslides slip develop with lower power capture, their education has an impact hydrometeorological regime of the area. Precipitation significantly replenish the landslide groundwater, thereby stimulate the process of sliding. On the formation and subsequent development of landslides also affected by the change of wet and dry periods.

The slopes of the western and north-western exposures in its hydrometeorological conditions most favorable for the formation and development of landslides. The role of vegetation in education landslides slip becomes more pronounced, and this role is twofold. On the one hand, arboreal and shrub vegetation fasten potential landslide mass within its distribution. On the other hand, large vegetable load potentially landslide strata increase their gravitational instability. The role of vegetation is rare, because the structural landslides developed mainly in zones of subalpine meadows and alpine crooked. Structural landslides slip significantly less landslides extrusion mass of the landslide material (about 100,000–15,000,000 cubic meters).

Geosystem, in which the most widespread structural landslides slip is a steep slopes (30–45°) Western and north-western exposure with subalpine motley-grass meadows on low-powered mountain-meadow soils and rare elfin low power meadow-forest soils on clay shales, siltstones and aleurolites Jurassic period.

Deluvial landslides (osovy or olivine) are developing everywhere, regardless of composing the slopes of indigenous breeds. Most commonly they spread on the slopes, folded limestone, marl and dolomites chalk. In education deluvial landslides equivalent is the role of groundwater, and atmospheric water. Necessary quantity of groundwater for education osow significantly less than the structural landslides about 0.01–0.02 L per second because the Cretaceous rocks are characterized by a lower density and easily damaged, and the influence of even a minimal amount of water. The appearance of the horizon waters between indigenous and sedimentary rocks leads to instant sliding of sedimentary thickness with the seizure of the moist layer of bedrock. For the formation of landslides enough alluvial-deluvial sediments power of 0.5–1 meters on slopes steeper than 20°.

The role of the exposition of the slopes is significant in the formation of deluvial landslides, with a surface character development. Landslides clearly confined to the most humid slopes of the western and north-western exposure. Least of all they are in the rain-shadow zone and dry submontane depressions. To activate deluvial landslides important change of dry and wet periods.

Vegetation also plays an important role in the formation of deluvial landslides. Landslides barely formed on tightly turf slopes. These types of landslides are the most widespread in the zone of alpine meadows and meadow slopes of the Pasture range. Power deluvial landslides vary from 10,000 to 100,000 cubic meters.

Geosystem, in which the most widely developed deluvial landslides are steep slopes (20–30°), mainly Western exposure with herb-gramineous plants with a low-powered mountain-meadow soils on limestone, mengele and dolomites chalk [4].

32.3 MONITORING OF LANDSLIDE PROCESSES

Currently in North Ossetia has registered dozens of current and stable landslide centers. In the basins of the rivers: Urukh observed 12 active and 15 of stable landslides, Ardon and Mamisondon – 10 active and 12 stable, Fiagdon – 8 active and 10 stable, Kambileevka rivers – 5 active and 7 stable, Gizeldon – 4 active and 5 stable [6, 8].

In connection with a considerable mass of remove material landslides have great destructive force and represent a danger. Most often exposed to landslides exposed Transcaucasian highway. In general, in the Mountainous Ossetia mass gathering powerful landslides celebrated in years 1960, 1975, 1984, 1985, 1987, 1990, 1997, 1999, 2002 [6, 7, 9].

It should be noted significant role of the technogenic load on the slopes activation of landslide processes. In particular, catastrophic progress in landslide Zinzer-1 in 1984 was connected with the construction and operation of the Transcaucasian highway [6, 7].

The impact of landslide processes can be divided into two types: direct threat of the destruction of objects, located in the zone of development of the landslide deposits and the threat of river channels overlap with the formation of temporary pond lakes displacement of tens to hundreds of thousands of cubic meters, at the breaking through of which is the formation of powerful debris flows, floods [6, 7].

Example of the destructive processes can serve activation of landslide slopes in 2002 in the towns of Upper Mizur, the Average Shed, Upper Sadon and others, there have been fixed the deformation of houses, which necessitated the resettlement of the population. Activation of the processes with overlapping of the rivers in 2002 was observed quite widely. Dammed lakes became the reason of formation of debris flows on the rivers Miramdon, Kazdon, Lamardon and other mountain rivers of the third order. Movement Miramdon the right-bank landslide was the main reason for the formation of a debris flow in a volume of around 50 thousand cubic meters with the destruction of a residential house in the village of Nizhniy Unal [10, 11].

32.4 DEBRIS FLOWS

The landscape conditions of formation of clearly distinguished three types: water mud micro mudflows, water mudstone and water stone.

Water stone mudflow down with a capacity of more than 100,000 cubic meters is the most dangerous. Their origin is dated to the high-altitude zone from 2600 to 3100 meters. Large capacity allows them to come down from the slopes of even a small slope (10°–25°). An important characteristic landscape mudflow basins is the type of surface sediments, as a source of hard component of a debris flow. Water stone mudflows formed from the moraines and fluvioglacial material pending mainly on the granites and Paleozoic granitoids. When there is a sufficient number of loose material in the upper reaches of the rivers, an important condition for the formation of debris flow become heavy or prolonged precipitation. For formation water stone mudflows required to rain more than 60 millimeters of rainfall and sharp change of the dry season – wet.

Exposition of the slopes is also important in the formation of various types of debris flows. The wet westerly and north-western slopes of the most favorable for their development, however, in general, exhibit confinement water stone mudslides are quite extensive.

Pockets of origin water stone mudflows are confined mainly to the belt of the alpine and subalpine meadows.

Geosystem, in which emanated water stone mudflows are the slopes of a small slope (20°), mainly western and eastern slopes with herb-gramineous vegetation of the alpine and subalpine meadows in mountain-meadow low thick soils on the moraine sediments overlying granite and Paleozoic granitoids.

The mud stone debris of medium power with the mass of the subject of debris material from 10,000 to 100,000 cubic meters are the most widespread. Their origin is timed to the heights of attitudes 1800–2600 meters and more. In connection with lower power mud stone mudflows for their coming, greater than for water stone mudflows, steepness of slopes – from 30°. Surface sediments, which are the basic material for the origin of mud stone mudflows – scree, placers, stone looses on clay shales, aleurolites and siltstones Jura.

For formation mud stone mudflows necessary loss of 50–60 millimeters of precipitation in one downpour, and the alternation of dry and wet periods. Source mud stone mudflows have a wide exhibit confinement. They are developed predominantly on wetter northern, western and north-western expositions slopes.

Vegetation has a significant stabilizing role for this type of debris flows for the reason that the root system of vegetation quite firmly clamps small and medium fragmental material. Pockets of origin mud stone mudflows are confined to rocky surfaces subalpine zone and partly to the high mountain curved forest.

Geosystems, which are formed from the mud stone mudflows is steep slopes (20–30°), western and north-western exposure with depression herb-gramineous alpine and subalpine vegetation on low-powered mountain-meadow soils on the loose fragmental sediments overlying clay shales, mudstones and siltstones.

Mud micro mudflows of small capacity with the mass of subject material not exceeding 10,000 cubic meters, developed everywhere, but most disseminate them widely celebrated in the lowlands below 1800–2000 meters. Low-power mudslides are formed on slopes steeper than 45°. Source of solid phase micro mudflows are alluvial-deluvial sediments on margele, limestones and dolomites of chalk. Precipitation from 30 to 40 mm for one rain can cause micro mudflows in geomorphic favorable conditions. Exhibition range, favorable for the development of micro mudflows is limited. They are solely the very moist western and very dry southern slopes due to rains. Unlike other types, micro mudflows widely developing and in the forest belt, especially in humid forests of lowland on the carbonate rocks.

Geosystem, in which developing mud micro mudflows is very steep (45°) and steep (30–45°) slopes of the western and southern exposures with beech-hornbeam forests on low-power of brown mountain-forest soils are alluvial-deluvial sediments overlaying cretaceous limestones and dolomites. [4].

32.5 MONITORING OF DEBRIS FLOWS

Debris flows are also widespread. Currently in North Ossetia recorded dozens of current and stabilize the landslide foci. In catastrophic mudflows occur every 10–15 years. Sat down with evacuation of more than 100,000 cubic meters belong to the category of "very powerful" and appear once in 5–10 years. Sat "medium power" with salt removal from 10000 to 100,000 cubic meters occur once in 2–3 years, "weak" and "very weak" sat down with removal of 900–10,000 cubic meters occur annually. In river basins: Urukh observed 12 active and 15 stable landslides. The largest and most destructive debris pockets are formed in river basins: Ardon – 2 very powerful, 8 medium power, and 9 weak; Zeidon – 4 very powerful, 7 average power and 3 weak; Karaygomdon – 2 very powerful, 2 medium power and 2 weak; Genaldon – 2 very powerful and 3 weak; Fiagdon – 5 average power and 2 weak; Terek – 2 very powerful, 4 medium power and 2 weak; Gizeldon – 2 very powerful, 2 medium power and 4 weak [6–8].

Mudflows have caused significant damage—destroy the industrial enterprises of resort and tourist objects, houses, power lines, bridges and highways—possible human casualties.

In the basin of the Sadon in July 1958 after heavy rains appeared mudflow, which washed away the mine dumps, destroyed the major part of the village Stroke and blocked the river bed, forming a dam lake. Some time later, the resulting lake burst and washed away on its way bridges, destroyed an automobile road and supports the monorail system, blocked by mud array of all the streets of the village Sadon.

In June 1964 in the village of Karman-Sindzikay river Ursdon came powerful mudflow, resulting in some 40 students, with the teacher and two military, which came to their aid.

In 1987 mudflow destroyed several houses in the village Nugzal, killed about 50 kilometers and destroyed 2-kilometer cloth of Transcaucasian highway. In 1989, the mud flow of the river Big Labagomdon destroyed the reinforced concrete bridge and killed a track road for 30 km In the river basin Zakkadon in the summer of 1987 mud stone mudflow caused erosion of the left bank with the seizure of the roadway [8].

In the night of 16 to 17 July 2000 the monthly norm of precipitation, causing the formation of a powerful mudslide in the region of the village Byron. The result was demolished bridge over the river Ardon and blocked cloth of Transcaucasian highway mud stone cone removal, clearing of which lasted for two weeks. In the night from 25 to 26 august 2000 on the Transcaucasian highway even came a new powerful mudflow, which in minutes blew only that the reconstructed bridge and destroyed two supports for high-voltage power line [10, 11].

Avalanches. According to landscape development conditions provided mainly three types of avalanches: chute, front and jumping.

Avalanche chute, displacement of 10,000 to 50,000 cubic meters, the most common type. Pitch range, spreading from 1800 to 2600 meters. These avalanches can be formed with the steepness of the slopes surface from 30 up to 45°. When such a steepness of the other types of avalanches are formed less.

The optimum weather conditions for the formation and disappearance of the gutter avalanches is a loss of snow precipitation by 50 to 100 millimeters longer average (30–40 mm) with average monthly temperature from 0 to 8°. When a sufficient amount of snow avalanches important change of snow, thaw, fen and blizzards. Most of avalanche danger is different leeward eastern, north-eastern and south-eastern slopes. The eastern slopes, warming in the first half of the day receive less heat and thereby retain large amounts of snow accumulation. The western slopes of the heat in the second half of the day and receive more heat, which contributes to the gradual melting and unloading of snow. In the spring, when increasing the inflow of solar energy, the most dangerous become the northern slopes, which have large volumes of snow accumulations to the beginning of the spring warming. On the southern expositions as well as on the western receiving more heat in the winter, snow accumulation and therefore they are less avalanches are in the spring.

Geosystem, which largely developed avalanche chute is slopes are steep (30–45°) and steep (20–30°) eastern and north-eastern slopes with pine-birch forests and birch elfin low power in the mountain-forest soils in granites, clay shales and sandstones.

Front avalanches capacity of about 50,000–100,000 cubic meters apply to the heights of 2600 and 3000 meters. For their formation requires the steepness of the slopes 45–60°. This is because the snow masses, coming off the front, overcoming a lot of resistance, lose a lot of snow. On the slopes of the small and medium steepness of the snow coming down vertically, may not reach the foot of the slope, therefore, does not form an avalanche. For the formation of a frontal avalanche you want more snow accumulations than for the gutter. They arise when solid precipitation on 80–120 millimeters longer average (40–60 mm) at a temperature of 3–15°. The most favorable exposure to the front avalanche – the northern, eastern and north-east, where the spring there is a massive front off the snow. Friendliness eastern slopes due to the fact that they occupy the position of the leeward slopes, on which accumulates a large amount of snow, necessary for formation of a frontal avalanches. Woody and shrub vegetation prevent outright front of avalanches. Deep subalpine tall grass contributes to slip with it snow masses, while the grass delay off the snow. However, the front avalanches usually go forested slopes, leaving "procesy." When it is destroyed or broken by pine and birch sags. Some experts avalanches believe that highland dwarf birch is the result of avalanches [12]. This type of avalanches is widely developed in the alpine, subalpine belts in high mountain curved forest.

Geosystem formation front of the avalanche is very steep (45–60°) and steep (30–45°) slopes, predominantly eastern and northern exposures with the alpine and subalpine and herb-gramineous plants with a low-powered mountain-meadow soils, granite and sandstones.

Bouncing avalanches are considered rare, although they are quite widespread in the highlands. By the volume of issued snow they are the most powerful (more than 100–000 cubic meters). A great power is not only due to large accumulation of snow at high altitudes, where they are formed at altitudes of 3000–3500 m, but also the fact that during the gathering (jump) avalanche does not lose snow on overcoming the way. These avalanches are formed on slopes steeper than 60° more often on the cliffs. They have great destructive power, high power and immediacy. For formation jumping avalanches required precipitation of more than 120 mm, with average monthly norm of 60–80 mm and temperature of 7 to 18°. They are formed on the slopes of different exposures, although most often found on the southern and eastern exposures leeward slopes. From the tops of the windward slopes are snow cornices, brings in the form of jumping avalanches to the bottom of the leeward slopes. This type of avalanches is more common in subnival zone.

Geosystems, which develop bouncing avalanches is very steep (60°) slopes of the cliffs, mainly eastern exposure with moss-lichen vegetation on fragmentary mining and peaty soils, glaciers and rocks of the granite and sandstones.

32.6 MONITORING OF AVALANCHES

Avalanches are widespread in the area of Lateral and Water dividing ranges and are formed on the slopes of the longitudinal depression. Avalanche here is characterized by the suddenness, periodicity, proceed with great speed, become more active in the winter and spring. Volumes of removal cones ranging 10,000–50,000 cubic meters, destruction (catastrophic) of individual buildings, roads, there may be casualties. In the basin of the river Zakkadon there are about 60 avalanche centers, which from December to April, down avalanches capacity of 800–7000 cubic meters. Fewer avalanche lesions observed in the basins of the rivers: Mamisondon – 35, Genaldon – 25, Ardon – 20, Karaygomdon – 20. Of them are quite powerful avalanches from 10,000 to 30,000 cubic meters descend into the valley of the river Ardon and destroy the Transcaucasian highway. Most widespread are the chute and front, less bouncing avalanche [1, 4, 5].

The main danger of avalanches is their suddenness and lightning. According to experts, with the impact of avalanches occur pressure 100–130 tons per square centimeter, as a result of break lines, collapsing buildings and bridges, killing forest areas [12].

Together with snow and ice avalanche is capable of carrying large ground masses, including blocks weighing up to 3000 tons, causing the formation of a mudflow most capable to destroy the city, town, kill cattle, people. So, in the XIX century was buried under an avalanche of village Zrug. In 1971 avalanche destroyed the village Mosca; in March 1967 destroyed a residential building with its citizens near the village of Nar and in January 1977 there was also destroyed two houses and three people died; February 14, 1932 in South Ossetia under the weight of snow was a village Aroshenda, which killed 112 people. It should be noted that the horses and cattle remained alive since, feeling the approach of danger, they went down to the river to drink. In the spring of 1992 under avalanches killed 30 people. In 1992, the avalanche had stopped the Transcaucasian highway even [11].

32.7 HAZARD ASSESSMENT

32.7.1 MASS ACTIVATION OF DANGEROUS PROCESSES IN 2002

Climatic anomalies contributed to the intensification of dangerous natural processes in 1992 and 1993, and by 1998 activity reduced. In 2001 recorded a record low level of activity. In 2002 there was a sharp spike in activity of hazardous processes, by their number, territorial coverage, and devastating effects. Under the existing criteria of assessment 2002 became the year's activities dangerous processes with catastrophic consequences. Rainfall 20 and 21 June 2002 amounted to 170 millimeters at a time, that in 2.7 times greater than the monthly average – 63 mm. The main events were the massive intensification of erosion processes, including the Kolka glacier collapse that led to the formation of a devastating ice-stone flow through the valley of the river Genaldon jumping below Karmadon gate in mud stone debris [13, 14].

The risk of dangerous processes in June 2002, in varying, covered the whole territory of the republic, however, to a greater extent affected the mountain administrative districts: Alagir, Suburban and the Iraf. Here catastrophic dimensions become mudflow, landslide and erosion processes [10, 11].

32.7.2 HIGH-ALTITUDE ZONE ZONING OF HAZARDOUS NATURAL PROCESSES

All natural processes in mountain areas have vertical zonalıty, therefore, the geographic spread of the dangerous natural processes is also a clear correlation with the absolute heights. So the greatest number of landslides have the Midlands – zone south of the Jurassic shale depression with numerous tectonic faults, fracture and overlap, folded easily decaying shales. The unloading material mudflows and avalanches also made the Midlands in the form of cones outstations. If we consider the anthropogenic factor, the residential zone, with its economic complex are nutrigenie hollow and valley middle, partially mountains and foothills.

Landslide processes are widespread in mountain areas of the republic, especially in the northern area of the Jurassic intermontane depression underskarp area of the Rocky ridge on the sections of the south intermontane depression and the main dividing ridge. The largest of them are: Hanikomski, Avsandurski, Kadatsky, Gulisky, Dzivgicky, Hanikomski – in the basin of the river Fiagdon; Kanicki, Dallagkomsky – in the basin of the river Mamisondon, Zaramagskiy – in the valley of the river Ardon; Ceycky – in the valley of the river Ceidon and others. Almost all of them are confined to regional faults and their cracks, with the depth of capture sliding slope up to 45–60 meters and more [14].

In addition, different types of landslides (structural plastic, plastic, mudflows, landslides cutting) developed in clay sediments of the lower sarmatian in the zone of the Terk and Sunzha ridges, where a rather abrupt their intensification was provoked by trimming the slopes during the construction of the road between the villages of Batako and Khurikau [7].

In the last decade sharply active landslide processes on the ledge of the right-bank alluvial-accumulative terrace of the river Terek in Mozdok district. Manifestations of large landslides with the threat of destruction of houses recorded in the area of the village of Oktyabrskoye, Kizlyar, Suhotskoe [6, 7].

On the territory of the Republic registered 355 of landslides. Their distribution in administrative districts is given in Table 32.1.

TABLE 32.1 Prevalence in the Territory of the Republic of North Ossetia-Alania Landslides

Administrative region, city	Area of the district, km^2	Number of land-slides	Affection (km^2 / %)			
			Total	Including:		
				Strong	Average	Weak
Mozdok district	1071	16	50/5	15/2	24/2	11/1

Kirovsky district	407	2	9/2	4/1	3/0,5	2/0,5
Right-bank area	442	3	20/4	6/1	10/2	4/1
Iraf district	1365	52	181/13	83/6	78/6	20/1
Digor district	593	15	77/13	27/4	35/6	15/3
Ardonsky district	377	0	0	0	0	0
Alagirsky district	2024	229	657/32	181/9	324/16	152/7
Suburban area	1440	26	97/6	21/1	55/4	21/1
The city of Vladikavkaz	281	12	23/8	6/2	15/5	2/1
In total for the RSO-Alania	7987	355	1114/14	1244/3	544/7	227/3

Note: Affection strong – struck more than 25% of the territory; average – struck 5–25% of the territory; weak – struck less than 5% of the territory.

Debris flows are also widespread. Within the North-Jurassic intermontane depression develops 30% of average power and size of mudflows. In the area South of the Jurassic shale depression operates up to 20% of debris flows. Maximum number of recent focus on West Mountain Ossetia, the minimum – in the Eastern part [8, 9].

According to the degree of debris flow hazards in Mountain Ossetia identified the following areas [16]:

1. The high mudflow risk – Lateral ridge and West (Digorskaya) part of the Main dividing range (26% of the whole territory of Mountainous Ossetia). There are mud pools, with a capacity of more than 100,000 cubic meters of solid material.
2. District middle of the debris flow hazard – The Main dividing range, the upper part of the river Fiagdon, North and South slate intermountain depression covers up to 27% of the whole territory of Mountainous Ossetia. Here there are boarded average power with removal of solid material from 10,000 to 100,000 cubic meters;
3. District weak debris flow hazard – Pastoral and Forested ridges. Covers 35% of the territory is Mountainous Ossetia. In the area there are traces of the passage of only a small mudflows capacity from 1000 to 10,000 cubic meters;
4. District nucleobases – wide parts of the longitudinal and cross river valleys (12% of the whole territory of Mountainous Ossetia), where the traces of the passage of mudflows are practically absent. [4]

Distribution of mudflow basins administrative districts of the Republic and affection of the territory of the mudflows are listed in Table 32.2.

TABLE 32.2 Prevalence in the Territory of the Republic of North Ossetia-Alania Landslides

Administrative region, city	Area of the district	Landslides	Affection (km^2 / %)			
			Total	Including:		
				Strong	Average	Weak
Mozdok district	1071	16	50/5	15/2	24/2	11/1
Kirovsky district	407	2	9/2	4/1	3/0.5	2/0.5
Right-bank area	442	3	20/4	6/1	10/2	4/1
Iraf district	1365	52	181/13	83/6	78/6	20/1
Digor district	593	15	77/13	27/4	35/6	15/3
Ardonsky district	377	0	0	0	0	0
Alagirsky district	2024	229	657/32	181/9	324/16	152/7
Suburban area	1440	26	97/6	21/1	55/4	21/1
The city of Vladikavkaz	281	12	23/8	6/2	15/5	2/1
In total for the Republic	7987	355	1114/14	1244/3	544/7	227/3

Note: Affection strong – struck more than 25% of the territory; average – struck 5–25% of the territory; weak – struck less than 5% of the territory.

Avalanches cover the whole of the alpine part of the Mountain Ossetia, starting from the heights of 1300 meters in Alagir gorge and with a mark of 1200 meters in the valley of the river, Fiagdon. In the Mountainous Ossetia recorded 495 existing avalanche sites: in the basin of the river Ardon – 300, in the basins of the Fiagdon, Gizeldon – 107 and in the Urukh river basin – 87. Avalanches are widespread in the area of the Lateral and the Main ridges, and are formed on the slopes of the longitudinal depression [15].

According to the degree of avalanche danger on the territory of Mountainous Ossetia identified the following areas [17]:

1. District most avalanche danger is high-mountainous part of the Watershed and Lateral ranges.
2. District significant avalanche hazard – massif of mount Kazbek.
3. Area with average avalanche danger – the upper reaches of the Gizeldon, Fiagdon, Urukh and Haznidon.
4. Area of low avalanche danger – average Gizeldon river, Fiagdon, Urukh and Haznidon.
5. District least avalanche danger – Gizeldon river mouths, Fiagdon, Urukh and Haznidon.

6. The area of potential avalanche danger – longitudinal depression between the ranges of the Rocky and Pasture.

32.8 PREVENTION AND PROTECTION

In the course of the research revealed that the mountain territory of the Republic of North Ossetia-Alania have a high degree of defeat of hazardous natural processes and requires implementation of measures for the protection from natural hazards.

In order to prevent and reduce the risk of dangerous natural processes important to have a guarantor support of the Republican authorities to observe the principle of rational use of nature, to carry out environmental protection measures and engineering protection of objects of social-economic infrastructure and natural territorial systems.

To increase the level of protection of territory and population of the Republic from the impact of hazardous natural processes authors were developed and offered some of the measures and activities for prevention and risk reduction.

32.8.1 ENGINEERING AND TECHNICAL PROTECTION

- Construction debris flow control chute and mud dam protection structures on the most dangerous waterways: Labagom, Casaicomdon, Karaughom.
- Organization and strengthening of the mine and construction of the dumps in the Upper Zgid and Upper Misure.
- Construction and rehabilitation of antiavalanche galleries along the Transcaucasian highway.
- Restriction of construction activities associated with the trimming of the mountain slopes, especially in the lower part of landslide-prone slopes.

32.8.2 CONSERVATION MEASURES

- Prohibition of the issuance of land allocated for the construction of dwelling houses and industrial objects without a preliminary assessment of exposure to areas exposed to hazardous processes.
- Non-admission of dumping on landslide slopes of wastewater.
- Protection of mountain forests, strengthening slope biotopes vegetation.

32.8.3 SCIENTIFIC RESEARCHES

- Activation of activity of the monitoring of exogenous processes of a Unified Environmental Monitoring RSO-Alania, with the involvement of resource programs of space monitoring.

- Development of GIS database and implementation of the scientific prediction of dangerous processes.

32.8.4 SUPPORT OF THE REPUBLICAN AUTHORITIES

- Development and implementation of programs on sustainable development of mountain territories in the framework of the program of strategic development of the RSO – Alania.
- To the leadership of North Ossetia-Alania to increase funding for research on the prediction and prevention of hazardous natural and anthropogenic processes on the territory of the Republic.
- Construction of additional roads for safe routes in order to ease the Transcaucasian highway and support the construction of Mamison tourist cluster.

KEYWORDS

- **climatic anomalies**
- **deluvial**
- **glacier**
- **granites**
- **landscape**
- **mountain systems**
- **shales**

REFERENCES

1. Khacajeva, F. M. (2009). Landscape Indication Natural Disasters in the Northern Caucasus (the Basin of the River Ardon), Monograph, Vladikavkaz, North-Ossetian State University, 160p. (in Russian)
2. Khacajeva, F. M. (2006). Landscape-Morphological Structure of the Land Slide Hearth, Bulletin of the Regional Branch of the Russian Geographical Society in the Republic of North Ossetia-Alania, *(9)*, 56–61p. (in Russian)
3. Khacajeva, F. M., & Tomaev, V. A. (2010). Geological and Geomorphological Factor of Development of Dangerous Processes in the Mountains of North Ossetia-Alania, Materials of the International Scientifically-Practical Conference, "Young Scientists in the Decision of Actual Problems of Science", 22–23 May (2010), Vladikavkaz, North Ossetian Institute of Humanities and Social Studies, 77–80. (in Russian)
4. Khacajeva, F. M. (1992). Landscape Analysis of Natural Disasters in the Northern Caucasus (the Basin of the River Ardon), Abstract of the Thesis of the Candidate of Geographical Sciences, St-Petersburg, 18p. (in Russian)

5. Khacajeva, F. M. (1996). Landscape Conditions of Formation Landslides, Avalanches and Mud Flows on the Example of the Ardon River Basin, The Bulletin of the Regional Branch of the Russian Geographical Society in the Republic of North Ossetia-Alania, *(1)*, 58–65. (in Russian)
6. Khacajeva, F. M., & Tomaev, V. A. (2010). Landslide Processes in Tectonically Active Zones of the Territory of the Republic of North Ossetia-Alania, Actual Problems of Modern Science and Education, Krasnoyarsk, Krasnoyarsk State University, 239–242. (in Russian)
7. Khacajeva, F. M., & Tomaev, V. A. (2010). Development of Land Slides in Areas Paleoseismo Dislocations Mountain Territories of North Ossetia-Alania, Proceedings of the VII International Scientific Conference "Sustainable Development of Mountain Territories in Conditions of Global Changes", 14–16 September (2010), Vladikavkaz, 14–18. (in Russian)
8. Khacajeva, F. M. (2005). Comprehensive Research of Natural Processes in the Conditions of North Ossetia, Natural Resources, North Ossetia-Alania, The Natural and Anthropogenic Catastrophes in the Republic of North Ossetia-Alania, Vladikavkaz, Project-Press, 95–106. (in Russian)
9. Khacajeva, F. M. (2010). The Development of Natural and Technological Disasters in Mountain Areas of the Republic of North Ossetia-Alania, Sustainable Development of Mountain Territories, Problems and Prospects, Vladikavkaz, North-Caucasian Mining and Metallurgical Institute, 44–48. (in Russian)
10. Bunkov, U. D. (2003). Information Report about Results of Inspection of Disastrous Floods and Manifestations of Exogenous Geological Processes on the Territory of the Republic of North Ossetia-Alania in 2002. Vladikavkaz, Territorial Fund of Geological Information, 43–46. (in Russian)
11. Dolgov, G. A., Bunkov, U. D., Drobyshev, V. N., et al. (1998). Information Report of the Territorial Center of Geological Monitoring of the Geological Environment on the State of the Geological Environment of the Republic of North Ossetia-Alania for (1992–1996) (Part 2 Monitoring of Exogenous Geological Processes), Vladikavkaz. Territorial Fund of Geological Information, 67–72. (in Russian)
12. Tushinckiy, G. K. (1963). Glaciers Snowfields and Avalanche, Monograph, Moscow, Publishing House Geografizdat, 312p. (in Russian)
13. Zalikhanov, Sh M. (1974). Avalanches and Development Prospects of the Mountains of North Ossetia, Monograph, Ordzhonikidze and Publishing House IR, 141. (in Russian)
14. Khacajeva, F. M. (2007). Map of Landslide Hazard, Atlas of North Ossetia, Vladikavkaz, Publishing House of the North-Ossetian State University, 7. (in Russian)
15. Steber, E. A. (1903). Glacier Landslides in the Headwaters of the River Genaldon, Vladikavkaz, Tersky Collection, 233–248. (in Russian)
16. Khacajeva, F. M. (2007). Map of Mudflow Hazards, Atlas of North Ossetia, Vladikavkaz, Publishing House of the North-Ossetian State University, 6. (in Russian)
17. Khacajeva, F. M. (2007). Avalanche Hazard Map, Atlas of North Ossetia, Vladikavkaz, Publishing House of the North-Ossetian State University, 8.(in Russian)

CHAPTER 33

AN EVALUATION OF THE BIOLOGICAL TREATMENT EFFECTIVENESS OF OIL POLLUTED SOILS FOR THE YAKUTIAN ARCTIC REGION

YULIYA S. GLYAZNETSOVA, IRAIDA N. ZUEVA, OLGA N. CHALAYA, and SARA H. LIFSHITS

Institute of Oil and Gas Problems Siberian Branch of the Russian Academy of Sciences, d. 1. Oktyabrskaya st., Yakutsk, 677890, Russia; E-mail: geochemlab@ipng.ysn.ru

CONTENTS

ABSTRACT

The experimental data on the study of oil pollution transformation features of permafrost soils has been concerned for an evaluation of oil polluted soil remediation with using biological preparations based on aboriginal hydrocarbon oxidizing micro flora. The complex of analytical methods had included FT-IR spectroscopy, adsorption column chromatography and gas chromatography-mass spectrometry as well as a geochemical approach for interpretation the obtained results.

33.1 INTRODUCTION

Recovery of hydrocarbon resources in the inclement conditions is a difficult technical problem including the obligation to guarantee the conditions for environmental management, industrial and environmental safety. Prior to the implementation of large-scale projects for hydrocarbons recovery it is important to assess the ecological status of the Arctic shelf to oil pollution.

Seaports, oil loading ships and storage capacities produce a significant contribution to environmental oil pollution. Bottom sediments of such objects are characterized by high level of oil hydrocarbons content. Thus the importance of monitoring of the coastal strip and the assessment of the environmental status of the areas due to anthropogenic activities leading to environmental change increases noticeably.

In the Arctic regions in contrast to regions with favorable climatic conditions the processes of biodegradation of petroleum hydrocarbons are much slower that predetermines the lower permafrost soils ability to cleanse itself. The presence of close overlying permafrost, low temperatures and short growing season – all this leads to lower rates of transformation of oil pollution.

Now day the environment Arctic Yakutia virtually little studied. The Republic of Sakha (Yakutia) is one of the largest territories in the Russian Federation. The Yakutia area is of 3103.2 thousand km. The Yakutia is located in the north-eastern part of the Asian continent in the basin of the Lena, Yana, Indigirka and Kolyma rivers lower reaches. The northern part of the Sakha Republic is washed by the cold of the Arctic seas: the Laptev and East Siberian. Oil and oil products are the most dangerous sources of contamination and represent environmental threat to the Arctic environment. They are delivered in Arctic regions along the rivers. Most of the oil depots are located along rivers and coastal areas of the Arctic seas. Oil pollution waters and coastal parts of the land occur as a result of accidental spills during transportation and during operation of tank farms.

Residual oil pollution in conditions of permafrost areas persisting for years becomes the source of oil hydrocarbons in river flow to the sea and its coastal part. Thus initially the local oil pollution can spread over a large area. The existence of permafrost soils, prolonged freezing water, widespread bogs and boggy territories are specific features of the vast Arctic regions [1, 2]. The soils of northern latitudes are a low biogenic as a little ability to heal itself. Oil spills penetrate to thaw zone and continue to spread laterally over this area being exposed to very slow transformation. Most of the spilled oil remains in the top 10 cm to the surface soil layer [3].

A priority of the northern territories environmental monitoring of oil objects is to minimize a damage as a result of human impact on natural ecosystems in the Arctic. Particular significance of research is to develop the best approaches for redress oil contaminated lands and building of effective oil destructors in extreme climatic conditions.

In our view the application of the organic geochemistry methods can promote the scientific support of some questions of monitoring and remediation of oil contaminated areas.

The key tasks of geochemical monitoring are search informative and reliable parameters for the identification of contamination of soils and bottom sediments with oil and oil products. These parameters may be useful for assess the quality of the treatment works and selection of the most effective destructors for arctic conditions.

For analyst it is clearly in determining the content of the "oil products" in the soils that the objects are the isolated extracts. The latest are a mixture of oil hydrocarbon compounds of anthropogenic, that is, proper oil pollution, and natural organic matter of soils. Each of these partials is characterized by a set of a large number of hydrocarbon compounds the content of which may vary over a wide range.

When monitoring the oil-contaminated areas the researcher gets the task of determining of oil pollution in the presence of a background native organic matter of soil or sediments. Much useful information about unusual and complexity of the task and possible approaches to some solutions can be found in number works [4–6]. Because of the complexity and diversity of chemical composition of different oil pollutants and specific composition of organic matter soils the question must be decided in each case in accordance with the specific task [6].

We believe in the petroleum geochemistry with a high degree of detail the processes of oil degradation had been studied both as for oil deposit conditions as according the results of the model experiments. Data on transformation processes of oil as a result of bacterial oxidation of hydrocarbons under aerobic conditions, the selectivity and the stage of these processes can be very useful for studying the features of degradation of oil pollution in time. The results can be the basis for creating new efficient oil destructors using indigenous micro flora in the development of methods to eliminate the oil spills.

The criteria for a correct assessment of the effectiveness of the developed oil destructors and control of remediation works of contaminated soils is an independent scientific task that requires special studies on transformation features of oil pollution in different conditions taking into account the influence of the natural background and secondary factors.

33.2 MATERIALS AND METHODOLOGY

Bitumoids of soil samples extracted with chloroform. Then the solution was filtered and distilled with using water bath. The extract was transferred to weighing bottles and brought to constant weight in a vacuum oven.

The chloroform extracts (bitumoids) studied by the infrared spectroscopy (FT-IR). The spectra were obtained on a FT-IR spectrometer "Protege 460" Company "Nicolet" in the broad range of 500–4000 cm^{-1}. Extracts were placed in the demountable cell with windows NaCl or KBr, the thickness of absorbing layer was 33 microns. The identification of absorption bands of the spectra had been made with the IR spectra of atlases and tables of wave numbers.

Group composition of the extracts: hydrocarbon components, benzene and alcohol-benzene resins were determined by adsorption column chromatography on silica gel. Asphaltenes precipitated from the extracts by excess of petroleum ether.

Gas chromatography-mass spectrometry (GC–MS) studies of the individual composition of the hydrocarbon fractions were carried out on a system that included an Agilent 6890 gas chromatograph with an interface to an Agilent 5973N high performance mass-selective detector. The chromatograph was equipped with a 30 m quartz capillary column with a diameter of 0.25 mm impregnated with an HP5 MS phase. Helium was used as a carrier gas; the flow rate was 1 mL/min; evaporator temperature 320 °C. The temperature rise was programmed in the temperature range 100–300 °C at a rate of 6 °C/min. The ionizing voltage of the source was 70 eV.

Objects of the study were the soil samples collected at the polluted territory of storage of fuels after one year of the diesel fuel spill. The background samples were taken at unpolluted areas at a distance far from the spill.

33.3 RESULTS AND DISCUSSION

As a practice of remediation of contaminated soils shows only biological methods based on stimulating the functional activity of hydrocarbon-oxidizing microorganisms provide more full recovery northern ecosystems. In this chapter, the experiment results of research on the degradation of oil-contaminated soil during bioremediation process are given. The soil clean up technology with using a biological product based on indigenous psichrotrop oxidizing micro flora isolated from oil contaminated permafrost soils of the Yakutia and is capable of developing positive at low temperatures (4 °C) in low aeration, moisture and lack of nutrient substances. The cultivation microorganisms in the medium containing oil substrate of the contaminated site allow tailoring them to its chemical composition and species of composition of the natural micro flora, which greatly accelerates the degradation of oil pollution.

The experiment was conducted under natural conditions in the territory of the storage of fuels located in the north-eastern Yakutia (Moma district) on the place of the emergency spill of diesel fuel. The total area of contamination set was 0.8 hectar. Permafrost soils characterize maximum temperature of –9 °C to –11 °C. In June the average temperature was of +6 °C to +14 °C. The experiment consisted of two stages. In the first stage in June samples of contaminated soils at the test sites were taken. Then soils of the test sites treatment were treated by biological preparation immobilized on mineral sorbents – zeolite and vermiculite. Immobilized microorganisms on the surface of minerals at the rate of 200 g/m^2 for zeolite and 50 g/m^2 for vermiculite were added to the contaminated soils. In the second stage on August the soil samples were again taken from these sites for the study of the degradation of diesel fuel and assessment of quality biological purification.

In addition, at both stages the control samples of contaminated sites were selected where the soil treatment was not carried out. This allowed comparing the effectiveness of action of biological product on degradation process of test samples and control samples without treatment. The results of study for the test permafrost soils contaminated with diesel fuel and background samples are shown in Table 33.1.

The data on natural background were used for assessing oil contamination of soil samples as for the degree of degradation of oil pollution. It was assumed that the composition of the contaminated samples after biological treatment should change in the direction to the composition of the background samples.

TABLE 33.1 Destruction of the Oil Pollution and Change the Chemical Composition of Bitumoids in the Process of Bioremediation of Permafrost Soils Contaminated by Diesel Fuel

Conditions of the experiment	Sample	Bitumoid content*, g/kg	Destruction of oil pollution, %	Group component composition of bitumoids, %		
				hydrocarbns	resins	asphaltenes
Background samples		0–1.516		6.23–13.79	58.22–69.34	21.94–32.70
Control samples without biological treatment						
June	1	25.823		92.64	6.87	0.49
August	2	21.247	17.80	81.99	13.23	4.77
Biological preparation and vermiculite						
before treatment - June	3	23.420		81.53	14.22	4.25
after treatment - August	4	2.450	89.53	23.05	65.29	11.66
Biological preparation and zeolite						
before treatment - June	5	18.350		80.72	16.37	2.91
after treatment - August	6	1.521	91.71	21.50	50.86	26.91

Biological preparation and zeolite						
before treatment - June	7	19.803		93.13	6.51	0.36
after treatment - August	8	2.875	85.48	17.05	71.14	11.81

* – oil product content was determined by the yield of the chloroform extract – chloroform bitumoid of the soil samples.

THE BACKGROUNDS SAMPLES

As can see from Table 33.1, the content of hydrocarbon compounds in the natural background samples varies from traces up to high values up to 1.516 g/kg. The peculiarities of the chemical structure of the background samples correspond to the bitumoids of the soil samples from the natural objects, previously studied [7, 8]. This is indicated by a close character of the IR spectra of samples and the data on group component composition with predominance of resins and asphaltenes over hydrocarbon compounds (see Table 33.1). At the same time according to the GC-MS low molecular weight alkanes $n\text{-}C_{12}\text{–}n\text{-}C_{15}$ had been detected. These hydrocarbons are not typical for alkanes row of organic matter of recent sediments and may indicate the presence of trace contamination by diesel fuel in the analyzed background samples. Such deviations from the natural background sample are typical of anthropogenic environment with oil-gas objects around. The analyzed samples can be considered as the background for the area with such objects.

On the first stage of the experiment (in June) in all samples (1, 3, 5, 7) of the contaminated sites, which were studied before the treatment them by the biological product the bitumoid content was equal to 18.350–25.823 g/kg (see Table 33.1). It corresponds to the high level of pollution on the classification [9]. The bitumoids are characterized by a close structural group composition with a predominance of hydrocarbons over compounds with oxygen-containing groups and bonds. It is indicated by the low value of absorption at the ranges 1100–1300, 1700–1740 and 3300 cm^{-1}. The infrared spectra of the extracts are typical for oil-contaminated areas. The group composition of them is also characteristic for oil contaminated soil samples. This is indicated by the dominance of hydrocarbon components 80.72–93.13% over the amount of asphaltenes and resins (see Table 33.1).

To the data on the GC/MS analysis the composition of saturated hydrocarbons of contaminated soil extracts corresponds to a diesel fuel which supports the predominance of relatively low molecular weight alkanes over high molecular (Table 33.2, samples 1, 2, 7). Hydrocarbon composition of the oil-contaminated sample is characterized by high content of normal alkanes, which are represented by homologues of $n\text{-}C_{12}\text{–}n\text{-}C_{29}$. 97% of them are relatively low molecular hydrocarbons of $n\text{-}C_{12}\text{–}n\text{-}C_{20}$, which is due the high value of their ratio to relatively high molecular weight homologues. Maximum distribution of n-alkanes is at n-$C_{15, 16}$, CPI is close to 1. Among the saturated hydrocarbons n-heptadecane (n-C_{17}) and n-octadecane (n-C_{18}) prevail over pristane (Pr) and phytane (Ph), the ratio of (Pr + Ph)/(n-C_{17} + n-C_{18}) is less than unity, which is typical for oil contamination.

TABLE 33.2 Change of Individual Composition of Saturated Hydrocarbons in the Process of Bioremediation of Permafrost Soils Contaminated by Diesel Fuel

	Sample				
Parameters	**1**	**2**	**7**	**8**	**Background sample**
	Control samples without biological treatment		**Biological preparation and zeolite**		
Conditions of the experiment	**Before treatment – June**	**After treatment – August**	**Before treatment – June**	**After treatment – August**	
n-C_{13}–n-C_{20}/n-C_{21}–n-C_{33}	34.13	26.57	33.65	1.19	0.59
max n-alkanes	n-$C_{16,17}$	n-$C_{15,16}$	n-$C_{15,16}$	n-$C_{17,18,27,29}$	n-$C_{27,29,31}$
CPI*	0.99	0.95	1.08	1.74	2.04
isoprenoids/n-alkanes	0.45	0.54	0.46	0.88	0.41
Pr/n-C_{17}	0.74	0.95	0.71	2.70	2.45
Ph/n-C_{18}	0.72	0.85	0.79	2.67	2.46
Pr+Ph/n-C_{17}+n-C_{18}	0.73	0.93	0.75	2.69	2.46

* Carbon preference index (CPI) ratio of peak heights or peak areas for odd- to even-numbered n-alkanes.

The results suggest that the composition of the soil samples analyzed nearly a year after the spillage of the diesel fuel there are no visible signs of biodegradation of contamination. This is dues by the very low ability of permafrost soils to clean itself in cold regions.

On the second stage (August), the results of studied of soil samples of the test sites over 2 months after treatment by the biological product showed decrease the residual oil hydrocarbons content. As result the destruction of oil contamination became 85.5–91.7% in comparison with 17.8% for the control sample without biological treatment (see Table 33.1).

High degree destruction of oil contamination is based on the results of the study the composition and chemical structure of selected extracts. As can be seen the infrared spectra (Fig. 33.1, sample 7) in the chemical structure of extracts of the initial soil sample hydrocarbon groups and bonds – long methylene chains (720 cm^{-1}) and aromatic hydrocarbons (750, 810 and 1600 cm^{-1}) prevail. Oxygen-containing groups and bonds are presented in small quantities (1170 and 1710 cm^{-1}). In the group composition of those samples (see Table 33.1) hydrocarbons are dominated, content of resins and asphaltenes is below 20%.

According to the group composition (see Table 33.1) it can be concluded that in the process of biodegradation the hydrocarbon part of pollution used at first. In case of the biological preparation in contaminated soil samples the hydrocarbon content in the extracts decreases from 80.72–93.13% to 17.05–23.05%. As can be expected after the biological treatment of samples the residual content of contamination has remarkably changed into the direction of increasing amount of asphaltenes and the resins especially (see Table 33.1). However, the FT-IR results showed that after treatment structural-group composition of extracts had became more close to the natural background sample and had differed from the typical residual oil contamination (Fig. 33.1, sample 8).

The data indicate a deep biochemical transformation of oil pollution and as result it didn't detect the presence of oil hydrocarbons in the studied samples by the IR-spectra. The type of the IR spectra of bitumoids corresponds to bitumoids of organic matter of recent sediments. Dominant absorption band 1700–1740 cm^{-1} indicates the large content of carbonyl groups, high absorption in 1170 cm^{-1} – the high amounts of ester bonds. Clearly expressed absorption in the range 3200–3600 cm^{-1} is associated with the presence of hydroxyl groups. In the chemical structure of bitumoids large content of long chain methylene (720, 730 and 1460 cm^{-1}) and low content of compounds with aromatic cycles (750 and 1600 cm^{-1}) had been detected. All of these features are characteristic of different types of permafrost soils of the Yakutia [8].

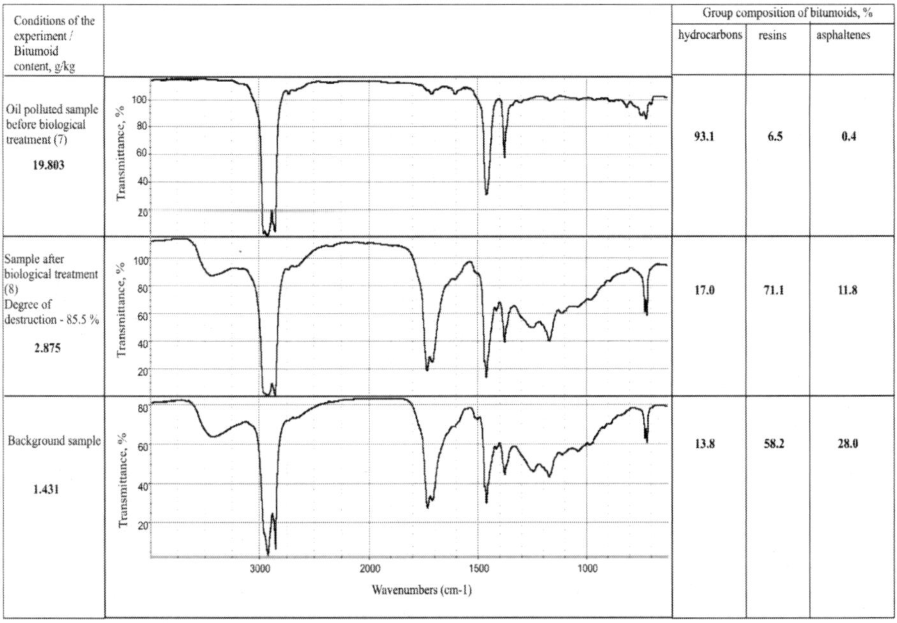

FIGURE 33.1 The infrared spectra of chloroform extracts of soils.

As result of the biological treatment of the contaminated samples there are significant changes in the composition of saturated acyclic hydrocarbons. The portion of n-alkanes reduced almost 10%, which were represented by the more long homologous series of n-C_{13} to n-C_{33} (see Table 33.2, sample 8; Fig. 33.2B). In their composition as compared to the initial oil contamination (Fig. 33.2A) has considerably increased the content relatively high molecular weight n-homologues n-C_{21}-n-C_{25} that is expressed in the decrease of the ratio of relatively low to relatively high molecular weight n-alkanes. The distribution of n-alkanes changed: it became the bimodal with maximum at n-$C_{17, 18 \text{ and }}$ n-$C_{27, 29}$. The redistribution between alkanes normal structure and isoprenoids changed: pristane and phytane prevailed over next eluted n-heptadecane and n-octadecane. For the sample 8 the coefficient of biodegradation (iC_{19}+iC_{20}/nC_{17}+nC_{18}) increased to 2.69 compared to 0.75 for the sample 7 (see Table 33.2).

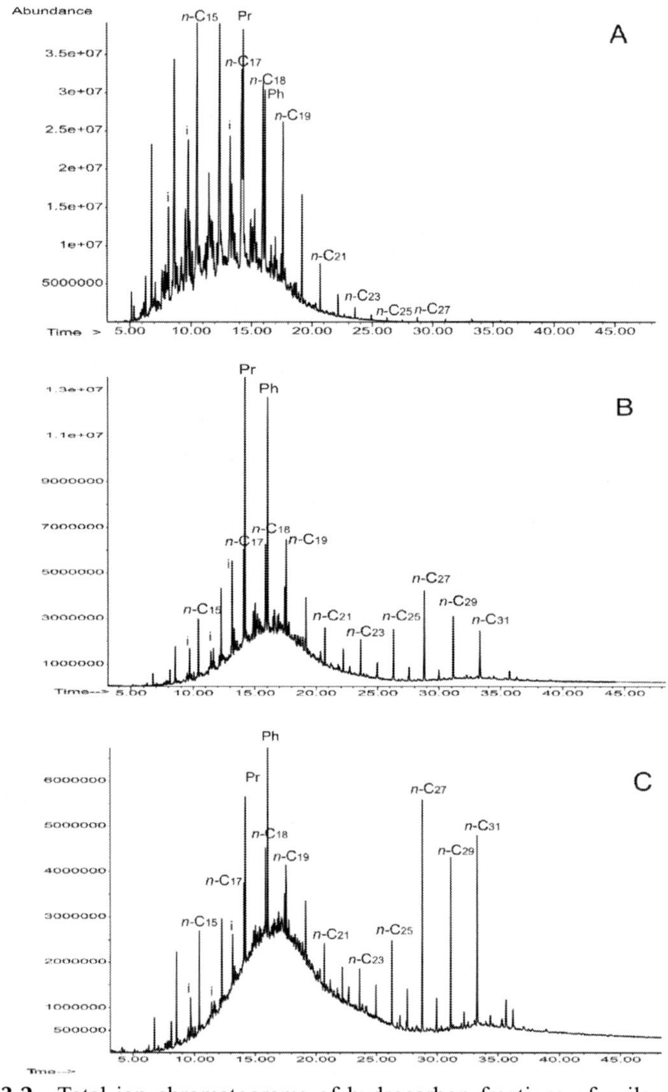

FIGURE 33.2 Total ion chromatograms of hydrocarbon fractions of soils extracts: A – sample 7 before biological treatment, B – sample 8 after biological treatment, C – background sample. i – isoprenoids.

33.4 CONCLUSIONS

The results showed that the use of a biological preparation specifically designed for contaminated permafrost soils on the basis of indigenous hydrocarbon-oxidizing microorganisms immobilized on natural sorbents could substantially reduce the residual oil content in the soil due to the high degradation of diesel fuel contamination. The destruction of oil pollution was from 85 to 92% of soil samples with high levels of contamination in comparison with 18% in the control samples.

As a result of the work on bioremediation revealed directed changes in the composition of oil pollution. It has been found reduction in the quantity of long methylene chains and an increase of oxygen-containing groups and bonds in the chemical structure of the extracts, the change of the ratios of normal alkanes and isoprenoids, a significant increase CPI and the coefficient of biodegradation (i-C_{19}+i-C_{20}/n-C_{17}+n-C_{18}). All this suggests a deep biochemical transformation of the oil pollution.

It has been revealed that after biological purification the chemical composition of the extracts of contaminated soil samples became substantially identical with that of background samples.

The studies showed the effectiveness of a biological product for cleaning oil-contaminated soils in cold Arctic regions.

ACKNOWLEDGEMENT

The authors thank microbiologist Erofeevskaya Larisa for the sample collection and all staff members of the laboratory Geochemistry for help during laboratory analysis of samples.

KEYWORDS

- **biological preparation**
- **microorganisms**
- **oil pollution**
- **soil**

REFERENCES

1. Yu, I. (1993). Natural and Technogenic Flows of Hydrocarbons in the Environment, Moscow State University, 208p. (in Russian).
2. Pikovskii, Yu I., & Puzanova, T. A. (2012). Ecological Problems of Oil Production in Russia, Fuel and Energy Complex of Russia, *1*, 34–37. (in Russian).
3. Gildeeva, I. M. (2003). The Effects of Seasonal Freezing and Thawing Processes on Soil Evolution of Oil Pollution in the Arctic [Shimansky, V. K. Ed] St. Peterburg, 118p. (in Russian).
4. White, D. M., Collins, D. S., Barnes, D., & Byard, H. (2004). Effect of Crude Oil Spill on Permafrost after 24 years in Interior Alaska, Procedure Cold Regions Engineering and Construction Conference American Society of Civil Engineers, Edmonton, May, 16–19.
5. White, D. M., & Irvine, R. L. (1996). The Bituminous Material in Arctic Peat, Implications for Analyses of Petroleum Contamination, J. Hazardous Materials, *49*, 81–196.

6. Bioremidiation of Petroleum Hydrocarbons in Cold Regions [Filler, D. M., Snape, I., Barnes, D. L. Eds (2008)] Cambridge University Press, 273p.
7. Zueva, I. N., Glyaznetsova, Yu S., Lifshits, S. H. et al. (2009). Research Methods of Surface Geochemical Hydrocarbon Fields of Natural and Technogenic Origin, Science and Education, *1*, 50–55. (in Russian).
8. Glyaznetsova, Yu S., Zueva, I. N., Chalaya, O. N., & Lifshits, S. H. (2010). Oil Pollution of Soils and Bottom Sediments at Territory of Yakutia (Composition, Distribution, Transformation), [Safronov, A. F. Ed] Yakutsk, Ahsaan, 160p. (in Russian).
9. Goldberg, V. M., Zverev, V. P., Arbuzov, A. I. et al. (2001). Technogenic Pollution of Natural Waters by Hydrocarbons, and its Environmental Consequences, Moscow, 94p. (in Russian).

INDEX

H

I

K

Q

R

T

U

V

W

X

Y

Z